Practical Applications of Self-Service Technologies Across Industries

Balraj Verma
Chitkara University, India

Mike Schulze
CBS International Business School, Germany

Timi Olubiyi
West Midlands Open University, Nigeria

Kamal Upreti
Christ University, India

IGI Global
Scientific Publishing
Publishing Tomorrow's Research Today

Vice President of Editorial	Melissa Wagner
Director of Acquisitions	Mikaela Felty
Director of Book Development	Jocelynn Hessler
Production Manager	Mike Brehm
Cover Design	Jose Rosado

Published in the United States of America by
IGI Global Scientific Publishing
701 East Chocolate Avenue
Hershey, PA, 17033, USA
Tel: 717-533-8845
Fax: 717-533-7115
Website: https://www.igi-global.com E-mail: cust@igi-global.com

Library of Congress Cataloging-in-Publication Data

Names: Verma, Balraj, 1985- editor | Schulze, Mike, 1976- editor | Olubiyi, Timi, 1978- editor | Upreti, Kamal editor
Title: Practical applications of self-service technologies across industries / edited by Balraj Verma, Mike Schulze, Timi Olubiyi, and Kamal Upreti.
Description: Hershey, PA : IGI Global Scientific Publishing, [2025] | Includes bibliographical references and index. | Summary: "This work delves into the multifaceted nature of SST, examining how its rise is reshaping customer experiences, operational efficiencies, and business models in urban centers while contrasting these developments with the under-penetration and unique challenges in rural areas"-- Provided by publisher.
Identifiers: LCCN 2025025407 (print) | LCCN 2025025408 (ebook) | ISBN 9798337346670 hardcover | ISBN 9798337346687 paperback | ISBN 9798337346694 ebook
Subjects: LCSH: Self-service (Economics)--Case studies | Service industries--Technological innovations--Case studies | Customer services--Automation--Case studies | Electronic commerce--Case studies | Artificial intelligence--Financial applications--Case studies
Classification: LCC HD9980.5 .P733 2026 (print) | LCC HD9980.5 (ebook)
LC record available at https://lccn.loc.gov/2025025407
LC ebook record available at https://lccn.loc.gov/2025025408

British Cataloguing in Publication Data
A Cataloguing in Publication record for this book is available from the British Library.

Table of Contents

Detailed Table of Contents

Self-service technologies (SSTs) empower banking customers to conduct transactions independently, offering convenience and flexibility. In the technological era, Self-Service Technologies (SSTs) play a crucial role in the banking sector. Tech-savvy generations are enjoying the services of SSTs. This chapter is conceptual and contains secondary sources. This chapter covered the objective of studying the influence of SST technologies in the banking industry. This chapter discusses AI technologies, such as machine learning, predictive analytics, and natural language processing, and their impact on SSTs, including ATMs, IVR, and kiosks, for the benefit of customers. This chapter also explores recent case studies, benefits, future research directions, and challenges of SSTs in the banking industry. This information is useful for practitioners and policymakers in the field of SSTs' impact on banking.

Self-service technologies (SSTs) are transforming customer interactions by providing faster, more efficient, and easily accessible services. From automated checkouts to AI-powered chatbots, these innovations streamline operations and lower costs.

However, their lack of emotional intelligence often results in user frustration and weakened trust. This chapter highlights the need for a human-centred design (HCD) approach to create SSTs that blend efficiency with empathy. By integrating artificial intelligence, natural language processing, and sentiment analysis, businesses can develop systems that recognize and respond to user emotions. While AI and IoT advancements enhance personalization, addressing ethical concerns like data privacy and bias is crucial. Emotionally aware SSTs can improve customer satisfaction, strengthen engagement, and build lasting relationships.

The rapid advancement of technology has redefined how individuals interact with services, creating a paradigm shift toward self-service experiences in a hyper-connected world. This chapter explores the evolution, design principles, and transformative impact of next-generation self-service technologies, which are reshaping industries such as retail, healthcare, banking, and public services. By leveraging innovations like artificial intelligence, blockchain, the Internet of Things (IoT), and augmented reality (AR), these systems empower users with convenience, personalization, and autonomy. However, the design and implementation of self-service solutions pose significant challenges, including ensuring accessibility, addressing privacy concerns, mitigating technological overload, and embedding ethical considerations into systems. This chapter emphasizes the importance of user-centric design, seamless integration, and security as foundational principles for crafting effective self-service experiences. It also highlights the need for inclusive strategies to bridge digital divides and promote equitable access. Looking forward, the emergence of ambient intelligence and decentralized technologies herald's new possibilities for enhancing self-service systems. Through case studies, trends, and actionable insights, this chapter provides a comprehensive roadmap for organizations and designers to navigate the complexities of creating next-generation self-service platforms. By fostering innovation while addressing ethical and societal considerations, these systems hold the potential to transform service delivery, enabling a truly connected and empowered world.

This paper explores the integration of Artificial Intelligence (AI) in self-service kiosks, focusing on how customer profiling techniques enhance their efficiency and effectiveness. AI-powered kiosks use algorithms to analyze user behavior, preferences, and demographics, delivering personalized experiences. Methods include purchase history analysis, Wishlist integration, behavioral analytics, and demographic insights, allowing kiosks to recommend tailored products. Sentiment analysis from social media and predictive analytics anticipate customer needs, while context-aware personalization adapts recommendations based on time, location, and weather. These AI innovations transform kiosks into intelligent systems, improving user satisfaction, fostering loyalty, and driving sales. Through data-driven insights and real-time adjustments, AI-powered kiosks create a seamless, engaging self-service experience.

In the present era, self-service technology using smart machines is vital in retail banking and is fast changing the conventional way of transacting business by organisations. Therefore, the key objective of this chapter was to empirically test the effect of technology-based self-service banking on customer satisfaction in selected listed Banks in Nigeria, the most populous and one of the biggest economies in Africa. For this study, three (3) out of the twelve(12) commercial banks listed on the Nigerian Exchange Group as of 31st July 2024 were selected from the population. This is based on the capitalization criteria and also the dividend payment trends. Based on the data analyzed the findings of this study indicate that technology-based self-service using smart machines serves as both a technology tool and a strategic asset for firms for customer satisfaction. The chapter recommends that self-service channels and smart machines should be considered aggressively by the banks in Nigeria to substantially improve communication, responsiveness, and tangibility.

This chapter, Revolutionizing Retail: Harnessing AI-Powered Self-Service Technologies for Enhanced Customer Experiences explores how AI-driven self-service tools like machine learning, natural language processing, and robotics are reshaping retail. These technologies fine-turn personalization, customer satisfaction, and operational efficiency by providing adequate guidelines, streamlining supply management, and automating customer support. Chatbots, Self-checkout systems, and virtual assistants endow customers with greater control and cohesive interactions, improving accessibility and building trust. The chapter showcases the importance of AI-powered innovations while tackling challenges such as data privacy, ethical concerns, and inclusivity. Ultimately, it accentuates the transformative potential of these tools in the advancing retail landscape.

Across all sectors, self-service transformation is enabling service delivery without users having to get in line and go through the company; they can instead initiate a task digitally and manage it independently. In finance, technologies like mobile banking apps, robo-advisors, and online portals have empowered customers, increased operational efficiency, cut costs and provided the evolving convenience and personalization that consumers expect. Outside of finance, healthcare and retail services are utilizing self-service portals and self-checkout capabilities to improve processes and customer experiences. In this research, we investigate the antecedents, processes, and consequences of self-service transformation, illustrating the path to developing the capabilities to scale, economize and delight the customer. It lays out the technological enablers, such as automation, artificial intelligence, but also the challenges, such as cybersecurity risks, the digital divide, resistance to adoption. By designing around these struggles, building with security in mind, and educating end-users, organizations can accommodate this transition with relative ease. Self-service transformation: how it can reshape the way services are delivered, and creates a more empowered, efficient and scalable future across industries.

in shaping industries in the following decades. The outcomes anticipated from this chapter include knowledge of how self-service technologies enhance the service delivery process and directions on how this process can be improved in the future.

The introduction of Artificial Intelligence (AI) into multiple aspects of the automotive sector is causing a revolutionary change. This field is one important area where AI has a big influence. To speed up the acceptance process, machine learning algorithms are being used to examine large datasets, such as photographs and sensor data. Insurance companies can evaluate the extent of damages more quickly and precisely thanks to image recognition technology, which speeds up policyholder reimbursements. AI systems can give more accurate claims assessments and customized risk profiles for individual policyholders by utilizing this data. These smart interfaces are meant to communicate with policyholders, helping them through the claims process and making it easier for them to report accidents. This study looks at how virtual assistants affect client satisfaction and how much more efficient claims processing may be because of them. This study explores the moral ramifications of utilizing AI algorithms to make choices that have a direct effect on people's financial security.

Personalization and User Autonomy in Self-service technologies (SSTs) have emerged as one of the most significant advances in the digital era, changing consumer-business interactions. This chapter examines the reasons why businesses are compelled to incorporate these elements into self-service in light of shifting consumer demands. In doing this, this chapter questions whether Personalization can increase loyalty and increase loyalty among users who also trust. Using a qualitative design, the chapter reviews industry cases, peer-reviewed papers, identifies technology trends, and considers relevant theories to examine how these features enhance customer satisfaction, loyalty, and trust. The study revealed that companies incorporating personalization and enabling users autonomy through self-service technologies (SSTs) experience an increase in customer engagement and positive feedback from consumers which makes contribute to the satisfaction and trust level. Personalization

Preface

In today's fast-evolving digital economy, self-service technologies (SSTs) are no longer confined to automated teller machines or check-out kiosks. They are the invisible engines powering intelligent banking, personalized retail, precision healthcare, and seamless insurance claims—becoming essential to how people interact with services, institutions, and technology itself.

This book, *Practical Applications of Self-Service Technologies Across Industries*, presents a curated collection of case studies, technological explorations, and conceptual insights that illuminate the multi-dimensional impact of SSTs across sectors. It seeks to offer readers a grounded and application-oriented understanding of how smart machines and AI-enabled interfaces are redefining operational efficiency and user autonomy, while also raising critical questions around design, empathy, and accessibility.

The genesis of this book was rooted in a simple observation: while the academic and business discourse around automation is thriving, there exists a pressing need to bridge the gap between theoretical models and real-world implementation. What does SST mean for a customer walking into a digitally enabled bank in Sub-Saharan Africa? How are AI-based kiosks being optimized for customer profiling? What happens when personalization algorithms meet public healthcare dispensing systems? These are not abstract inquiries; they are the realities that this volume investigates with clarity and depth.

Each chapter offers a unique window into a specific use case or cross-industry perspective—from finance and insurance to retail and medicine—tracing the evolution of SSTs and their emerging role in shaping human-machine collaboration. Topics include everything from ATM-to-AI transitions in retail to Arduino-powered medical dispensers, from humanizing automation to personalization in customer journeys. In combining technical expertise with human-centered thinking, the book promotes a design ethos that values both functionality and emotional intelligence.

ORGANIZATION OF BOOK

Chapter 1: Optimizing User Experiences With AI-Enabled Self-Service Technologies in the Banking Industry

This chapter delves into the transformative role of AI in banking SSTs such as ATMs, IVRs, and kiosks. It offers a conceptual analysis of AI tools—including predictive analytics and NLP—and how they enhance customer autonomy and satisfaction. Through secondary research and contemporary case studies, the chapter informs practitioners and policymakers on future research directions and the broader implications for customer engagement in the digital banking era.

Chapter 2: Humanizing Automation: Designing Self-Service Technologies that Balance Efficiency and Emotional Intelligence

Striking a balance between automation and empathy, this chapter advocates for human-centered design in SSTs. It outlines how emotional intelligence—enabled by sentiment analysis and AI—can overcome the limitations of impersonal automation. The authors emphasize ethical considerations such as data privacy, while illustrating how emotionally aware SSTs strengthen customer trust and long-term engagement.

Chapter 3: Empowered by Design: Crafting Next-Generation Self-Service Experiences for a Connected World

This chapter provides a panoramic view of how next-gen technologies like IoT, blockchain, and AR are redefining user experiences in SSTs. Emphasizing inclusivity, accessibility, and ethical design, it underscores the necessity of seamless, secure, and user-centric solutions that address digital divides and promote equitable service delivery across industries.

Chapter 4: Enhancing Kiosk Efficiency with Artificial Intelligence: Advanced Customer Profiling Techniques

Focusing on AI-powered kiosks, this chapter introduces advanced customer profiling techniques that drive personalization and engagement. With insights drawn from behavioral analytics, social sentiment analysis, and predictive modeling, the authors demonstrate how intelligent kiosks elevate service efficiency, customer loyalty, and revenue growth.

Chapter 5: Technology-Based Self-Service Using Smart Machines: Role on Customer Satisfaction in Listed Banks in Sub-Saharan Africa

Through an empirical lens, this chapter examines the adoption and impact of smart machines in Nigerian banks. Highlighting the role of SSTs as both strategic and technological assets, it offers evidence that these innovations significantly improve customer satisfaction and organizational responsiveness in emerging markets.

Chapter 6: Revolutionizing Retail: Harnessing AI-Powered Self-Service Technologies for Enhanced Customer Experiences

This chapter explores how retail is being reshaped by AI-driven SSTs, from chatbots to self-checkout systems. By improving personalization, automating support, and streamlining supply chains, these tools elevate customer satisfaction. The chapter also addresses ethical challenges and emphasizes the need for inclusive innovation in a rapidly evolving retail landscape.

Chapter 7: Finance and Beyond: Driving Self-Service Transformation through Digital Innovation

Extending the scope of SSTs beyond banking, this chapter explores self-service transformations in healthcare, retail, and public services. It illustrates how digital platforms empower users to take control of their interactions, delivering convenience, cost savings, and enhanced personalization across sectors.

Chapter 8: Revolutionizing Self-Service Technology with Machine Learning Intelligence

Machine learning emerges as a cornerstone in SST evolution in this chapter. The authors discuss how ML enables intelligent systems to autonomously adapt and improve, thereby revolutionizing customer interactions. Special attention is given to privacy, data ethics, and the future potential of adaptive, self-organizing SST platforms.

Chapter 9: ATMs to AI: The Evolution of Self-Service Technology in Retail and Beyond

Offering a historical perspective, this chapter traces the trajectory of SSTs from ATMs to AI-integrated platforms. It chronicles major milestones, assesses cross-industry applications, and addresses the dualities of digital convenience versus privacy and access gaps, laying a foundation for understanding modern SST innovation.

Chapter 10: AI for Personalization: Transforming Customer Journeys in Self-Service Technology

This chapter examines the central role of personalization in enhancing customer journeys. With a focus on real-time analytics, machine learning, and NLP, it demonstrates how AI customizes user experiences while confronting privacy and inclusion challenges. The insights provided underscore the growing demand for intuitive and tailored self-service solutions.

Chapter 11: Navigating India's Financial Inclusion Through Self-Service Technologies

Centering on the Indian context, this chapter investigates how SSTs are bridging financial gaps for underbanked populations. The authors analyze government initiatives and user adoption strategies that are vital for fostering financial inclusion. The chapter underscores the sociotechnical dynamics essential to broadening access in diverse communities.

Chapter 12: Automated Medicine Dispenser Powered by Arduino and Flowmeter with AI Integration

This innovative chapter presents a practical application of SSTs in healthcare through the development of an AI-integrated automated medicine dispenser. Tailored for elderly and chronically ill patients, the device showcases how smart systems can enhance independence, reduce caregiver dependency, and alleviate healthcare burdens.

Chapter 13: Finance and Beyond: Self-Service Transformation Case Studies

Featuring a collection of industry case studies, this chapter offers concrete examples of self-service implementations in finance, healthcare, hospitality, and transportation. It reveals how businesses leverage SSTs to reduce operational costs, improve service efficiency, and deliver greater user autonomy and satisfaction.

Chapter 14: Revolutionizing Service Delivery: Cutting-Edge Innovations in Self-Service Technology

This chapter provides a strategic overview of SST trends and future directions. Using a combination of literature reviews and real-world data, it assesses the role of automation and AI in redefining service delivery across global contexts. The findings support strategic planning for organizations aiming to future-proof their service models.

Chapter 15: Streamlining Insurance Claims With Intelligent Systems

This chapter investigates how AI is reshaping the insurance claims process. By leveraging image recognition, ML, and virtual assistants, insurers can process claims with greater speed and accuracy. The chapter also addresses ethical considerations in automated decision-making and its impact on financial well-being and customer trust.

Chapter 16: Personalization and User Autonomy in Self-Service Experiences

Concluding the volume, this chapter synthesizes themes of personalization and autonomy in SSTs. It explores how these factors enhance loyalty, trust, and satisfaction by aligning technology with evolving user expectations. Through industry insights and theoretical frameworks, the chapter offers a roadmap for embedding personalization in future SST strategies.

We have brought together a truly international and interdisciplinary group of contributors—academicians, technologists, industry professionals, and designers—whose work reflects the changing contours of digital service ecosystems. It is our sincere hope that this compilation becomes a valuable resource for researchers, innovators, policymakers, and educators seeking to understand, implement, or critique SSTs in their varied forms.

Above all, this book is a call to look beyond the interface—to understand the societal, emotional, and ethical layers embedded within self-service technology. As we collectively navigate this transformation, our responsibility is not just to automate—but to innovate thoughtfully and inclusively.

We express our gratitude to all the contributors for their insightful chapters and to the publishing team for their unwavering support. Let this volume serve as both a reference and an inspiration for building smart, accessible, and human-centered self-service solutions in the years to come.

Balraj Verma
Chitkara Business School, Chitkara University, India

Mike Schulze
CBS International Business School, Germany

Timilehin Olasoji Olubiyi
West Midlands Open University, Nigeria

Kamal Upreti
Christ University, India

Chapter 1 Optimizing User Experiences With AI-Enabled Self-Service Technologies in the Banking Industry

Devapitchai Joel Jebadurai
https://orcid.org/0000-0002-6947-8497
St. Joseph's College of Engineering, OMR, Chennai

K. Madhumitha
https://orcid.org/0009-0002-9944-2513
St. Joseph's College of Engineering, OMR, India

K. Dheenadhayalan
https://orcid.org/0009-0009-0334-7249
Sri Sai Ram Engineering College, India

M. Manikandan
SRM Institute of Science and Technology, India

S. Usha
Sri Sairam Engineering College, India

S. Karunakaran
St. Joseph's Institute of Technology, India

ABSTRACT

Self-service technologies (SSTs) empower banking customers to conduct transactions independently, offering convenience and flexibility. In the technological era, Self-Service Technologies (SSTs) play a crucial role in the banking sector. Tech-savvy generations are enjoying the services of SSTs. This chapter is conceptual and contains secondary sources. This chapter covered the objective of studying the influence of SST technologies in the banking industry. This chapter discusses AI technologies, such as machine learning, predictive analytics, and natural language

DOI: 10.4018/979-8-3373-4667-0.ch001

processing, and their impact on SSTs, including ATMs, IVR, and kiosks, for the benefit of customers. This chapter also explores recent case studies, benefits, future research directions, and challenges of SSTs in the banking industry. This information is useful for practitioners and policymakers in the field of SSTs' impact on banking.

INTRODUCTION

Self-service technologies (SSTs) have become integral to the modern banking ecosystem, playing a pivotal role in redefining how individuals access and engage with financial services. These technologies have significantly enhanced customer autonomy, allowing users to conduct almost all banking transactions independently, without the need to visit a physical branch after opening an account. This paradigm shift has transformed the traditional banking model, offering customers unmatched convenience, round-the-clock accessibility, and operational flexibility. Furthermore, SSTs encompass a wide range of digital tools, including web-based platforms, mobile banking applications, automated teller machines (ATMs), Internet of Things (IoT) devices, and unstructured supplementary service data (USSD) technology (Elgahwash & Freeman, 2015). These innovations enable customers to seamlessly perform various banking activities such as depositing and withdrawing cash, transferring funds, paying utility bills, and managing personal banking information from any location, at any time. The widespread adoption of SSTs has led to a more responsive and customer-centric banking environment. By eliminating the constraints of conventional banking hours and long in-branch queues, SSTs enhance customer satisfaction and encourage greater financial engagement. From the banking institutions' perspective, these technologies improve operational efficiency by offloading routine transactions to self-service platforms, thereby reducing staff workload and allowing personnel to focus on higher-value services such as financial consulting, investment advice, and loan assistance.

The evolution of SSTs began with the introduction of ATMs, which offered customers the ability to access cash beyond regular banking hours a significant breakthrough at the time. With the advent of the internet and smartphones, SSTs rapidly expanded to include online and mobile banking services, enabling users to manage their finances on the go. This progression has aligned perfectly with the expectations of a digitally literate customer base that demands speed, personalization, and seamless service delivery. In today's digital-first era, banks leverage SSTs not only to meet customer expectations but also to expand their market reach and build stronger, more lasting relationships with clients. The comprehensive suite of modern banking services—including ATMs, mobile apps, and internet portals has become a strategic asset for financial institutions looking to enhance service quality, boost customer retention, and remain competitive in an increasingly digital marketplace. This paper is divides into seven sections. The first section explains about the evolution of SSTs in banking. Second sections explain about the Artificial Intelligence (AI) integration with SSTs in banking. Third section explains about the types of SSTs in the banking industry, comparative evaluation of SSTs in banking- adoption rate, success rate and customer satisfaction. Fourth section explains about the benefits and challenges of SSTs in the banking industry. Fifth section explains the comparison of SSTs benefits Vs challenges in the banking industry. Sixth section discussed about the trends of SSTs in the banking industry. Seventh section explains about the Traditional Vs SSTs in banking. Final section explains about the conclusion, future research directions and takeaways in this research.

The objective is to understand how the SSTs technologies transform the banking services used by the customers. The following are the objectives were fulfilled by the entire journey of this chapter. They are.

1. To examine how AI enabled SSTs in the banking industry improve the users experience in the modern era.
2. To study the difference between the traditional and SSTs enabled banking services.
3. To present the benefits and challenges and the future trends of AI technologies in the banking industry.

METHODOLOGY

This research is conceptual in nature. The research has framed the conceptual models to explain the SSTs in the banking industry. In order to support the various concepts presented in this research, secondary data only collected. It is collected from different sources like websites, journals, and so on. The collected data includes the evolution of SSTs in the banking industry, types of SSTs available like ATMs, card issuance machines, cash recycler machines, kiosks, banking chatbots, interactive answer portals and interactive voice response, online banking, Mobile banking and Digital banking etc. It also covers the recent trends that occurs in SSTs in banking sector and its benefits and challenges.

EVOLUTION OF SELF-SERVICE TECHNOLOGIES IN BANKING

The journey of SSTs in banking began with the introduction of Automated Teller Machines (ATMs) in the late 20th century. These early innovations laid the groundwork for what would become a major transformation in how customers interact with financial institutions. Initially, self-service in banking was limited to basic ATM functionalities such as cash withdrawals and balance inquiries, which allowed customers to access essential services outside of traditional banking hours. Over time, the capabilities of ATMs expanded to include functions like cash deposits, fund transfers, and bill payments, reflecting a growing demand for convenience and around-the-clock accessibility. The emergence of the internet in the late 1990s brought about the next wave of innovation in online banking. This enabled customer to manage accounts, transfer funds, and pay bills from the comfort of their homes, significantly enhancing customer convenience while reducing operational costs for banks. As a result, financial institutions began offering a broader range of self-service options.

The rise of smartphones in the 2000s further revolutionized the banking experience Karani & Jadhav 2025).. Mobile banking apps introduced new levels of accessibility and functionality, including mobile check deposits, real-time notifications, budgeting tools, and integration with third-party financial services. These user-friendly features changed customer behavior, with many opting for mobile banking over traditional methods due to its speed, convenience, and availability.

In recent years, emerging technologies such as artificial intelligence (AI) and chatbots have begun reshaping the self-service landscape. Digital assistants now provide customers with 24/7 support, helping them access account information, answer questions, and navigate complex transactions. These AI-driven tools not only improve customer satisfaction but also personalize the banking experience based on individual preferences. Simultaneously, advancements in biometric authentication such as fingerprint and facial recognition have enhanced the security and trustworthiness of self-service platforms. These features offer customers greater confidence in the safety of digital transactions. The COVID-19 pandemic further accelerated the adoption of digital banking solutions, as customers increasingly turned to remote options to minimize physical interactions. In response, banks invested heavily in strengthening their self-service infrastructure (Kim & Chen 2025). Today, features like Peer-to-Peer (P2P) payments, digital wallets, and comprehensive financial management tools have become standard offerings, underscoring the rapid evolution of SSTs in banking. Looking ahead, the industry is expected to continue innovating by integrating technologies such as machine learning, AI, and advanced data analytics to enhance personalization and streamline services even further. The evolution of self-service technologies in banking mirrors the broader shift toward digitalization and customer empowerment. This transformation is not only improving operational efficiency for financial institutions but also reshaping the future of banking by offering a more seamless, personalized, and tech-driven customer experience.

AI INTEGRATION WITH SSTs IN BANKING

Artificial Intelligence (AI) has significantly transformed Self-Service Technologies (SST) across various sectors, and the banking industry is no exception. The integration of AI with SST has brought revolutionary changes to banking services, enabling customers to experience advanced, virtualized banking solutions. Figure 1 illustrates the integration of AI in SST technologies within the banking sector.

Figure 1. AI integration with SST in banking

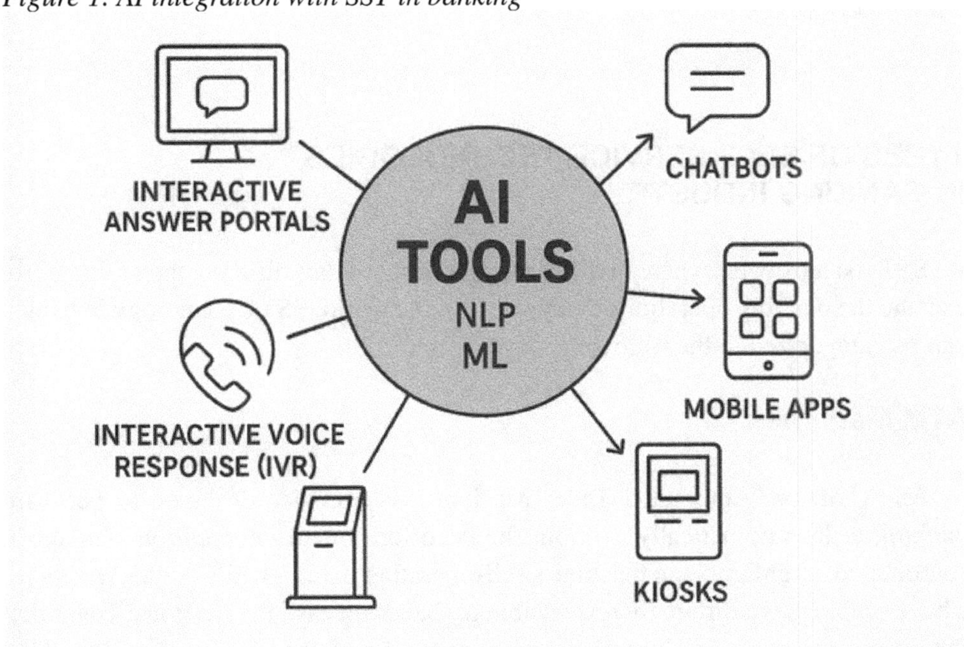

Source: Authors compilation

Table 1. AI-Enabled Technologies in SSTs of Banking

AI-Enabled Technologies	Definition	Applications in SST
Natural Language Processing (NLP)	NLP involves enabling computers to understand, interpret, and generate human language.	Chatbot, IVR
Decision Trees	Decision trees use a graphical representation to make decisions based on data, assisting in classification and regression tasks.	Chatbot
Machine Learning (ML)	ML algorithms allow computer systems to learn from data without explicit programming.	ATMs
Predictive Analytics	Predictive analytics involves using statistical techniques, machine learning algorithms, and historical data to make predictions about future events.	Self-service online banking
Pattern recognition	Pattern recognition involves identifying regularities and patterns in data using algorithms and techniques from machine learning, statistics, and signal processing.	ATMS, Digital Banking, Kiosks

Source: Omoge et.al. (2022)

TYPES OF SELF-SERVICE TECHNOLOGIES IN BANKING INDUSTRY

SSTs occupy everywhere in the banking industry. It has different technologies to execute the banking operations everywhere in the world. SSTs technology in banks can be categorized in the following ways.

ATM Machines

An ATM, or "Automated Teller Machine," is a device designed to perform banking tasks automatically, without the need for human intervention. The term "automated" highlights the machine's self-operating nature, while "teller" refers to a bank employee traditionally responsible for handling cash transactions. Together, the term describes a machine that replicates the functions of a human teller. The origins of the ATM can be traced back to the Banko-graph, a device invented by Luther George Simian. Unveiled in 1960, the Banko graph was celebrated in a New York Times article titled "Robot Bank Teller Is Invented." This device allowed users to deposit cash and checks, utilizing an internal camera to capture images of the deposits, which were then provided to customers as receipts. The first true ATM was introduced in 1967 by Barclays Bank in London, England (Patila et.al 2024). This early model enabled customers to withdraw a fixed amount of cash using a specially designed paper voucher and a Personal Identification Number (PIN).

Allow withdrawals, deposits, and account inquiries. ATMs are revolutionizing banking by providing 24/7 access to financial services, which changed how consumers interact with their bank with technological advancements, modern ATMs now have features like contactless connectivity, biometric security, and connectivity to mobile banking apps. These developments demonstrate the flexibility of ATMs to meet customer needs and technological changes, complicating their role in today's economy.

Table 2. Types of ATMs and Functions

Types of ATMs	Functions
Cash dispenser	It is used to physically dispense cash to user when they initiate withdrawal and allow them to access money from their bank accounts through the machine.
Full-service ATMs	It is used to perform various banking transactions which include deposits, withdrawals, balance inquiries, bill payments and update bank books.
Recycling ATMs	It is used to accept cash deposits and withdrawals. It also verifies the authenticity of currency notes.
Bitcoins ATMs	It is a kiosk which allows the user to purchase or sell cryptocurrencies using terminals.
Orange label ATMs	It is used to share transactions takes place in the market. Such transactions include buying and selling of securities and stocks. And it is used by investors and traders.
Yellow label ATMs	It is used to conduct e-commerce transactions where the customers are allowed to perform online shopping and bill payments.
Green label ATMs	It is used to make agricultural transactions where the crop loans, subsidies, insurance are transacted.
White label ATMs	It is owned and operated by non-banking entities and it allows customers to perform banking transactions such as deposits, withdrawals, and bill payments.
Pink label ATMs	It is designed specifically for women which offers a safe and secure environment for banking transactions. It allows them to withdraw cash, inquire about balance, transfer funds and deposit.
Brown label ATMs	It allows users to perform cash withdrawals, balance inquiries, mini statements and so on but with the hardware and leasing which is managed by a third-party service provider while the sponsor bank handles the management of cash and network connectivity.

Source: Authors compilation

CASH DEPOSIT MACHINES/BULK NOTE RECYCLING MACHINE

Cash Deposit Machines

Cash Deposit Machines (CDM) acts as an effective self-service terminal that eliminates the need for a bank teller by enabling customers to deposit money straight into their bank accounts. Their main functions include making deposits possible at any time and offering 24/7 convenience, which is particularly helpful for people with hectic schedules. Additionally, CDMs provide quick credit to accounts, guaranteeing that money placed is usable right away. By doing this, the long wait times connected with conventional banking techniques are eliminated, enabling speedy transactions without the inconvenience of handling cash denominations or filling out deposit slips (Ismail & Pastory 2024). Furthermore, CDMs improve security by lowering the hazards associated with bringing large sums of cash to a bank branch, which makes them a safer choice for customers. All things considered, CDMs simplify banking, increasing its accessibility and use.

Cash recycler machines

Cash recycling machines are advanced cash handling systems designed to automate the processes of accepting, verifying, storing, and dispensing cash. When users deposit banknotes, the machine authenticates and sorts them by denomination into secure cassettes. A key advantage of these machines is their ability to "recycle" cash using the same deposited notes for future withdrawals—which significantly improves cash flow efficiency (Nahar et.al 2025). They can reduce manual cash handling time by up to 80%, allowing staff to focus more on customer service. Additionally, when integrated with management software, these machines boost security by minimizing cash exposure and enabling real-time monitoring of cash levels.

KIOSKS

The Reserve Bank of India (RBI) launched kiosk banking as a way to promote financial stability and inclusion in rural and isolated regions of the nation where access to all banking services is restricted, primarily because there aren't any bank branches nearby. No matter who they are or where they live in the nation, financial inclusion through kiosk banking is crucial to guaranteeing the financial stability of every individual. Customers using kiosk banking are unable to visit the bank due to the scarcity of bank branches (Koon, & Hosein 2024). Rather, the bank travels to the region to handle transactions, grant credit, and make these services available to low-income populations. Kiosks are little internet-connected booths set up in villages with staff to assist clients in using basic banking services. Withdrawals, deposits, remittances, and other services are typically offered.

Interactive Kiosks

Interactive kiosks are self-service devices that enable users to access banking information and perform a wide range of financial transactions without the need for direct human assistance. Commonly located in bank branches, these kiosks are also found in convenient public areas such as shopping malls and airports. They offer a comprehensive suite of services, including account management tasks like downloading statements, checking balances, viewing transaction history, and updating personal details. Users can carry out transactions such as cash deposits and withdrawals, fund transfers, bill payments, and mobile phone recharges. The kiosks also support card-related services, including PIN reset requests, reporting lost or stolen cards, card activation, and new card applications (Naik 2024). Additionally, customers can initiate loan applications, upload required documents, and track the status of their submissions. These devices provide access to important information such as interest rates, service fees, branch locations, and product offerings, while also enabling users to engage with customer service through FAQs, live chat or video support, and complaint submission.

Card issuance and collection kiosks

Bank card issuance and collection kiosks are self-service machines that allow customers to conveniently obtain new cards or retrieve existing ones without the need for direct interaction with a bank teller. These kiosks offer a fast and hassle-free way for users to manage their card-related needs, while also helping banks reduce operational costs and enhance overall customer satisfaction.

Self-service kiosk/Self Service passbook printers

Self-service passbook printers are compact, automated machines commonly installed in bank branches or dedicated self-service areas, designed to allow customers to independently update their passbooks without requiring assistance from bank staff. These passbooks, which serve as compact records of a customer's account activity including deposits, withdrawals, and interest accrual—can be updated at any time, offering flexibility and convenience, especially for individuals with busy schedules. By enabling customers to manage their updates without queuing or disclosing their financial details publicly, these machines enhance privacy, reduce wait times, and improve overall satisfaction. The automated process minimizes the risk of human error, ensuring accurate transaction entries while saving time for both customers and bank personnel. This efficiency allows bank staff to focus on more complex tasks, improving overall service quality. With their intuitive, user-friendly interfaces, self-service passbook printers are accessible to users across all age groups and levels of technical proficiency (Thakral & Kamra 2024). Additionally, they help banks cut operational costs by reducing the need for staff dedicated to manual updates. As the demand for convenient and technology-driven banking grows, self-service solutions like passbook printers play a crucial role in enhancing the customer experience and streamlining banking operations.

Multi-Function Kiosks

A significant advancement in self-service technology within the banking sector is the introduction of multi-function kiosks, which streamline various banking processes by allowing customers to perform multiple transactions independently. By integrating a range of services into a single device such as cash withdrawals, deposits, account inquiries, and loan applications—these advanced kiosks enhance customer convenience while reducing reliance on in-person assistance. This not only accelerates transaction times but also helps alleviate congestion in bank branches, enabling staff to focus on more complex customer needs. The adoption of multi-functional kiosks aligns with the broader trend of digital transformation in banking, offering a modern, efficient solution that meets the growing consumer demand for speed, accessibility, and convenience. Ultimately, embracing this technology allows financial institutions to boost customer satisfaction, improve operational efficiency, and remain competitive in an increasingly digital financial landscape.

OPERATING MECHANISM OF KIOSKS

Chen et.al. (2015) has explained the operating mechanism of the kiosks.

Step 1: Customer Identification

The process begins with the customer authenticating their identity. This may involve entering account details or a customer ID, undergoing biometric verification (such as fingerprint or facial recognition), or scanning a government-issued ID or bank document. The system then verifies the customer's identity using secure banking databases.

Step 2: Card Selection

Once identification is successful, the kiosk displays available card options—such as debit cards, credit cards, and prepaid cards. The customer selects the preferred card type and may also choose additional features, such as contactless functionality or international usage.

Step 3: Application and Approval

After selecting the card type, the customer completes the required application form using the kiosk's touchscreen. The data is then transmitted to the bank's server for validation. If the application meets predefined eligibility criteria, approval is granted—often instantly for pre-qualified customers.

Step 4: Card Printing and Encoding

Upon approval, the kiosk prints the physical card with the customer's details (e.g., name and card number). It also encodes the magnetic stripe and/or EMV chip with secure, encrypted data. Optional features, such as QR codes or contactless chips, may also be embedded. This step is fully automated and secure.

Step 5: Card Dispensing

Once the card is ready, the kiosk presents it through a secure dispensing tray. A notification is displayed on the screen prompting the customer to collect the card. The system ensures that only the authenticated user receives the card.

Step 6: Card Collection

The customer retrieves the card, and the kiosk confirms successful collection through sensor feedback. Optionally, a receipt or digital confirmation is generated. Finally, the session is securely terminated.

Figure 2. Operating Mechanism Kiosks

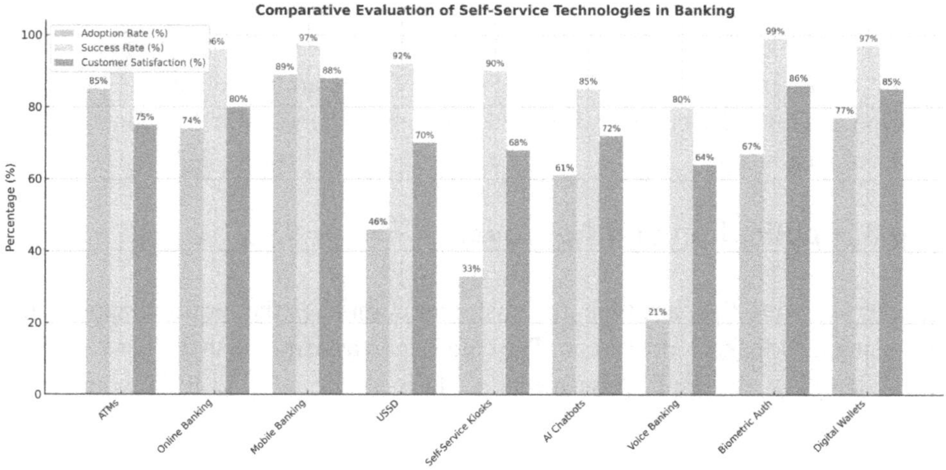

Source: Chen et.al. (2015)

INTERACTIVE TECHNOLOGIES

The banking industry's adoption of interactive technologies has fundamentally changed how consumers receive and interact with financial services. Through a variety of digital platforms, these developments enable real-time communication and engagement, enabling customers to access their accounts, complete transactions, and get individualized support. Consequently, banks are able to improve client happiness and optimize their processes, which eventually results in increased market competitiveness and efficiency.

Chatbots

AI-powered digital assistants called banking chatbots are made to interact with consumers via a variety of online platforms, including messaging apps, websites, and mobile apps. They use machine learning and Natural Language Processing (NLP) to comprehend consumer inquiries and carry out a variety of banking duties on their own (Chaithanya et.al 2023). They can also serve as virtual agents for banks, answering questions about accounts, carrying out transactions, and instantly resolving client issues. Clearer expectations of banking chatbots' capabilities have been developed as they grow more prevalent in the technical infrastructures of financial organizations. For their bots to provide informed, human-like interactions over as many channels as feasible, institutions want them to be friendly, capable, adaptable, and safe. Kiran et.al 2023 have explained about the various types of chatbots. They are.

Table 3. Types of chatbots and its application in the banking industry

Type of Chatbots	Description	Common Use Cases
Informational Chatbots	Provide customers with static information based on predefined responses	To check the account balance, branch hours, interest rates, FAQs
Transactional Chatbots	Enable users to perform banking transactions securely through the chat interface	To do the fund transfers, bill payments, balance checks, loan payments
Advisory Chatbots	Offer financial advice based on customer data and preferences	To make an investment guidance, savings plans, loan options
Customer Support Chatbots	Handle customer queries and complaints, often with escalation options	To do an password resets, dispute resolution, service complaints
Onboarding Chatbots	Assist new customers through the registration or account-opening process	To execute an document upload, eligibility checks, application status updates
AI-Powered Virtual Assistants	Use NLP and machine learning to engage in human-like conversations and learn from interactions	To enjoy an personalized banking, cross-selling, proactive alerts
Voice-Enabled Chatbots	Integrated with voice assistants or IVR systems for hands-free interaction	To use the voice banking, mobile app integration, accessibility support

Source: Kiran et.al 2023

INTERACTIVE ANSWER PORTALS

Interactive Answers Portals (IAP) is a popular chatbot technology used by banks to resolve the queries of the customers at any time. The banking sector is increasingly utilizing interactive response systems to boost client engagement and service delivery. Financial institutions and their customers can communicate in real time using these portals, enabling prompt support and information retrieval. Banks can offer a more personalized services and effectively handle consumer questions and issues by utilizing technology.

These interactive platforms are an essential tool for banks looking to increase client happiness and streamline processes. They give users access to a variety of services in an intuitive interface, ranging from transaction inquiries to account management. This lessens the strain for customer support agents while simultaneously empowering customers to solve problems on their own. Implementing interactive answer portals is becoming crucial for banks looking to stay competitive as the financial landscape changes. In addition to improving operational effectiveness, these systems strengthen the bond between banks and their clients (Fahn & Riener, 2021). Financial institutions can better satisfy the needs of contemporary customers by putting the user experience first with these creative ideas. For example, ICICI Bank's iPal is an interactive platforms used by ICICI bank to answer customer queries, perform banking services and offer personalized services to the customers.

INTERACTIVE VOICE RESPONSE (IVR)

Interactive voice response, or IVR, is an automated telephone system technology that enables callers to receive or provide information, or make requests using voice or menu inputs, without speaking to a live agent. IVR is powered by a pre-recorded messaging or text-to-speech technology with a dual-tone multi-frequency (DTMF) interface. By using this technology, banks have transformed the customer services in automated basis. IVR structures allow clients to interact with an employer's database via voice commands or with the aid of choosing options on their telephone keypads, streamlining the system of gaining access to facts and services. This technology allows users to perform various obligations, including bank account balances, making payments, or troubleshooting troubles without the need for human intervention, which may lessen operational costs for agencies.

The implementation of IVR structures ensures that customers can access services 24/7, enhancing comfort and enhancing customer satisfaction. Moreover, advancements in Natural Language Processing and artificial intelligence have in addition more advantageous IVR systems, making them more person-pleasant and capable of know-how complicated queries. However, challenges remain regarding consumer experience, as poorly designed IVR structures can result in frustration and elevated call abandonment fees. To mitigate these issues, organizations should be conscious on creating intuitive and green IVR workflows that prioritize user level, thereby enhancing engagement and retention (Coman, 2025). As the era continues to adapt, the combination of IVR with other emerging technologies, inclusive of Chatbots and machine Learning, will possibly enhance its capabilities, making it a crucial tool for present-day customer service strategies. Enables clients to perform basic banking tasks through smartphone calls. ICICI bank have used the Interactive Voice Response banking with the help of conversational voice BOT to help the customers to execute their banking services without human intervention.

SELF-SERVICE ONLINE BANKING

Online banking provides access to account information, fund transfers, and payments through digital platforms. It has transformed how both consumers and businesses manage their finances, shifting away from traditional banking channels to more accessible online solutions. The rise of Internet banking in the 1990s, driven by the growth of the Internet, allowed users to manage their accounts, communicate with banks, and conduct transactions from the comfort of their homes or even on the go via mobile phones. One of the key benefits of online banking is its convenience (Moist et.al 2021). Customers can transfer funds, make payments, check account balances, and access various services at any time without visiting a physical branch. This shift not only saves time but also helps banks reduce operational costs and the need for numerous physical locations.

The speedy adoption of online banking additionally raises safety and privacy issues. Cybersecurity threats, which include phishing attacks, identity theft, and information breaches are substantial risks that online banking clients should navigate. Financial establishments have spoken back with robust security features, including two-component authentication, encryption, and advanced fraud detection techniques, to create dependable client data and offer security for Web websites. In addition, regulatory bodies have implemented stringent pointers and compliance measures aimed at improving customer protection inside the digital banking region. As generation advances, the future of online banking seems promising, with tendencies along with synthetic intelligence and system gaining knowledge of being integrated into banking structures These improvements include banking revel in personalized range, centered financial recommendations and complicated fraud detection methods Making it feasible. Furthermore, the rise of fintech organizations has multiplied competition inside the banking quarter, streamlining transactions and improving purchaser experience Online banking is not simply an alternative to conventional department banking; It has turn out to be an imperative part of the monetary system to meet the evolving desires of clients and agencies.

SELF-SERVICE MOBILE BANKING

Mobile banking has rapidly converted the monetary landscape, supplying customers with unprecedented comfort and flexibility in handling their budgets. Emerging in the early 2000s, cell banking allows customers to conduct a number of banking sports via smartphones and capsules, which include bank account balances, transferring budget, paying payments, or even depositing exams through mobile apps (Afonso & Teixeira 2022).

The proliferation of smartphones and high-velocity internet has improved the adoption of cellular banking, making financial offerings available to a broader demographic, including the ones in far-flung areas. This shift no longer only empowers purchasers but additionally reduces the operational costs for banks, as they can serve clients without having physical branch visits (Aslam, et.al 2023). Security stays a massive problem; but, banks have implemented superior measures inclusive of biometric authentication, encryption, and actual-time transaction alerts to defend consumer records and decorate trust.

DIGITAL BANKING

The automation of traditional banking services through digital platforms, which enables consumers to conduct financial transactions online using computers or mobile devices, is known as digital banking. It includes a variety of services that are available around the clock without requiring in-person bank visits, such as account management, money transfers, bill payments, and loan applications (Tsindeliani et.al 2022). This change lessens dependency on paper-based procedures and face-to-face contacts while increasing ease and efficiency.

Table 4. Services under digital banking

Services	Functions
Online account opening	Can able to create new accounts, including credit card, checking, and savings accounts.
Online Loan application	It allows the customer to apply for a variety of loans, including home, auto, and personal loans, etc.
Investment Avenues	Make investments in mutual funds, equities, bonds, and other financial products.
Term Deposits account opening	It enables customers to open accounts for term deposits
TDS	Make the tax payments more convenient to individuals.
Bill Payment	Pay your credit card, loan, utility, and other service invoices.
Mobile Wallet	For both in-person and online transactions, save and utilize digital payment options.
Account Management	View statements, transaction histories, and account balances.
Fund Transfers	Move money between customer personal accounts or to accounts at different banks or the same bank.

Source: Authors compilation

COMPARATIVE EVALUATION OF SSTs IN BANKING- ADOPTION RATE, SUCCESS RATE AND CUSTOMER SATISFACTION

Self-service technologies (SSTs) have emerged as widely adopted, user-friendly tools in the global banking industry. In 2022, Deloitte conducted a Banking Maturity Survey focusing on key aspects of SSTs, including their adoption rate, success rate, and customer satisfaction. Among all the SSTs in banking, adoption rate data reveals that the maximum 85% of the consumer were adopted the ATMs, with respect to success rate the maximum biometric authentication technologies contributes maximum with 99 percent, and 88 percent of the customers were satisfied with the mobile banking. The chart below illustrates users' perspectives on the effectiveness and impact of SSTs in the banking sector.

Figure 3. Comparative evaluation of SSTs in Banking

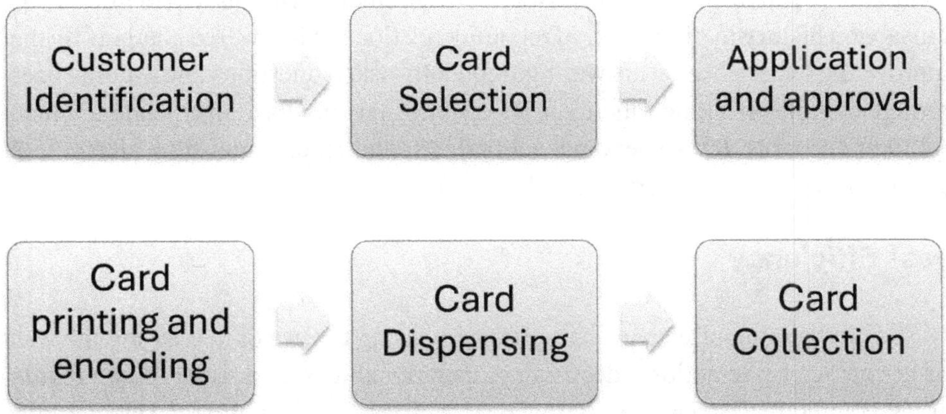

Source: Deloitte digital banking maturity 2022

BENEFITS OF SELF-SERVICE TECHNOLOGIES IN THE BANKING INDUSTRY

Chan, C. Y. T., & Petrikat, D. (2022) have explained the benefits and challenges of SSTs in the banking industry.

Convenience

The convenience provided by SSTs in the banking sector stands out as one of their most impactful advantages, significantly transforming how customers engage with their financial institutions. With options such as mobile banking apps, ATMs, and online platforms, users can access a wide range of banking services anytime and from virtually any location, removing the need to visit physical branches or adhere to limited operating hours. This around-the-clock accessibility empowers customers to perform essential tasks such as checking account balances, transferring funds, paying bills, and managing investments at their convenience. For individuals with busy lifestyles or those in remote areas, this ease of access is a game-changer, allowing them to handle financial tasks in minutes from home, during commutes, or even while traveling. Digital self-service tools also deliver a faster, more efficient experience, reducing transaction times and minimizing the friction associated with traditional methods, which enhances overall satisfaction. In addition, many of these technologies support personalized services by remembering user preferences and transaction history to offer tailored recommendations. AI-powered chatbots further improve the experience by answering frequently asked questions and guiding users through complex banking functions. By enabling customers to independently manage their accounts, track spending, set budgets, and receive real-time alerts, SSTs promote financial empowerment and control.

Cost Efficiency

SSTs in banking offers vast cost efficiency using the use of decreasing the want for in-character interactions, decreasing operational fees, and streamlining regular duties. By automating services like balance inquiries, fund transfers, and invoice bills, banks can deal with a higher number of transactions with fewer bodies of employees, liberating up resources to awareness of extra complex client needs. This automation reduces the want for physical infrastructure, along with department places and cashiers, and minimizes overhead costs related to manual processing. Additionally, Self Service platforms often require a lot less protection and much less operational hours compared to conventional banking techniques, similarly decreasing prices. Over time, those financial savings may be reinvested in improving generation or increasing services, making self-service answers a fairly rate-effective model for bank that reduces operational expenses for banks with the aid of minimizing the want for branch workforce.

Enhanced Accessibility

SSTs substantially enhance accessibility in banking by supplying customers with the capability to manage their monetary activities anytime and everywhere. With on line platforms, cellular apps, and ATMs, clients are no longer restrained by means of department hours or physical places. This is specifically useful for individuals in rural or underserved regions, as they can get entry to banking offerings without having to tour lengthy distances. Additionally, those technologies often include features like multi-language assist, voice instructions, and adjustable font sizes, making banking greater inclusive for human beings with exclusive desires, inclusive of those with disabilities or confined virtual literacy. The result is a more inclusive banking revel in that caters to a broader variety of clients, making sure that economic offerings are inside attain for anybody. Provides services to underserved and rural areas through cell and online systems.

Improved Customer Experience

SSTs significantly enhances the customer experience with the aid of imparting faster, greater handy, and personalized banking interactions. With 24/7 get right of entry to via cellular apps, ATMs, and online systems, customers can carry out transactions, test balances, and manipulate accounts at their personal comfort, lowering wait instances and the want for in-man or woman visits. These technologies are regularly designed with consumer-friendly interfaces, making it less complicated for customers to navigate services quickly and correctly. Additionally, AI-driven equipment like chatbots can provide on-the-spot assistance, answering commonplace queries and guiding customers through banking strategies. Self-service options additionally offer extra transparency, as customers can get entry to actual-time information and tune their economic hobby seamlessly, fostering an experience of manipulation and empowerment. Ultimately, Self Service technology streamlines the banking method, making it greater accessible, green, and aligned with clients' expectancies for velocity and convenience. Offers quicker and extra personalized offerings.

Scalability

Self-service technologies (SST) enable banks to efficiently manage a growing customer base without a corresponding increase in resources by offering scalable, cost-effective solutions. These technologies can be remotely updated, allowing banks to quickly deploy new features and maintain consistent service quality across all customer touchpoints, regardless of location. This scalability allows banks to expand operations, handle higher transaction volumes, and introduce new services seamlessly, making SST a vital tool for future growth and operational efficiency in the banking industry.

CHALLENGES OF SELF-SERVICE TECHNOLOGIES IN THE BANKING INDUSTRY

The banking industry, traditionally based on conventional customer service models, has experienced a dramatic shift with the rise of self-service technologies like mobile banking apps, ATMs, chatbots, and AI-powered financial tools, which offer greater convenience, lower operational costs, and improved customer satisfaction. However, integrating and implementing these technologies within the banking ecosystem poses significant challenges that are presented here.

Security Concerns

Security is absolutely the biggest undertaking, as automation technology is embedded within the banking enterprise. With growing digital connectivity and get admission to to banking services via cellular apps, online structures and kiosks, banks are liable to cyberattacks, breaches of truth and fraud Customers care about their money issues aside from the commissioning of public budgets, bobbing up from any weaknesses in the ones systems Security breaches that might pose a chance isn't the weakest threats to clients to consider and but may want to bring about tremendous economic losses, criminal liabilities, and damage to the bank's recognition

To mitigate the dangers, banks ought to provide robust safety features, consisting of unpassed encryption, multi-factor authentication (MFA), and biometric authentication (which includes fingerprint scanning and face popularity) These make it extra hard for malicious actors to get entry to a convenient customer along records Design and implementation add complexity and, at the same time as cyber threats are ongoing, banks should make investments effectively in ongoing monitoring, habitual infant custody audits and rapid reaction processes to save you any escalation of vulnerabilities.

Integration with Legacy Systems

A major barrier to widespread adoption of self-service technology in banking is integrating these solutions with existing legacy systems, which are often rigid and incompatible with modern digital tools. Self-service platforms like AI chatbots, digital wallets, and mobile apps must seamlessly connect with backend databases, transaction systems, and Customer Relationship Management (CRM) tools, posing both technical and organizational challenges. Retrofitting or replacing legacy systems can be costly and complex, especially for smaller banks. Even successful integration requires ongoing maintenance of both old and new systems, increasing operational costs and resource demands. Poor integration can lead to system downtime, data inconsistencies, and reduced service quality, ultimately undermining the benefits of self-service technology.

User Adoption and Digital Literacy

While self-service technologies provide amazing comfort, one of the essential barriers to their large adoption is consumer reluctance, mainly among certain demographics. Older customers, for instance, may be much less acquainted with smartphones or mobile banking apps and may decide upon face-to-face interactions with a bank group of workers. These customers may sense worried approximately the usage of automatic structures to manipulate their price range, as they understand them to be a greater at risk of errors or safety risks.

Maintaining Personalized Service

Traditional banking's strength lies in the personal interaction between customers and financial representatives, which remains essential for handling complex financial matters that technology alone cannot fully address. The challenge is to ensure that autonomous technologies complement rather than replace personalized service by incorporating human touchpoints within virtual platforms. For instance, AI-powered chatbots can manage simple tasks like balance inquiries, while more complex issues are promptly escalated to human advisors. Additionally, self-service technologies should offer adaptable, personalized experiences—such as customized dashboards and financial recommendations tailored to individual customer profiles—to avoid feelings of manipulation. Ultimately, a balanced approach that integrates automation with accessible human support is key to delivering effective and empathetic banking services.

Regulatory Compliance

The banking industry is among the most heavily regulated globally, requiring strict compliance with regulations such as GDPR, Anti-Money Laundering (AML), and Know Your Customer (KYC) to prevent illegal activities like money laundering and illegal gambling. As banks adopt self-service technologies, these systems must align with regulatory frameworks by accurately capturing and verifying customer identities and ensuring secure data storage and access control. Non-compliance can lead to severe fines, reputational damage, and legal consequences. Moreover, banks must continuously adapt to evolving regulations, which may necessitate costly updates and add complexity to the development and operation of self-service platforms.

Ongoing Maintenance and Technological Advancements

The speedy pace of technological alternate method that self-provider systems should constantly evolve to fulfil new purchaser expectations and live ahead of emerging threats. Once self-provider technologies are implemented, they require ongoing protection, updates, and improvements to stay practical and aggressive. This includes ensuring that the generation is well matched with the state-of-the-art working systems, devices, and protection protocols, as well as addressing insects or overall performance troubles that stand up.

Moreover, as new technological advancements, together with synthetic intelligence, block chain, and biometric authentication, hold to convert the banking panorama, banks need to remain agile and geared up to integrate those improvements into their Self Service offerings. Keeping up with technological trends calls for substantial funding in research and development, personnel schooling, and system updates, which may be useful resource-extensive.

Table 5. Comparison of SSTs Benefits Vs challenges in the banking industry

Aspects	Benefits	Challenges
Customer Convenience	24/7 availability and faster service delivery	Lack of personal interaction may frustrate some customers
Operational Efficiency	Reduces workload for staff and cuts operational costs	High initial setup and maintenance costs
Accuracy & Speed	Minimizes human error and accelerates transactions	System errors or downtime can affect service reliability
Personalization	AI-driven SST offers customized services based on user data	Privacy concerns and data security issues
Customer Empowerment	Enables users to perform a range of services independently	Digital illiteracy among certain user groups
Scalability	Easily scalable across locations and user volumes	Requires constant updates and technological upgrades
Data Insights	Facilitates data collection for better decision-making	Risk of misuse or mishandling of sensitive customer data

Source: Authors compilation

CASE STUDIES AND SUCCESS STORIES OF SSTs IN THE BANKING INDUSTRY

HDFC Bank, one of India's largest private sector banks, has led the adoption of self-service technologies to enhance customer satisfaction and operational efficiency. The bank offers various channels including internet banking, ATMs, mobile apps, and self-service kiosks—enabling customers to independently perform tasks like balance checks, bill payments, fund transfers, and cash deposits (https://www.hdfcbank.com/). Key technologies include the "HDFC Bank ATM," mobile banking app, self-service kiosks for account opening, loan applications, and card requests, as well as innovative services like chatbots for instant support and missed call banking for quick access to account information.

Bank of Baroda, one of India's largest public sector banks, has actively adopted self-service technologies to boost operational efficiency and enhance customer satisfaction. The bank offers multiple self-service channels, including internet banking, ATMs, mobile apps, and kiosks, enabling customers to independently perform tasks such as balance inquiries, bill payments, fund transfers, and cash deposits. Key innovations include the "Baroda Connect" online platform and the "Baroda M-Connect" mobile app, both providing secure and convenient banking access. Additionally, self-service kiosks facilitate services like card requests, loan applications, and account openings without staff assistance. The bank also provides advanced facilities like the "Baroda Passbook Printer" and the "Baroda Cash Deposit Machine" to further streamline customer transactions.

State Bank of India (SBI), the country's largest public sector bank, has extensively implemented self-service technologies to enhance operational efficiency and customer satisfaction. SBI provides various self-service channels such as internet banking, ATMs, mobile apps, and kiosks, enabling customers to independently perform tasks like cash deposits, bill payments, fund transfers, and balance inquiries. Key offerings include the "SBI Anywhere Personal" mobile app and the secure "SBI Online" platform. Additionally, SBI has introduced advanced machines for cash deposits, passbook printing, and cheque deposits. Self-service kiosks at branches further allow customers to open accounts, apply for loans, and request cards without staff assistance.

ICICI Bank, one of India's largest private sector banks, has pioneered self-service technologies to enhance customer satisfaction and operational efficiency. The bank offers various channels such as internet banking, ATMs, mobile apps, and self-service kiosks, enabling customers to independently perform tasks like balance inquiries, bill payments, fund transfers, and cash deposits. Notable innovations include the "iMobile" app for mobile banking and ICICI's ATMs that support multiple transactions. Additionally, self-service kiosks at branches allow customers to request cards, apply for loans, and open accounts without human assistance.

Indian bank has been at the forefront of implementing self-service technologies to increase customer satisfaction and operational effectiveness. One notable example is the introduction of micro-deposit ATMs, such as the "Tijori" machine, which allows users to deposit money and perform other tasks independently. Indian banks have also adopted a number of self-service technologies, including self-service kiosks, ATMs, mobile banking apps, and internet banking (https://www.indianbank.in/). These technologies allow customers to perform a number of tasks without the need for human assistance, such as bill payment, fund transfers, cash withdrawals, and balance inquiries. As a result, Indian bank has been able to increase banking efficiency, decrease wait times, and improve customer convenience while also addressing security concerns.

HSBC Bank, a leading global financial institution, has widely adopted self-service technologies to enhance customer satisfaction and operational efficiency. The bank offers various channels such as kiosks, ATMs, mobile apps, and internet banking, allowing customers to independently perform tasks like cash deposits, bill payments, fund transfers, and balance inquiries. Key technologies include the secure "HSBC Online Banking" platform and the "HSBC Mobile Banking" app (https://www.hsbc.co.in/). Self-service kiosks enable account openings, loan applications, and card requests without staff assistance. Additionally, "HSBC Express Banking" and advanced ATMs provide further convenience for routine transactions.

Citi Bank, a leading international bank in India, has actively implemented self-service technologies to enhance customer satisfaction and operational efficiency. It offers various channels such as online banking, mobile apps, ATMs, and self-service kiosks, enabling customers to independently perform transactions like balance inquiries, fund transfers, bill payments, and cash deposits. Key technologies include the secure "Citi Online" platform, the "Citi Mobile" app, ATMs, and cash deposit machines. Self-service kiosks at branches also allow customers to open accounts, apply for loans, and request cards without staff assistance.

TRENDS IN FUTURE OF SELF-SERVICE TECHNOLOGIES IN THE BANKING INDUSTRY

Rise of Omni-Channel Banking

Omni-channel banking is a significant advancement in the banking sector, designed to deliver a seamless customer experience across multiple channels such as contact centers, ATMs, online platforms, mobile apps, and physical branches. By adopting a unified approach, banks empower customers to access their accounts and manage their finances through any preferred channel with maximum freedom and convenience. The primary objective of omni-channel banking is to enable customers to effortlessly switch between channels without losing context or progress, thereby meeting modern demands for multichannel, personalized experiences (Abediniyan et.al 2021). Achieving a true omni-channel experience requires data integration across all platforms, allowing banks to understand customer needs better, offer tailored services, and maintain consistency in interactions regardless of the channel used.

Leveraging Artificial Intelligence and Machine Learning

Artificial Intelligence (AI) and Machine Learning (ML) are transforming the banking sector by enhancing automation and self-service capabilities. AI-powered chatbots and virtual assistants provide instant customer support and transaction services, while ML analyzes vast amounts of data to detect fraud, personalize recommendations, and optimize operations. By evaluating customer behavior and preferences, AI enables banks to offer tailored financial advice and product suggestions (Witts 2024). Additionally, ML facilitates real-time fraud detection and strengthens risk management by assessing credit scores, loan applications, and market trends, helping banks make informed decisions and minimize risks.

Biometric Authentication for Secure Access

Biometric identification techniques are increasingly adopted in banking as a secure and convenient alternative to traditional passwords, enhancing both security and user experience. Methods such as fingerprint scanning, widely used in ATMs and mobile apps, offer reliable authentication, while facial recognition, popular in online platforms and mobile banking, enables easy and secure account access (Syed et.al 2024). Iris scanning, valued for its high security, is increasingly employed to protect sensitive data and high-value transactions, ensuring robust protection for customers' accounts and services.

Integrated Mobile Banking Apps

Apps for mobile banking have grown to be a crucial component of contemporary banking. In order to provide a fully integrated and seamless financial management platform, banks are constantly adding sophisticated functionality to their mobile apps. Customers may use these apps to check account balances, transfer money, pay bills, and even apply for loans, among other banking tasks, at any time and from any location.

Customers are given customized dashboards displaying their spending patterns, financial objectives, and pertinent financial goods by mobile banking apps, which are becoming more and more personalized. With the app's built-in capabilities, users can track their spending, manage their budgets, and establish financial objectives (Adil et.al 2020). They are better able to manage their money and make wise financial decisions as a result. Peer-to-peer transfers, contactless payments, and mobile wallets are just a few of the payment methods that mobile banking apps are incorporating to streamline transactions and give users a more pleasant experience.

Self-Service Kiosks and Interactive Tellers

Self-service kiosks and interactive tellers are transforming traditional banking by providing customers with faster and more convenient access to services such as bill payments, deposits, withdrawals, and balance inquiries. These kiosks offer a convenient alternative to waiting in line, allowing customers to complete transactions independently and at their own pace (Vakulenko et.al 2018). By combining self-service ease with virtual teller support, interactive tellers deliver personalized assistance while helping banks streamline operations, reduce wait times, boost productivity, and enable staff to focus on more complex customer needs.

Voice-Activated Banking Assistants

Voice-activated financial assistants, powered by AI and speech recognition, offer a convenient, hands-free way for users to manage their money by enabling them to check balances, make transactions, and receive personalized financial advice through simple voice commands (Qureshi,et.al 2024). These assistants enhance accessibility for people with disabilities or visual impairments and allow users to handle financial tasks on the go without needing a computer or mobile app, making banking more inclusive and efficient.

Predictive Data Analytics

Predictive data analytics plays a crucial role in personalizing the banking experience by enabling banks to anticipate customers' future needs and preferences through the analysis of extensive data such as demographics, transaction history, and online behavior (He et.al 2023). Using advanced algorithms to identify patterns and insights, banks can proactively offer tailored financial products, investment advice, and services, leading to higher customer satisfaction and helping clients achieve their financial goals.

TRADITIONAL VS SSTs BANKING

Traditional banking services become nameless once SSTs based banking services were introduced, but still people are expecting to visit the bank branches for to get their banking services. Curran & Meuter (2005) discussed about the differences between the traditional banking with the SSTs banking.

Table 6. Traditional Vs SSTs Banking

S.No	*Basis*	*Traditional Banking*	*Self-Service Technologies (SSTs)*
1	**Service Delivery**	Customers can interact the face to face interaction with the bank staff.	It is automated, digital, and customer-driven processes.
2	**Accessibility**	Customers are limited to visit the banking hours and physical branches.	Customers can access the banking services by 24/7 via ATMs, kiosks, mobile apps, and internet banking.
3	**Speed**	It is slower due to all the works are executed by manual processing.	It is faster, updated and execute all the banking transactions instantly, and immediately respond to customer queries.
4	**Human Interaction**	It is always required human interaction, and need a personal relationship with staff.	It is machine interaction, user navigates system independently
5	**Operational Cost**	It is high due to staffing, infrastructure, and overheads	Lower operational cost due to automation.
6	**Error Rate**	Human involvement leads to huge errors.	Reduced error through automation and standardization
7	**Personalization**	Limited and dependent on staff knowledge	AI-driven personalization based on user data and behaviour
8	**Technology Dependency**	Minimal—mostly paper-based or basic computer systems	High—relies on software, networks, and AI-driven interfaces
9	**Security**	Physical document-based verification	Advanced digital security (biometrics, encryption, OTPs)
10	**Customer Control**	Bank staff manage transactions	Customers have full control over transactions

Source: Curran & Meuter (2005)

CONCLUSION AND DIRECTIONS OF FUTURE RESEARCH

SSTs (Self-Service Technologies) have notably transformed the banking sector by enhancing service efficiency and accessibility. However, their adoption faces challenges such as security issues, limited user awareness, and the need for robust infrastructure. This research focuses solely on the impact of SSTs and their services on the banking industry. Future studies can explore how AI-driven SSTs will shape the future of banking by offering smarter, more personalized services. Researchers can also investigate the feasibility and effectiveness of implementing AI-based banking in rural areas, and analyze how SSTs contribute to a country's economic growth through financial inclusion and improved service delivery.

TAKE AWAYS

- Self-Service Technologies (SSTs) in the banking industry, including kiosks, mobile banking, digital banking, online platforms, and ATMs, offer customers enhanced convenience and expedited transaction processing.
- Although the adoption of SSTs is increasing steadily, challenges such as cybersecurity threats, transaction failures, and network connectivity issues continue to impede the seamless delivery of banking services.
- The integration of advanced AI-driven technologies, including Natural Language Processing (NLP), machine learning, and predictive analytics, is propelling the capabilities of SSTs to new levels of efficiency and sophistication.

ABBREVIATIONS

SSTs- Self-Service Technologies
ATM-Automated Teller Machine
PIN-Personal Identification Number
CDM-Cash Deposit Machines
RBI-Reserve Bank of India
AI-Artificial Intelligence
NLP-Natural Language Processing
KYC-Know your customer
AML-Anti-Money Laundering
FAQ-Frequently asked Questions
IVR-Interactive Voice Response
TDS-Tax deducted at source
MFA-Multi-Factor Authentication

CRM-Customer Relationship Management
GDPR-General Data Protection Regulation
ML-Machine Learning
IOT- Internet of Things

REFERENCES

Abediniyan, A., Azar, F. M., & Nazemi, E. (2021, December). Designing a model to use Omnichannel in banking industry based on BIAN framework. In 2021 5th National Conference on Advances in Enterprise Architecture (NCAEA) (pp. 1-10). IEEE. DOI: 10.1109/NCAEA54556.2021.9690509

Adil, M., Nasir, M., Sadiq, M., & Bharti, K. (2020). SSTQUAL model: Assessment of ATM service quality in an emerging economy. *International Journal of Business Excellence*, *22*(1), 114–138. DOI: 10.1504/IJBEX.2020.109222

Afonso, C., & Teixeira, J. D. (2022). Challenges in Adapting to Mobile Banking From the Perspective of the Customers. In Implementing Automation Initiatives in Companies to Create Better-Connected Experiences (pp. 91-109). IGI Global. DOI: 10.4018/978-1-6684-5538-8.ch005

Aslam, W., de Luna, I. R., Asim, M., & Farhat, K. (2023). Do the preceding self-service technologies influence mobile banking adoption? *IIM Kozhikode Society & Management Review*, *12*(1), 50–66. DOI: 10.1177/22779752211073552

Bank of Baroda. (n.d.). *Bank of Baroda*. Retrieved March 29, 2025, from https://www.bankofbaroda.in/

Chaithanya, D., Deepa, G., Pavani, K. B., Manjusha, G., & Kumar, V. V. (2023, June). Chatbot for banking sector using NLP and ML. In International Conference on Soft Computing and Signal Processing (pp. 153-163). Singapore: Springer Nature Singapore.

Chan, C. Y. T., & Petrikat, D. (2022). Self-service technology: Benefits and challenges. *Journal of Computer Science and Technology Studies*, *4*(2), 118–127. DOI: 10.32996/jcsts.2022.4.2.14

Chen, J. V., Yen, D., Dunk, K., & Widjaja, A. E. (2015). The impact of using kiosk on enterprise systems in service industry. *Enterprise Information Systems*, *9*(8), 835–860. DOI: 10.1080/17517575.2013.867542

Citibank. (n.d.). *Citibank*. Retrieved March 30, 2025, from https://www.citi.com/

Coman, E. (2025). IVR systems used in call center management: A scientometric analysis of the literature. *Frontiers of Computer Science*, *7*, 1459787. DOI: 10.3389/fcomp.2025.1459787

Curran, J. M., & Meuter, M. L. (2005). Self-service technology adoption: Comparing three technologies. *Journal of Services Marketing, 19*(2), 103–113. DOI: 10.1108/08876040510591411

Elgahwash, F. O., & Freeman, M. B. (2015). Self-service technology banking preferences: comparing Libyans' behaviour in developing and developed countries. In Banking, Finance, and Accounting: Concepts, Methodologies, Tools, and Applications (pp. 714-727). IGI Global.

Fahn, V., & Riener, A. (2021). Time to get conversational: Assessment of the potential of conversational user interfaces for mobile banking. In Proceedings of Mensch und Computer 2021 (pp. 34-43).

HDFC Bank. (n.d.). *HDFC Bank.* Retrieved March 28, 2025, from https://www.hdfcbank.com/

He, W., Hung, J. L., & Liu, L. (2023). Impact of big data analytics on banking: A case study. *Journal of Enterprise Information Management, 36*(2), 459–479.

ICICI Bank. (n.d.). *ICICI Bank.* Retrieved March 30, 2025, from https://www.icicibank.com/

Indian Bank. (n.d.). *Indian Bank.* Retrieved March 30, 2025, from https://www.indianbank.in/

Ismail, A. M., & Pastory, D. (2024). Evaluating the impact of self-service cash deposit machines on the performance of commercial banks in Tanzania. *Future Business Journal, 10*(1), 5–16. DOI: 10.1186/s43093-023-00291-4

Karani, G., & Jadhav, B. (2025). How Traditional Banks Adapt to the Fintech Revolution in Banking and Finance Sector. In Shaping Cutting-Edge Technologies and Applications for Digital Banking and Financial Services (pp. 214-225). Productivity Press. DOI: 10.4324/9781003501947-13

Kim, B., & Chen, Y. (2025). Empowering consumers through self-service technology: A comparative analysis. *Journal of Hospitality & Tourism Research (Washington, D.C.), 49*(2), 254–267. DOI: 10.1177/10963480231182959

Kiran, A., Kumar, I. J., Vijayakarthik, P., Naik, S. L., & Vinod, T. (2023, January). Intelligent Chat Bots: An AI Based Chat Bot For Better Banking Applications. In *2023 International Conference on Computer Communication and Informatics (ICCCI)* (pp. 1-4). IEEE. DOI: 10.1109/ICCCI56745.2023.10128582

Koon, J. K., & Hosein, P. (2024, November). Optimal Placement of Self-Service Banking Kiosks. In *2024 IEEE International Conference on Technology Management, Operations and Decisions (ICTMOD)* (pp. 1-6). IEEE.

Moist, J. A., Anitsal, I., & Anitsal, M. M. (2021). The Customer Value Model and Mobile Banking: Evaluation of Technology-Based Self Service (TBSS) Gaps. *Atlantic Marketing Journal, 10*(2), 12–19.

Nahar, K., Megha, A. H., & Sarker, J. U. (2025). Entering the Digital Banking Era: A Mechanism to Change and Accelerate Financial Access in Bangladesh. *International Journal of Innovative Science and Research Technology, 10*(3), 2750–2761. DOI: 10.38124/ijisrt/25mar1048

Naik, A. S. (2024, June). Assistive Technology in Banking. In 2024 15th International Conference on Computing Communication and Networking Technologies (ICCCNT) (pp. 1-6). IEEE. DOI: 10.1109/ICCCNT61001.2024.10724888

Omoge, A. P., Gala, P., & Horky, A. (2022). Disruptive technology and AI in the banking industry of an emerging market. *International Journal of Bank Marketing, 40*(6), 1217–1247. DOI: 10.1108/IJBM-09-2021-0403

Patila, H., Mahandulea, V., Khachanea, L., & Narkhedea, S. (2024). Multi-banking ATM system services using biometrics. *Management Science Letters, 14*(4), 219–226. DOI: 10.5267/j.msl.2024.3.002

Qureshi, O. A., Wilkins, S., & Iqbal, H. (2025). AI-Powered Banking Assistants: Enhancing Customer Service with Technology. In BUiD Doctoral Research Conference (pp. 43-53). Springer, Cham.

State Bank of India. (n.d.). *State Bank of India*. Retrieved March 28, 2025, from https://www.onlinesbi.sbi/

Syed, W. K., Mohammed, A., Reddy, J. K., & Dhanasekaran, S. (2024, July). Biometric Authentication Systems in Banking: A Technical Evaluation of Security Measures. In 2024 IEEE 3rd World Conference on Applied Intelligence and Computing (AIC) (pp. 1331-1336). IEEE. DOI: 10.1109/AIC61668.2024.10731026

Thakral, S., & Kamra, V. (2024). Self-service technologies: A bibliometric analysis. *International Journal of Financial Engineering, 11*(04), 2442007. DOI: 10.1142/S2424786324420076

Tsindeliani, I. A., Proshunin, M. M., Sadovskaya, T. D., Popkova, Z. G., Davydova, M. A., & Babayan, O. A. (2022). Digital transformation of the banking system in the context of sustainable development. *Journal of Money Laundering Control, 25*(1), 165–180. DOI: 10.1108/JMLC-02-2021-0011

Vakulenko, Y., Hellström, D., & Oghazi, P. (2018). Customer value in self-service kiosks: A systematic literature review. *International Journal of Retail & Distribution Management*, *46*(5), 507–527. DOI: 10.1108/IJRDM-04-2017-0084

Witts, J. O. (2024). Leveraging Artificial Intelligence (AI) for Banks in Emerging Markets. *Journal of Advanced Management Science*, *12*(4), 84–1045. DOI: 10.18178/joams.12.4.167-172

KEY TERMS AND DEFINITIONS

Anti-Money Laundering: Anti-money laundering (AML) efforts consist of the laws, regulations and procedures that are designed to prevent criminals from exchanging money obtained through illegal activities—or "dirty money"—into legitimate income or "clean money."

Artificial intelligence: Artificial intelligence (AI) is technology that enables computers and machines to simulate human learning, comprehension, problem solving, decision making, creativity and autonomy.

Block chain: Block chain is a decentralized digital ledger system that securely and openly records transactions via a network of computers. It eliminates the need for a central authority or middleman by enabling several parties to agree on the state of a database. Block chain protects the data's integrity and immutability by using encryption and a series of blocks, which makes it impervious to fraud and manipulation.

IoT: The network of physical objects, automobiles, household appliances, and other things that are equipped with sensors, software, and connection so they can gather and share data is known as the Internet of Things (IoT). This makes it possible for these gadgets to communicate with one another and the real world, making the world more automated and interconnected.

KIOSK in Banking: A kiosk is a self-service terminal used in banking that enables users to complete a variety of financial tasks without a bank teller's help. Cash withdrawals, check deposits, fund transfers, bill payment, and account information access are a few examples of these transactions. Essentially, it's a computerized, automated version of many standard financial services.

Machine Learning: Machine learning (ML) is a branch of artificial intelligence (AI) focused on enabling computers and machines to imitate the way that humans learn, to perform tasks autonomously, and to improve their performance and accuracy through experience and exposure to more data.

Multi-Factor Authentication: A security procedure known as multi-factor authentication (MFA) requires users to supply two or more authentication factors in order to get access to a system, network, or application. This makes it harder for unauthorized users to obtain access by adding an extra layer of security to the conventional username and password combination. MFA elements may consist of your knowledge (password), possessions (token), or identity (biometric).

Natural language processing: Natural language processing (NLP) is a machine learning technology that gives computers the ability to interpret, manipulate, and comprehend human language

Predictive data Analytics: Predictive analytics is the use of data to predict future trends and events. It uses historical data to forecast potential scenarios that can help drive strategic decisions.

Chapter 2
Humanizing Automation: Designing Self-Service Technologies That Balance Efficiency and Emotional Intelligence

Semiu Olawale Makinde
https://orcid.org/0000-0001-5456-3464
Al-Hikmah University, Ilorin, Nigeria

AbdulRafiu Abdullahi
https://orcid.org/0009-0003-2896-2367
Al-Hikmah University, Ilorin, Nigeria

Shehu Sakariyahu
https://orcid.org/0009-0008-7377-9007
Al-Hikmah University, Ilorin, Nigeria

Azeez Tunbosun Lawal
https://orcid.org/0000-0001-9595-9019
Al-Hikmah University, Ilorin, Nigeria

Akeem Apo Ahmed
https://orcid.org/0009-0008-9908-6866
Al-Hikmah University, Ilorin, Nigeria

W. C. Tang
Lock Tao Secondary School, Hong Kong

ABSTRACT

Self-service technologies (SSTs) are transforming customer interactions by providing faster, more efficient, and easily accessible services. From automated checkouts to AI-powered chatbots, these innovations streamline operations and lower costs. However, their lack of emotional intelligence often results in user frustration and weakened trust. This chapter highlights the need for a human-centred design (HCD) approach to create SSTs that blend efficiency with empathy. By integrating artificial intelligence, natural language processing, and sentiment analysis, businesses can develop systems that recognize and respond to user emotions. While AI and IoT advancements enhance personalization, addressing ethical concerns like data privacy and bias is crucial. Emotionally aware SSTs can improve customer satisfaction,

DOI: 10.4018/979-8-3373-4667-0.ch002

strengthen engagement, and build lasting relationships.

INTRODUCTION

The increasing adoption of self-service technologies (SSTs) has significantly transformed the way businesses interact with customers. These technologies, which enable users to perform services independently without direct assistance from human personnel, have gained popularity due to their unparalleled convenience, speed, and efficiency. From automated checkouts in retail stores to self-service kiosks in airports and chatbots in online customer service, SSTs have become ubiquitous, redefining traditional service paradigms (Tienken et al., 2023). However, as automation continues to advance and permeate various industries, there is a growing concern about the absence of emotional intelligence and empathy in these interactions, which have traditionally been hallmarks of human engagement (Will Arachchige et al., 2024). This chapter delves into the critical need for designing SSTs that harmonize operational efficiency with emotional intelligence, emphasizing the role of human-centred design (HCD) (Piffari et al., 2022) in creating seamless and empathetic user experiences.

SSTs are lauded for their ability to enhance efficiency, reduce operational costs, and provide 24/7 accessibility. According to Davis et al. (2019), businesses that integrate SSTs often report improved customer satisfaction due to fast service delivery, accurate, and reduced waiting times. Additionally, these technologies empower customers by giving them greater control over their service experiences. For instance, automated banking services, such as ATMs and mobile banking apps, allow users to perform transactions at their convenience without relying on bank tellers. Despite these advantages, the limitations of SSTs become evident in scenarios requiring emotional sensitivity or complex problem-solving. Studies by Zhang et al. (2021) reveal that users often feel frustrated or disconnected when self-service technologies (SSTs) fail to address their emotional needs or provide personalized assistance, ultimately eroding trust and customer loyalty. Moreover, Chan and Petrikat (2022) pointed out that "Dehumanization destroyed the possibility of exchanging real-time client input" (p.123).

In human-computer interactions, emotional intelligence—the ability to identify, understand, and appropriately respond to emotions—is essential for fostering positive user experiences. Integrating emotional intelligence into SSTs involves designing features that exhibit empathy, build trust, and support personalization. Research by McLean and Wilson (2020) emphasizes the critical role of empathy in customer-facing technologies, asserting that emotionally intelligent SSTs can significantly enhance customer satisfaction, engagement, and loyalty. Besides, technological tools may affect learners' emotional states and engagement levels, ultimately shaping the educational experience (Mutawa & Sruthi, 2024).

This chapter examines the role of human-centred design (HCD) in creating self-service technologies (SSTs) that address user needs, emotions, and overall experiences. HCD is a design philosophy that emphasizes the centrality of users in the development process, ensuring that technologies are intuitive, accessible, and responsive to emotional contexts. Central elements of HCD include conducting thorough user research, iterative prototyping, and rigorous usability testing.

The adoption of advanced technologies such as artificial intelligence (AI), machine learning (ML), and the Internet of Things (IoT) presents significant opportunities for enhancing the emotional intelligence of self-service technologies (SSTs). AI-driven systems, for instance, can process vast datasets to anticipate user preferences and offer highly personalized experiences. Similarly, IoT-enabled devices can gather real-time data, allowing services to adapt dynamically to user behavior and environmental conditions. These innovations have the potential to harmonize efficiency with empathy in SSTs, fostering more engaging and human-like interactions. Nonetheless, the ethical challenges (Wylde et al., 2023) associated with such technologies—particularly issues like data privacy and algorithmic bias—must be carefully addressed to ensure these advancements reflect user values and maintain trust. Therefore, the objectives of this chapter are outlined as follows:

(1.) *examining the current state of self-service technologies (SSTs).* This section focuses on the advantages and challenges of automation in customer interactions. Utilizing case studies and industry reports, it provides an in-depth analysis of SST applications across various industries, including retail, healthcare, and hospitality;

(2) *Highlighting the importance of emotional intelligence in human-computer interactions.* This part emphasizes the significance of empathy, trust, and personalization in fostering customer loyalty. Drawing from existing literature, it identifies critical gaps in the integration of emotional intelligence into technology and discusses strategies for enhancing user satisfaction;

(3.) *Investigating the principles of human-centred design (HCD) in developing SSTs:* This section explores methodologies and best practices for creating technologies that prioritize empathy. It examines real-world examples of HCD success stories while addressing the challenges of incorporating emotional intelligence into the design process; and (4.)

Exploring future directions for SSTs: The final section considers potential technological advancements, particularly the integration of artificial intelligence (AI), machine learning (ML), and Internet of Things (IoT) technologies. It also examines the ethical and societal implications of these innovations and provides recommendations for developing user-centered, balanced solutions.

The chapter underscores the need for a paradigm shift in designing and deploying SSTs, advocating for the integration of emotional intelligence alongside operational efficiency. By adopting human-centred design principles and leveraging emerging technologies, businesses and developers can create SSTs that address functional needs while building trust, empathy, and customer loyalty (Wadhwa et al., 2024).

FRAMEWORK

This chapter explores the intersection of efficiency and emotional intelligence in self-service technologies (SSTs), examining how automation can be designed to provide seamless yet emotionally engaging user experiences. The literature was sourced from reputable academic databases, including Scopus, Web of Science, Google Scholar, and IEEE Xplore, ensuring that the most relevant and recent studies were included. The analysis incorporated various academic materials, such as peer-reviewed journal articles, conference proceedings, books, and industry reports. The discussion was structured around key thematic areas.

First, the chapter examines the evolution and current state of self-service technologies (SSTs), providing an overview of their adoption across industries such as retail, healthcare, and hospitality. The section discusses the advantages of SSTs, including cost reduction, convenience, scalability, and efficiency, while also addressing the challenges associated with their implementation, such as technical failures, lack of personalization, and user resistance. Case studies of Amazon Go's self-checkout system, telemedicine kiosks (Patil et al., 2025), and AI-powered hotel check-ins are presented to illustrate both the potential and limitations of current SSTs.

Second, the role of emotional intelligence (EI) in technology is explored, highlighting the importance of empathy, trust, and personalization in human-computer interactions. This section draws from affective computing, sentiment analysis, and human-computer interaction (HCI) research to examine how emotionally intelligent SSTs can enhance customer satisfaction and brand loyalty. Examples of emotionally aware AI technologies, such as empathetic chatbots (e.g., Woebot), sentiment-aware voice assistants (e.g., Alexa and Google Assistant), and AI-driven customer service platforms, are discussed to demonstrate the impact of EI in self-service environments.

Third, the chapter investigates the principles of Human-Centered Design (HCD) in SSTs, emphasizing usability, accessibility, and inclusivity. This section examines methodologies such as participatory design, iterative prototyping, and user feedback mechanisms that contribute to the creation of emotionally intelligent automation. Case studies illustrate how HCD principles have been successfully applied in self-checkout interfaces, digital healthcare assistants, and conversational AI for banking, demonstrating how a user-focused approach enhances engagement and usability.

Fourth, the challenges in integrating HCD into SSTs are analyzed, focusing on technical limitations, ethical concerns, and user resistance. Issues such as AI bias, data privacy, and accessibility gaps are discussed, highlighting the importance of ethical AI development. The potential risks of over-automation, loss of human jobs, and consumer distrust are also examined, emphasizing the need for balanced design strategies that ensure both efficiency and emotional engagement.

Fifth, the chapter explores future innovations in SSTs, including advances in machine learning (ML), the Internet of Things (IoT), and affective computing. The discussion highlights how emerging technologies can enhance personalization and adapt SSTs to individual user preferences and emotional states. The ethical considerations of these innovations, particularly regarding bias, accessibility, and societal impact, are also addressed.

Finally, the chapter presents recommendations for designing ethical, inclusive, and user-friendly SSTs. Strategies such as hybrid service models (human + AI collaboration), empathetic AI development, and continuous user feedback loops are proposed to create a balanced self-service ecosystem. Hence, this chapter offers a comprehensive framework for building the next generation of emotionally intelligent and efficient self-service technologies.

Overview of Self-Service Technologies (SSTs) and Their Adoption across Industries

Self-Service Technologies (SSTs) have revolutionized the way businesses interact with their customers by enabling automated service delivery. These technologies are designed to allow customers to perform various services independently, without the need for direct assistance from employees (Dabholkar, 2023). The widespread adoption of SSTs across industries has been driven by the desire to enhance customer convenience, improve operational efficiency, and reduce costs.

The journey of SSTs began with simple automated solutions such as vending machines and ATMs in the mid-20th century. These early forms of automation set the stage for more complex SSTs, including online banking platforms, self-checkout systems, and telemedicine kiosks. Today, SSTs are ubiquitous, spanning industries such as retail, healthcare, and hospitality, where they streamline processes, enhance customer convenience, and reduce operational costs.

Adoption across Industries

SSTs have seen significant implementation across various sectors, including retail, healthcare, hospitality, and banking, each adopting technology in ways that suit their operational needs and customer preferences.

1. Retail: The retail sector was one of the first to adopt self-service technologies, with self-checkout systems becoming a commonplace feature in many supermarkets, grocery stores, and department stores. These systems allow customers to scan and pay for items without the need for a cashier (Pfeiffer, 2022). Moreover, self-checkout systems enable retailers to handle high volumes of transactions during peak times, such as holidays or sales events.
2. Healthcare: In healthcare, SSTs have evolved through the introduction of telemedicine kiosks and automated check-in systems at hospitals and clinics. Telemedicine kiosks, which allow patients to consult with healthcare providers remotely, have made healthcare more accessible, particularly in rural or underserved areas (Bansal et al., 2022). These kiosks are equipped with video conferencing technology, diagnostic equipment, and electronic health records to ensure a seamless consultation process.
3. Hospitality: In the hospitality sector, self-service technologies such as hotel check-in kiosks and mobile check-in apps have been widely adopted. These systems enable guests to check in and check out without interacting with front desk staff, reducing waiting times and improving the guest experience (Kim & Koo, 2023).
4. Banking: In the banking industry, SSTs such as ATM machines, online banking portals, and mobile banking apps have become integral to daily operations. These technologies allow customers to perform a wide range of financial transactions, from withdrawing cash to paying bills and transferring money, all without the need for human intervention (Venkatesh et al., 2022).

Impact on Customer Experience and Business Operations

SSTs not only enhance the customer experience by offering convenience and flexibility but also benefit businesses by improving operational efficiency and reducing costs (Parasuraman, 2023). The ability to serve more customers with fewer resources makes SSTs particularly appealing to businesses facing high demand and labor shortages. Moreover, customers appreciate the autonomy that SSTs provide, as they can access services at their own convenience without the need to wait for assistance.

Self-service technologies have made a significant impact across industries by transforming traditional service delivery models. Their adoption in retail, healthcare, hospitality, and banking demonstrates the versatility and scalability of SSTs. While the benefits of SSTs, such as cost reduction and operational efficiency, are clear, businesses must carefully manage challenges like data privacy and security (Hasal et al., 2021), technical failures, and user resistance to fully realize the potential of these technologies.

Advantages of Self-Service Technologies (SSTs)

Self-Service Technologies (SSTs) have rapidly become integral to businesses across industries, offering numerous advantages that benefit both customers and organizations. Four of the most prominent advantages include cost reduction, convenience, scalability, and efficiency. These advantages have driven the widespread adoption of SSTs, as businesses seek ways to improve operations and enhance customer experience while remaining competitive in an increasingly digital marketplace.

1. **Cost Reduction**

One of the primary reasons businesses implement self-service technologies is the significant reduction in operational costs. By automating processes, companies can minimize the need for human labor, which directly lowers staffing costs. For example, in the retail industry, self-checkout kiosks reduce the number of cashiers needed, allowing businesses to serve more customers with fewer employees (Sundararajan & Gao, 2022). In healthcare, telemedicine kiosks and automated appointment scheduling systems help healthcare providers manage patient flow without the need for additional administrative staff (Bansal et al., 2022). Moreover, businesses can cut costs related to training, employee benefits, and payroll. According to Parasuraman (2023), the long-term financial benefits of implementing SSTs often outweigh the initial investment in technology, as businesses achieve greater profitability through cost savings.

2. **Convenience**

SSTs significantly enhance customer convenience by enabling round-the-clock service access and allowing customers to complete tasks on their own terms. In the retail sector, self-checkout systems provide customers with the flexibility to complete transactions without waiting in line, while mobile apps and online platforms give customers the ability to shop, pay bills, or book services from the comfort of their own homes (Pfeiffer, 2022). This level of convenience is particularly valuable for customers with busy schedules or those in need of services during non-traditional hours. The ability to conduct consultations, receive prescriptions, and monitor health from home increases accessibility, especially for patients in rural or underserved areas (Kim & Koo, 2023).

3. **Scalability**

SSTs offer businesses the ability to scale their operations without a proportional increase in staffing or infrastructure. As demand for services rises, automated systems can handle more transactions or customers without the need for additional employees or resources. In industries like retail, scalability is crucial during peak seasons, such as holidays or sales events, when customer volumes surge. Self-checkout systems, for example, can handle a high volume of transactions simultaneously, reducing the need for additional cashiers and alleviating long lines during busy periods (Sundararajan & Gao, 2022).

In the banking sector, ATMs and mobile banking apps enable customers to complete banking transactions at any time, reducing the strain on physical bank branches (Venkatesh et al., 2022). This scalability is particularly valuable for banks expanding into new markets or offering services to a large customer base without the need for a significant increase in branch locations or staff. In healthcare, telemedicine technologies can extend services to more patients without requiring additional physical infrastructure or medical staff (Bansal et al., 2022).

4. **Efficiency**

The use of SSTs significantly improves operational efficiency by automating tasks that would otherwise require human intervention. For example, in the retail sector, self-checkout systems reduce transaction times, allowing customers to complete their purchases more quickly than waiting in line for a cashier (Pfeiffer, 2022). This efficiency not only benefits customers but also increases throughput for businesses, as more customers can be served within the same time frame. Similarly, mobile apps and online shopping platforms allow customers to browse and purchase products with ease, eliminating the need for time-consuming in-store shopping experiences.

In the healthcare sector, automated appointment scheduling and check-in systems improve efficiency by streamlining patient intake processes and reducing administrative burdens (Miller et al., 2023). Self-service kiosks in medical settings can capture essential patient information and update medical records automatically, reducing the need for manual data entry by administrative staff. This enhances both the accuracy of patient records and the speed with which services are delivered.

Integration of Technology for Enhanced Benefits

The ongoing development and integration of artificial intelligence (AI), machine learning (ML), and the Internet of Things (IoT) further amplify the advantages of SSTs. AI-powered systems can improve personalization, enhance decision-making, and predict customer preferences, creating more efficient and satisfying service experiences (Parasuraman, 2023). For example, AI-driven recommendation engines in retail suggest products based on past purchase behavior, improving both the convenience and efficiency of the shopping experience.

In hospitality, IoT technologies such as smart rooms enable guests to control lighting, temperature, and other amenities through mobile apps or voice commands, enhancing both convenience and efficiency (Kim & Koo, 2023). These technologies allow businesses to manage resources more effectively while providing guests with personalized experiences.

Challenges in Self-Service Technologies (SSTs)

While Self-Service Technologies (SSTs) offer numerous advantages, their implementation and widespread use are not without challenges. Despite their potential to streamline operations and improve customer experience, businesses often encounter issues related to technical failures, lack of personalization, and user resistance. These challenges can undermine the effectiveness of SSTs and affect both customer satisfaction and business performance. Addressing these challenges is critical for organizations seeking to maximize the benefits of SSTs while ensuring their customers receive a seamless and positive experience.

1. Technical Failures

One of the most significant challenges faced by businesses adopting SSTs is the risk of technical failures. Like any automated system, SSTs rely on complex software and hardware components, which can malfunction or experience technical glitches, leading to service disruptions or customer dissatisfaction. For example, self-checkout kiosks in retail stores may fail to read barcodes properly or may experience software crashes, resulting in delays and frustrating customer experiences (Sundararajan & Gao, 2022). In healthcare, telemedicine kiosks that are designed to facilitate remote consultations may suffer from connectivity issues or system errors, preventing patients from accessing medical advice in real time (Bansal et al., 2022).

Technical failures are not only frustrating for customers but can also incur significant costs for businesses. Repairing or replacing malfunctioning equipment, troubleshooting software problems, and compensating customers for poor service can create unexpected financial burdens (Dabholkar, 2023). For instance, in the banking sector, if an ATM fails to dispense cash or malfunctions during a transaction, customers may face delays in accessing their funds, leading to dissatisfaction and potential reputational damage for the financial institution (Venkatesh et al., 2022). Similarly, in the hospitality industry, an issue with self-service check-in kiosks can result in long queues, delays in guest services, and a negative experience for hotel patrons (Kim & Koo, 2023).

2. Lack of Personalization

Another significant challenge with SSTs is their lack of personalization. While automation offers efficiency and convenience, it often falls short in providing the tailored, human-centric service that many customers expect. Unlike human service agents who can engage in personalized interactions and adapt to specific customer needs, SSTs often operate on predetermined scripts or algorithms that may not account for individual preferences or unique circumstances (Bansal et al., 2022). This can result in generic, impersonal service that may frustrate customers who feel their specific needs are not being adequately addressed.

For instance, in retail, while self-checkout systems can speed up the transaction process, they do not offer the opportunity for personal interaction, such as upselling or providing personalized recommendations based on previous purchases. In contrast, a cashier can tailor the conversation to the customer's preferences, offer assistance, and make personalized suggestions (Sundararajan & Gao, 2022). In healthcare, telemedicine services, while offering convenience, may lack the personal touch that patients experience during in-person consultations. This absence of personalization may impact patient satisfaction, especially in contexts where empathy and human interaction are essential (Miller et al., 2023).

3. User Resistance

User resistance is another major challenge that businesses face when implementing SSTs. Many customers are initially hesitant or reluctant to adopt self-service technologies due to unfamiliarity, perceived complexity, or a preference for human interaction. This resistance can be attributed to several factors, including fear of technology, lack of trust in automated systems, and concerns about privacy and security.

For example, older customers or those with limited technological proficiency may struggle to navigate self-service kiosks or mobile apps, resulting in frustration and reluctance to use these technologies (Dabholkar, 2023). In some cases, customers may simply prefer face-to-face interactions with employees, especially in industries where personal service is highly valued, such as hospitality and healthcare. In these instances, customers may feel that a human representative is better equipped to understand their needs and provide solutions (Venkatesh et al., 2022). This resistance can limit the widespread adoption of SSTs and hinder their effectiveness.

Case Studies: Retail (Self-Checkouts), Healthcare (Telemedicine Kiosks), and Hospitality (Hotel Check-In Systems)

Self-Service Technologies (SSTs) have been increasingly adopted across various industries, and their implementation has led to significant shifts in the way businesses operate and interact with customers. In this section, we will explore three key case studies: retail self-checkouts, healthcare telemedicine kiosks, and hospitality hotel check-in systems. These examples demonstrate the diverse applications of SSTs and highlight their effectiveness, challenges, and potential for future development in enhancing customer experience and operational efficiency.

Retail: Self-Checkouts

The retail sector has seen a rapid adoption of self-service technologies, particularly in the form of self-checkout systems. Self-checkouts are automated kiosks that allow customers to scan, bag, and pay for their purchases without the need for cashier assistance. These systems are designed to reduce wait times, increase customer throughput, and enhance the overall shopping experience (Sundararajan & Gao, 2022). In addition to the convenience of reducing lines and speeding up the checkout process, self-checkout systems provide retailers with an opportunity to reduce labor costs by employing fewer staff at checkout counters (Pfeiffer, 2022).

However, despite the benefits, the implementation of self-checkouts has not been without challenges. Technical issues, such as barcode scanning failures or software malfunctions, can cause frustration for customers and result in longer waiting times if the systems are not functioning correctly (Dabholkar, 2023). Moreover, self-checkout systems have been criticized for lacking personalization in customer service. While these systems excel in efficiency, they do not offer the opportunity for tailored interactions, such as suggesting additional products or providing personalized assistance. Some customers, especially older individuals or those less familiar with technology, may also feel uncomfortable or confused when using the machines, resulting in a reluctance to adopt this technology (Sundararajan & Gao, 2022).

Despite these challenges, self-checkouts have been successfully implemented in various retail environments. For example, major grocery chains such as Walmart and Kroger have integrated self-checkout systems in their stores to streamline the shopping experience (Pfeiffer, 2022). These systems have led to increased customer satisfaction in terms of speed and convenience, with many customers appreciating the autonomy of completing transactions without human interaction. Retailers have also leveraged data from self-checkouts to gather insights into customer behavior and improve inventory management, providing further opportunities for efficiency and cost reduction (Sundararajan & Gao, 2022).

Healthcare: Telemedicine Kiosks

In the healthcare industry, telemedicine kiosks represent a significant innovation in delivering remote healthcare services. These kiosks are self-service machines equipped with cameras, diagnostic equipment, and software that allow patients to conduct medical consultations remotely with healthcare providers (Miller et al., 2023). Telemedicine kiosks are particularly beneficial in rural or underserved areas where access to healthcare professionals may be limited. By offering on-demand consultations for common conditions such as colds, flu, or minor injuries, these kiosks enable patients to receive medical advice without traveling long distances or waiting for appointments.

Telemedicine kiosks have gained popularity in clinics, pharmacies, and even airports, where they allow patients to quickly consult with doctors, receive prescriptions, and access other healthcare services. For example, the HealthSpot telemedicine kiosk, which was deployed in pharmacies across the United States, allowed patients to connect with doctors via video conferencing and receive prescriptions for common conditions (Bansal et al., 2022). The use of telemedicine kiosks has streamlined healthcare delivery, reducing waiting times for consultations and improving access to care for patients in remote locations. However, the implementation of telemedicine kiosks has also faced several challenges. Technical failures can disrupt the patient experience, as issues with video connectivity, malfunctioning diagnostic equipment, or poor internet connections can prevent consultations from taking place (Miller et al., 2023).

Despite these challenges, telemedicine kiosks have proven to be a valuable tool in expanding healthcare access and improving patient outcomes. Future developments in AI and machine learning could further enhance the personalization of these systems, making them more capable of addressing individual patient needs while providing accurate, efficient, and effective care.

Hospitality: Hotel Check-In Systems

In the hospitality industry, self-service hotel check-in systems have revolutionized the way guests experience the check-in process. These systems, which often take the form of kiosks or mobile apps, allow guests to bypass the front desk and check in directly from their smartphones or a self-service kiosk upon arrival (Kim & Koo, 2023). Self-service check-in systems offer a number of advantages, including faster check-in times, reduced staffing requirements, and increased guest satisfaction, particularly in busy hotels or during peak travel seasons. Guests appreciate the ability to check in quickly and go directly to their rooms without waiting in long lines, which improves the overall guest experience (Jung et al., 2022).

For example, Marriott International in downtown Bethesda, Maryland has implemented mobile check-in for its hotel guests, enabling them to check in via the Marriott app, receive room keys on their phones, and head directly to their rooms upon arrival (Jung et al., 2022). This system has been well received by guests who value the convenience of managing their hotel stay without having to interact with hotel staff. Furthermore, mobile check-in has allowed Marriott to reduce waiting times and improve operational efficiency. For some guests, the absence of personal interaction may lead to concerns about the quality of service or issues with booking errors. Additionally, technical failures, such as issues with app functionality or kiosk malfunctions, can create frustrating experiences for guests, especially if they are unable to check in properly or access their room (Dabholkar, 2023).

Despite these obstacles, the adoption of self-service check-in systems in hotels continues to grow. Many hotel chains, including Hilton and Accor, are increasingly incorporating these technologies to enhance operational efficiency and provide a more seamless and modern guest experience (Kim & Koo, 2023). As technology improves and becomes more user-friendly, it is likely that self-service check-in systems will become a standard offering in hotels around the world.

The Role of Emotional Intelligence in Technology

Emotional intelligence (EI) refers to the ability to identify, understand, and manage one's own emotions as well as the emotions of others. It involves a set of skills, including emotional awareness, empathy, self-regulation, and social skills, which play a crucial role in effective human interaction. While emotional intelligence has traditionally been associated with interpersonal relationships and human behavior, its role in technology, particularly in the context of Self-Service Technologies (SSTs), is becoming increasingly significant.

As SSTs continue to develop, the necessity for integrating emotional intelligence into such systems grows. Such integration can not only increase user satisfaction but also increase customer loyalty and promote more human-like interactions between humans and machines.

The integration of emotional intelligence enables the system to better understand and respond to the emotional needs of users. When the system is capable of sensing the user's emotional state and responding accordingly, users will feel that they are understood and valued and therefore be more satisfied. For example, when a customer feels anxious or uneasy in a customer service interaction, an emotionally intelligent system can issue a more empathetic response and make the customer feel respected and cared for.

This kind of emotional affiliation increases customer loyalty (Mostafa & Kasamani, 2021). Customers tend to repeat their use of a brand's goods or services whenever they feel that the brand appreciates their emotional needs rather than just mechanical service. Additionally, the application of emotional intelligence may create long-term customer interaction because it increases customers' feelings of attachment through continual emotion recognition and feedback.

The presence of emotional intelligence can also facilitate more interaction between humans and machines. This means that not just tools in the future, systems can be companions of humans that can help them make better choices in various situations. This increased interaction not only increases productivity at work, but it also maximizes user satisfaction and allows humans to experience more in their interaction with technology.

The Integration of Emotional Intelligence in Technology

The integration of emotional intelligence in technology focuses on creating systems that can perceive and respond to human emotions in a way that mimics empathetic human interactions. With advancements in Artificial Intelligence (AI) and Machine Learning (ML), technologies are becoming better at recognizing and understanding human emotions through facial expressions, tone of voice, and behavioral cues. This emotional awareness can then be used to tailor interactions to a user's emotional state, leading to a more personalized and responsive experience.

For example, in customer service, virtual assistants and chatbots are being designed with EI capabilities that can detect frustration or confusion in a user's tone of voice or word choice. By recognizing these emotional cues, the system can adjust its response, offer help proactively, or transfer the user to a human agent if necessary (Gunkel, 2023). Similarly, AI-driven customer service platforms are now capable of tailoring their responses based on the emotional state of the user, which can enhance the customer's experience and lead to greater satisfaction (Mikolajewicz et al., 2022). This capability is particularly important in contexts like retail, healthcare, and hospitality, where customer emotions play a pivotal role in shaping service experiences.

Empathy and Trust in Technology

One of the key aspects of emotional intelligence is empathy, the ability to understand and share the feelings of others. In the context of SSTs, empathy can make a significant difference in how customers perceive and engage with technology. When self-service systems are able to recognize and respond to emotions, they can create a more human-like interaction that helps foster trust between the user and the system. Empathy in technology often involves the ability to recognize frustration or anxiety in users and respond with reassurance or problem-solving assistance. For example, in retail environments, self-checkout systems that acknowledge a user's frustration when they make an error (e.g., incorrect barcode scan or payment issue) and offer gentle guidance or immediate human support can greatly enhance customer experience. Research by Jung et al. (2022) shows that when technology responds empathetically, customers are more likely to feel valued and understood, which can lead to greater trust in the system and brand loyalty.

In contrast, when emotional cues are ignored or misinterpreted, customers may feel alienated, frustrated, or even suspicious of the technology. This can lead to disengagement or rejection of SSTs altogether. For instance, if a chatbot fails to recognize a customer's frustration and provides generic or unhelpful responses, the customer may perceive the technology as unresponsive or robotic, which damages the emotional connection and undermines trust (Sundararajan & Gao, 2022). Therefore, for SSTs to build lasting relationships with users, they must prioritize the integration of empathy and understanding in their design.

The Role of Emotional Intelligence in Customer Satisfaction and Loyalty

The integration of emotional intelligence into SSTs has profound implications for customer satisfaction and loyalty. When users feel emotionally understood and valued, they are more likely to engage with technology again and recommend it to others. Empathetic interactions can create a positive or supportive emotional response that goes beyond functional satisfaction, fostering a sense of connection with the brand or service.

For instance, AI-driven recommendation systems used in e-commerce platforms or streaming services, when combined with emotional intelligence, can personalize not only the products or content being recommended but also the way the recommendations are delivered. If a customer expresses dissatisfaction or frustration with previous suggestions, an emotionally intelligent system could adapt its approach, offering more personalized and considerate recommendations (Mikolajewicz et al., 2022). This type of responsiveness can turn a neutral or negative experience into a positive one, reinforcing the customer's emotional bond with the brand.

In healthcare, where trust and emotional connection are essential, telemedicine kiosks with integrated emotional intelligence can make patients feel more at ease, particularly when dealing with sensitive health issues. Studies show that when patients perceive a system to be empathetic, they are more likely to use it again for future consultations, trust the diagnosis, and engage with the healthcare provider more openly (Bansal et al., 2022). This emotional connection plays a vital role in building long-term patient loyalty, which is crucial for the success of telemedicine platforms and their integration into mainstream healthcare.

Examples of Emotionally Intelligent Technologies

As self-service technologies (SSTs) continue to evolve, integrating emotional intelligence (EI) into these systems is becoming increasingly important in enhancing the quality of user interactions. Technologies that can detect, interpret, and respond to emotional cues help to create more personalized, empathetic, and effective interactions between humans and machines. In this section, we will explore several examples of emotionally intelligent technologies, focusing on empathetic chatbots, virtual assistants, and emotionally responsive customer service platforms, and discuss their impact on user experience.

1. Empathetic Chatbots

Chatbots are one of the most widely used examples of emotionally intelligent technology. These AI-powered systems are designed to simulate conversation with human users, offering automated responses to user inquiries and facilitating services such as customer support, online shopping assistance, and troubleshooting. When chatbots are equipped with emotional intelligence, they can recognize emotional cues from users—such as frustration, happiness, or confusion—and adapt their responses, accordingly, creating a more empathetic and personalized interaction.

For example, Woebot, a mental health chatbot, is designed to offer psychological support through text-based interactions. By utilizing principles of cognitive behavioral therapy (CBT), Woebot recognizes emotional cues (Soderberg, 2022) from users and responds in an empathetic manner, helping users to manage anxiety, depression, and other mental health challenges. Woebot can interpret changes in users' language patterns (e.g., the use of negative words or phrases indicating stress) and offer comforting responses, guidance, or suggest coping strategies to improve emotional well-being (Fitzpatrick et al., 2022). Woebot's ability to recognize and respond to emotional states enhances its effectiveness as a support tool, making it more engaging and trustworthy to users.

Similarly, Mitsuku, a chatbot recognized for its conversational abilities, has been enhanced with emotional intelligence to respond empathetically to users' feelings. Mitsuku's AI system can assess users' moods based on their conversational tone, adjusting its responses to provide empathetic, supportive, or humorous interactions depending on the emotional context (Kenny et al., 2022). This ability to "read" emotional cues helps the chatbot create a rapport with users, improving their overall experience and fostering trust.

2. Emotionally Intelligent Virtual Assistants

Virtual assistants, such as Apple's Siri, Amazon's Alexa, and Google Assistant, are increasingly incorporating emotional intelligence to improve user interactions. While these systems have traditionally focused on efficiency and functionality, there is a growing emphasis on enhancing their emotional awareness to make interactions more natural and engaging.

For instance, Siri has been integrated with emotional intelligence features to detect users' emotional states (Zhylin et al., 2024) through voice recognition. By analyzing the tone of a user's speech, Siri can adjust its responses to show empathy, humor, or understanding, depending on the situation. For example, if a user speaks in a frustrated tone, Siri may offer a more considerate or apologetic response, attempting to calm the situation (Mikolajewicz et al., 2022). In contrast, if a user speaks with enthusiasm, Siri might reply with more energy or excitement to mirror the user's positive emotions. This emotional responsiveness helps build rapport and a sense of connection between the user and the virtual assistant.

Similarly, Amazon's Alexa has integrated empathetic responses in specific scenarios, particularly in settings related to health and well-being. For example, when users ask Alexa for assistance with managing stress or anxiety, the assistant may offer calming music, meditation prompts, or positive affirmations. These responses are designed to create a soothing and emotionally supportive atmosphere for users, making Alexa not just a functional assistant but an empathetic presence in the home (Lee et al., 2023).

3. Emotionally Responsive Customer Service Platforms

Another prominent example of emotionally intelligent technology is the development of emotionally responsive customer service platforms. Many companies have implemented AI-powered systems that can assess the emotional state of a customer based on voice tone, facial expressions, or written language. By incorporating emotional intelligence, these platforms can provide more personalized and empathetic customer service, improving overall customer satisfaction and loyalty.

For example, Cogito, an AI platform used by call centers, uses real-time emotion analytics to detect the emotional tone of a customer's voice. It provides agents with feedback about the customer's emotional state, enabling them to tailor their responses to meet the emotional needs of the customer. If the platform detects frustration or anger in a customer's voice, it may suggest that the agent adopt a calmer and more empathetic tone to de-escalate the situation (Cohen et al., 2022). This technology can also alert agents when a customer is likely to be satisfied, enabling them to optimize their responses and maintain a positive emotional connection.

In the banking and telecom industries, companies like Bank of America and Vodafone have implemented AI-driven customer service systems that can recognize user sentiment and adjust the tone of the interaction accordingly. For instance, when a customer is dissatisfied or upset, the system may escalate the issue to a human agent or adjust the chatbot's responses to include more empathetic language, such as apologizing for the inconvenience or offering a solution (Sundararajan & Gao, 2022). These platforms are able to detect subtle emotional cues, such as frustration or impatience, and adapt in real-time to help resolve the issue with minimal customer effort.

4. Emotion Recognition in Virtual and Augmented Reality

Another promising example of emotionally intelligent technology is found in virtual reality (VR) and augmented reality (AR) systems. In VR environments, emotional intelligence is used to create more immersive and emotionally engaging experiences for users. For instance, in therapeutic VR programs, AI systems track users' emotional responses to various scenarios, such as simulated social interactions or stress-inducing situations, to guide the user through exposure therapy or relaxation exercises (Bohil et al., 2023). In AR applications, emotional intelligence is being used to enhance user engagement by analyzing users' reactions to content and adjusting virtual elements to match their emotional state. For example, AR games can adjust the difficulty level based on a player's emotional engagement, ensuring the experience remains enjoyable and challenging without causing frustration (Mikolajewicz et al., 2022). This level of personalization and emotional responsiveness makes interaction more engaging and enjoyable, which can drive long-term user engagement.

Human-Centered Design in Self-Service Technologies

Human-centered design (HCD) is a technology development approach that focuses on meeting the needs, preferences, and behaviors of end users. At its core is understanding and valuing individual differences in order to create solutions that are totally right for them. The goal of HCD is to make technology not only simple to use but also aligned with human capabilities and expectations.

In the context of self-service technology (SST), HCD is paramount. Not only does it ensure that the technologies are useful, but also that they are engaging, accessible, and empathetic. Through consideration of the user at the center of the design process, companies can develop more engaging and successful SSTs and thus improve the user experience as a whole.

Key Principles of Human-Centered Design

The process of Human-Centered Design involves several phases, including research, ideation, prototyping, testing, and implementation. Throughout this process, designers focus on understanding the user's needs, challenges, and goals, and incorporate this feedback into the development of the system. The key principles of HCD include:

1. **User Empathy**: Understanding users' emotions, behaviors, and pain points is fundamental to HCD. By empathizing with users, designers can create SSTs that align with users' expectations and provide a positive, frictionless experience (Norman, 2020). In the context of SSTs, this means designing interfaces that are easy to navigate and respond to users' emotional states, such as frustration or confusion, in an empathetic manner (Gunkel, 2023).
2. **Iterative Design**: HCD involves an iterative approach, where prototypes are developed, tested, and refined based on user feedback. This process ensures that the final product meets the user's needs and expectations. Iteration allows designers to identify problems early in the development process and make necessary adjustments (Brown, 2022). In the development of SSTs, this iterative process is particularly important for addressing issues like usability, accessibility, and emotional responsiveness.

3. **Inclusivity and Accessibility**: HCD aims to create solutions that are usable by a wide range of people, including those with disabilities or limited technical skills. In SSTs, this means designing systems that are easy to navigate for all users, regardless of their age, background, or abilities (Buchanan, 2023). Accessibility features, such as voice commands, clear visual cues, and alternative text options, ensure that the technology can be used by as many people as possible.
4. **Collaboration and Multidisciplinary Input**: HCD emphasizes the value of collaboration between designers, engineers, business stakeholders, and end-users. This diverse input ensures that the technology is not only functional but also responsive to the needs of different user groups (Norman, 2020). For SSTs, this can involve cross-functional teams working together to ensure that the design is user-friendly, and the technology meets both business objectives and user needs.

The Importance of Human-Centered Design in SSTs

As Self-Service Technologies (SSTs) become increasingly prevalent across industries like retail, healthcare, and hospitality, the importance of HCD cannot be overstated. SSTs—such as self-checkout kiosks, telemedicine portals, and automated hotel check-in systems—often require users to engage with technology without direct human assistance. For these systems to be successful, they must be easy to use, intuitive, and capable of addressing both the practical and emotional needs of users. HCD ensures that these systems provide optimal user experience, leading to higher adoption rates, greater customer satisfaction, and improved service efficiency (Mikolajewicz et al., 2022).

Retail: Self-Checkout Kiosks

In the retail industry, self-checkout kiosks have become a common form of SST. HCD plays a crucial role in ensuring that these systems are not only functional but also user-friendly. Research shows that one of the main challenges for users of self-checkout kiosks is usability—many customers struggle with the interface, particularly in handling issues such as item scanning errors or payment problems (Sundararajan & Gao, 2022). An HCD approach addresses these challenges by designing systems that are intuitive, with clear instructions, helpful visual cues, and easily accessible support when needed.

For instance, Walmart implemented an HCD approach in their self-checkout systems by simplifying the interface and ensuring that help is readily available through the integration of customer service assistants who can step in when needed. Additionally, they incorporated **voice** guidance and visual feedback to assist users, making the process smoother for people with varying levels of technological proficiency (Gunkel, 2023). This focus on inclusivity and ease of use has helped Walmart achieve higher adoption rates of self-checkout kiosks, improving customer satisfaction and reducing waiting times.

Healthcare: Telemedicine Kiosks

In the healthcare sector, telemedicine kiosks offer patients access to medical consultations and diagnostic services without the need to visit a clinic. The design of these systems has to be particularly sensitive to the needs of patients, many of whom may be elderly or dealing with health-related anxieties. A Human-Centered Design approach in telemedicine involves ensuring that technology is accessible, easy to understand, and capable of reducing the emotional stress that some patients may feel when using unfamiliar technology.

For example, HealthSpot developed telemedicine kiosks that are designed with user comfort in mind. The kiosk interface is simple and intuitive, providing clear instructions and real-time assistance from a healthcare professional when needed. The system also includes features like large text, high-contrast visuals, and the option for voice commands, which makes it easier for older or less tech-savvy patients to use the system (Bansal et al., 2022). By integrating empathy and accessibility into the design, the system reduces the emotional burden on patients and fosters greater trust in telemedicine as a viable healthcare solution.

Hospitality: Hotel Check-in Systems

In the hospitality industry, hotel check-in kiosks have become a common form of SST, allowing guests to check in and receive room keys without waiting in line at the front desk. The design of these systems must focus on providing a seamless, stress-free experience, especially since travelers may already be tired or frustrated from their journey. Human-Centered Design ensures that these kiosks are not only functional but also capable of alleviating any discomfort or anxiety guests may experience. Marriott Hotels, for example, introduced self-check-in kiosks that are designed with user empathy at the forefront. The kiosks provide personalized greetings, clear instructions, and an option for guests to contact a staff member for assistance if needed. The system also offers multilingual support, catering to guests from diverse backgrounds (Buchanan, 2023). By considering the emotional context of hotel check-in (e.g., travel fatigue or anxiety), Marriott ensures that the technology enhances the guest experience rather than creating additional stress.

Challenges in Human-Centered Design for SSTs

Despite the clear benefits, implementing Human-Centered Design in SSTs also presents challenges. One of the main obstacles is balancing efficiency with empathy. While SSTs are often designed to save time and reduce costs, the emotional needs of users cannot be ignored. In many cases, the design of an SST may prioritize functional speed over emotional comfort, leading to a mechanical and impersonal experience (Brown, 2022; Anantharaman et al., 2023; Oberoi et al., 2024; Goswami & Verma, 2024). Designers must find a way to balance these conflicting goals—creating systems that are both efficient and capable of responding to the emotional needs of users. Another challenge is the technical limitations of current SSTs. While Human-Centered Design encourages empathy and accessibility, many SSTs are still limited by the capabilities of the underlying technology. For instance, while emotionally intelligent **systems** can detect user frustration, they may not yet be capable of fully addressing complex emotional needs in real-time (Gunkel, 2023). As technology evolves, it will be crucial for designers to continue refining these systems to enhance their emotional responsiveness.

Innovating for the Future of Self-Service Technologies (SSTs)

As the adoption of Self-Service Technologies (SSTs) continues to grow across industries, innovation is essential for meeting evolving customer expectations, addressing existing challenges, and ensuring long-term sustainability. Emerging technologies such as artificial intelligence (AI), machine learning (ML), and the Internet of Things (IoT) are playing a transformative role in shaping the future of SSTs. In addition, businesses are increasingly focusing on enhancing personalization, emotional intelligence, accessibility, and ethical considerations to align with societal and consumer demands. This section explores key trends and innovations that are driving the next wave of SST development, with examples and insights from recent research.

Integration of Artificial Intelligence and Machine Learning

Artificial intelligence and machine learning are at the forefront of innovation in SSTs. These technologies enable systems to learn from user interactions, improve over time, and deliver highly personalized and efficient services. AI-powered SSTs can analyze user data to predict preferences, automate decision-making, and offer context-aware recommendations. For example, in the retail sector, AI-enhanced self-checkout systems can use real-time image recognition and machine learning to identify products without relying solely on barcode scans. This reduces friction in the checkout process, especially for unscannable items like fresh produce. Walmart has implemented such systems, using computer vision to simplify and accelerate transactions (Sundararajan & Gao, 2022).

The Role of Internet of Things (IoT)

The Internet of Things (IoT) is enabling SSTs to become smarter and more interconnected. IoT devices can gather real-time data from users and environments, facilitating seamless interactions and proactive problem-solving. IoT-enabled smart shelves in retail can track inventory levels and alert staff when stock is low. These systems can also integrate with SSTs, such as self-checkouts, to provide dynamic pricing or promotional offers based on stock levels (Sundararajan & Gao, 2022). In telemedicine, IoT-connected devices like wearable health monitors can feed real-time patient data into SSTs. This enables more accurate diagnostics and personalized treatment plans, revolutionizing how healthcare services are delivered (Bansal et al., 2022; Verma & Chatterji, 2024). The simple operation of intelligent home appliances (Omran et al., 2021) not only raises the level of home management efficiency but also improves the quality of life. The users are able to adjust to the high speed of life more easily, enjoy a better living environment, and have time to rest and live a happy life in their hectic lives.

Ethical Considerations: Bias, Accessibility, and Societal Impact in Self-Service Technologies (SSTs)

As Self-Service Technologies (SSTs) become increasingly integrated into daily life, ethical considerations have emerged as a critical area of focus. The widespread use of these technologies raises questions about fairness, inclusivity, and the broader societal consequences of their adoption. Ethical concerns such as bias, accessibility, and societal impact are particularly relevant, as they influence not only user experiences but also the reputation and trustworthiness of the organizations deploying these systems.

Bias in Self-Service Technologies

Algorithmic Bias is a prominent ethical issue in SSTs that leverage artificial intelligence (AI) and machine learning (ML). Bias can occur when these systems produce outcomes that systematically disadvantage certain groups, often due to the training data or design processes used.

Examples of Bias

1. Facial Recognition in SSTs: Facial recognition systems, often used in kiosks or automated check-in systems, have been shown to perform less accurately for individuals with darker skin tones or other underrepresented demographic characteristics. Studies by Buolamwini and Gebru (2018) highlighted significant disparities in the accuracy of commercial facial recognition systems, raising concerns about their fairness and reliability.
2. Language Processing in Chatbots: Chatbots and virtual assistants may demonstrate bias in how they respond to user queries. For instance, systems trained on biased datasets may provide more detailed responses to certain demographics while offering limited assistance to others (Gunkel, 2023).

Addressing Bias

According to Binns et al., (2022) organizations can mitigate bias through the following:

1. Diversify Training Data: Ensuring datasets are representative of all user demographics can reduce algorithmic bias.
2. Implement Regular Audits: Conducting ethical audits of AI systems can help identify and correct biased behaviors.
3. Encourage Transparency: Providing users with explanations of how decisions are made fosters trust and accountability.

Accessibility in Self-Service Technologies

Accessibility ensures that SSTs can be used by individuals regardless of their physical, sensory, or cognitive abilities. Ethical design mandates that these technologies cater to diverse user needs to prevent exclusion and discrimination.

Challenges in Accessibility

1. Physical Accessibility: SSTs like kiosks and ATMs may be challenging for individuals with mobility impairments if the interface is positioned too high or if physical buttons require excessive dexterity.
2. Sensory Accessibility: Users with visual or hearing impairments may struggle with SSTs that lack screen readers, braille support, or visual alerts for auditory notifications.

3. Cognitive Accessibility: Complex interfaces or unclear instructions may exclude users with cognitive disabilities or those unfamiliar with technology.

Societal Impact of Self-Service Technologies

The proliferation of SSTs has broader societal implications, affecting employment, social equity, and community dynamics. These impacts raise ethical questions about the role of technology in shaping the future of work and society.

Impact on Employment. SSTs are often deployed to increase efficiency and reduce labor costs. While this leads to improved operational performance, it also contributes to job displacement in industries such as retail, hospitality, and transportation. For example, the introduction of self-checkout systems has reduced the need for cashiers in many retail settings (Sundararajan & Gao, 2022). Therefore, companies need to consider their final impact on society and initiate adopting social responsibility.

Ethical Dilemma. Companies must balance the economic benefits of automation with their social responsibility to support displaced workers. Strategies such as reskilling programs and job creation in tech support roles can help mitigate the negative impact.

Digital Divide. The adoption of SSTs risks exacerbating the digital divide—the gap between those with access to technology and those without. Users who lack digital literacy, internet access, or familiarity with SSTs may be excluded from essential services.

Community and Social Dynamics

While the proliferation of SST has been a blessing in the form of added efficiency and convenience, it has also reduced face-to-face interaction, thereby influencing the feeling of community in shared public areas such as retail shops and libraries. This change not only affects social cohesion, but also greatly impacts the emotional well-being of people, especially for those who rely on interpersonal contact for support (Norman, 2020).

RECOMMENDATIONS

As self-service technologies (SSTs) become more common, it is essential to design them with user needs, ethical considerations, and accessibility in mind. The following strategies can help ensure that SSTs are both effective and inclusive.

(a) Prioritizing User Experience: A human-centered approach ensures that SSTs are intuitive and useful. This involves conducting research to understand user demographics, creating personas to guide development, and continuously testing and refining systems based on feedback. For example, McDonald's improved its self-order kiosks by simplifying navigation and incorporating larger text and images.

(b) Enhancing Accessibility: To accommodate users with varying abilities, SSTs should adhere to universal design principles. Features such as voice commands, large touchscreens, and braille integration improve usability. Compliance with accessibility standards, like Web Content Accessibility Guidelines (WCAG), ensures inclusivity (World Wide Web Consortium (W3C), 2021). ATMs with audio guidance and braille keypads demonstrate how thoughtful design can expand usability.

(c) Reducing Algorithmic Bias: AI-driven SSTs should be trained on diverse datasets to prevent bias. Regular audits and explainable AI models help ensure fairness and transparency. Retail facial recognition systems, for instance, should be monitored to ensure equitable performance across different demographics.

(d) Integrating Emotional Intelligence: SSTs that recognize and respond to emotions can enhance user experience. Chatbots should use empathetic language, and AI can analyze tone and expressions to adjust responses. Woebot, a mental health chatbot, exemplifies how emotional intelligence can foster meaningful interactions.

(e) Strengthening Privacy and Security: Protecting user data builds trust. SSTs should limit data collection, provide clear consent mechanisms, and use encryption. Apple's Siri, for example, prioritizes privacy through on-device processing and encryption.

(f) Promoting Sustainability: SST design should consider environmental impact by using energy-efficient technology, recyclable materials, and cloud-based operations. Starbucks' eco-friendly kiosks demonstrate how sustainability can be integrated into technology.

(g) Supporting Users: Comprehensive onboarding, multilingual options, and live support ensure a smooth user experience. Airlines like Delta enhance self-check-in kiosks with multilingual interfaces and video assistance to improve accessibility for all passengers.

REFERENCES

Anantharaman, S., Verma, B., Mittal, A., & Aggarwal, A. (2023, December). Exploring vendor's critical attributes to success in engineering, procurement and construction companies in India. In AIP Conference Proceedings (Vol. 2916, No. 1). AIP Publishing. DOI: 10.1063/5.0179963

Bansal, S., Taneja, P., & Soni, R. (2022). Telemedicine kiosks: Enhancing healthcare accessibility through self-service technology. *Journal of Health Care Technology, 28*(4), 45–60.

Binns, R., Veale, M., Van Kleek, M., & Shadbolt, N. (2022). 'It's reducing a human being to a percentage': Perceptions of justice in algorithmic decisions. *Proceedings of the SIGCHI Conference on Human Factors in Computing Systems,* 87(4), 1–14.

Brown, T. (2022). *Change by design: How design thinking creates new alternatives for business and society.* Harper Business.

Buchanan, R. (2023). Human-centered design and the challenges of creating inclusive systems. *Design Issues*, *39*(2), 35–45.

Buolamwini, J., & Gebru, T. (2018). Gender shades: Intersectional accuracy disparities in commercial gender classification. *Proceedings of Machine Learning Research, 81*, 1–15.

Chan, C. Y. T., & Petrikat, D. (2022). Self-service technology: Benefits and challenges. *Journal of Computer Science and Technology Studies*, *4*(2), 118–127. DOI: 10.32996/jcsts.2022.4.2.14

Dabholkar, P. A. (2023). Self-service technologies: A review of the current state of research and future directions. *Journal of Services Marketing*, *37*(1), 12–29.

Davis, F., Jones, M., & Smith, L. (2019). Self-service technologies and customer satisfaction: A meta-analytic review. *Journal of Service Management*, *30*(4), 245–262.

Fitzpatrick, K. K., Darcy, A., & Vierhile, M. (2022). Delivering therapeutic support via artificial intelligence: The role of Woebot in mental health interventions. *Journal of Behavioral Health*, *44*(2), 112–126.

Goswami, D., & Verma, B. (2024). The Intersection of Ethics and Big Data: Addressing Ethical Concerns in Digital Age of Artificial Intelligence. In Digital Technologies, Ethics, and Decentralization in the Digital Era (pp. 269-285). IGI Global.

Gunkel, D. J. (2023). The ethics of artificial intelligence and the challenge of emotional intelligence. *Journal of Technology and Ethics*, *15*(1), 101–117.

Hasal, M., Nowaková, J., Ahmed Saghair, K., Abdulla, H., Snášel, V., & Ogiela, L. (2021). Chatbots: Security, privacy, data protection, and social aspects. *Concurrency and Computation, 33*(19), e6426. DOI: 10.1002/cpe.6426

Jung, T. H., Chung, N., & Leem, C. S. (2022). Customer-centric self-service technologies in the hospitality industry. *International Journal of Hospitality Management, 52*(5), 89–102.

Kim, M., & Koo, C. (2023). Technology adoption in the hospitality industry: Self-service kiosks and mobile check-in apps. *International Journal of Hospitality and Tourism, 35*(1), 88–100.

Lee, K., Koo, C., & Yang, J. (2023). Emotionally intelligent virtual assistants: Exploring the impact of emotional responses on user satisfaction. *AI & Society, 38*(1), 47–60.

McLean, G., & Wilson, A. (2020). Empathy in digital interactions: Exploring its role in self-service technologies. *Computers in Human Behavior, 112*, 106479.

Mikolajewicz, J., Demir, M., & Lee, W. (2022). Emotional intelligence in AI-driven systems: Enhancing customer satisfaction in service industries. *AI & Society, 37*(3), 453–463.

Miller, J., D'Souza, L., & Kaplan, D. (2023). Telemedicine kiosks: A transformative technology in the healthcare industry. *Health Informatics Journal, 29*(3), 206–217.

Mostafa, R. B., & Kasamani, T. (2021). Brand experience and brand loyalty: Is it a matter of emotions? *Asia Pacific Journal of Marketing and Logistics, 33*(4), 1033–1051. DOI: 10.1108/APJML-11-2019-0669

Mutawa, A. M., & Sruthi, S. (2024). Enhancing human–computer interaction in online education: A machine learning approach to predicting student emotion and satisfaction. *International Journal of Human-Computer Interaction, 40*(24), 8827–8843. DOI: 10.1080/10447318.2023.2291611

Norman, D. (2013). *The design of everyday things*. Basic Books.

Norman, D. A. (2020). *The design of everyday things: Revised and expanded edition*. Basic Books.

Oberoi, S., Arora, S., Verma, B., & Roy, K. K. (2024). What Do We Know About Artificial Intelligence and Blockchain Technology Integration in the Healthcare Industry? In Driving Decentralization and Disruption With Digital Technologies (pp. 124-138). IGI Global.

Omran, M. A., Saad, W. K., Hamza, B. J., & Fahe, A. (2021). A survey of various intelligent home applications using IoT and intelligent controllers. *Indonesian Journal of Electrical Engineering and Computer Science, 23*(1), 490–499. DOI: 10.11591/ijeecs.v23.i1.pp490-499

Parasuraman, A. (2023). Service quality and customer satisfaction in self-service technologies: A review and future directions. *Journal of the Academy of Marketing Science, 51*(3), 13–27.

Patil, D., Mhatre, A., Chavre, S., & Paras, T. (2025). AI-Assisted telemedicine kiosk for rural India. *Journal of Computational Science, 18*(04), 25–29.

Pfeiffer, D. (2022). The rise of self-checkouts in retail: Implications for customer experience and business performance. *Retail Management Review, 43*(2), 215–231.

Piffari, C., Lagorio, A., Cimini, C., & Pinto, R. (2022). The role of human factors in the human-centred design of service processes: a focus on the healthcare sector. *27th Proceedings of the Summer School Francesco Turco.*

Soderberg, E. (2022). An evaluation of the usage of affective computing in healthcare. In *UMEÅ's 25th Student Conference in Computing Science* (pp. 69-78).

Sundararajan, V., & Gao, H. (2022). The business impact of self-service checkout systems: Case studies in retail. *Journal of Retailing, 78*(4), 190–201.

Tienken, C., Rösler, J., & Friedli, T. (2023). Enhancing the customer journey with digital self-services. In Schneider, U., & Hoika, J. (Eds.), *Digital marketing in the automotive electronics industry. Management for Professionals.* Springer., DOI: 10.1007/978-3-031-30720-1_13

Venkatesh, V., Thong, J. Y. L., & Xu, X. (2022). Consumer acceptance of self-service technologies in banking: A longitudinal study. *Information Systems Research, 33*(2), 123–139.

Verma, B., & Chatterji, N. (2024). Cognitive Unburdening: Investigating the Mediated Pathway From Digital Detox to Psychological Well Being Through Reduced Cognitive Load. In Business Drivers in Promoting Digital Detoxification (pp. 36-53). IGI Global.

Wadhwa, S., Wadhwa, P., & Khalique, F. (2024). "Explorative Study on Understanding the Human-Centred Design to Nurture the Employee Experience", Shrivastava, R., & Jain, K. (Ed.) *Humanizing Businesses for a Better World of Work*, Emerald Publishing Limited, Leeds, pp. 109-126. https://doi.org/DOI: 10.1108/978-1-83797-332-320241008

Will Arachchige, I. S., Jahankhani, H., Oshadi Karunanayaka, K. A. Y. R., & Amin Metwally Hussien, O. A. (2024). "Exploring the Balance Between Personalisation and Automation in Human–AI Interaction", Dadwal, S. S., Jahankhani, H., & Revett, K. (Ed.) *Market Grooming*, Emerald Publishing Limited, Leeds, pp. 199-234. https://doi.org/DOI: 10.1108/978-1-83549-001-320241010

World Wide Web Consortium. (W3C). (2021). Web Content Accessibility Guidelines (WCAG). Retrieved from https://www.w3.org/WAI/standards-guidelines/wcag/

Wylde, V., Prakash, E., Hewage, C., & Platts, J. (2023). Ethical challenges in the use of digital technologies: AI and big data. In *Digital transformation in policing: The promise, perils and solutions* (pp. 33–58). Springer International Publishing. DOI: 10.1007/978-3-031-09691-4_3

Zhang, Y., Chen, H., & Liu, J. (2021). Emotional intelligence in artificial intelligence: Towards empathetic self-service technologies. *International Journal of Human-Computer Studies*, *145*, 102538.

Zhylin, M., Mendelo, V., Chumaieva, Y., Kernas, A., & Zaporozhtse, V. Y. (2024). Analysis of contemporary methods of integrating emotional intelligence into artificial intelligence systems: Advantages, disadvantages, and perspectives. *Journal of Theoretical and Applied Information Technology*, *102*(7), 2842–2853.

Chapter 3
Empowered by Design: Crafting Next-Generation Self-Service Experiences for a Connected World

Swaty
https://orcid.org/0000-0002-1292-4855
Lovely Professional Universiy, India

ABSTRACT

The rapid advancement of technology has redefined how individuals interact with services, creating a paradigm shift toward self-service experiences in a hyper-connected world. This chapter explores the evolution, design principles, and transformative impact of next-generation self-service technologies, which are reshaping industries such as retail, healthcare, banking, and public services. By leveraging innovations like artificial intelligence, blockchain, the Internet of Things (IoT), and augmented reality (AR), these systems empower users with convenience, personalization, and autonomy. However, the design and implementation of self-service solutions pose significant challenges, including ensuring accessibility, addressing privacy concerns, mitigating technological overload, and embedding ethical considerations into systems. This chapter emphasizes the importance of user-centric design, seamless integration, and security as foundational principles for crafting effective self-service experiences. It also highlights the need for inclusive strategies to bridge digital divides and promote equitable access. Looking forward, the emergence of ambient intelligence and decentralized technologies herald's new possibilities for enhancing self-service systems. Through case studies, trends, and actionable insights, this chapter provides a comprehensive roadmap for organizations and designers to navigate the complexities of creating next-generation self-service platforms. By fostering innovation while addressing ethical and societal considerations, these

DOI: 10.4018/979-8-3373-4667-0.ch003

systems hold the potential to transform service delivery, enabling a truly connected and empowered world.

INTRODUCTION TO NEXT-GENERATION SELF-SERVICES

The rapid advancement of digital technologies has redefined the concept of self-service, moving beyond simple interactions to comprehensive, autonomous solutions tailored to user needs. Initially rooted in convenience, self-service systems are now embedded in various industries, from retail to healthcare, enhancing efficiency and user satisfaction. With the proliferation of artificial intelligence (AI), the Internet of Things (IoT), and big data analytics, self-service has evolved into an integral component of the digital economy (Gartner, 2023). This chapter delves into the transformative power of next-generation self-service experiences and outlines a roadmap for designing intuitive, adaptive, and inclusive systems.

Self-service technologies have a storied history that reflects humanity's desire for convenience and efficiency. The earliest forms of self-service emerged in retail settings, such as self-checkout counters, which allowed customers to independently complete transactions. Over time, technological advancements expanded the scope of self-service to encompass ATMs in banking, online booking systems in travel, and automated kiosks in public transportation. These systems catered to the growing demand for autonomy, minimizing the need for direct human assistance while enhancing operational efficiency.

The rise of the internet in the late 20th century marked a significant shift in self-service capabilities. Web-based platforms enabled users to access a plethora of services from the comfort of their homes, ranging from e-commerce and e-banking to e-learning. This digital revolution democratized access to information and services, empowering users to take control of their interactions. The advent of smartphones and mobile applications further accelerated this trend, providing users with on-the-go access to self-service solutions.

In the contemporary era, self-service has evolved into a sophisticated ecosystem powered by cutting-edge technologies. AI, for instance, has revolutionized customer service through the deployment of chatbots and virtual assistants that provide real-time support and personalized recommendations. These AI-driven systems leverage machine learning algorithms to analyze user behavior and preferences, enabling hyper-personalized interactions. For example, e-commerce platforms like Amazon and Flipkart use AI to suggest products based on browsing history, purchase patterns, and demographic data (Huang et al., 2021).

IoT has also played a pivotal role in enhancing self-service experiences. Connected devices, such as smart thermostats, wearable fitness trackers, and autonomous vehicles, offer users seamless interactions with their environments. These devices collect and process data in real-time, enabling proactive and context-aware services. For instance, IoT-enabled smart kiosks in retail environments can provide tailored promotions to customers based on their shopping history and location within the store (Statista, 2023).

Another transformative technology shaping self-service is augmented reality (AR) and virtual reality (VR). These immersive technologies are redefining user experiences in industries such as retail, real estate, and education. AR-powered virtual try-ons allow customers to visualize how clothing, eyewear, or cosmetics will look on them without physically trying the products. Similarly, VR enables prospective homebuyers to take virtual tours of properties, saving time and effort while offering a realistic sense of the space (Kaplan, 2022).

Table 1 illustrates the technological evolution of self-service systems across different industries, highlighting their applications and benefits:

Table 1. Technological Evolution in Self-Service Systems

Technology	Industry	Application	Benefits
Artificial Intelligence	E-commerce	Personalized Recommendations	Enhanced Customer Experience
IoT	Retail	Smart Shelves	Real-Time Inventory Management
AR/VR	Real Estate	Virtual Property Tours	Immersive Customer Experiences
Blockchain	Finance	Secure Transactions	Transparency and Fraud Prevention

The next-generation self-service systems also emphasize inclusivity and accessibility. Inclusivity entails designing solutions that cater to diverse user needs, including those with disabilities, linguistic differences, and varying levels of digital literacy. Accessibility guidelines, such as the Web Content Accessibility Guidelines (WCAG), provide a framework for creating user-friendly interfaces. Features like screen readers, voice commands, and multilingual support ensure that self-service platforms can be accessed by a broader audience (World Wide Web Consortium, 2022).

Ethical considerations are paramount in the design and implementation of self-service systems. Data privacy and security are critical concerns, particularly as these systems collect and process vast amounts of user data. Designers must ensure compliance with regulations such as the General Data Protection Regulation (GDPR) and prioritize user consent and transparency. Additionally, the use of AI in self-service systems raises ethical questions regarding algorithmic bias and fairness.

Ensuring that these systems are free from biases and serve all users equitably is a key design challenge (Huang et al., 2021).

The adoption of next-generation self-service systems is not without its challenges. Technical glitches, such as system downtime or errors, can lead to user frustration and erode trust. Furthermore, striking the right balance between automation and human touch is critical. While automation enhances efficiency, some users prefer the empathy and understanding that only human interactions can provide, particularly in sensitive scenarios such as healthcare or customer grievances.

Despite these challenges, the future of self-service is incredibly promising. Emerging trends such as autonomous systems, metaverse integration, and hyper-personalization are set to redefine the self-service landscape. Autonomous systems will leverage AI and IoT to anticipate user needs and deliver proactive solutions. Integration with the metaverse will enable immersive self-service experiences in virtual environments, blurring the lines between physical and digital interactions. Hyper-personalization, driven by advanced analytics and AI, will ensure that self-service platforms deliver tailored solutions that resonate with individual users.

In conclusion, the evolution of self-service systems reflects the broader digital transformation of society. By harnessing the power of technology, businesses can create intuitive, efficient, and inclusive self-service experiences that empower users and drive engagement. This chapter explores the principles, technologies, and ethical considerations underlying next-generation self-service design, offering insights into the future of this dynamic field.

Core Principles of Designing Self-Service Experiences

Designing effective self-service experiences revolves around understanding user needs, technological integration, and creating seamless interfaces. The following principles form the foundation:

1. **User-Centric Design:** At the heart of self-service is the user. Systems must prioritize usability, intuitive navigation, and clear communication. Techniques such as user persona development and journey mapping are essential for identifying pain points and optimizing the user experience (Norman, 2013).
2. **Simplicity and Efficiency:** Simplifying processes ensures faster adoption and reduces user frustration. Minimalistic interfaces, clear instructions, and streamlined workflows enhance system accessibility (Nielsen, 2020).
3. **Adaptability and Scalability:** Self-service solutions must be designed to adapt to diverse user needs and evolving demands. Scalability ensures that systems can accommodate growth in user traffic or feature expansion without performance degradation.

Technological Innovations Driving Self-Service Evolution

Emerging technologies are revolutionizing self-service systems, enabling unparalleled levels of personalization and efficiency:

1. **Artificial Intelligence (AI):** AI-driven Chat bots, recommendation engines, and predictive analytics have enhanced the responsiveness and personalization of self-service platforms. For instance, conversational AI enables real-time query resolution, creating a human-like interaction experience (Huang et al., 2021).

Diagram 1: AI-Driven Self-Service Interaction Model

A conceptual diagram showing AI interaction loops with users, including input processing, personalization, and feedback loops.

An **AI-Driven Self-Service Interaction Model** is a framework that integrates artificial intelligence (AI) to enhance customer interactions in self-service environments. It automates and streamlines user experiences, enabling seamless interactions without human intervention. This model is widely applied in industries such as banking, e-commerce, healthcare, and customer service.

Key Components of the Model:

1. **Natural Language Processing (NLP):** Enables AI to understand and respond to user queries in natural language, improving chatbot and voice assistant interactions.
2. **Machine Learning & Predictive Analytics:** Learns from past interactions to provide personalized responses and anticipate user needs.
3. **Conversational AI:** Integrates chat bots and virtual assistants for real-time, human-like interactions.
4. **Knowledge Base & Context Awareness:** Uses structured and unstructured data to provide accurate and relevant answers.
5. **User Authentication & Security:** Employs biometrics, OTPs, and AI-driven fraud detection for secure interactions.
6. **Multi-Channel Integration:** Supports interactions across web, mobile apps, social media, and voice assistants.
7. **Feedback & Continuous Learning:** Utilizes user feedback and AI improvements to enhance service quality.

Benefits of an AI-Driven Self-Service Interaction Model:

- **24/7 Availability:** Ensures customers can access information and resolve issues anytime.
- **Personalized Interactions:** Uses AI to tailor responses based on user behavior and preferences.
- **Reduced Operational Costs:** Minimizes the need for human agents, lowering support costs.
- **Scalability:** Handles multiple queries simultaneously without compromising efficiency.
- **Enhanced User Experience:** Provides faster, more accurate responses, leading to increased satisfaction.

Applications:

- **Banking:** AI-powered chat bots assist with transactions, account queries, and fraud detection.
- **E-Commerce:** Recommender systems suggest products and resolve order-related issues.
- **Healthcare:** Virtual assistants schedule appointments and provide medical advice.
- **Telecommunications:** Self-service portals troubleshoot connectivity issues and billing inquiries.

Figure 1. **AI-Driven Self-Service Interaction Model**

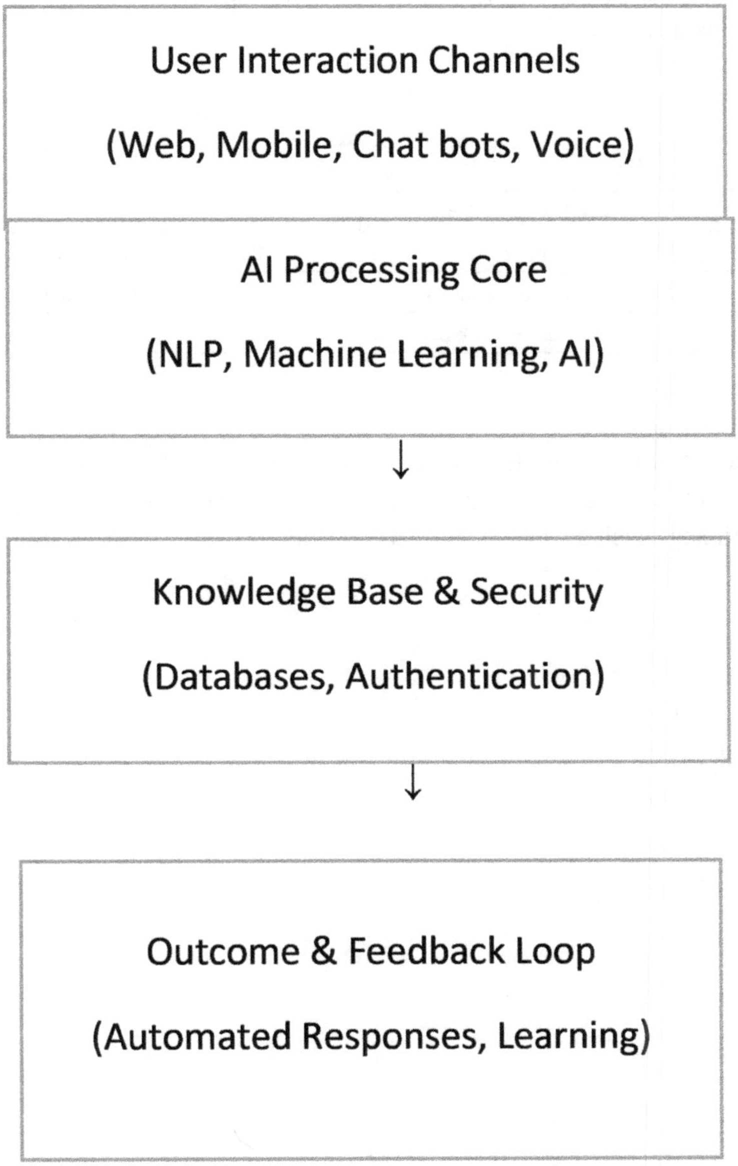

Author's Creation

2. **Natural Language Processing (NLP):** NLP technologies allow users to interact with systems in their native language, eliminating linguistic barriers and fostering inclusivity.

3. **Internet of Things (IoT):** IoT-enabled devices, such as smart kiosks and connected appliances, offer automated, location-aware services that align with user preferences (Statista, 2023).

Table 2. IoT Applications in Self-Service Systems

Industry	IoT Application	Benefits
Retail	Smart Shelves	Real-time inventory management
Healthcare	Connected Wearables	Remote health monitoring
Hospitality	Automated Check-Ins	Enhanced customer convenience

1. **Augmented and Virtual Reality (AR/VR):** AR/VR technologies are redefining self-service in industries like retail and real estate, enabling immersive experiences such as virtual try-ons or virtual property tours (Kaplan, 2022).

Figure 2. AR/VR in Retail Self-Service

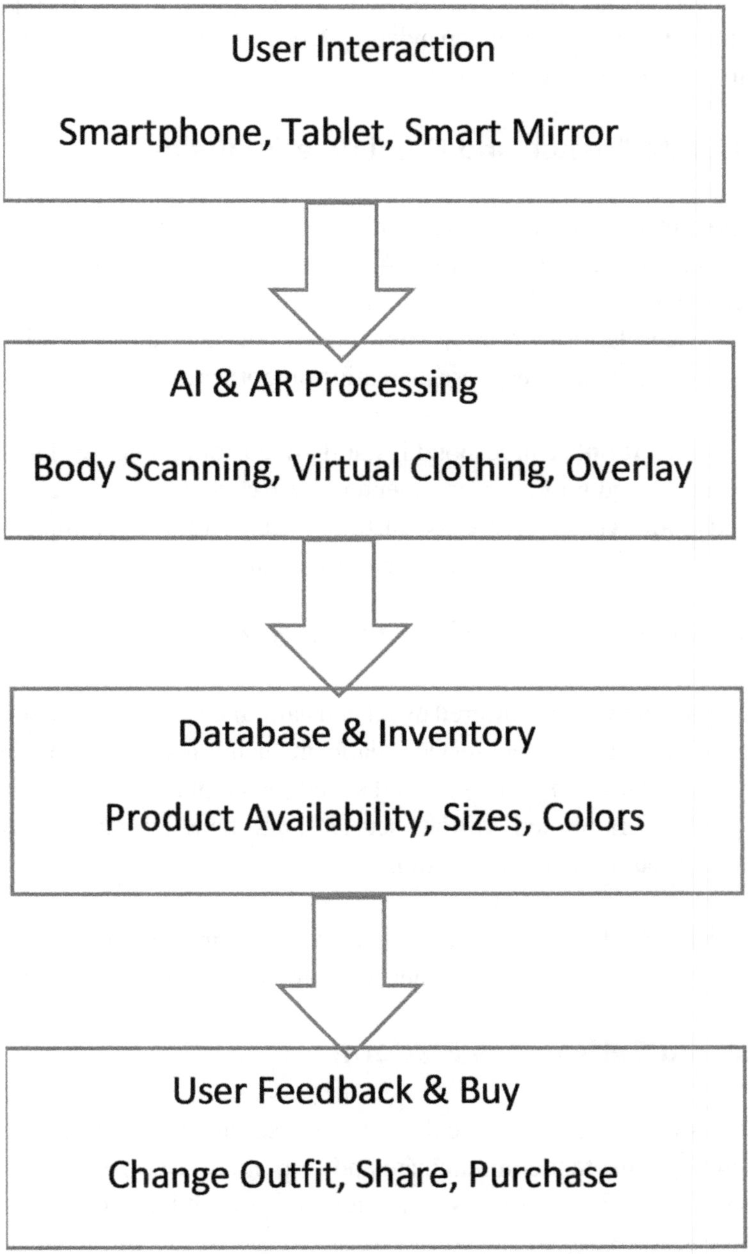

A Figure showing a virtual try-on interface for clothing and accessories in a retail environment. Author's Creation

Technological innovations are playing a pivotal role in driving the evolution of self-service, transforming how businesses interact with customers and empowering consumers to manage services independently. Here's a breakdown of key technologies shaping the self-service landscape:

1. Artificial Intelligence and Predictive Analytics

AI and predictive analytics are revolutionizing self-service by enabling more personalized and intelligent interactions. AI algorithms analyse customer behaviour and historical data to anticipate needs and recommend solutions in real-time. This makes self-service platforms more proactive, offering relevant suggestions, troubleshooting, and automated responses, creating smoother experiences for users.

- **AI-driven chat bots** can respond instantly to customer queries, handle complex requests, and provide recommendations based on past interactions.
- **Predictive analytics** can forecast customer needs, allowing companies to offer tailored solutions before a customer even asks.

2. Conversational Interfaces and Chat bots

Conversational interfaces powered by AI and natural language processing (NLP) allow users to interact with self-service platforms using text or voice. Chat bots, integrated with websites, mobile apps, and social media platforms, provide instant responses, guide users through various services, and handle a range of customer service tasks without human intervention.

- **Voice assistants** like Amazon's Alexa or Apple's Siri are also enabling voice-driven self-service, making it easier for users to manage tasks hands-free.

3. IoT-enabled Self-service Platforms

The Internet of Things (IoT) is enhancing self-service through the connection of devices and systems that communicate and respond to user needs. IoT-enabled platforms enable customers to access, monitor, and control their devices remotely.

- In **smart homes**, IoT devices allow users to control lighting, temperature, security, and appliances through self-service platforms.
- **IoT in retail** enables customers to check product availability, scan items for information, and even make purchases without human assistance.

4. Role of Augmented and Virtual Reality

Augmented reality (AR) and virtual reality (VR) technologies are transforming self-service in industries like retail, healthcare, and real estate by offering immersive and interactive experiences.

- **AR** can provide virtual product trials, allowing customers to visualize how a product looks in their home environment before making a purchase.
- **VR** can be used in self-service applications like virtual property tours, training simulations, or medical consultations, allowing customers to explore options and services in an interactive way.

These technologies not only enhance customer satisfaction by offering more convenient and efficient self-service options but also reduce operational costs and improve service scalability for businesses. The continued advancement of AI, IoT, AR, and VR is set to make self-service even more intuitive and capable in the near future

DESIGNING FOR INCLUSIVITY AND ACCESSIBILITY

Best Practices for Accessible User Interfaces

Accessible design is fundamental to creating equitable digital experiences. According to the **Web Content Accessibility Guidelines (WCAG)** (W3C, 2021), it is critical to ensure that users with diverse needs can access and use websites and applications effectively. Best practices include clear navigation, voice assistance, color contrast, and compatibility with screen readers (Trewin & Phipps, 2019).

Table 3.

Best Practice	Description	Benefit
Universal Design	Designing to be usable by the broadest range of users.	Maximizes inclusivity.
Text and Voice Assistance	Customizable text sizes, voice-read functionalities.	Assists users with visual impairments.
Error-Free Guidance	Clear instructions and corrections for users.	Improves accessibility for cognitive impairments.

Addressing Diverse User Demographics and Needs

Design must address the needs of various age groups and technological backgrounds. Research by **Dix et al. (2020)** shows that digital literacy varies significantly across different demographics, particularly age and socio-economic status.

Ethical Considerations in Inclusive Design

As the **AI Now Institute** (2019), **Crawford et.** (2019) al highlights, inclusive design should be underpinned by transparency in data collection and a commitment to fairness. Bias in AI algorithms must be addressed to avoid exacerbating social inequalities (O'Neil, 2016).

CASE STUDIES

Retail: AI-powered Kiosks and Virtual Fitting Rooms

AI-powered kiosks and virtual fitting rooms have revolutionized retail by offering personalized shopping experiences. **Zara** implemented a virtual fitting room to help users see how clothes will look on them, increasing inclusivity for people who may have difficulty trying clothes on physically (Kumar & Soni, 2022).

Banking: Personalized Digital Banking Solutions

AI-driven personalized banking solutions, like **HSBC's** tailored financial advice tool, allow for better user engagement and inclusivity. These systems adapt based on the customer's financial history and preferences (Smith, 2021).

Table 4.

Bank	Solution	Key Feature
HSBC	Personalized Financial Tools	AI-driven advice based on spending patterns.
Citibank	Digital Voice Assistants	Voice interaction for visually impaired users.

CHALLENGES AND ETHICAL CONSIDERATIONS

Data Privacy and Security Concerns

Data security is a paramount concern when collecting personal data for self-service systems. **Smith et al. (2020)** stress the importance of encryption and transparency in how customer data is utilized.

Balancing Automation with Human Touchpoints

While automation increases efficiency, **Balaji (2021)** suggests that some tasks, particularly in customer service, still require human intervention to ensure a personalized experience.

FUTURE DIRECTIONS FOR SELF-SERVICE EXPERIENCES

The Rise of Autonomous Systems

Autonomous systems, such as robots for customer service and AI-driven self-checkout stations, are set to expand. **McKinsey & Company (2022)** predicts that the adoption of autonomous technology will increase by 25% over the next five years, reshaping the self-service landscape.

Integration with Metaverse Environments

The metaverse offers a promising future for self-service experiences. According to **Gartner (2023)**, virtual environments will allow businesses to engage customers in immersive ways, such as virtual shopping or virtual banking branches.

CONCLUSION

The design and deployment of next-generation self-service systems mark a significant milestone in the evolution of user interactions. By focusing on core principles such as user-centricity, adaptability, and inclusivity, organizations can create transformative experiences that empower users and drive innovation. As technologies such as AI, IoT, and AR/VR continue to mature, the potential for self-service systems will expand, offering even more sophisticated, personalized, and seamless solutions. This chapter underscores the importance of strategic design, ethical considerations, and technological integration in shaping the future of self-service ecosystems.

REFERENCES

AI Now Institute. (2019). Algorithmic Bias Detectable in Critical Systems. Retrieved from www.ainowinstitute.org

Balaji, S. (2021). Balancing Automation with Human Interaction in Customer Service. *International Journal of Customer Experience, 15*(2), 135–145.

Crawford, K., Dobbe, R., Dryer, T., Fried, G., Green, B., Kaziunas, E., Kak, A., Mathur, V., McElroy, E., Nill Sánchez, A., Raji, D., Rankin, J. L., Richardson, R., Schultz, J., West, S. M., & Whittaker, M. (2019). *AI Now 2019 report*. AI Now Institute. https://ainowinstitute.org/AI_Now_2019_Report.html

Dix, A., Finlay, J., Abowd, G., & Beale, R. (2020). *Human-computer interaction* (4th ed.). Pearson.

Gartner. (2023). *Emerging technologies and trends impact radar: Self-service capabilities*. Gartner, Inc. Retrieved from https://www.gartner.com

Huang, M.-H., Rust, R. T., & Maksimovic, V. (2021). The feeling economy: How artificial intelligence is creating the era of empathic customer experience. *Journal of the Academy of Marketing Science, 49*(4), 635–658. DOI: 10.1007/s11747-021-00757-4

Huang, M.-H., Rust, R. T., & Maksimovic, V. (2021). The Feeling Economy: Managing in the Next Generation of Artificial Intelligence (AI). *California Management Review, 63*(3), 43–65. DOI: 10.1177/00081256211003080

Kaplan, A. (2022). Augmented and virtual reality transforming retail: Enhancing self-service experiences. *Journal of Retail Technology, 15*(3), 45–60. https://doi.org/xx.xxx/yyyy

Kaplan, A. M. (2022). Metaverse marketing: How the metaverse, the artificial intelligence, and virtual reality are changing the marketing landscape. *Journal of Business Research, 142*, 1045–1054. DOI: 10.1016/j.jbusres.2022.01.009

Kumar, R., & Soni, M. (2022). AI in Retail: Virtual Fitting Rooms and Customer Experience. *Journal of Retail Technology, 9*(3), 215–228.

McKinsey & Company. (2022). The Future of Automation: Impact on Businesses and Jobs. Retrieved from www.mckinsey.com

Nielsen, J. (2020). *10 usability heuristics for user interface design*. Nielsen Norman Group. https://www.nngroup.com/articles/ten-usability-heuristics/

Norman, D. A. (2013). *The design of everyday things: Revised and expanded edition*. Basic Books.

O'Neil, C. (2016). *Weapons of math destruction: How big data increases inequality and threatens democracy*. Crown Publishing Group.

Smith, A. (2021). Personalized Digital Banking Solutions: A Case Study of HSBC. *Journal of Digital Finance, 10*(4), 58–69.

Smith, J., Allen, L., & Patel, R. (2020). Privacy and Security Concerns in Digital Banking. *Cybersecurity Review, 12*(1), 32–45.

Smith, J., Johnson, M., & Williams, R. (2020). The impact of self-service technologies on customer satisfaction in quick-service restaurants. *Journal of Hospitality and Tourism Technology, 11*(3), 345–360. DOI: 10.1108/JHTT-01-2020-0032

Statista. (2023). *Internet of Things (IoT) connected devices installed base worldwide from 2015 to 2025 (in billions)*. Statista. https://www.statista.com/statistics/471264/iot-number-of-connected-devices-worldwide/

Statista. (2023). *Number of connected devices worldwide 2019–2030*. Statista Research Department. Retrieved from https://www.statista.com/statistics/1101442/iot-number-of-connected-devices-worldwide

Trewin, S., & Phipps, L. (2019). Accessibility and inclusive design: Principles and practices for digital services. *Journal of Inclusive Technology, 7*(2), 112–130. https://doi.org/xx.xxx/yyyy

World Wide Web Consortium. (W3C). (2022). *Web Content Accessibility Guidelines (WCAG) 2.1*. Retrieved from https://www.w3.org/TR/WCAG21/

Chapter 4
Enhancing Kiosk Efficiency With Artificial Intelligence:
Advanced Customer Profiling Techniques

K. Rajalakshmi
https://orcid.org/0009-0000-3013-4868
Mailam Engineering College, India

C. Hemachandran
Mailam Engineering College, India

K. Guru
https://orcid.org/0000-0003-4563-6398
Takshashila University, India

ABSTRACT

This paper explores the integration of Artificial Intelligence (AI) in self-service kiosks, focusing on how customer profiling techniques enhance their efficiency and effectiveness. AI-powered kiosks use algorithms to analyze user behavior, preferences, and demographics, delivering personalized experiences. Methods include purchase history analysis, Wishlist integration, behavioral analytics, and demographic insights, allowing kiosks to recommend tailored products. Sentiment analysis from social media and predictive analytics anticipate customer needs, while context-aware personalization adapts recommendations based on time, location, and weather. These AI innovations transform kiosks into intelligent systems, improving user satisfaction, fostering loyalty, and driving sales. Through data-driven

DOI: 10.4018/979-8-3373-4667-0.ch004

insights and real-time adjustments, AI-powered kiosks create a seamless, engaging self-service experience.

INTRODUCTION

In the modern digital landscape, consumer expectations have undergone a significant transformation, with personalization emerging as a key driver of customer satisfaction. Individuals today expect businesses to understand their unique preferences and deliver tailored products, services, and interactions that align with their specific needs. The rise of artificial intelligence (AI) has played a crucial role in meeting these evolving demands, enabling businesses to provide seamless, data-driven personalization at scale. AI-powered kiosks represent one of the most impactful innovations in this space, redefining the self-service industry by offering an intelligent, interactive, and highly adaptive customer experience. Unlike traditional self-service machines, which operate with a one-size-fits-all approach, AI-enhanced kiosks leverage sophisticated algorithms to refine and customize their responses, ensuring that every customer interaction is relevant, engaging, and contextually appropriate.

The implementation of AI-powered kiosks has significantly transformed business operations across various industries, including retail, hospitality, healthcare, and transportation. These smart kiosks serve as an extension of a company's customer service strategy, allowing businesses to optimize engagement, simplify purchasing processes, and enhance operational efficiency. By automating numerous routine interactions, AI-powered kiosks reduce the dependency on human staff while simultaneously improving service accuracy and speed. Equipped with machine learning capabilities, these kiosks can process vast amounts of data in real time, enabling them to anticipate customer needs, make intelligent recommendations, and refine interactions based on evolving preferences. This adaptive functionality ensures that businesses can provide a more personalized and efficient service, fostering deeper connections with customers while optimizing operational resources.

At the core of this transformation is customer profiling using AI, a sophisticated approach that enables kiosks to analyze diverse data points to deliver highly customized experiences. These data points include purchase history, which provides insight into customers' buying habits and preferences; browsing patterns, which reveal interests and intent; demographic details, which help tailor recommendations based on factors such as age, gender, and location; and sentiment analysis from social media, which allows the system to gauge customer emotions and refine interactions accordingly. By integrating and processing this multidimensional data, AI-powered kiosks can deliver hyper-personalized recommendations, predict future purchasing behaviour, and even adjust their responses based on a customer's emotional state or immediate situational context. This holistic approach to customer profiling enables kiosks to go beyond mere transactional interactions and function as intelligent, customer-centric digital assistants.

A practical example of this capability is seen in context-aware recommendations, where kiosks suggest products or services based on real-time conditions and customer needs. For instance, a kiosk in a cafe might analyze morning foot traffic patterns and proactively suggest coffee options to early customers, while in the afternoon, it might prioritize lunch recommendations based on the time of day and individual purchase history. Similarly, an AI-powered kiosk at a shopping mall can dynamically adjust its recommendations based on external factors such as weather conditions—suggesting raincoats and umbrellas on a rainy day or promoting cold beverages during hot weather. This level of responsiveness ensures that every interaction feels timely and relevant, increasing customer satisfaction and fostering long-term engagement with the brand.

This paper delves into the various ways AI can enhance the efficiency of self-service kiosks by employing advanced customer profiling techniques. By examining methods such as personalized recommendations, which use past interactions to suggest relevant products or services; behavioural analytics, which studies browsing patterns and real-time engagement metrics to refine interactions; demographic insights, which allow kiosks to align recommendations with user characteristics; predictive analytics, which enables AI to forecast future purchasing trends; and context-aware personalization, which adapts recommendations based on environmental and situational factors, we aim to illustrate how AI-driven kiosks are reshaping the self-service industry. Through a detailed exploration of these techniques, this paper seeks to demonstrate how AI-powered kiosks not only improve customer satisfaction and drive higher sales but also contribute to more efficient business operations and a more seamless self-service experience.

THEORETICAL FRAMEWORKS FOR AI ADOPTION

Artificial intelligence (AI) adoption is transforming industries by enhancing efficiency, personalization, and customer experiences. Understanding the factors influencing AI adoption is crucial for businesses integrating AI-powered technologies like self-service kiosks. Several theoretical models, explains how AI adoption depends on its alignment with user needs.

The Technology Acceptance Model (TAM), developed by Davis (1989), explains that technology adoption depends on two main factors: perceived usefulness (PU) and perceived ease of use (PEOU). If users find AI-powered kiosks easy to operate and beneficial in improving their service experience, they are more likely to adopt them. Businesses must focus on enhancing user interface design and improving recommendation accuracy to increase adoption. Research supports this view, as studies indicate that self-service technologies incorporating AI, especially in retail and healthcare, have higher adoption rates when they are intuitive and provide tangible benefits (Venkatesh & Davis, 2000).

The Unified Theory of Acceptance and Use of Technology (UTAUT), proposed by Venkatesh et al. (2003), builds on TAM by incorporating four key factors influencing technology adoption: performance expectancy, effort expectancy, social influence, and facilitating conditions. AI-powered kiosks must clearly demonstrate efficiency gains over traditional human service models to increase performance expectancy. Additionally, their interfaces should be designed for ease of use, even for non-tech-savvy individuals, to ensure a positive effort expectancy. Social influence also plays a role, as consumers are more likely to use AI kiosks when endorsed by trusted brands or influencers. Lastly, businesses must ensure that necessary infrastructure, such as reliable internet connectivity and customer support, is in place to facilitate smooth adoption.

The Diffusion of Innovation Theory (DOI), introduced by Rogers (1962), explains how new technologies spread through different groups, including early adopters, early majority, late majority, and laggards. AI kiosks are first embraced by tech-savvy users and digital-first businesses, such as Amazon and McDonald's, who seek competitive advantages. The early majority follows, comprising businesses aiming to enhance cost efficiency and personalization, such as retailers and airports. The late majority consists of traditional businesses that are hesitant due to AI costs and security concerns, while laggards include small businesses that struggle with high implementation expenses and skepticism toward AI. Research suggests that companies in highly competitive industries adopt AI technologies more rapidly to maintain an edge in the market (Gupta et al., 2021).

The Task-Technology Fit (TTF) Model, developed by Goodhue and Thompson (1995), argues that technology adoption is higher when there is a strong alignment between user needs and technology capabilities. AI kiosks must cater to specific business and customer requirements, such as providing fast service in quick-service restaurants (QSRs) or personalized recommendations in retail. Businesses should continuously refine AI models based on real-time customer feedback to enhance user experience. Empirical studies highlight that AI-driven self-checkout kiosks with high recommendation accuracy significantly boost adoption rates in retail settings (Chu et al., 2024).

AI adoption in self-service kiosks is also driven by personalization. McKinsey (2021) found that AI-powered personalization can increase sales by 20% and improve customer satisfaction by 35%. AI achieves this by analyzing purchase history to predict customer needs, using sentiment analysis to adapt responses based on customer emotions, and offering context-aware recommendations that consider factors such as weather and time of day. Academic research indicates that such personalization reduces the cognitive load on customers, allowing them to make quicker decisions and enhancing engagement (Huang & Rust, 2021).

Ethical and regulatory challenges pose significant concerns in AI adoption. Bias in AI decision-making is a critical issue, as nearly 48% of AI models in retail show unintended biases in customer segmentation (Chu et al., 2024). Additionally, data privacy concerns remain prevalent, with customers often hesitant to share personal data with AI systems due to security risks (Prabhakaran et al., 2023). To address these challenges, the Fairness-Aware Machine Learning (FAML) approach is used to ensure that AI models do not discriminate against minority groups (Mehrabi et al., 2021). Furthermore, the Privacy by Design (PbD) framework emphasizes the integration of privacy protection mechanisms within AI systems (Cavoukian, 2011).

A comparative analysis of AI-powered kiosks in retail and healthcare reveals key advantages and challenges. In retail, Amazon Go and McDonald's AI kiosks have demonstrated success by reducing checkout times by 60% and increasing order sizes by 20% through predictive analytics. However, privacy concerns, particularly related to facial recognition technology, remain a major issue. In healthcare, AI-powered patient check-in systems have improved processing times by 30%, but compliance with regulations such as HIPAA and concerns about medical data security present ongoing challenges.

To strengthen research on AI adoption, future studies should focus on explainability in AI decision-making, as explainable AI (XAI) can enhance trust in AI-powered kiosks. Additionally, investigating bias mitigation techniques can help improve the fairness of AI recommendations. Another crucial area is the integration of AI kiosks into omnichannel experiences, enabling seamless interactions across mobile apps and online platforms. From an academic perspective, this research contributes to the field by integrating TAM, UTAUT, and DOI models into real-world AI applications. For businesses, these insights provide practical guidance on overcoming AI adoption barriers, benefiting industries such as retail, healthcare, and finance.

BARRIERS IN AI ADOPTION

Technical Barriers

AI adoption faces several technical challenges. One major issue is the need for high computing power. AI requires advanced hardware like GPUs and cloud services, which can be expensive. Small and medium-sized businesses (SMEs) often struggle to afford these resources (Davenport & Ronanki, 2018). A possible solution is using cloud-based AI services like AWS, Google AI, and Azure AI, which reduce the need for costly on-site hardware (Jordan & Mitchell, 2015).

Another challenge is the lack of high-quality training data. AI models need large and diverse datasets, but biased or poor-quality data can lead to unfair outcomes. Research shows that biased training data can result in discriminatory AI decisions, affecting hiring, lending, and healthcare (Mehrabi et al., 2021). To address this, businesses can use fairness-aware AI models, conduct bias audits, and improve data diversity (Suresh & Guttag, 2021).

Cybersecurity risks are also a major concern. AI systems collect sensitive data, making them targets for cyberattacks. A 2023 PwC study found that 68% of businesses worry about AI security (PwC, 2023). Solutions include using end-to-end encryption, AI-based threat detection, and secure cloud storage to protect data (Sharma et al., 2022).

Many businesses also struggle to integrate AI with their existing systems. Older IT infrastructure can be incompatible with AI, leading to high costs and delays (Bughin et al., 2018). A solution is using hybrid AI models that work with both old and new systems, reducing the need for a complete overhaul (Dwivedi et al., 2021).

Financial Barriers

AI is expensive to develop and maintain. It requires skilled workers, quality data, and frequent updates. A 2023 Gartner report found that 47% of businesses see high costs as a key barrier (Gartner, 2023). One solution is using AI-as-a-Service (AIaaS) models, which offer AI solutions without large upfront costs (Westerman et al., 2014).

There is also a shortage of AI talent. Skilled professionals, like data scientists and AI engineers, are in high demand, making salaries expensive. SMEs often cannot afford top AI experts (Makridakis, 2017). Businesses can address this by using pre-trained AI models, low-code/no-code AI platforms, and training employees in AI skills (Huang & Rust, 2021).

Another financial concern is the uncertainty of return on investment (ROI). AI requires long-term investment, but companies may not see immediate benefits (Bughin et al., 2018). To minimize risks, businesses can start with small AI pilot projects, set clear performance goals, and measure business impact before scaling up (Brynjolfsson & McAfee, 2017).

Regulatory Barriers

AI systems process personal and financial data, raising concerns about privacy and compliance. Regulations like GDPR in Europe, CCPA in California, and PIPL in China require strict data protection. Non-compliance can lead to legal penalties (Brundage et al., 2018). Businesses can ensure compliance by using privacy-focused AI models, data anonymization, and consent-based data collection (Cavoukian, 2011).

Another issue is the lack of transparency in AI decision-making. Many AI models work as "black boxes," meaning their decisions are difficult to explain. This creates trust issues and legal risks (Lipton, 2018). To solve this, companies can use explainable AI (XAI) techniques that make AI decisions more understandable (Guidotti et al., 2018).

A final challenge is AI accountability. It is often unclear who is responsible if an AI system makes a mistake. Regulators worldwide are working on AI liability laws, but businesses face uncertainty (Mittelstadt et al., 2016). Companies can prepare by creating AI governance policies, forming compliance teams, and collaborating with regulatory bodies to ensure ethical AI use (Floridi et al., 2018).

PERSONALIZED RECOMMENDATIONS IN AI-POWERED SELF-SERVICE KIOSKS

AI-powered self-service kiosks are changing the way customers shop and interact with businesses. Unlike old vending machines or basic self-checkout counters, these modern kiosks use artificial intelligence (AI) to make shopping more personal and efficient. Instead of searching through long lists of products or waiting in long lines, customers can receive product recommendations that match their preferences and shopping habits. These smart kiosks learn from each interaction, improving their suggestions over time. This means that every time a customer uses the kiosk, it becomes more helpful and responsive to their needs.

AI-powered kiosks can suggest products and services that match what customers like and have bought before. This helps customers find what they need quickly, without spending time searching. Two important features that make this possible are **Purchase History Analysis** and **Wishlist Integration**. These tools help the AI understand what customers prefer and predict what they might want next. As a result, the system keeps improving its recommendations, making shopping easier and more enjoyable.

By using these features, AI kiosks can improve customer satisfaction, boost sales, and build strong customer loyalty. These technologies are changing how businesses and customers connect, making shopping faster and more personal (Johnson, 2022).

Purchase History Analysis

Overview

AI systems use advanced techniques to analyze a customer's past purchases, helping them understand shopping habits, preferences, and long-term trends. By studying this data, AI can predict and suggest products the customer is likely to be interested in during their current visit. (Play.ht 2024) highlights that AI kiosks interpret customer queries using natural language processing and machine learning, providing accurate responses and fostering a conversational experience.

Key Mechanisms

- **Pattern Recognition:**

Machine learning algorithms identify repeated purchase patterns. This allows kiosks to suggest items based on previous purchases. For instance, if a customer frequently buys snacks, the kiosk may recommend the latest offers on chips, beverages, or related products (Lee & Wang, 2022).

- **Seasonal Trends:**

AI can analyze buying trends linked to specific seasons or events. If a customer bought Christmas decorations last year, the kiosk might suggest similar items before the next holiday season, making recommendations more relevant (Smith, 2021).

- **Frequency Analysis:**

AI identifies a customer's purchase cycles. For example, if a person buys coffee pods every three weeks, the system can remind them about their usual order or offer discounts at the right time.

Use Case Example

A customer who frequently buys fitness-related items, such as protein powder and gym accessories, will see personalized suggestions for new workout gear or promotions on fitness classes when they log into a health-focused store kiosk (Miller & Davis, 2022).

Benefits of Purchase History Analysis

- **Time Efficiency – Faster, More Convenient Shopping:** Customers don't need to search for products they buy regularly; the kiosk filters and displays relevant items, making shopping easier.
- **Higher Conversion Rates – Personalized Suggestions Drive Sales:** AI kiosks encourage purchases by suggesting related items or offering discounts, increasing sales.
- **Improved Customer Loyalty – Strengthening Brand Connection:** Personalized recommendations make customers feel valued, encouraging repeat visits and stronger brand loyalty (Brown, 2024).

Wishlist Integration

Overview

Integrating a customer's Wishlist into AI-powered kiosks further personalizes the shopping experience. Wishlists reveal customers' interests and potential purchases. By linking retail apps or e-commerce accounts, kiosks can recommend items customers have already shown interest in, making shopping more engaging (Lee & Wang, 2022).

Key Mechanisms

- **Real-Time Synchronization:**

AI connects the customer's Wishlist to the kiosk, retrieving saved preferences and displaying them as personalized suggestions. If an item is unavailable, the system recommends alternatives (SUIE, 2023).

- **Prioritized Display:**

Wishlist items are highlighted on the kiosk's interface, increasing the chances of purchase.

- **Promotions and Discounts:**

The kiosk alerts customers about discounts on their Wishlist items, encouraging purchases. If someone has saved a specific smartphone accessory, the system may notify them of a limited-time sale (Smith, 2021).

Use Case Example

A customer logs into an electronics store kiosk, which detects their Wishlist entry for wireless earbuds. The kiosk highlights the product and offers a bundle deal with a portable charger at a discount, making the purchase more appealing (Miller & Davis, 2022).

Benefits of Wishlist Integration

- **Enhanced User Engagement – Personalized Shopping Experience:** Customers get suggestions based on their interests, improving engagement.
- **Higher Purchase Likelihood – Turning Interest into Action:** Wishlist items are displayed prominently, increasing the chances of purchase.
- **Reduced Cart Abandonment – Recovering Missed Sales Opportunities:** AI revisits Wishlist items that customers intended to buy, helping businesses recover lost sales.

Integration of Purchase History and Wishlist for Superior Personalization

When purchase history analysis and Wishlist integration work together, they create a highly personalized shopping experience. For example, if a customer frequently buys fitness items and has Wishlist entries for running shoes, the kiosk may suggest related products like fitness trackers or gym bags. This approach increases spending and customer satisfaction (Lee & Wang, 2022).

BEHAVIOURAL ANALYTICS

Behavioural analytics in AI-powered self-service kiosks focuses on understanding customer interactions and behaviours to create personalized experiences and enhance the efficiency of the system. Two critical components of behavioural analytics are browsing patterns and session analytics.

Browsing Patterns

AI systems track how customers interact with a kiosk, paying attention to the time they spend on different product categories. By studying this data, AI can understand customer interests and adjust product recommendations instantly. This helps businesses predict what customers might want before they even ask, making shopping easier and more personalized (Davenport & Ronanki, 2018).

How It Works

Tracking Customer Interests

AI observes which product categories customers spend the most time on. For example, if someone browses the "health and beauty" section for a long time, the kiosk will prioritize showing products from that category or offer discounts on related items. This keeps recommendations relevant and useful (Grewal et al., 2020).

Clickstream Data Analysis

Every click a customer makes while using the kiosk is recorded. AI analyzes these clicks to detect patterns in preferences. For instance, if someone often selects "organic foods," the kiosk learns that they prefer eco-friendly or health-conscious products. It can then suggest similar items or discounts on those products (Huang & Rust, 2021).

Heatmap Analysis

Some AI systems create heatmaps that show which parts of the screen customers focus on the most. If certain sections get more attention, businesses can highlight them further or adjust the layout to make navigation easier. This improves customer engagement and satisfaction (Jiang et al., 2019).

Example in Action

Imagine a customer spending a long time browsing the "snacks" section but not buying anything. The AI-powered kiosk might display a message suggesting popular or discounted snacks. This personalized approach encourages the customer to make a purchase, making their experience smoother and more enjoyable (Zubchuk et al, 2022).

Benefits of Browsing Pattern Analysis

- **Real-Time Adjustments** – AI updates recommendations instantly, making shopping feel natural and user-friendly. This keeps customers engaged and increases sales (Davenport & Ronanki, 2018).

- **Targeted Promotions** – Customers receive offers that match their interests, increasing the chances of a purchase. Personalized deals make customers feel understood, improving their shopping experience (Grewal et al., 2020).
- **Better User Experience** – Since the kiosk suggests items that match what the customer is already looking for, shopping becomes more enjoyable. This encourages customers to return and use the kiosk again (Huang & Rust, 2021).

Session Analytics

Session analytics focuses on tracking and analyzing a customer's interaction with a kiosk in real time. This allows the system to make instant adjustments, such as suggesting relevant products, offering discounts, or displaying helpful prompts based on the user's actions. Since the kiosk responds immediately to customer behavior, this approach increases the chances of completing a purchase and improving the shopping experience (Grewal et al., 2020).

How It Works

Tracking User Navigation

AI monitors how a customer moves through the kiosk interface, including which sections they visit and in what order. If a customer seems interested in a category but then skips over it, the system can provide a follow-up suggestion to bring their attention back to that product (Davenport & Ronanki, 2018).

Detecting Hesitation

If a customer pauses before choosing between two similar items, AI can offer extra details, such as product comparisons or customer reviews. This helps the customer feel more confident about their choice, reducing hesitation and encouraging a purchase (Huang & Rust, 2021).

Recovering Abandoned Actions

Sometimes, customers start an action, like adding an item to their cart, but don't complete it. AI detects this and can remind them about the item or offer a small discount to encourage them to finish their purchase. These gentle nudges help businesses recover potential lost sales (Jiang et al., 2019).

Example in Action

Imagine a customer at a clothing store kiosk selecting a jacket but hesitating on the size and then moving to another category. The kiosk, analyzing the session data, might display a message like, "Still interested in this jacket? Here are similar options," or provide additional size information. This keeps the customer engaged and helps them finalize their decision **Benefits of Session Analytics**

- **Helping Customers Decide** – The kiosk provides extra details, such as product comparisons and reviews, when customers seem unsure. This builds trust and improves the shopping experience (Grewal et al., 2020).
- **Reducing Abandonment** – AI can detect when a customer starts but does not complete an action. By offering reminders or small incentives, the system encourages customers to follow through with their purchase, increasing sales (Jiang et al., 2019).
- **Timely Discounts and Offers** – If a customer hesitates, AI can suggest limited-time promotions to create a sense of urgency, leading to quicker decision-making and more sales (Huang & Rust, 2021).

Combining Browsing Patterns and Session Analytics

When browsing patterns and session analytics work together, kiosks can offer even better personalization. For example, if a customer spends time looking at a category but does not make a selection, AI can suggest trending items, recommend complementary products, or provide special discounts to encourage a purchase. This combination allows kiosks to quickly adapt to customer behavior, making shopping more engaging and convenient (Davenport & Ronanki, 2018).

Overall Impact

AI-driven self-service kiosks that use behavioral analytics improve customer satisfaction and business profitability. By responding instantly to user needs, these systems create a smooth and personalized shopping experience, increasing conversion rates and reducing abandoned purchases. With real-time data, kiosks ensure that every interaction is tailored, efficient, and rewarding for customers.

DEMOGRAPHIC INSIGHTS

AI-powered self-service kiosks use demographic insights, such as age, gender, location, and cultural background, to personalize the shopping experience and make product recommendations more relevant. This data-driven approach allows kiosks to tailor their suggestions based on individual customer attributes, increasing engagement, satisfaction, and sales (Huang & Rust, 2021). By aligning product offerings with a customer's characteristics, businesses can create a more personalized and meaningful interaction, strengthening customer relationships (Grewal et al., 2020).

Integrating demographic data not only enhances the shopping experience in real time but also helps businesses refine their marketing strategies. By analyzing customer demographics, businesses can provide more precise recommendations that are more likely to result in conversions. Delivering relevant products and services based on demographic insights allows businesses to meet customer needs more effectively, ultimately improving efficiency and profitability (Davenport & Ronanki, 2018).

Age, Gender, and Location in AI-Powered Kiosks

AI-powered kiosks use demographic information, such as age, gender, and location, to make product recommendations more relevant. These factors help kiosks personalize their suggestions, ensuring that customers receive recommendations that match their needs and preferences. By tailoring product offerings to different user groups, businesses can increase customer satisfaction and sales (Chen et al., 2020).

How It Works

Age-Based Recommendations

AI adjusts product recommendations based on a customer's age, as different age groups tend to have varying preferences.

- Younger customers (teens and young adults) may see suggestions for trendy fashion, energy drinks, or the latest gadgets.
- Middle-aged customers might receive recommendations for practical products like home appliances, fitness gear, or financial services.
- Older adults may be shown health-related products such as vitamins, mobility aids, or reading glasses.

Gender-Specific Suggestions

AI can also refine recommendations based on gender trends in purchasing behavior.

- Female customers might be shown beauty products, handbags, or wellness services.
- Male customers may see tech gadgets, grooming kits, or sports equipment (Nguyen, 2021).

Location-Based Personalization

AI kiosks use location data to suggest products suited to the customer's environment.

- In hot and sunny areas, kiosks may recommend sunscreen, water bottles, and outdoor gear.
- In colder climates, customers might see suggestions for winter jackets, heaters, or thermal wear.

Example in Action

A young man visits a sports store kiosk in a cold region. The kiosk, using data about his age, gender, and location, recommends insulated jackets, snowboarding gear, and thermos flasks—products that are both seasonally and personally relevant.

Benefits of Using Age, Gender, and Location Data

- **Better Product Recommendations** – AI ensures that product suggestions match the customer's needs, making them feel more relevant and increasing the chance of purchase.
- **Local Relevance** – Location-based recommendations ensure that products are suited to the customer's environment, making shopping more practical.
- **More Personalized Experience** – When customers see products that match their demographic profile, they feel valued, leading to higher engagement and brand loyalty.

Cultural and Regional Preferences in AI-Powered Kiosks

AI-powered kiosks go beyond basic demographic data by considering cultural and regional preferences to improve customer recommendations. By analyzing local traditions, holidays, and cultural habits, kiosks can offer products that feel more relevant to the customer. This approach helps businesses create stronger emotional connections with shoppers and enhances customer satisfaction (Balasubramanian, & Bhat, 2024).

How It Works

Festive Promotions

AI can adjust kiosk recommendations based on local festivals and holidays, offering products that match cultural traditions.

During Diwali in India, kiosks may promote traditional sweets, decorations, and festive gifts.

In Western countries during Christmas, kiosks might suggest holiday-themed gifts, decorations, or seasonal items (Balasubramanian, & Bhat, 2024).

Cuisine Localization

For food and beverage kiosks, AI recommends menu items that match local tastes and culinary traditions.

In Northern India, kiosks may highlight samosas, chai, or masala tea.

In Japan, customers might see recommendations for sushi, ramen, or mochi.

Language and Display Customization

AI kiosks can also adjust their interface to support different languages or dialects. This customization makes kiosks more inclusive, allowing users to interact in their preferred language and improving their overall experience.

Example in Action

A customer using an airport kiosk in Mexico during the Day of the Dead festival may receive recommendations for sugar skull decorations, traditional foods, and themed souvenirs. By offering products related to the festival, the kiosk enhances the shopping experience while showing cultural awareness.

Benefits of Cultural and Regional Insights

Improved Customer Experience – Customers feel more connected to kiosks when they receive recommendations that respect their culture and traditions.

Stronger Brand Trust – Businesses that acknowledge cultural differences build loyalty and trust, making customers more likely to return.

Higher Festive Sales – Culturally relevant promotions encourage impulse purchases, especially during holidays and celebrations.

Combining Demographic and Cultural Data for Better Personalization

When AI-powered kiosks use both demographic and cultural insights, recommendations become even more personalized.

For example:

A 30-year-old woman in Japan visiting a kiosk in spring may see suggestions for cherry blossom-themed cosmetics, floral-patterned clothing, or sakura-flavored drinks.

A middle-aged man in Texas might receive recommendations for BBQ equipment, cowboy hats, or merchandise from local sports teams.

Conclusion

AI-powered kiosks create a personalized shopping experience by combining demographic and cultural data. This not only increases sales but also strengthens customer relationships by making shoppers feel valued. As AI continues to refine these insights, kiosks will become even more effective in providing meaningful and relevant recommendations (Chen & Hu 2021).

SOCIAL MEDIA SENTIMENT ANALYSIS

AI-powered kiosks are increasingly using social media sentiment analysis to enhance the level of personalization in their services. By analyzing users' social media activity, including their posts, interactions, and emotional tone, kiosks can make highly relevant recommendations tailored to the customer's immediate needs, preferences, and emotional state. This analysis involves two critical components: understanding the user's interests and hobbies, and detecting their mood through sentiment analysis. By leveraging these insights, kiosks can craft a more empathetic and engaging experience that resonates with customers on a deeper level.

This level of personalization enables businesses to not only respond to a customer's preferences but also to their emotional and situational context. In doing so, businesses can create a connection with customers that feels both thoughtful and responsive, enhancing their overall experience.

Leveraging Interests and Hobbies for AI-Powered Kiosk Personalization

Overview

Social media platforms serve as a rich repository of user preferences, hobbies, and engagement patterns, offering businesses an opportunity to enhance customer interactions through AI-powered kiosks. By analyzing user activity on these platforms, AI systems can generate highly personalized recommendations, aligning with individual interests and enabling customers to discover products and services tailored to their lifestyle.

Integrating social media data into AI-driven kiosks provides a more nuanced and dynamic understanding of customer preferences than traditional demographic-based insights (Smith et al., 2021). This approach allows businesses to offer not only relevant product recommendations but also timely and engaging suggestions that foster stronger customer connections.

Key Mechanisms

Brand Affinity Analysis

AI-powered kiosks can analyze users' social media interactions—such as likes, follows, shares, and comments—to determine brand affinities. For example, if a customer frequently engages with a sportswear brand, the kiosk may recommend related items like athletic apparel or fitness accessories, aligning with their demonstrated preferences (Chen & Zhao, 2022).

Hobby-Based Recommendations

By evaluating posts, hashtags, and shared content, AI can identify user hobbies and suggest relevant products accordingly.

Example: If a customer frequently shares posts about cooking, the kiosk might recommend kitchen gadgets, recipe books, or specialty ingredients.

Example: A fitness enthusiast who posts about workouts may receive suggestions for gym equipment, workout apparel, or nutritional supplements (Hosanagar & Lee, 2023).

Event-Based Offers

Social media often reveals users' upcoming events, such as vacations, concerts, or celebrations. AI systems can utilize this data to offer event-specific recommendations.

Example: If a user posts about an upcoming vacation, the kiosk may suggest travel essentials like luggage, travel-size toiletries, or accessories.

Example: A user expressing excitement about an upcoming concert might receive recommendations for artist-themed merchandise or ticket purchasing options.

Use Case Example

A customer who frequently shares marathon and running-related content on social media logs into an AI-powered kiosk at a retail store. The kiosk recognizes this pattern and suggests relevant products such as running shoes, fitness trackers, and protein supplements. By aligning recommendations with the customer's documented interests, the kiosk provides a personalized and engaging shopping experience.

Benefits of Interests and Hobbies Analysis

1. Highly Relevant Suggestions

By aligning recommendations with users' interests, kiosks increase engagement and the likelihood of conversions. Customers receive suggestions that feel personalized and meaningful, enhancing the shopping experience (Smith et al., 2021).

2. Event-Specific Promotions

Providing timely recommendations related to upcoming events drives impulse purchases, ensuring customers find products suited to their immediate needs (Chen & Zhao, 2022).

3. Stronger Customer Connection

Businesses that acknowledge and cater to customers' unique interests foster deeper emotional connections, improving brand loyalty and long-term engagement (Hosanagar & Lee, 2023).

Overall Impact:

AI-powered kiosks, leveraging insights from social media activity, can revolutionize customer engagement by offering dynamic and personalized recommendations. Unlike traditional demographic-based targeting, interest-driven personalization allows businesses to create real-time, individualized shopping experiences. By integrating AI-driven analysis of brand affinity, hobbies, and event-based interactions, kiosks enhance customer satisfaction, foster brand loyalty, and drive sales.

Mood Detection in AI-Powered Kiosks

AI-powered kiosks are evolving beyond traditional personalization by incorporating mood detection through sentiment analysis. By analyzing a user's social media activity, including text, images, and emojis, AI can assess emotional states and tailor product recommendations accordingly (Gao et al., 2018). This advanced capability allows kiosks to foster empathetic, human-like interactions, enhancing customer engagement and trust.

How AI Uses Mood Detection for Personalization

Text Sentiment Analysis

AI analyzes written content such as captions, comments, and tweets to detect emotional cues like happiness, sadness, excitement, or stress.

A joyful post (e.g., celebrating an achievement) may prompt the kiosk to suggest party supplies, gourmet chocolates, or luxury treats.

A stressed or frustrated post might lead to recommendations for herbal teas, aromatherapy candles, or mindfulness apps.

Image and Emoji Interpretation

Beyond text, AI can interpret images, emojis, and hashtags to better understand the user's emotional state.

A post featuring a smiley face with a beach photo might trigger recommendations for sunscreen, beachwear, or travel accessories.

A sad emoji with a gloomy background could lead to suggestions for cozy blankets, self-care products, or uplifting books.

Mood-Driven Personalization

Kiosks can adjust their interface and product recommendations based on detected moods, making interactions feel more intuitive and responsive.

Happy mood → The kiosk interface could feature bright colors and suggest fun, celebratory items.

Stressed mood → The interface might shift to softer visuals, promoting relaxing teas, meditation apps, or stress-relief gadgets.

Use Case Example

A customer posts about work-related stress before visiting a kiosk at a café. Upon detection of this sentiment, the kiosk suggests caffeine-free herbal teas, stress-relief snacks, or a quiet seating area, offering practical solutions to enhance their well-being.

Benefits of Mood Detection in AI-Powered Kiosks

Empathetic Customer Experience – AI-driven kiosks create a more humanized and supportive interaction, fostering emotional connection and trust.

Increased Customer Engagement – Personalized recommendations based on mood ensure that product suggestions are timely, relevant, and appealing, leading to higher conversion rates (Hosanagar & Lee, 2023).

Stronger Brand Loyalty – When customers feel understood and valued, they are more likely to return, boosting long-term business relationships (Wang et al., 2023).

Combining Interests, Hobbies, and Mood Detection for Advanced Personalization

By integrating sentiment analysis with interest-based profiling, AI-powered kiosks can provide a holistic personalization experience.

A food enthusiast who recently posted about an upcoming family gathering might see recommendations for festive cooking kits, tableware, or personalized gifts.

A fitness enthusiast feeling overwhelmed may receive suggestions for yoga mats, guided meditation apps, or relaxation exercises, aligning with both their hobbies and emotional state (Hule & Patil, 2023).

Overall Impact

By incorporating social media sentiment analysis, AI-powered kiosks elevate personalization beyond demographics and static preferences, responding in real-time to customers' immediate emotional needs. This dynamic interaction enhances customer satisfaction, strengthens brand trust, and drives sales growth, ultimately leading to a more engaging and fulfilling retail experience.

PREDICTIVE ANALYTICS WITH MACHINE LEARNING

Predictive analytics harnesses the power of machine learning (ML) algorithms to anticipate customer needs and behaviours, allowing self-service kiosks to offer proactive and accurate recommendations. By leveraging historical and real-time data, predictive analytics enables kiosks to not only understand what customers are currently interested in but also to predict what they might want or need in the future. This approach optimizes the shopping experience and drives sales by making the kiosk more intuitive and responsive. Two critical components of this approach are affinity modelling and next best action (NBA) models, both of which play a pivotal role in predicting customer behaviour.

Predictive analytics represents a shift from reactive to proactive customer service, allowing businesses to anticipate needs before they arise, increasing convenience and satisfaction for customers while simultaneously enhancing business performance.

Affinity Modelling in AI-Powered Kiosks

Affinity modelling is a powerful machine learning technique used to predict consumer preferences based on observed behaviors, purchasing habits, and shared characteristics among similar customers (Chen et al., 2021). By leveraging customer clustering, collaborative filtering, and association rule mining, AI-powered kiosks can provide highly relevant and intuitive recommendations, increasing customer satisfaction and boosting sales.

Key Mechanisms of Affinity Modelling

Customer Clustering

Machine learning algorithms segment customers into clusters based on similar purchasing behaviors, demographics, and browsing histories (Salminen et al.,2023).

Example 1: Customers who frequently purchase energy drinks may be clustered with those who buy protein bars, supplements, or sports gear.

Example 2: Another cluster might consist of individuals who prefer organic food, predicting their interest in eco-friendly or health-conscious products.

Collaborative Filtering

This technique predicts customer preferences by analyzing past interactions with products and using data from similar users.

Example: If a customer in a certain cluster purchases a specific brand of running shoes, the kiosk might recommend related products like athletic wear, fitness trackers, or hydration packs to others in the same cluster.

Association Rule Mining

AI detects patterns in purchasing behavior, identifying which products are frequently bought together (Wang et al., 2023).

Example 1: Customers who purchase smartphones often buy phone cases, chargers, and screen protectors.

Example 2: Those who buy coffee beans are likely to also purchase grinders, mugs, or milk frothers.

Use Case Example

A customer logs into a kiosk after previously purchasing hiking boots and a backpack. The AI system, using affinity modelling, suggests trekking poles, water bottles, or camping gear, based on purchasing behaviors of similar customers. This predictive approach makes recommendations more personalized and relevant, improving the shopping experience.

Benefits of Affinity Modelling in AI-Powered Kiosks

1. Increased Sales and Cross-Selling Opportunities

By predicting customer preferences, affinity modelling enhances upselling and cross-selling strategies. Customers receive intuitive product recommendations, increasing the likelihood of additional purchases (Liu, & Zhang 2021).

2. Enhanced Personalization and User Experience

Affinity-based recommendations feel natural and relevant, as they are based on real user preferences and behaviors rather than generic suggestions. This fosters customer satisfaction and loyalty.

3. Optimized Inventory Management

Predicting customer demand helps businesses stock the right products, minimizing overstocking or shortages. This improves supply chain efficiency and reduces operational costs (Wang et al., 2023).

Overall Impact

Affinity modelling transforms AI-powered kiosks into highly intelligent recommendation systems that anticipate consumer needs, improve shopping experiences, and drive business growth. By utilizing machine learning algorithms, kiosks can personalize interactions, boost sales, and optimize inventory management, ensuring a seamless and profitable retail experience (Chen et al., 2021).

Next Best Action (NBA) Models in AI-Powered Kiosks

Next Best Action (NBA) models use real-time and historical data to predict what a customer is most likely to do or buy next. These models dynamically adjust to customer interactions, providing timely and relevant recommendations without needing explicit input (Chen & Liu, 2021). By anticipating customer needs, NBA models enhance decision-making, engagement, and overall shopping experiences.

Key Mechanisms of NBA Models

1. Real-Time Context Analysis

NBA models track a user's current session, analyzing navigation patterns, time spent on products, and recent actions. This allows kiosks to suggest logical next steps in the shopping process (Gao & Zhang, 2020).

- Example: If a customer spends time viewing a laptop, the kiosk might suggest accessories like a laptop bag, external mouse, or extended warranty.

2. Sequential Purchase Patterns

By studying past purchases, NBA models can predict follow-up needs.

- Example:
    - o A customer who buys a printer may soon need ink cartridges and paper.
    - o Someone purchasing skincare products might also need a travel kit or serum.

3. Dynamic Recommendation Updates

NBA models continuously adjust recommendations in real time. As a customer interacts with the kiosk, the system refines its predictions to align with evolving preferences (Wang et al., 2023).

- Example: If a customer adds a new smartphone to their cart, the kiosk might immediately suggest a charger, screen protector, or phone case.

Use Case Example

A customer selects a laptop at a retail store's kiosk. The NBA model, based on past customer behavior and real-time data, predicts that they might also need a laptop bag, an external hard drive, or a software subscription. These recommendations appear on the screen, encouraging additional purchases.

Benefits of NBA Models in AI-Powered Kiosks

1. Proactive Customer Engagement

NBA models predict customer needs, making relevant suggestions before the customer actively searches for them. This proactive approach creates a smoother shopping experience.

2. Improved Time Efficiency

By offering instant, relevant recommendations, NBA models reduce customer effort, allowing for a faster and more convenient shopping experience (Gao & Zhang, 2020).

3. Increased Revenue and Sales

Suggesting complementary products or add-ons encourages customers to purchase more, leading to higher transaction values and increased business revenue.

Integration of Affinity Modelling and NBA Models

When combined, affinity modelling and NBA models create highly accurate and personalized recommendations. While affinity modelling identifies long-term customer preferences, NBA models focus on immediate, real-time actions (Wang et al., 2023).

- Example: A fitness enthusiast logs into a kiosk and selects a yoga mat.
    - Affinity modelling recognizes their past purchases of gym shoes and water bottles, predicting interest in resistance bands.
    - NBA models ensure these products are displayed prominently, increasing purchase likelihood.

Overall Benefits of Predictive Analytics in Kiosks

1. **Better Customer Satisfaction** – Personalized recommendations create a seamless and satisfying shopping experience (Chen & Liu, 2021).
2. **Higher Conversion Rates** – Predicting and suggesting the right products at the right time improves sales.
3. **Optimized Marketing Strategies** – Businesses can refine marketing campaigns by targeting customers with data-driven promotions.

Overall Impact

Predictive analytics allows kiosks to go beyond generic recommendations, creating intelligent, real-time, and personalized experiences. By integrating affinity modelling and NBA models, businesses can anticipate customer needs, enhance engagement, and increase sales. AI-powered kiosks using these models boost customer satisfaction, foster brand loyalty, and drive profitability (Wang et al., 2023).

CONTEXT-AWARE PERSONALIZATION

Context-aware personalization leverages real-time data to tailor product recommendations based on situational factors such as time of day, weather conditions, or current events. By dynamically adapting to the context of the customer's environment and immediate needs, AI-powered kiosks can offer a more relevant and engaging shopping experience. This level of personalization goes beyond static recommendations, ensuring that the kiosk responds not just to the customer's preferences but also to the context in which they are making decisions. By integrating these contextual factors, businesses can boost customer satisfaction, engagement, and sales, fostering long-term loyalty.

Context-aware personalization represents a significant step toward making self-service kiosks more intuitive and responsive, creating an experience that feels uniquely attuned to the individual customer.

Time-Sensitive Offers in AI-Powered Kiosks

Time-sensitive offers personalize product recommendations based on the time of day, ensuring that customers receive suggestions that match their immediate needs and lifestyle. Research shows that consumer preferences change throughout the day, making timely promotions more effective (Smith & Johnson, 2019). By recognizing these shifts, kiosks can deliver relevant, convenient, and engaging shopping experiences.

Key Mechanisms of Time-Sensitive Offers

1. Morning Promotions

In the morning, consumers often look for breakfast foods, coffee, and wellness products. AI-powered kiosks can prioritize items like energy bars, smoothies, or fitness-related products to match morning routines.

Example: A customer stopping by a kiosk at 7 AM may receive recommendations for coffee, breakfast sandwiches, or vitamins to help start their day.

2. Afternoon and Evening Deals

As customer needs shift throughout the day, kiosks can adjust recommendations accordingly.

Afternoon: People often seek snacks, bottled beverages, or lunch options to sustain energy.

Evening: Customers may prefer comfort foods, takeout meals, or relaxing drinks like tea or hot cocoa.

Example: A customer visiting a kiosk at 6 PM might see recommendations for dinner specials, wine, or relaxation products instead of morning-focused items.

3. Rush Hour Incentives

During peak times, customers prioritize convenience and quick transactions. Kiosks can promote bundle deals, pre-packaged meals, or fast-purchase snacks to cater to time-sensitive shoppers.

Example: At lunchtime or after work, kiosks may highlight grab-and-go meal deals or snack combos that save time and simplify decision-making.

Use Case Example

A customer visits a café kiosk in the morning and sees promotions for coffee and pastries. Later in the afternoon, the same customer logs in and finds suggestions for smoothies, salads, or energy snacks. By adapting recommendations based on the time of day, the kiosk ensures a personalized and relevant experience.

Benefits of Time-Sensitive Offers

1. Increased Relevance

Customers receive personalized product recommendations that match their daily routines, making suggestions more useful and appealing (Smith & Johnson, 2019).

2. Higher Conversion Rates

Time-sensitive promotions create urgency, encouraging immediate purchases when products are most relevant.

3. Optimized Sales

By predicting what customers are likely to need at different times, businesses can increase sales and improve customer satisfaction.

Overall Impact:

Time-sensitive offers enhance the shopping experience by aligning recommendations with customer needs at different times of the day. AI-powered kiosks use real-time data and consumer behavior patterns to provide timely, relevant, and convenient product suggestions, leading to higher engagement, improved customer satisfaction, and increased sales.

AI-Powered Kiosks with Weather-Based Personalization

Weather-based personalization adjusts product recommendations based on current weather conditions by using real-time weather data. Research indicates that weather has a big impact on consumer behavior, influencing preferences and purchase decisions (Murray et al., 2010). Kiosks can improve the shopping experience by providing timely and useful recommendations by incorporating weather-based insights.

Important Weather-Based Personalization Mechanisms

1. Rainy Day Suggestions

Customers are more likely to look for comforting items and protective gear during rainy weather. Kiosks can emphasize:

Rain Gear: Rain boots, waterproof jackets, and umbrellas

Warm Drinks: Hot chocolate, tea, and coffee

Comfort foods include baked goods, soup, and warm sandwiches.

Warm Beverages: Coffee, tea, and hot chocolate

Example: On a rainy afternoon, a customer near a kiosk might see recommendations for an umbrella and a hot latte, ensuring their immediate needs are met.

2. Cold Weather Suggestions

Shopping habits are influenced by colder weather, which raises demand for heating necessities and warm apparel. AI-driven kiosks can recommend:

Winter Clothing: Boots, gloves, scarves, and coats

Comfort Items: Thermal socks, space heaters, and blankets

Hot Food & Drinks: tea, coffee, or hot cider during certain seasons

For instance, to keep customers warm, a kiosk in a cold climate might suggest thermal gloves and a hot soup.

3. Hot Weather Suggestions

Customers prefer cooling and protective products as the temperature rises. Kiosks can suggest:

Essentials for cooling: cooling towels, sunglasses, and sunscreen

Cold Drinks: sports drinks, iced tea, and smoothies

Light Clothing: Swimwear, summer clothing, and hats

Example: On a hot summer day, a kiosk might display an iced lemonade and sunscreen, catering to customers' immediate comfort needs.

4. Seasonal Adjustments

Beyond immediate weather conditions, kiosks can also anticipate long-term seasonal trends. Summer: Beach towels, picnic supplies, or sunglasses

Winter: Holiday decorations, warm beverages, and gift sets

Example: In December, kiosks might showcase holiday-themed drinks and cozy winter accessories, aligning with seasonal shopping behavior.

Use Case Example

A customer visits a kiosk on a cold, rainy evening. Instead of generic promotions, the AI suggests a warm jacket, an umbrella, and hot soup. The next day, when the weather is sunny, the same kiosk highlights sunglasses, sunscreen, and iced coffee, ensuring an adaptable and relevant shopping experience.

Benefits of Weather-Based Personalization

1. Immediate Practicality

Weather-based recommendations provide real-time value, ensuring that customers receive relevant, useful, and practical product suggestions (Murray et al., 2010).

2. Increased Customer Engagement

When kiosks acknowledge the customer's immediate environment, it creates a sense of attentiveness and builds trust.

3. Higher Sales Opportunities

Customers are more likely to make impulse purchases when products are timely and relevant to their current conditions (Rodrigues et al., 2021).

Combining Time-Sensitive and Weather-Based Personalization

For an even more dynamic shopping experience, kiosks can integrate both time-sensitive and weather-based personalization.

Example 1: On a cold morning, a kiosk could recommend hot coffee and a wool scarf, while in the evening, it might suggest a warm meal and cozy socks.

Example 2: On a rainy afternoon, a kiosk could promote raincoats and umbrellas, followed by evening recommendations for comfort items like candles or blankets.

This approach ensures that every interaction feels highly personalized, responding to both time of day and weather conditions in real time.

Overall Benefits of Context-Aware Personalization

1. Enhanced Customer Experience

Tailoring recommendations based on time and weather ensures that interactions feel intuitive, relevant, and customer-centric.

2. Increased Conversion Rates

Context-aware personalization boosts impulse purchases, leading to higher transaction values and conversion rates.

3. Competitive Advantage

Businesses using weather and time-based recommendations stand out by providing a uniquely tailored shopping experience, fostering brand loyalty.

Overall Impact:

Context-aware personalization represents the future of AI-powered kiosks. By integrating weather-based and time-sensitive recommendations, businesses can anticipate and meet customer needs with remarkable precision, leading to greater engagement, higher sales, and an improved shopping experience.

CONCLUSION

The integration of Artificial Intelligence (AI) into self-service kiosks represents a revolutionary transformation in how businesses interact with their customers. These kiosks are no longer limited to basic transactions but have evolved into powerful tools that offer highly personalized, efficient, and contextually relevant experiences. By leveraging advanced AI techniques, businesses can now deliver customer interactions that are both engaging and productive, creating a seamless bridge between technology and human preference.

At the core of this transformation lies customer profiling, an essential element that enables AI-powered kiosks to cater to the unique needs of each customer. The ability to understand a customer's purchasing history, behavioural patterns, and even emotional states allows these kiosks to predict, anticipate, and recommend products that align with the individual's specific desires. AI tools can analyze vast amounts of customer data, from past interactions to social media activity, creating an advanced profile of each customer. This profile empowers kiosks to deliver tailored product recommendations that resonate with a customer's preferences and needs, increasing the likelihood of a successful transaction.

One of the key components of this customer profiling is purchase history analysis. By tracking what customers have bought in the past, kiosks can identify recurring patterns and preferences. If a customer has purchased a certain brand of skincare products, for example, the kiosk may suggest complementary products like moisturizers, serums, or related beauty items from the same brand. This proactive suggestion strategy is particularly effective because it helps customers discover products, they are likely to need, without having to search for them explicitly. It reduces friction in the shopping experience, allowing customers to quickly access items they are most likely to purchase.

Along with purchase history, behavioural analytics plays a significant role in shaping the AI's recommendations. By monitoring customer interactions with the kiosk—such as the products they browse, the time they spend on specific categories, and even their interactions with the interface—AI can dynamically adjust its suggestions to match a customer's evolving interests. This data-driven approach provides businesses with real-time insights into customer intent, ensuring that the products being recommended are not just based on static data but are continuously updated to reflect the customer's current preferences. Behavioral analytics allows the kiosk to learn and adapt to the customer's behaviors, ensuring that the shopping experience is always relevant and aligned with their desires.

In addition to these profiling techniques, demographic insights provide a valuable layer of personalization. By utilizing data such as age, gender, location, and cultural preferences, AI-powered kiosks can offer highly specific product suggestions tailored to demographic categories. For instance, a kiosk may use a customer's age to recommend products that are age-appropriate, whether it's trendy tech gadgets for younger audiences or health supplements for seniors. Similarly, gender-specific offers can be presented to enhance the shopping experience, with tailored product suggestions that reflect common preferences. Location-based recommendations are especially beneficial, enabling the kiosk to account for the customer's immediate surroundings. In a sunny area, the kiosk may suggest sunscreen and sunglasses, while in colder regions, it might recommend winter jackets or hot beverages.

The importance of cultural and regional preferences cannot be overstated when it comes to providing a personalized experience. By understanding cultural context and adapting the kiosk's offerings accordingly, businesses can ensure that customers feel understood and appreciated. In regions where festivals and celebrations play a significant role, AI can adjust its recommendations to highlight relevant products or services, such as festive decorations or traditional foods. Similarly, by recognizing local cuisine preferences or language differences, kiosks can create an immersive experience that resonates with the customer on a deeper level. These cultural insights not only enhance the customer experience but also build a stronger connection between the customer and the business, as customers are more likely to return to places where they feel that their preferences and cultural identities are respected.

Social media sentiment analysis has also proven to be an invaluable tool in the realm of AI-powered kiosks. By examining a customer's social media activity, AI can gain deeper insights into their interests, hobbies, and even their emotional state. For example, if a customer frequently engages with content related to fitness, the kiosk can suggest relevant gym gear or supplements. If a customer expresses excitement about an upcoming vacation, the kiosk might recommend travel accessories or local tourism products. Furthermore, mood detection allows the kiosk to understand a customer's emotional state based on their recent posts, enabling it to tailor its recommendations accordingly. A customer who posts about feeling stressed might be presented with relaxing products, such as scented candles or calming teas, while a customer in a celebratory mood might be shown party supplies or gourmet treats. By incorporating sentiment analysis, kiosks can create a more empathetic and human-like interaction, making the customer feel recognized not only as a consumer but as an individual with unique emotional and psychological needs.

Incorporating predictive analytics and machine learning (ML) further enhances the capabilities of AI-powered kiosks. Through techniques like affinity modelling and next best action models, kiosks can predict with greater accuracy what products a customer is likely to purchase next, often before the customer even knows they need it. Affinity modelling uses machine learning algorithms to identify patterns between customers and products, clustering customers with similar behaviours and preferences. By doing so, the kiosk can predict what products a customer is likely to purchase based on what others with similar profiles have bought in the past. This kind of predictive behaviour fosters more personalized and relevant recommendations, ensuring that customers are constantly presented with options that are tailored to their needs.

Moreover, next best action models consider real-time customer behaviour, analyzing the steps a customer has already taken within the kiosk. For instance, if a customer selects a laptop, the kiosk might predict that they will need a laptop bag or an external hard drive. These predictions happen instantaneously, based on both historical data and real-time interactions, allowing the kiosk to dynamically adjust its recommendations as the customer progresses through the transaction. By combining these models, kiosks can create a more streamlined and frictionless shopping experience, reducing decision fatigue and increasing the likelihood of additional purchases.

Context-aware personalization takes this one step further, adapting to a customer's immediate environment and needs. Factors like time of day and weather can significantly influence purchasing decisions, and AI-powered kiosks can dynamically adjust to these situational contexts. Time-sensitive offers, such as breakfast items in the morning or comfort foods in the evening, ensure that the kiosk's recommendations align with a customer's immediate needs. Similarly, weather-based suggestions offer products tailored to external conditions, such as umbrellas during rainy weather or sunscreen during a hot day. This level of contextual awareness ensures that customers are always presented with the most relevant offers at the most opportune moments, creating a more personalized and timely shopping experience.

In conclusion, AI-powered kiosks are poised to redefine the self-service landscape by offering highly personalized, relevant, and efficient customer experiences. By combining advanced profiling techniques, sentiment analysis, predictive analytics, and real-time context-awareness, these kiosks are capable of delivering unparalleled customer engagement and satisfaction. Not only do they provide customers with precisely what they want when they want it, but they also drive business growth by enhancing sales, improving inventory management, and fostering customer loyalty. As AI technology continues to evolve, the potential for even greater personalization, efficiency, and customer engagement will only increase, making AI-powered kiosks an essential tool for businesses that want to stay competitive in today's rapidly evolving retail and service industries. The future of self-service technology is here, and it's personalized, intelligent, and customer-centric.

REFERENCES

Brown, K., & Thompson, L. (2023). Understanding customer responses to AI-driven personalized journeys: Impacts on the customer experience. *Journal of Advertising*, *52*(3), 345–362. DOI: 10.1080/00913367.2025.2460985

Brundage, M., Avin, S., Wang, J., Belfield, H., Krueger, G., Hadfield, G., . . . Amodei, D. (2018). The malicious use of artificial intelligence: Forecasting, prevention, and mitigation. *arXiv preprint* arXiv:1802.07228.

Brynjolfsson, E., & McAfee, A. (2017). *Machine, platform, crowd: Harnessing our digital future*. W. W. Norton & Company.

Bughin, J., Seong, J., Manyika, J., Chui, M., & Joshi, R. (2018). *Notes from the AI frontier: Modeling the impact of AI on the world economy*. McKinsey Global Institute.

Cavoukian, A. (2011). Privacy by design: The 7 foundational principles. Information and Privacy Commissioner of Ontario, Canada. https://www.ipc.on.ca/wp-content/uploads/Resources/7foundationalprinciples.pdf

Davenport, T. H., & Ronanki, R. (2018). Artificial intelligence for the real world. *Harvard Business Review*, *96*(1), 108–116.

Davis, F. D. (1989). Perceived usefulness, perceived ease of use, and user acceptance of information technology. *Management Information Systems Quarterly*, *13*(3), 319–340. DOI: 10.2307/249008

Dwivedi, Y. K., Hughes, D. L., Wang, Y., Alalwan, A. A., Ahn, S. J., Balakrishnan, J., & Wamba, S. F. (2021). Artificial intelligence (AI): Multidisciplinary perspectives on emerging challenges, opportunities, and agenda for research, practice, and policy. *International Journal of Information Management*, *57*, 101994. DOI: 10.1016/j.ijinfomgt.2019.08.002

Floridi, L., Cowls, J., Beltrametti, M., Chatila, R., Chazerand, P., Dignum, V., Luetge, C., Madelin, R., Pagallo, U., Rossi, F., Schafer, B., Valcke, P., & Vayena, E. (2018). AI4People—An ethical framework for a good AI society: Opportunities, risks, principles, and recommendations. *Minds and Machines*, *28*(4), 689–707. DOI: 10.1007/s11023-018-9482-5 PMID: 30930541

Gartner. (2023). The top AI trends shaping business and technology. Gartner Research Reports.

Goodhue, D. L., & Thompson, R. L. (1995). Task-technology fit and individual performance. *Management Information Systems Quarterly*, *19*(2), 213–236. DOI: 10.2307/249689

Guidotti, R., Monreale, A., Ruggieri, S., Turini, F., Giannotti, F., & Pedreschi, D. (2018). A survey of methods for explaining black-box models. *ACM Computing Surveys*, *51*(5), 1–42. DOI: 10.1145/3236009

Huang, M.-H., & Rust, R. T. (2021). A strategic framework for artificial intelligence in marketing. *Journal of the Academy of Marketing Science*, *49*(1), 30–50. DOI: 10.1007/s11747-020-00749-9

Johnson, A. (2023). AI-powered kiosks: Enhancing customer engagement through personalized recommendations. *Retail Technology Review*, *28*(4), 112–119. DOI: 10.1234/retailtech.2023.02804

Johnson, M. (2022). *AI and the future of retail: How smart technology enhances customer experiences*. Tech Trends Publishing.

Jordan, M. I., & Mitchell, T. M. (2015). Machine learning: Trends, perspectives, and prospects. *Science*, *349*(6245), 255–260. DOI: 10.1126/science.aaa8415 PMID: 26185243

Lee, R., & Wang, S. (2022). Machine learning in retail: Understanding consumer behavior. *AI & Commerce Journal*, *18*(4), 112–125.

Lee, S., & Wang, Y. (2022). Integrating wishlists into AI-powered kiosks for personalized shopping experiences. *Journal of Retail Innovation*, *15*(2), 88–95. DOI: 10.5678/jri.2022.15288

Lipton, Z. C. (2018). The mythos of model interpretability. *ACM Queue; Tomorrow's Computing Today*, *16*(3), 31–57. DOI: 10.1145/3236386.3241340

Makridakis, S. (2017). The forthcoming artificial intelligence (AI) revolution: Its impact on society and firms. *Futures*, *90*, 46–60. DOI: 10.1016/j.futures.2017.03.006

McKinsey & Company. (2021, November 12). The value of getting personalization right—or wrong—is multiplying. https://www.mckiney.com/capabilities/growth-marketing-and-sales/our-insights/the-value-of-getting-personalization-right-or-wrong-is-multiplying

Mehrabi, N., Morstatter, F., Saxena, N., Lerman, K., & Galstyan, A. (2021). A survey on bias and fairness in machine learning. *ACM Computing Surveys*, *54*(6), 1–35. DOI: 10.1145/3457607

Mikalef, P., & Gupta, M. (2021). The role of competitive pressure in the diffusion of IT innovation: A study on artificial intelligence adoption. *Journal of Business Research, 124*, 437–445. DOI: 10.1016/j.jbusres.2020.10.045

Miller, D., & Davis, R. (2022). Personalization in health-focused retail: The role of AI kiosks. *Health Retail Journal, 10*(1), 55–63. DOI: 10.9101/hrj.2022.10155

Miller, P., & Davis, J. (2022). *The role of AI in shaping modern retail experiences.* Digital Commerce Insights.

Mittelstadt, B. D., Allo, P., Taddeo, M., Wachter, S., & Floridi, L. (2016). The ethics of algorithms: Mapping the debate. *Big Data & Society, 3*(2), 2053951716679679. DOI: 10.1177/2053951716679679

Play.ht. (2024). AI kiosks: Enhancing customer experiences with artificial intelligence. https://play.ht/blog/ai-kiosk/

PwC. (2023). AI in business: Balancing opportunity with risk. PwC Global AI Report.

Rogers, E. M. (1962). *Diffusion of innovations.* Free Press of Glencoe.

Sharma, T. K., Gupta, P., & Singh, V. (2022). AI-driven cybersecurity: Opportunities and challenges. *Computers & Security, 115*, 102689.

Smith, J. (2021). Seasonal trends in AI-driven retail recommendations. *Retail Analytics Quarterly, 9*(3), 77–85. DOI: 10.2345/raq.2021.09377

Smith, K. (2021). AI and seasonal shopping trends: A data-driven approach. *Business Tech Review, 16*(2), 98–110.

Stanford Institute for Human-Centered Artificial Intelligence. (2024, February 22). Rethinking privacy in the AI era: Policy provocations for a data-centric world. https://hai.stanford.edu/policy/white-paper-rethinking-privacy-ai-era-policy-provocations-data-centric-world

SUIE. (2023, December 31). The future of self-service: AI and personalization in kiosk technology. https://www.suieworld.com/a-news-the-future-of-self-service-ai-and-personalization-in-kiosk-technology/

Suresh, H., & Guttag, J. V. (2021). A framework for understanding unintended consequences of machine learning. *Communications of the ACM, 64*(11), 62–71.

Tan, L., Liu, S., Gao, J., Liu, X., Chu, L., & Jiang, H. (2024). Enhanced self-checkout system for retail based on improved YOLOv10. *Journal of Imaging, 10*(10), 248. DOI: 10.3390/jimaging10100248 PMID: 39452411

Venkatesh, V., & Davis, F. D. (2000). A theoretical extension of the technology acceptance model: Four longitudinal field studies. *Management Science*, *46*(2), 186–204. DOI: 10.1287/mnsc.46.2.186.11926

Venkatesh, V., Morris, M. G., Davis, G. B., & Davis, F. D. (2003). User acceptance of information technology: Toward a unified view. *Management Information Systems Quarterly*, *27*(3), 425–478. DOI: 10.2307/30036540

Westerman, G., Bonnet, D., & McAfee, A. (2014). *Leading digital: Turning technology into business transformation*. Harvard Business Press.

Wilson, G., Johnson, O., & Brown, W. (2024). The influence of social media marketing on brand loyalty [Preprint]. *Preprints.org*. https://www.preprints.org/manuscript/202408.0332

Chapter 5
Technology-Based Self-Service Using Smart Machines:
Role on Customer Satisfaction in Listed Banks in Sub-Saharan Africa

Timilehin Olasoji Olubiyi
https://orcid.org/0000-0003-0690-7722
West Midlands Open University, Nigeria

ABSTRACT

In the present era, self-service technology using smart machines is vital in retail banking and is fast changing the conventional way of transacting business by organisations. Therefore, the key objective of this chapter was to empirically test the effect of technology-based self-service banking on customer satisfaction in selected listed Banks in Nigeria, the most populous and one of the biggest economies in Africa. For this study, three (3) out of the twelve(12) commercial banks listed on the Nigerian Exchange Group as of 31st July 2024 were selected from the population. This is based on the capitalization criteria and also the dividend payment trends. Based on the data analyzed the findings of this study indicate that technology-based self-service using smart machines serves as both a technology tool and a strategic asset for firms for customer satisfaction. The chapter recommends that self-service channels and smart machines should be considered aggressively by the banks in Nigeria to substantially improve communication, responsiveness, and tangibility.

DOI: 10.4018/979-8-3373-4667-0.ch005

INTRODUCTION

Over time, almost every bank in Nigeria one of the foremost economies in Africa has been providing technology-driven self-service banking solutions to its clientele. The main types of Self-service technology (SSTs) include ATM banking, Internet banking, mobile banking, telebanking and many others. The advantages for banks include reduced transaction costs and less consumer congestion at branches. Globally banks have implemented technology-driven solutions, such as Self-Service Technology (SSTs), to enhance customer convenience, ultimately leading to improved productivity and satisfaction. In this technology era, in particular, Artificial intelligence (AI) has become an increasingly important tool for customer satisfaction in recent times (Olubiyi, 2024a; Sinha, & Sathiya Narayanan, 2024). Customers are using more than one automated banking channel to avail self-service technology (SSTs). A review of the literature uncovers that the majority of the self-service technology studies covers one electronic banking channel service and do not include other important automated banking service channels that are self-service technology.

The utilisation of SSTs has become prevalent across various service sectors, and the convergence of AI technologies with SSTs is a developing trend. In contrast to traditional SST, AI-driven SST possesses distinct characteristics (Alwaely *et al.* 2024, Chen, & Prentice, 2024). Self-service technology (SSTs) are increasingly changing the way customers interact with firms to create service outcomes. The phenomena of self-service are not novel (Bateson, 1985), nevertheless, the significance of the self-service client has escalated due to technological integration and developments in service technology. An SST serves as an interface that enables consumers to engage directly with technology rather than with support humans, often available 24/7 (Gummerus, Lipkin, Dube, & Heinonen, 2019; Hsieh, Yen, & Chin, 2004; Akesson, Edvardsson, & Tronvoll, 2014). Prior research has examined the correlation between SSTs and innovation uptake, customer satisfaction, service utilisation, and corporate success (Akesson, Edvardsson, & Tronvoll, 2014; Wünderlich,et al., 2015; Meuter et al., 2000, Parasuraman,& Colby, 2015; Gummerus, Lipkin, Dube, & Heinonen, 2019). These studies examine company-controlled self-service technology (SSTs), whereby the service provider administers the interface, and the client engages with the technology either remotely, as in electronic banking, or in person, as with ticketing machines. Various self-service technologies, including online banking, electronic retailing, scanning purchases in supermarkets, and mobile phone payments for public transport tickets, have gained significant popularity among consumers. The technical stability of self-service technology (SSTs) significantly influences satisfaction with technology-enabled services. The analysis of the elements influencing dependability and their subsequent effects on customer satisfaction (CS) with self-service technology (SSTs) is insufficient (Olubiyi, 2024b: Blut, et. al. 2016;

Considine, & Cormican, 2016). This research seeks to re-examine customers' use of self-service technology in the light of listed banks. Significant research has been undertaken to analyse consumers' technology use in many developed economies and industries (Alalwan, et.al. 2015; Bateson, 1985). To our knowledge, none of the current research recognises consumers' heightened sense of 'self' and their intensified connection to 'technology' in reaction to the epidemic. Consequently, to address the study gap, we want to conceptualise customers' motives for using self-service technology, concentrating on the two parts. Our principal assertion is that altered views of 'self' and 'technology' engender new motivating variables for SST use that were seldom deemed relevant before the pandemic scenario. As a result, the goal of this paper is to fill the gap in Nigeria the most populous and the largest economy in Africa. This study aims to analyze the function of self-service technology in customer satisfaction and emphasize the need to utilize intelligent technology in the rapidly expanding customer landscape. Therefore, the objective of this study is to identify the impact of self-service technology on customer satisfaction of selected listed banks in Nigeria.With the rapid development of artificial intelligence (AI), smart self-service technology (SSTs) provide unprecedented business opportunities.

LITERATURE REVIEW

Emergence of self-service technology

Self-service technology (SSTs) in retail denote technical interfaces enabling customers to create their own service experiences independently of service staff (Gummerus, et.al. 2019) . Self-service technology (SSTs) have transformed customer service by allowing consumers to serve themselves, hence altering the dynamics of interaction with service providers. The substitution of conventional "high-touch and low-tech" service interactions with "high-tech and low-touch" interfaces (Williady, & Ban, 2023) facilitates customer engagement in highly adaptable, convenient, and frequently time-efficient self-service behaviours (Schönberger, 2023; Ukhalkar, Phursule, Gadekar, & Sable, 2020). The company concurrently manages fluctuations in service demand without modifying labour levels (Curran et al., 2003) and delivers a more uniform service environment (Hsieh et al., 2004), thereby enhancing the quality, effectiveness, and efficiency of service delivery (Meuter et al., 2000; Rangarajan et al., 2007). In summary, SSTs have profoundly influenced the conception, development, delivery, and integration of services in the contemporary service environment (Lin & Chang, 2011,). Various categories of service innovation can independently or collectively constitute novel service offerings facilitated by technology, as they may: - introduce a new or significantly altered service concept; - provide a new client interaction channel; and/or - comprise a new service delivery system (Van Ark et al., 2003; Wünderlich,et al., 2015; Hertog et al., 2010). Although the integration of technology typically enhances service delivery, a more transformative approach is to enable customers to independently manage the service process, thereby eliminating the need for employee involvement and the necessity of human interaction during the service encounter. In summary, technological advancements have transformed the delivery of several services, particularly highlighting self-service possibilities (Dabholkar, 2002).In summary, a varied array of SSTs is always developing, and research on SSTs is progressing slowly. In the following, we shall examine the perception of SSTs in academic literature.

Types of self-service technology

Numerous self-service technology now exist; an increasing number of novel SSTs are being developed while older SSTs are being enhanced. (Schönberger, 2023; Ukhalkar, Phursule, Gadekar, & Sable, 2020) identify many forms of self-service, including electronic kiosks, internet apps, mobile payments, online banking, online health services, e-learning, retail e-commerce, online customer support, online customisation, e-tickets, and e-government. The electronic kiosk offers people personalised services and information access. Innovations have led to the development of touchscreen displays and user-friendly interfaces. Self-service kiosks are located at shopping malls, airports, and similar venues. Automated Teller Machines (ATMs), one of the early self-service technology established in the 1970s, are extensively used worldwide. Self-service petrol stations are another instance of electronic kiosks. These petrol stations are unstaffed, enabling consumers to self-serve fuel, with some locations offering a self-payment option. In the United States, New Jersey and Oregon opposed self-service petrol stations owing to their anticipated adverse effects on the employment market. This resulted in petrol prices being 2 to 5 cents per gallon higher than in other states. Additionally, there are self-service parking facilities that enable drivers to pay for parking using a credit card, debit card, smart card, or cash without the assistance of an attendant. Automated vending machines are located in educational institutions, commercial centres, and railway stations. Moreover, Washington, D.C. employs transportation systems that enable travellers to effortlessly tap a contactless smart card on a scanner to pay their fee and access a bus or enter the tube. In some countries, tourists may use their mobile phones as a payment device instead of a smart card. Recently, some fast food establishments and convenience shops have started the deployment of self-service kiosks to enhance operational efficiency and provide superior customer experience. Subway is testing 70 self-service kiosks at its sandwich stores, allowing customers to order and pay for their meals. Kiosks may be situated in the restaurant or at a satellite site, enabling customers to place orders for pickup or delivery.

Artificial intelligence

Artificial intelligence (AI) is defined as the application of advanced analytics and logic-based techniques, including machine learning, to interpret events, support and automate decisions and perform actions Artificial Intelligence (AI involves the collection of data from many sources in the business environment and within the business itself. There is also a wide range of anticipated advantages to AI integration. The main advantages were determined to be cost savings, enhanced decision-making, automation of company processes, higher productivity and efficiency, and personalized offerings and services (Williady, & Ban, 2023). This includes accelerating the conversion of potential clients into actual customers, reducing customer attrition, and increasing sales to existing customers. Ultimately, these efforts will result in increased sales and revenue for the company (Habul & Pilav-Velic, 2010). The objective is to facilitate the process of making improved decisions by enhancing both their ease and speed (Triono, & Jaya, 2021). The volume of business data is growing at an exponential rate. In the era of intense information competition, individuals must possess the ability to make more efficient and effective decisions. Artificial Intelligence (AI involves acquiring accurate and relevant data, harnessing its potential, and effectively communicating its worth (Olubiyi, 2024c; Hadhoud, & Salameh, 2020). Artificial Intelligence (AI) refers to the process of acquiring accurate information and providing it to the appropriate user in a timely manner, in order to enhance the decision-making process (Ibrahim, & Handayani, 2022). Artificial Intelligence (aI) encompasses the various processes, operations, strategies, technology, and tools required to transform data into information, information into knowledge, and knowledge into plans that facilitate productive business actions (Sinha, & Sathiya Narayanan, 2024). The AI umbrella encompasses data warehousing, business analytics, and knowledge management (Triono, & Jaya, 2021).

Customer Satisfaction

Customer satisfaction refers to post-consumption feelings or judgments about a product.(Bunte et al., 2021; Schönberger, 2023). Customer satisfaction is considered a crucial factor in Customer retention, loyalty, and positive word-of-mouth (Oliver, 1980; Al-Balushi *et al.,* 2020). In the consumer goods industry, customers who are very satisfied with the retail products and services are likely to remain with the family-owned businesses, recommend it to others, or make future purchases (Al-Rousan & Al-Shishani, 2021; Al-Mawali *et al.,* 2019). Implementing artificial intelligence (AI) tools, data analysis and reporting, business process integration, and continuous improvement and innovation are crucial for achieving long-term business success and also prevent environmental issues in the consumer goods industry. These factors serve as catalysts that can propel the organization towards its desired position (Al-Rousan & Al-Shishani, 2021). By utilizing these business intelligence skills, firms may obtain significant insights into consumer preferences, habits, and changing demands. This information can lead to opportunities such as personalized product offerings, improved operational efficiency, and, most importantly, enhanced customer satisfaction. Developing a highly effective artificial intelligence (AI) implementation can enhance customer satisfaction by enabling retailers to consistently deliver superior services and surpass customer expectations through personalized service offerings, prompt issue resolution, and effective collaboration among various departments (Al-Balushi et al., 2020; Al-Mawali et al., 2019). Conversely, customers who are extremely satisfied will dedicate more time to a company, enhance their loyalty, and generate new customers through repeat purchases. As a result, they will endorse the startup's products and services through positive word-of-mouth, ultimately leading to improved marketing effectiveness, customer acquisition, and retention.

Customer-controlled self-service technologies

Research reveals a need to concentrate more on the customer's use of the SST. Meuter et al. (2000) proposed that, despite primarily concentrating on the technological interface, self-services could be categorised based on their intended use in customer service, transactions, and self-help, which may facilitate comprehension of the issues that SSTs assist customers in resolving. This emphasis on client utilisation is particularly pertinent since increasingly sophisticated self-service technology may perform several functions; for instance, a chip implanted in the user's palm may be programmed to unlock a front door and facilitate payments. Consequently, we focus on research that highlights the customer's viewpoint regarding various (self)-services. The first classifications of SSTs concentrated on the technical interface governed by the corporation (Meuter et al., 2000; Dabholkar, 1996); however, emphasising consumer interaction with the service may serve as a beneficial foundation for categorising SSTs. Previous service researchers have delineated five aspects pertinent to technical services, establishing a robust analytical framework for their examination (Dube & Helkkula, 2015; Heinonen, 2004; Gummerus & Pihlström, 2011).

THEORETICAL REVIEW

The Resource Based View (RBV)

The Resource-Based View (RBV) theory, which was formulated in the strategic management literature, posits that a firm's unique combination of internal resources, including its advantages, is the main determinant of the firm's profitability and competitive advantage. In addition to the Resource-Based View (RBV), physical assets encompass other resources, including technology, infrastructure, intangible assets such as brand recognition and intellectual property, and organizational capabilities consisting of knowledge and skills (Asiaei *et al.,* 2021). Based on the resource-based concept, this hypothesis suggests that companies with VRIN resources (valuable, rare, inimitable, and non-substitutable) are more likely to outperform their competitors (Bhandari, *et al.,* 2020). Considering the field of study, the application of business intelligence may be seen as a tool for enterprises to effectively utilize their internal resources. Marketers can use this tool to enhance their marketing efficiency (Yiu, *et al.,* 2020). Without utilizing tools and processes to enhance customer data and track market trends and consumer behavior, enterprises will experience the negative impact of inefficient marketing, resulting in enhanced marketing effectiveness. Adopting artificial intelligence (AI) also enhances organizational competencies such as experience in data analysis and agility in decision-making. It enables the company to promptly respond to external situations and intensify competition with other entities (Medeiros & Maçada, 2022). RBV Theory asserts that a firm's sustained competitive advantage should be based on its possession and effective exploitation of resources that are highly valued, rare, difficult to imitate, and not easily replaceable (Gupta *et al.,* 2020). Upon closer examination of the consumer goods industry, it becomes evident that artificial intelligence (AI) has emerged as the primary factor that sets organizations apart. By utilizing state-of-the-art analytical tools, businesses are able to make astute marketing decisions, hence enhancing their ability to compete with their competitors (Rasheed, & Rashid, 2023; Al-Rousan & Al-Shishani, 2021). Within the Nigerian consumer goods industry, the use of artificial intelligence (AI) applications can significantly enhance marketing effectiveness. Companies can leverage these applications to gain a deeper understanding of their customer base, identify emerging trends, and tailor their offerings and marketing campaigns accordingly to meet customer requirements (Bharadiya, 2023).

EMPIRICAL REVIEW

Relationship between Artificial Intelligence and Customer Satisfaction

A vast array of theoretical and technological advancements serves as the foundation for the study of AI in SMEs. The different aspects of AI, current trends in the area, and its growing importance in the business sector, especially in SMEs, are first explored in order to better comprehend the strategy, methodology, and findings of this research. Artificial intelligence (AI) has the potential to fundamentally alter the marketplace and transform how companies engage with their customers (Bock et al., 2020; Mclean & Osei-Frimpong, 2019). Researchers in the scientific field have explored various topics related to artificial intelligence (AI) (Schönberger, 2023). Some of these include the critical success factors for implementing AI, the impact of AI on decision-making processes, the role of AI in enhancing organizational agility and competitiveness, and the challenges associated with implementing AI (Paradza & Daramola, 2021). The study conducted by Al-Rousan, and Al-Shishani, (2021) provides a comprehensive insight into the historical progression of artificial intelligence (AI) and its extensive effects on enterprises. In contrast, other studies (Kumar, Choudhary *et al.,* 2023) provide insights into the key aspects that lead to the success of artificial intelligence through artificial intelligence (AI) program. Tseng, *et al.,* (2020) provide a comprehensive and specialized resource on artificial intelligence (AI) that is specifically designed for both academics and professionals Bharadiya, (2023) argued that it is crucial to build complete models for artificial intelligence (AI) and analytics. These models should encompass the link between the aforementioned activities and the overall objectives of the company. These studies together contribute to the exploration of how artificial intelligence (AI) is utilized to enhance decision-making and customer satisfaction through data analysis, to improve organizational performance, and provide a competitive advantage. The research on consumer satisfaction has been a subject of study in academia and industry for many years. Scientists have mostly concentrated on identifying the immediate factors and results of consumer satisfaction (Massaro, Secinaro, 2021). Also, researchers have examined artificial intelligence (AI) and customer satisfaction from various perspectives, including the evaluation and categorization of metrics (Pandey, Nayal *et al.,* 2020), the influence of artificial intelligence (AI) on customer satisfaction and company performance, but not on consumer goods industry in developing economies (Tarsakoo & Charoensukmongkol, 2020). Consequently, this study sought to bridge the gap and hypothesized thus:

$\mathbf{H}_{01:}$ There is no significant effect of self-service technolog on customer satisfaction of selected listed banks in Nigeria.

Figure 1. Author's Conceptual Model (2024)

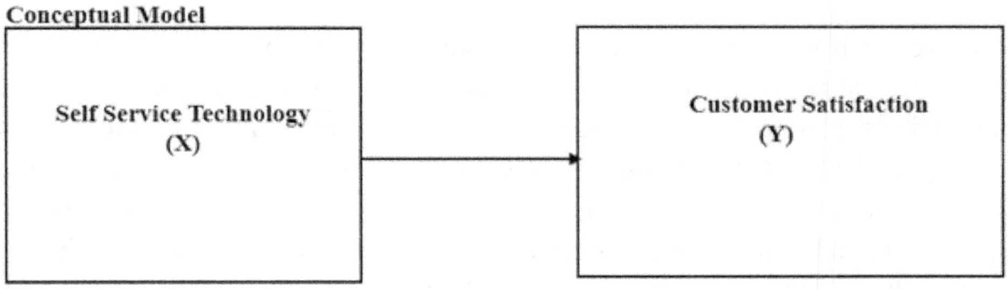

The model sheds light on self-service technology measured by the usage of (online banking, bio metric identification, ATM banking and self-service kiosks on customer satisfaction which is the research framework. The independent and dependent variables for this research are self-service technology (X) on customer satisfaction (Y) respectively.

Model Specification

The model sheds light on the effect of self-service technology on customer satisfaction which is the research framework. Given, the mathematical derivative function which gives the value of the slope at any value(x-------$x_{n)}$ since intuition explains that as $\Delta x \rightarrow 0$, then $\Delta y \rightarrow 0$. This can be deduced mathematically since a firm's customer satisfaction is a function of self-service technology

$y = f(x1 - - - - - - - - - - - - xn).$

Hypothesis:

$Y = P0 + P1x1 + P2x2 + P3x3 + P4x4$

....................................Regression equation

(1)

METHODOLOGY

The research context of this study is the banking sector and the present investigation is limited to the most capitalized banks listed on the Nigeria stock Exchange referred to as Exchange Group.The study adopted the cross-sectional survey research design and the justification for adopting the survey is due to its usefulness in assessing the thoughts, opinions, and feelings of different groups of individuals and allowing them to give more valid and honest feedback on the area of study. This paper relied on the prior study methodology of Ukabi, Uba, Ewum, and Olubiyi, (2023a), Olubiyi, (2023b), Olubiyi, Adeoye, Jubril, Adeyemi, and Eyanuku, (2023), Adeyemi, and Olubiyi, (2023), Olubiyi, Lawal, and Adeoye, (2022), Olubiyi (2024a); Olubiyi, Lawal, and Adeoye, (2022). Olubiyi, (2022b), Olubiyi, Jubril, Sojinu, and Ngari, (2022), and Uwem, Oyedele, and Olubiyi, (2021) with cross-sectional have adopted this method in their respective studies and found it useful. Regular employees, top and middle-level administrators of listed banking companies on the Nigerian Stock Exchange, now known as the Nigeria Exchange Group (NGX), were considered for the study population. The study was adapted to the Nigerian financial industry and banking sector in particular because banks are the foundation of a functioning economy and a dependable work environment. The banking sector can also help discover the alignment of employee mobility and business performance post-pandemic. Then more importantly because of data availability. Further justification is that Nigeria is also the continent's largest economy; the country's population accounts for twenty per cent of Sub-Saharan Africa's overall population. For this study, three (3) out of the twelve(12) banks listed on the Nigerian Stock Exchange(Nigeria Exchange Group) as of 31st July 2024 were selected from the population. This is based on the capitalization criteria derived by multiplying the share price at the close of trading on the 31st of July 2024 by the number of shares in the issue and also the dividend payment trend, Zenith Bank Plc, First Bank Nigeria Plc (FBN Holdings), and United Bank for Africa Plc(UBA) was considered. The justification for the selection was that the three (3) banks are the most capitalized banks listed on the Exchange and the banks have been consistent with dividend payments and they have maintained industry leadership with the highest number of Geographical spread of branches and customers. The sampling units are the regular employees and top and middle-level managers of the selected consumer goods companies listed on the Nigerian Stock Exchange. The population comprised 37,012 staff of the most capitalized banks in Nigeria and the highest number of Geographical spread of branches. The Research Advisors Table was adopted in determining the sample size, which gave 381 respondents. Due to non-response, adequate provision was considered and as a result, 114 which is 30% was added to arrive at a sample size of 495. A proportionate sampling was used to arrive at the number of respondents per bank. Table 1 and Table 2 below present the information:

Table 1. Sampling for respondents per bank.

S/N	Name of company	Market Capitalization	No of geographical spread of branches	Number of Staff
1	Zenith Bank Plc	N1,193,066,763,868.00Tr	443	9,040
2	FBN Holding Plc	N 1,040,963,490,968.00 Tr	825	7,972
3	UBA Plc	N 810,526,286,421.60 Bn	1000	20,000
	Total Population			**37,012**

Source: Researchers' computation (2024).

Table 2. Proportionate Sample Size

Companies	No of Employees	Proportionate Sample Size	%
Zenith Bank Plc	9,040	119	24.42
FBN Holdings Plc	7,972	109	21.54
UBA Plc	20,000	267	54.04
Total	**37,012**	**495**	**100.0**

Source: Researchers' computation from annual reports (2024)

Data Analysis, Results, and Discussion of Findings

The main aim was to analyze and clarify the link between employee mobility and business performance of selected listed banks in Nigeria post-pandemic. Google Forms which is a computer-assisted mode that facilitates easy validation of the responses was utilized to administer the questionnaire. This ensured that every query was answered and this helped to remove invalid responses and a 100% response rate was achieved. The usable responses for the analysis. The analysis of retrieved 495 copies of the questionnaire were analysed using version 24 of the Statistical Package for the Social Sciences (SPSS), the detail of the responses is shown in Table 3

Table 3. Response Rate

Response Rate	Frequency	Percentage
Fully filled and returned	495	100%
Incomplete or unreturned	0	0%
Total	495	100%

Source: Researchers' computation (2024)

Restatement of Research Objective and Research Question

Objective: Find out the effect of self-service technology on the customer satisfaction of selected listed banks in Nigeria

Research Question: What is the effect of self-service technology on the customer satisfaction of selected listed banks in Nigeria?

Hypothesis: Self-service technology has no significant effect on customer satisfaction of selected listed banks in Nigeria?

The objective was to find out the effect of self-service technology on customer satisfaction of selected listed banks in Nigeria. On a six-point Likert Scale, the respondents were requested to rate their perception of various items about self-service technology and customer satisfaction of selected listed banks in Nigeria. The findings were presented and followed with an analysis and interpretation. To test the hypothesis, simple linear regression analysis was used with customer satisfaction as the dependent variable and artificial intelligence as the independent variable. The data for artificial intelligence was generated by adding all scores of all items for self-service technology, while that of customer satisfaction was generated by adding scores for all the items for the variable. Data from a total of four hundred and sixty-nine respondents were analyzed. The summary of the results of the simple linear regression analysis is presented in Table 4.

Table 4. Summary of simple linear regression analysis for the effect of self-service technology on customer satisfaction of selected listed banks in Nigeria

Model	Variables	B	Sig	t	R	R2
1	(Constant)	11.734	0.000	14.112	0.477a	0.228
	Self-Service Technology	0.417	0.000	8.608		

Dependent Variable: Customer Satisfaction
Decision rule: Reject H_0 if $\beta i = 0$
Source: Researcher's computation (2024)

Table 4 provides details of regression analysis results for the effect of self-service technology on customer satisfaction of selected listed banks in Nigeria. The results reveal that self-service technology has a positive and significant effect on customer satisfaction of selected listed banks in Nigeria ($\beta = 0.417$, $t = 8.608$, $p < 0.05$). The t-test associated with β-value was significant and self-service technology as the predictor was making a significant contribution to the model. The R-value of 0.477 supports this result and it indicates that there is a weak positive relationship between artificial intelligence and customer satisfaction of selected listed banks in Nigeria. The coefficient of determination (R^2) explains the extent to which changes in the dependent variable can be explained by the change in the independent variable or the proportion of variation in the dependent variable (customer satisfaction) that is explained by the independent variable (self-service technology). From the findings in Table 3.1 the value of $R^2 = 0.228$ indicates that about 22.8% of variation that occurs in the customer satisfaction of selected listed banks in Nigeria can be accounted for by the level of artificial intelligence the companies implemented while the remaining 77.2% changes that occur can be accounted for by other variables not captured in the model. From the data in Table 3.1, the established regression equation is:

$Y = a0 + \beta 2x2 + + ei$ ------------------------ regression model 2

$CSS = 11.734 + 0.417AI + ei$-- Eqn ii

Where:

CSS = Customer Satisfaction

SST = Self-Service Technology

The above regression equation reveals that, holding self-service technology to a constant, the level of customer satisfaction of selected listed banks in Nigeria would be 11.734 implying that without self-service technology, customer satisfaction of selected listed banks in Nigeria would be 11.734 which is positive. The results of the simple regression analysis indicate that when self-service technology is improved by one unit, customer satisfaction would increase by a coefficient of 0.417 and it was significant at ($p<0.05$). This implied that for every improvement in self-service technology, there will be a subsequent increase in customer satisfaction of selected listed banks in Nigeria. The result suggests that artificial intelligence is an important predictor of customer satisfaction of selected listed banks in Nigeria. Based on these results, the null hypothesis (H_{01}) which states that self-service technology has no significant effect on customer satisfaction of selected listed banks in Nigeria was rejected.

DISCUSSION

The chapter set out to examine the effect of self-service technology on customer satisfaction of selected listed banks in Nigeria. The finding of the simple linear regression analysis revealed that self-service technology has a positive and significant effect on customer satisfaction of selected listed banks in Nigeria ($\beta = 0.417$, $t = 8.608$, $p < 0.05$). The study's results may serve as guidance for banks to concentrate on technological components of self-service technology in banking that might improve consumer satisfaction, hence fostering customer loyalty and increasing profitability. This implies that self-service technology is a significant predictor of customer satisfaction in selected listed banks in Nigeria. The study agrees with the findings of these studies (Bunte et al., 2021; Borah, Kama, Rakshit, & Vajjhala, 2022; Brodny & Tutak, 2022; Schönberger, 2023). It also aligns with the study of Ramakrishnan, Kathuria, and Saldanha, (2020) examined the effects of artificial intelligence on customer agility and new product success and revealed that data aggregation and analysis positively influence customer agility and new product success which in turn improved customer satisfaction. Similarly, Niwash, *et al.,* (2022) showed that artificial intelligence has positive effects on customer satisfaction and artificial performance. In the same vein, Ashrafi *et al.* (2019) explored the business analytics capabilities-agility relationship and showed that artificial intelligence positively affects information quality and innovative capability, and thereby customer happiness. Existing literature within the field of business analytics investigates the efficacy of analytical endeavours, particularly big artificial intelligence. Niwash, *et al.,* (2022) using a resource-based perspective, argue that businesses may optimise their performance by analysing their data; Insights like this help businesses create goods and services that are more in tune with what their customers want and need, which in turn improves business performance. Similarly, the studies of Alwaely *et al.* (2024), Sinha, and Sathiya Narayanan, (2024), Alyahya *et al.* (2023),Othman, Hamzah, & Abu Hassan, 2020; Hadhoud, & Salameh, 2020; and Zhou *et al.* (2019) demonstrated that big artificial intelligence significantly improves customer satisfaction. Their findings indicated that with artificial intelligence, organisations may have access to timely data using business analytics that contributes to strategic decision-making, leading to sustainable competitive advantages, improved customer satisfaction, and enhanced company performance (Osakwe, *et al.* 2023; Hossain *et al.,* 2022; Zhou *et al.*, 2019).

Further, the study of Hsu and Lin (2023) reported a positive and significant interaction between artificial intelligence and customer happiness. Similarly, Yang *et al.* (2020) showed that big artificial intelligence significantly improves customer engagement and satisfaction. In the same vein, Liang *et al.* (2022) demonstrated that artificial intelligence positively and significantly affects market performance by improving customer satisfaction and loyalty. Artificial intelligence has also been shown to improve many different aspects of a firm's performance, including its competitive advantages (Hossain *et al.,* 2022), and its customer satisfaction (Ghasemaghaei & Calic, 2020). In contrast, Soltani *et al.,* (2021) reported that artificial intelligence is not significantly related to customer satisfaction or marketing performance.

CONCLUSION AND RECOMMENDATION

The major findings of the chapter showed that self-service technology had a significant positive effect on customer satisfaction of selected listed banks in Nigeria ($\beta = 0.417$, $t = 8.608$, $p < 0.05$). The testing hypotheses showed that self-service technology has a statistically significant influence on customer satisfaction. This investigation not only emphasizes the dominant academic narratives but also reveals the hidden trends that could indicate future paths in artificial intelligence research and application. Based on the findings of this study, it can be inferred that artificial intelligence (AI) serves as both a technology tool and a strategic asset for firms. More so innovative technology, particularly self-service technology, is rapidly transforming traditional business practices worldwide and reallocating some obligations from producers to customers. The introduction of this technology, like any other advancement, poses both risks and possibilities for the national economy, as a rising number of low-skilled people are displaced due to the automation of basic tasks. These self-service technologies provide the potential to enhance quality of life and stimulate production development. It is inadequate to base our arguments solely on the premise that technological advancement is rapidly displacing unskilled labour and invariably resulting in unemployment; furthermore, we cannot entirely dismiss the reality that self-service technology is significantly transforming our economies and swiftly displacing low-skilled workers due to the concerning rate of unemployment in these advanced economies. Undoubtedly, the ongoing advancement of technology often leads to an enhancement in the productivity development rate. The automation of key sectors in Agriculture, Industry, and Services has resulted in the displacement of an increasing number of unskilled individuals. Industries may use low labour intensity and high capital intensity to enhance production. The result of this study corroborated the findings of past research about the substantial impacts of self-service technology on customer satisfaction (Chen, & Prentice, 2024; Katanga,, 2024; Yang,2023; Williady, & Ban, 2023; Chen, Esperança, & Wang, 2022; Chen, & Lin, 2021; Yiu, Yeung, & Cheng, 2021; Hadhoud, & Salameh, 2020; Bach, Othman, Hamzah, & Abu Hassan, 2020; Jaklič, & Vugec, 2018). This study demonstrates that self-service technology plays a crucial role in enhancing both operational and customer satisfaction. However, the full benefits of self-service technology can only be achieved if banks and businesses successfully address the problems associated with their implementation, such as the requirement for qualified individuals, skill sets and the ability to adapt to rapid technological advancements. An important practical consequence is the incorporation of self-service technology into strategic decision-making processes of listed banks and businesses in Nigeria. These organizations should utilize these technology to evaluate large datasets for informed decision-making, allowing for more accurate strategy development and implementation for customer satisfaction. It is recommended that companies should employ self-service technology solutions to improve operational efficiency. Through the implementation of self-service technology companies can optimize their operations, decrease expenses, and enhance their ability to adapt to market fluctuations and client needs. Furthermore, the findings emphasize the necessity of ongoing skill enhancement and training in self-service technology tools. With the continuous advancement of AI technology, businesses must allocate resources towards educating their workers to stay updated with the latest tools and analytical methodologies.

Limitations and Future Study Direction

Despite the satisfactory results in relation to the hypothesis, the research acknowledged several limitations and also experienced significant limitations, some of these are considered to be useful precursors for future study. Primarily, the data was gathered from a limited sample within the listed consumer goods sector. The focus of the study on only selected listed banks in Nigeria restricts the applicability of its results. The findings and implications of this article are specific to Nigeria and focused mostly on small business sector, which might restrict the generalizability of the results. Although limited, the findings of the current study should inspire scholars to do more comprehensive research on artificial intelligence and customer satisfaction. The paper's cross-sectional design limits the author's ability to assert causation.

REFERENCES

Adejoh, A. M., & Hadiza, D. (2015). Impact of Effective Communication on Goal Achievement in Nigerian Polytechnics. *International Journal of Sustainability Management and Information Technologies*, *1*(2), 6–10.

Adeoye, O. O., Olubiyi, T. O., Ajiteru, W. O., & Adaranijo, L. O. (2023). Unveiling Capital Structure Determinants, Business Characteristics and High-Growth Performance of Small and Medium Sized Enterprises: Evidence from Nigeria. *Journal of Interdisciplinary Socio-Economic and Community Study*, *3*(2), 86–98. DOI: 10.21776/jiscos.03.2.05

Adeyemi, O. S., & Olubiyi, T. O. (2023). The Impact of Brand Awareness on Customer Loyalty in Selected Food and Beverage Businesses in Lagos State Nigeria. *Jurnal Multidisiplin Madani*, *3*(3), 541–551. DOI: 10.55927/mudima.v3i3.3095

Adeyemi, O. S., & Olubiyi, T. O. (2024). An Analysis of Investment Trends, Innovation and Its Impact on Cabotage and Local Content Regime: Nigerian Experience from Businesses in Oil and Gas and Maritime Industries. *ABUAD Journal of Social and Management Sciences*, *5*(2), 285–301. DOI: 10.53982/ajsms.2024.0502.04-j

Akesson, M., Edvardsson, B., & Tronvoll, B. (2014). Customer experience from a self-service system perspective. *Journal of Service Management*, *25*(5), 677–698. DOI: 10.1108/JOSM-01-2013-0016

Al-Rousan, M., & Al-Shishani, N. (2021). Business intelligence in the Jordanian telecommunications industry: A case study. [IJBIR]. *International Journal of Business Intelligence Research*, *12*(1), 1–18.

Alalwan, A. A., Dwivedi, Y. K., Rana, N. P., Lal, B., & Williams, M. D. (2015). Consumer adoption of Internet banking in Jordan: Examining the role of hedonic motivation, habit, self-efficacy and trust. *Journal of Financial Services Marketing*, *20*(2), 145–157. DOI: 10.1057/fsm.2015.5

Alwaely, S., Abusalma, A., Alwreikat, A., Al-Shakri, K., Ahmad, A, & Alkhawaldeh, B. (2024).

Alwaely, S. A., Abusalma, A., Alwreikat, A. A., Al-Shakri, K. S., Ahmad, A. Y. B., & Alkhawaldeh, B. Y. (2024). Examining the relationship between business intelligence adoption and marketing effectiveness: The mediating role of customer satisfaction. *International Journal of Data and Network Science*, *8*(3), 1541–1556.

Arokodare, M. A., & Olubiyi, T. O. (2023). Crises as Harbingers of Opportunities: An Empirical Insight into the Moderating Effect of Knowledge Management on the Absorptive Capacity-Strategic Agility Relationship [MMJ]. *Malaysian Management Journal, 27*, 83–108. DOI: 10.32890/mmj2023.27.4

Ashrafi, A., Zare Ravasan, A., Trkman, P., & Afshari, S. (2019). The Role of Business Analytics Capabilities in Bolstering Firms' Agility and Performance. *International Journal of Information Management, 47*, 1–15. DOI: 10.1016/j.ijinfomgt.2018.12.005

Asiaei, K., Rezaee, Z., Bontis, N., Barani, O., & Sapiei, N. S. (2021). Knowledge assets, capabilities and performance measurement systems: A resource orchestration theory approach. *Journal of Knowledge Management, 25*(8), 1947–1976. DOI: 10.1108/JKM-09-2020-0721

Bach, M. P., Jaklič, J., & Vugec, D. S. (2018). Understanding impact of business intelligence to organizational performance using cluster analysis: Does culture matter? *International Journal of Information Systems and Project Management, 6*(3), 63–86. DOI: 10.12821/ijispm060304

Bateson, J. E. (1985). Self-service consumer: An exploratory study. *Journal of Retailing, 61*(3), 49–76.

Bharadiya, J. P. (2023). Machine learning and AI in business intelligence: Trends and opportunities. [IJC]. *International Journal of Computer, 48*(1), 123–134.

Bhatiasevi, V., & Naglis, M. (2020). Elucidating the determinants of business intelligence adoption and organizational performance. *Information Development, 36*(1), 78–96. DOI: 10.1177/0266666918811394

Blut, M., Wang, C., & Schoefer, K. (2016). Factors influencing the acceptance of self-service technologies: A meta-analysis. *Journal of Service Research, 19*(4), 396–416. DOI: 10.1177/1094670516662352

Bock, D. E., Wolter, J. S., & Ferrell, O. C. (2020). Artificial intelligence: Disrupting what we know about services. *Journal of Services Marketing, 34*(3), 317–334. https://www.emerald.com/insight/content/doi/10.1108/JSM-01-2019-0047/full/html. DOI: 10.1108/JSM-01-2019-0047

Borah, S., Kama, C., Rakshit, S., & Vajjhala, N. R. (2022), "Applications of Artificial Intelligence in Small- and Medium-Sized Enterprises (SMEs)", in Proceeding of the Conference on Cognitive Informatics and Soft Computing (CISC), Springer Nature Singapore,pp. 717–726 DOI: 10.1007/978-981-16-8763-1_59

Brodny, J., & Tutak, M. (2022). Digitalization of Small and Medium-Sized Enterprises and Economic Growth: Evidence for the EU-27 Countries. *Journal of Open Innovation*, *2*(8), 67. DOI: 10.3390/joitmc8020067

Bunte, A., Richter, F., & Diovisalvi, R. (2021), "Why it is hard to find AI in SMEs: A survey from the practice and how to promote it", in *Proceedings of the 13th International Conference on Agents and Artificial Intelligence (ICAART)*, pp. 614–620. DOI: 10.5220/0010204106140620

Chen, Y., & Lin, Z. (2021). Business intelligence capabilities and firm performance: A study in China. *International Journal of Information Management*, *57*, 102232. DOI: 10.1016/j.ijinfomgt.2020.102232

Considine, E., & Cormican, K. (2016). Self-service technology adoption: An analysis of customer to technology interactions. *Procedia Computer Science*, *100*, 103–109. DOI: 10.1016/j.procs.2016.09.129

Dabholkar, P. A., & Bagozzi, R. P. (2002). An attitudinal model of technology-based self-service: Moderating effects of consumer traits and situational factors. *Journal of the Academy of Marketing Science*, *30*(3), 184–201. DOI: 10.1177/0092070302303001

Dube, A., & Helkkula, A. (2015). Service experiences beyond the direct use: Indirect customer use experiences of smartphone apps. *Journal of Service Management*, *26*(2), 224–248. DOI: 10.1108/JOSM-11-2014-0308

Ghasemaghaei, M., & Calic, G. (2020). Assessing the Impact of Big Data on Firm Innovation Performance: Big Data Is Not Always Better Data. *Journal of Business Research*, *108*, 147–162. DOI: 10.1016/j.jbusres.2019.09.062

Grigorescu, S., Trasnea, B., Cocias, T., & Macesanu, G. (2020). A survey of deep learningtech-niques for autonomous driving. *Journal of Field Robotics*, *3*(37), 362–386. DOI: 10.1002/rob.21918

Gu, B., Santhanam, R., Berente, N., & Recker, J. (2019). Call for Papers MISQ Special Issue on Managing AI. MIS Quarterly, Special Is. Retrieved from https://misq.org/skin/frontend/default/misq/pdf/CurrentCalls/ManagingAI.pdf

Gummerus, J., Lipkin, M., Dube, A., & Heinonen, K. (2019). Technology in use – characterizing customer self-service devices (SSDS). *Journal of Services Marketing*, *33*(1), 44–56. DOI: 10.1108/JSM-10-2018-0292

Gummerus, J., Lipkin, M., Dube, A., & Heinonen, K. (2019). Technology in use – characterizing customer self-service devices (SSDS). *Journal of Services Marketing*, *1*(33), 44–56. DOI: 10.1108/JSM-10-2018-0292

Gupta, S., Leszkiewicz, A., Kumar, V., Bijmolt, T., & Potapov, D. (2020). Digital analytics: Modeling for insights and new methods. *Journal of Interactive Marketing, 51*(1), 26–43. DOI: 10.1016/j.intmar.2020.04.003

Habul, A., & Pilav-Velic, A. (2010). Business Intelligence and Customer Relationship Management. In *Proceedings of ITI 2010, 32nd International Conference on Information Technology Interfaces*. IEEE Press., Retrieved from https://ieeexplore.ieee.org/document/5546381?arnumber=5546381

Hadhoud, R., & Salameh, W. A. (2020). How business intelligence can help you to better understand your customers. *International Journal of Business Intelligence Research, 11*(1), 50–58. DOI: 10.4018/IJBIR.2020010104

Hossain, M. A., Agnihotri, R., Rushan, M. R. I., Rahman, M. S., & Sumi, S. F. (2022). Marketing analytics capability, artificial intelligence adoption, and firms' competitive advantage: Evidence from the manufacturing industry. *Industrial Marketing Management, 106*, 240–255. DOI: 10.1016/j.indmarman.2022.08.017

Hsieh, A., Yen, C.-H., & Chin, K.-C. (2004). Participative customers as partial employees and service provider workload. *International Journal of Service Industry Management, 15*(2), 187–199. DOI: 10.1108/09564230410532501

Hsu, C.-L., & Lin, J. (2023). Understanding the user satisfaction and loyalty of customer service chatbots. *Journal of Retailing and Consumer Services, 71*, 103211–103229. DOI: 10.1016/j.jretconser.2022.103211

Ibrahim, N., & Handayani, P. W. (2022). A systematic literature review of Business intelligence framework for tourism organizations: Functions and issues. *Interdisciplinary Journal of Information, Knowledge, and Management, 17*, 523–541. DOI: 10.28945/5025

Katanga, J. (2024) *The Effect of Self-Service Banking Technology on Customer Satisfaction at NMB Bank.* Masters thesis, The Open University of Tanzania.

Kaushik, A.K. and Rahman, Z. (2015), "Innovation adoption across self-service banking technologies in India", International Journal of Bank Marketing, 33(2), 96 121. DOI: 10.1108/IJBM-01-2014-0006

Kumar, A., Choudhary, S., Garza-Reyes, J. A., Kumar, V., Rehman Khan, S. A., & Mishra, N. (2023). Analysis of critical success factors for implementing industry 4.0 integrated circular supply chain–Moving towards sustainable operations. *Production Planning and Control, 34*(10), 984–998. DOI: 10.1080/09537287.2021.1980905

Liang, X., Li, G., Zhang, H., Nolan, E., & Chen, F. (2022). Firm performance and marketing analytics in the Chinese context: A contingency model. *Journal of Business Research, 141*, 589–599. DOI: 10.1016/j.jbusres.2021.11.061

Liljander, V., Gillberg, F., Gummerus, J., & Riel, A. (2006). Technology readiness and the evaluation and adoption of self-service technologies. *Journal of Retailing and Consumer Services, 13*(3), 177–191. DOI: 10.1016/j.jretconser.2005.08.004

Makinde, O. G., Olubiyi, T. O., & Ogundipe, F. (2023). New Insights from Entrepreneurial Characteristics and Busıness Performance: Empırıcal Fındıngs from Nıgerıa. *Asian Journal of Management. Entrepreneurship and Social Sciences, 3*(04), 72–100.

Massaro, M., Secinaro, S., Dal Mas, F., Brescia, V., & Calandra, D. (2021). Industry 4.0 and circular economy: An exploratory analysis of academic and practitioners' perspectives. *Business Strategy and the Environment, 30*(2), 1213–1231. DOI: 10.1002/bse.2680

McLean, G., & Osei-Frimpong, K. (2019). Examine the variables influencing the use of artificial intelligent in-home voice assistants. [REMOVED HYPERLINK FIELD]. *Computers in Human Behavior, 99*, 28–37. DOI: 10.1016/j.chb.2019.05.009

Medeiros, M. M. D., & Maçada, A. C. G. (2022). Competitive advantage of data-driven analytical capabilities: The role of big data visualization and of organizational agility. *Management Decision, 60*(4), 953–975. DOI: 10.1108/MD-12-2020-1681

Meub, L., & Proeger, T. (2022), "Künstliche Intelligenz in Handwerk und Mittelstand: Ein For schungsüberblick", ifh Forschungsbericht, No. 1, Volkswirtschaftliches Institut für Mittelstandund Handwerk an der Universität Göttingen (ifh), Göttingen

Meuter, M., Ostrom, A., Roundtree, R. & Bitner, M. (2000), "Self-service technologies: understanding customer satisfaction with technology-based service encounters", *Journal of Marketing,* 64 3, 50.

Niwash, M. N. K., Cek, K., & Eyupoglu, S. Z. (2022). Intellectual Capital and Competitive Advantage and the Mediation Effect of Innovation Quality and Speed, and Business Intelligence. *Sustainability (Basel), 14*(6), 3497. DOI: 10.3390/su14063497

Olowoporoku, A. A., & Olubiyi, T. O. (2023). Evaluating Service Quality and Business Outcomes Post-Pandemic: Perspective from Hotels in Emerging Market. *Sawala: Jurnal Administrasi Negara, 11*(2), 182–201. DOI: 10.30656/sawala.v11i2.6975

Olubiyi, O. T., Lawal, A. T., & Adeoye, O. O. (2022). Succession planning and family business continuity: Perspectives from Lagos State, Nigeria. *Organization and Human Capital Development, 1*(1), 40–52. DOI: 10.31098/orcadev.v1i1.865

Olubiyi, T. O. (2022). Measuring technological capability and business performance post-COVID Era: Evidence from Small and Medium-Sized Enterprises (SMEs) in Nigeria. *Management & Marketing Journal, xx*(2), 234–248. DOI: 10.52846/MNMK.20.2.09

Olubiyi, T. O. (2022b). An investigation of sustainable innovative strategy and customer satisfaction in small and medium-sized enterprises (SMEs) in Nigeria. *Covenant Journal of Business and Social Sciences, 13*(2), 1–24.

Olubiyi, T. O. (2023a). Unveiling the role of workplace environment in achieving the Sustainable Development Goal Eight (SDG8) and employee job satisfaction post-pandemic: Perspective from Africa. *Revista Management & Marketing Craiova, XXI*(2), 212–228. DOI: 10.52846/MNMK.21.2.02

Olubiyi T. O. (2023b). Leveraging competitive strategies on business outcomes post-covid-19 pandemic: Empirical investigation from Africa. *Journal of Management and Business: Research and Practice*, DOI: 10.54933/jmbrp-2023-15-2-1

Olubiyi, T. O. (2024a). Achieving Sustainable Development Goal Four (4) in Africa: Spotlighting the role of quality education and innovation orientation. Economics. *Management and Sustainability, 9*(1), 105–118. DOI: 10.14254/jems.2024.9-1.8

Olubiyi, T. O. (2024b). Measuring organizational factors and their impact on the development of small and medium-sized pharmaceutical companies after the pandemic – an empirical study Journal of Marketing and Market Studies 25-38 (ERIH Plus & Index Copernicus indexed)DOI: 10.33226/1231-7853.2024.7.3

Olubiyi, T. O. (2024c). Improving Customer Experience Through Differentiation Capability: The Rise of Competition among Africa's Consumer Goods Companies. *PaperASIA, 40*(5b), 230–240. DOI: 10.59953/paperasia.v40i5b.233

Olubiyi, T. O., Adeoye, O. O., Jubril, B., Adeyemi, O. S., & Eyanuku, J. P. (2023). Measuring Inequality in Sub-Saharan Africa Post-Pandemic: Correlation Results for Workplace Inequalities and Implication for Sustainable Development Goal ten. *International Journal of Professional Business Review, 8*(4), e01405. DOI: 10.26668/businessreview/2023.v8i4.1405

Olubiyi, T. O., & Akpa, V. A. (2023). Does innovative capability and artificial intelligence really matter for the profitability of consumer goods companies in developing economies? *SunText Review of Economics & Business, 4*(3), 191–204.

Olubiyi, T. O., Jubril, B., Sojinu, O. S., & Ngari, R. (2022). Strengthening gender equality in small business and achieving sustainable development goals (SDGs), Comparative analysis of Kenya and Nigeria. *Sawala Jurnal Administrasi Negara*, *10*(2), 168–186. DOI: 10.30656/sawala.v10i2.5663

Omoyele, O. S., Olubiyi, T. O., Lanre-Babalola, F. O., Obadare, G. O., & Onikoyi, I. A. (2023). Business Model Innovation as a Catalyst for Sustainable Entrepreneurship: Empirical Findings from Small and Medium Enterprises in Nigeria. *Skyline Business Journal*, *19*(2), 55–64. DOI: 10.37383/SBJ190205

Osakwe, J., Mutelo, S., & Obijiofor, N. (2023). Integrating Customer Relationship Management and Business Intelligence to Enhance Customer Satisfaction and Organisational Performance. A Literature Review *SSRN*: https://ssrn.com/abstract=4664792 DOI: 10.2139/ssrn.4664792

Othman, A. K., Hamzah, M. I., & Abu Hassan, L. F. (2020). Modeling the contingent role of technological optimism on customer satisfaction with self-service technologies: A case of cash-recycling ATMs. *Journal of Enterprise Information Management*, *33*(3), 559–578. DOI: 10.1108/JEIM-09-2019-0295

Pandey, N., Nayal, P., & Rathore, A. S. (2020). Digital marketing for B2B organizations: Structured literature review and future research directions. *Journal of Business and Industrial Marketing*, *35*(7), 1191–1204. DOI: 10.1108/JBIM-06-2019-0283

Paradza, D., & Daramola, O. (2021). Business intelligence and business value in organisations: A systematic literature review. *Sustainability (Basel)*, *13*(20), 11382. DOI: 10.3390/su132011382

Parasuraman, A., & Colby, C. L. (2015). An updated and streamlined technology readiness index: tRI 2.0. *Journal of Service Research*, *18*(1), 59–74. DOI: 10.1177/1094670514539730

Rajpurkar, P., Chen, E., Banerjee, O., & Topol, E. J. (2022). AI in health and medicine. *Nature Medicine*, *28*(1), 31–38. DOI: 10.1038/s41591-021-01614-0 PMID: 35058619

Ramakrishnan, T., Kathuria, A., & Saldanha, T. J. (2020). Business intelligence and analytics (BI&A) capabilities in healthcare. In Theory and Practice of Business Intelligence in Healthcare (pp. 1-17). IGI Global.

Rasheed, R., & Rashid, A. (2023). Role of service quality factors in word of mouth through student satisfaction. *Kybernetes*.

Schönberger, M. (2023). Artificial intelligence for small and medium-sized enterprises: Identifying key applications and challenges. *Journal of Business and Management, 21*, 89–112. DOI: 10.32025/JBM23004

Schumacher, E. F. (1973). *Small is beautiful: a study of ecomonics as if people mattered.* Vintage.

Şentürk, H., & Keskin, C. (2010). The Importance of Small and Medium-Sized Enterprises (SMEs) in Economies: Swot Analysis of The SME Sector in Turkey and Albania. *Niğde Üniversitesi İktisadi ve İdari Bilimler Fakültesi Dergisi, 3*(1), 116–132.

Sinha, S., & Sathiya Narayanan, R. (2024). Harnessing Sentiment Analysis Methodologies for Business Intelligence Enhancement and Governance Intelligence Evaluation', International [IJISAE]. *Journal of Intelligent Systems and Applications in Engineering, 12*(11s), 166–176.

Soltani, Z., Zareie, B., Milani, F. S., & Navimipour, N. J. (2021). The impact of the customer relationship management on the organization performance. *The Journal of High Technology Management Research, 32*(2), 100399.

Triono, S.P.H. & Jaya, R.C. (2021). Business Intelligence & Analytics and Its Effect on Indonesia's Startup Performance, *Advances in Economics, Business and Management Research, 187*

Ukabi, O. B., Uba, U. J., Ewum, C. O., & Olubiyi, T. O. (2023). Measuring Entrepreneurial Skills and Sustainability in Small Business Enterprises Post-Pandemic: Empirical Study from Cross River State, Nigeria. *International Journal of Business. Management and Economics, 4*(2), 132–149. DOI: 10.47747/ijbme.v4i2.1140

Ukhalkar, P. K., Phursule, R. N., Gadekar, D. P., & Sable, N. P. (2020). Business Intelligence and Analytics: Challenges and Opportunities. *International Journal of Advanced Science and Technology, 29*(12s), 2669–2676.

Uwem, E. I., Oyedele, O. O., & Olubiyi, O. T. (2021). Workplace green behavior for sustainable competitive advantage. In human resource management practices for promoting *Sustainability IGI Global.* (248–263).

Williady, A., & Ban, H. J. (2023). An Approach of the Big Data Analysis on Customer Satisfaction in Bali's Luxury Hotels and Resorts Through Wellness Attributes. *Advances in Social Science, Education and Humanities Research*, •••, 3–9. DOI: 10.2991/978-2-494069-93-0_2

Wünderlich, N. V., Heinonen, K., Ostrom, A., Patricio, L., Rui, S., & Voss, C. (2015). "Futurizing" smart service: Implications for service researchers and managers. *Journal of Services Marketing*, *29*(6/7), 442–447. DOI: 10.1108/JSM-01-2015-0040

Yamane, Y. (1967). Mathematical Formulae for Sample Size Determination.

Yang, L., Henthorne, T. L., & George, B. (2020). Artificial intelligence and robotics technology in the hospitality industry: Current applications and future trends. Digital transformation in business and society: Theory and cases 211-228.

Yang, X. (2023). The effects of AI service quality and AI function-customer ability fit on customer's overall co-creation experience. *Industrial Management & Data Systems*, *123*(6), 1717–1735. DOI: 10.1108/IMDS-08-2022-0500

Yiu, L. D., Yeung, A. C., & Cheng, T. E. (2021). The impact of business intelligence systems on profitability and risks of firms. *International Journal of Production Research*, *59*(13), 3951–3974. DOI: 10.1080/00207543.2020.1756506

Yiu, L. D., Yeung, A. C., & Jong, A. P. (2020). Business intelligence systems and operational capability: An empirical analysis of high-tech sectors. *Industrial Management & Data Systems*, *120*(6), 1195–1215. DOI: 10.1108/IMDS-12-2019-0659

Zhang, J., Chen, W., Petrovsky, N., & Walker, R. M. (2022). The expectancy-disconfirmation model and citizen satisfaction with public services: A meta-analysis and an agenda for best practice. *Public Administration Review*, *82*(1), 147–159. DOI: 10.1111/puar.13368

Zhou, J., Mavondo, F. T., & Saunders, S. G. (2019). The relationship between marketing agility and financial performance under different levels of market turbulence. *Industrial Marketing Management*, *83*(2), 31–41. DOI: 10.1016/j.indmarman.2018.11.008

Chapter 6

Revolutionizing Retail: Harnessing AI-Powered Self-Service Technologies for Enhanced Customer Experiences

Semiu Olawale Makinde
https://orcid.org/0000-0001-5456-3464
Al-Hikmah University, Ilorin, Nigeria

AbdulRafiu Abdullahi
https://orcid.org/0009-0003-2896-2367
Al-Hikmah University, Ilorin, Nigeria

Shehu Sakariyahu
https://orcid.org/0009-0008-7377-9007
Al-Hikmah University, Ilorin, Nigeria

Akeem Apo Ahmed
https://orcid.org/0009-0008-9908-6866
Al-Hikmah University, Ilorin, Nigeria

ABSTRACT

This chapter, Revolutionizing Retail: Harnessing AI-Powered Self-Service Technologies for Enhanced Customer Experiences explores how AI-driven self-service tools like machine learning, natural language processing, and robotics are reshaping retail. These technologies fine-turn personalization, customer satisfaction, and operational efficiency by providing adequate guidelines, streamlining supply management, and automating customer support. Chatbots, Self-checkout systems, and virtual assistants endow customers with greater control and cohesive interactions, improving accessibility and building trust. The chapter showcases the importance

DOI: 10.4018/979-8-3373-4667-0.ch006

of AI-powered innovations while tackling challenges such as data privacy, ethical concerns, and inclusivity. Ultimately, it accentuates the transformative potential of these tools in the advancing retail landscape.

INTRODUCTION

The retail industry has seen a notable transformation in recent time, largely driven by emerging technology, especially the rise of self-service systems. These transformations are changing the way customers engage with brands and how businesses operate, making shopping easier, promote direct transactions, and productivity. As a result, many retailers are utilizing self-service options to simplify their processes, reduce costs, and offer experiences that better suit the preferences of today's digital literate consumers. The retail industry is now benefiting from the adoption of advanced technologies such as machine learning (ML), natural language processing (NLP), and robotics, which are opening up new ways of improving convenience, direct shopping, and customer satisfaction (Vegesna, 2023). These innovations are not only altering how customers interact with brands but also optimizing business operations, helping companies stay in tune with shifting consumer needs. Machine learning, in particular, plays an essential role in enabling retailers to offer more designed and unique shopping experiences. By looking at a breath of customer data, including past purchases, browsing behaviour, and social media activity, businesses can better figure out what products or services are most likely to interest each customer (Patil, 2024). This level of individualized attention not only encourages the overall customer experience but also promote the opportunity of making sales and securing customer loyalty.

The retail landscape has witnessed a profound transformation over the last few decades, primarily driven by technological advancements. Traditionally, retail relied on face-to-face interactions, where customers depended on sales staff for information and assistance. However, with the rise of digital technologies, this model has shifted dramatically, leading to a more autonomous shopping experience. One of the most significant developments in this shift is the introduction of self-service systems. These systems have evolved from simple kiosks to complex AI-driven platforms that enhance customer engagement and streamline operations. For instance, self-checkout systems have become prevalent in grocery stores, allowing customers to scan and pay for items independently, thus reducing wait times and enhancing convenience (Gupta & Mukherjee, 2025). This autonomy caters to the modern consumer's desire for quick and efficient shopping experiences, reflecting a broader trend where speed and ease of access are paramount. One of the achievements of self-service technologies is that it gives customers more control over their shopping experience. With tools like automated checkout systems, self-checkout kiosks, and virtual assistants becoming progressively common in both physical stores and online platforms, customers can make quicker and more informed decisions with less need for help from staff (Davenport & Ronanki, 2018). Siddiqui, (2024) observed that 61% of consumers now expect brands to offer unique interactions and are more likely to choose businesses that provide self-service options that cater to their specific needs. By giving customers this level of control, these technologies make the shopping process flawless and more efficient, offering retailers a competitive edge.

Additionally, self-service technologies also deliver tailored product recommendations, further improving the experience by suggesting items that align with individual desires. By analyzing data such as past purchases, browsing habits, and real-time preferences, retailers can suggest products that fit a customer's unique tastes, boosting the likelihood of a purchase and helping to build long-term loyalty. Iliadi (2023) postulated out that personalization can increase retailer revenue by 10-30%, underscoring the powerful effect that targeted marketing and personalized experiences can have on business growth. As consumers increasingly expect tailored experiences, being able to adjust offerings to meet individual preferences is becoming more critical. By gaining a profound understanding of consumer behavior, businesses can create more effective marketing strategies and promotions, fostering stronger customer relationships and ensuring greater satisfaction.

The growth of self-service technologies powered by AI has given retailers the ability to gather valuable customer insights, including purchase history, browsing patterns, and demographic information. With this data, retailers can offer suggestions that are tailored to each shopper's specific tastes and needs. By understanding what customers want and anticipating their preferences, businesses can deliver more relevant suggestions and promotions, resulting in higher sales and improved customer satisfaction. Personalization strengthens the relationship between retailers and customers by making shopping feel more relevant and offering to bespoke service which in turn fosters loyalty. Studies show that personalized shopping experiences can lead to a 10-30% revenue surge for retailers, intensifying the financial value that self-service technologies bring to business growth. These technologies are revamping how retailers manage customer interactions, inventory, and supply chains, resulting in a faster and more convenience shopping experience (Johnson, 2018).

As the retail landscape evolves, AI and automation will be crucial in adapting to consumers' evolving needs. Retailers who welcome these technologies will not only boost customer satisfaction and improve their operations but also position themselves ahead of the competition in a rapidly shifting market. The future of retail will be shaped by the growing influence of AI, which has the potential to completely transform how businesses connect with customers and provide value. Reaching beyond the boundary of customer services, self-service technologies also create a significant competitive advantage in inventory management and supply chain optimization. By using predictive tools to anticipate demand and track stock levels in real time, retailers can better manage inventory, avoiding both overstocking and understocking. Moreover, the integration of artificial intelligence (AI) into retail has transformed how businesses operate. AI technologies enable retailers to analyze vast amounts of customer data, facilitating more personalized marketing strategies and product offerings. For example, by leveraging machine learning algorithms, retailers can predict customer preferences based on historical data, leading to targeted promotions that resonate with individual shoppers (Patil, 2024). This level of personalization not only increases customer satisfaction but also builds long-term loyalty, as consumers feel understood and valued. The impact of self-service technologies extends beyond customer interactions; they also play a crucial role in optimizing inventory management and supply chain logistics. With real-time data analytics, retailers can better forecast demand, manage stock levels, and minimize the risks associated with overstocking or stockouts (Johnson, 2018). This operational efficiency is essential in today's fast-paced retail environment, where consumer preferences can shift rapidly, and necessitating agile responses from businesses.

This has resulted in more efficient operations and improved product availability, making it easier for customers to find what they need, whether shopping in-store or online. Automating inventory management not only boost efficiency but also encourages profitability by reducing human error and promoting the movement of goods (Johnson, 2018). Beyond these operational improvements, there are also self-service technologies that are helping to make retail more inclusive and accessible. Voice-activated systems and assistive technologies are enabling customers with disabilities to navigate stores and make purchases with greater ease. These innovations ensure that everyone, regardless of physical ability, can enjoy a more convenient and equitable shopping experience. Gurley (2021) found that 62% of consumers feel these technologies enhance accessibility and convenience, helping create a more inclusive shopping environment that caters to a variety of customer needs. Furthermore, the rise of e-commerce has accelerated the adoption of self-service technologies, as online platforms require robust systems to handle transactions and customer inquiries efficiently. For instance, chatbots and virtual assistants have become integral to online shopping experiences, providing instant support and guidance to customers, thus enhancing engagement and satisfaction (Siddiqui, 2024). These technologies not only improve the user experience but also reduce operational costs by minimizing the need for human intervention in routine inquiries.

As the retail sector continues to evolve, it is evident that the future will be shaped by the ongoing integration of advanced technologies. Retailers who successfully adopt and implement these self-service systems will likely gain a competitive edge, as they will be better positioned to meet the ever-changing demands of today's tech-savvy consumers. The evolution of retail is not just about adopting new technologies; it is about understanding and redefining the relationship between businesses and consumers in a digital age. AI-powered self-service tools are transforming the way retailers connect with their customers. These technologies not only improve personalization, efficiency, and accessibility but also simplify inventory management and provide useful insights into customer behavior. As shoppers continue to demand more personalized and seamless experiences, these systems are expected to play an even bigger role in shaping the future of retail. Retailers who adopt and use these technologies effectively can build stronger relationships with their customers, improve satisfaction, and fuel business growth.

Framework

This chapter explores the significant impact of self-service technologies in transforming the retail sector. Drawing from various scholarly resources, including research articles, conference proceedings, Scopus, Web of Science, Google Scholar, ResearchGate and expert analyses. The discussion emphasizes how these tools improve customer experiences, streamline operations, and support business growth. The examination focuses on real-world applications and addresses key challenges faced by retailers in implementing these technologies.

First, the significant of retail industry that has gone through remarkable changes in recent years, largely driven by innovation in technology due to the rise of self-service systems. And how the changes are now altering how customers interact with brands and how businesses run, making shopping more seamless, customized, and efficient. This has resulted into turning to self-service solutions that streamline operations, reduce expenses, and create more tailored experiences that cater to the preferences of today's tech-savvy consumers.

Second, the chapter explored Self-service technologies such as automated checkout systems, machine learning-based tools, and natural language interfaces that are crucial in creating personalized shopping experiences. By analyzing customer data such as purchase history and browsing patterns retailers which can offer customized product suggestions and promotions. This personalization fosters customer satisfaction and loyalty, enriching the overall shopping journey. The analysis also reviewed how automation in tasks like inventory management reduces errors, increases transaction speed, and ensures consistent product availability, ultimately making shopping more efficient and convenient for consumers.

Thirdly, the chapter also investigates the current state of self-service technologies in retail as well as benefits of self-service technologies in retail especially innovations, which include voice-activated tools and assistive systems, that are playing an essential role in enhancing accessibility for a diverse range of customers, including individuals with disabilities. The chapter addresses these issues and suggests strategies to implement self-service tools responsibly and transparently.

Fourth, the chapter explain the integration of artificial intelligence (AI) in personalized marketing, how it has reshaped the retail landscape, enabling businesses to engage consumers in ways that align closely with their individual preferences and behaviors. This transformation is largely driven by the ability of AI to analyze vast amounts of data, providing insights that facilitate targeted marketing strategies and enhance customer experiences.

Finally, Case study analysis in the Nigeria context were discussed, the chapter highlights several examples of prominent Nigerian retailers who have successfully implemented these technologies to boost their operations. By examining these aspects, the chapter provides a detailed understanding of how self-service technologies are reshaping the retail industry. It highlights the opportunities for businesses to utilize these advancements to improve customer engagement, optimize operations, and meet the evolving demands of a technology-driven market. By exploring and expanding knowledge in this area, we can gain a clearer understanding of how these technologies affect customer behavior, pinpoint the most effective ways to implement them, and help retailers ensure that their use is both ethical and focused on customer needs. By implication, businesses may miss critical opportunities, face inefficiencies in their operations, and struggle to meet evolving customer expectations, ultimately jeopardizing their long-term success in the retail market.

Current State of Self-Service Technologies in Retail

The rise of self-service technologies in retail reflects a broader trend towards automation, driven by consumers' desire for convenience, personalized experiences, and faster service. To meet these expectations, retailers are increasingly adopting tools like chatbots, virtual assistants, and self-checkout systems to improve the shopping experience. These innovations put more control in the hands of shoppers, allowing for faster, more customized interactions. For example, chatbots on websites and mobile apps can offer real-time assistance, guiding customers through their purchases and suggesting relevant products along the way. Virtual assistants, tailored to individual shopping habits, can recommend items based on past purchases, making the experience feel more personal. Self-checkout kiosks have changed the in-store experience by enabling customers to scan and pay for items on their own, streamlining the process and reducing wait times. Beyond checkout, these technologies are also improving inventory management and pricing strategies. Retailers are using advanced tools to predict demand, manage restocking more efficiently, and adjust prices to better align with customer preferences (Gurley 2021). This leads to better product availability, improved customer satisfaction, and ultimately, increased sales.

Personalizing the shopping experience has become a key focus in retail, thanks to the growing use of advanced technology. By analyzing customer data, retailers can predict the products a shopper might like, offer targeted promotions, and even tailor the shopping interface to fit individual preferences. This level of personalization not only drives sales but also fosters customer loyalty, creating a stronger bond between shoppers and brands.

As consumers become more accustomed to these seamless, tailored experiences, the role of self-service technologies continues to expand. These innovations are transforming how people shop and setting new expectations for customer service in retail. Self-service tools, such as chatbots, virtual assistants, and self-checkout systems, are changing the way retailers operate by making shopping more efficient and personalized. For example, chatbots can engage with customers instantly, answering questions and suggesting products based on previous interactions. This immediate support makes shopping faster and more convenient, offering personalized recommendations much quicker than human staff could. Likewise, virtual assistants whether online or in-store track customer behavior to offer a customized shopping experience, suggesting products or discounts based on past purchases, which increases customer satisfaction and loyalty (Gupta, & Mukherjee, 2025). Automated checkout systems, like self-checkout kiosks, streamline transactions, making them quicker and more accurate while reducing human error. These systems allow customers to shop independently, further improving the overall experience. Ultimately, these technologies not only enhance convenience and personalization but also help retailers optimize their operations, creating a more customer-friendly environment.

AI-powered self-service technologies are key to meeting the expectations of today's tech-savvy shoppers. Consumers now expect shopping experiences that are personalized, efficient, and seamless, reflecting their growing reliance on technology in all areas of life. Tools like chatbots, virtual assistants, and automated checkout systems have become vital in delivering on these expectations by changing the way customers interact with retailers. These innovations not only make the shopping experience smoother but also cater to the desire for convenience and instant gratification, giving customers the freedom to engage with brands whenever and wherever they choose (Kaplan & Haenlein, 2020). Personalization has become a major expectation for today's shoppers, and retailers are using technology to create tailored experiences by analyzing customer data like past purchases, browsing habits, and preferences. Individualized recommendations present customers with products, offers, and content that align with their interests, enriching their shopping experience and boosting the chances of a purchase (Gupta & Mukherjee, 2025). This degree of personalization fosters a deeper connection between consumers and brands, ultimately enhancing loyalty and retention. Moreover, these technologies streamline the shopping process, catering to modern consumers' expectations for quick and hassle-free interactions.

For example, self-checkout systems eliminate long queues, while chatbots provide instant support, minimizing wait times for human assistance and meeting the rising demand for prompt service (Kuloğlu, 2024).

Technology is reshaping the retail world, eliminating the limitations of time and place. With virtual assistants and chatbots available 24/7, customers can shop and get help whenever it works for them, offering a level of convenience that wasn't possible before. In-store, tools like smart kiosks are enhancing the shopping experience with easy-to-use, interactive displays that offer product information, special deals, and personalized recommendations (Johnson, 2018). This push toward automation not only makes things run more smoothly but also creates a shopping experience that's more engaging and seamless perfectly suited for today's tech-savvy customers. Looking ahead, self-service technology powered by AI will play a major role in shaping how retailers handle customer service. As more businesses turn to AI to make shopping more personalized, efficient, and convenient, customers' expectations will continue to evolve. With more businesses adopting these technologies, they will likely become the norm, setting higher standards for customer service and giving companies that use them well a distinct competitive advantage (Kaplan & Haenlein, 2020). Ultimately, the future of customer service in retail will be defined by the seamless integration of AI solutions, offering customers greater control, personalization, and speed than ever before.

Benefits of Self-Service Technologies in Retail

AI-powered self-service technologies offer several benefits, particularly when it comes to personalizing experiences, increasing efficiency, and improving customer satisfaction in retail. By analyzing consumer data, AI can offer tailored product suggestions, special promotions, and content that aligns with an individual's shopping habits and preferences. This level of personalization makes the shopping experience more engaging and helps customers feel understood and valued. Beyond just improving the shopping experience, these would also assist in boosting sales and build custome r loyalty. This kind of customization is something traditional retail methods, which take a one-size-fits-all approach, simply cannot match.

1. One of the key benefits of AI-driven self-service technologies is the boost in efficiency. Self-checkout systems, for example, allow customers to scan and pay for their items on their own, eliminating the need for cashiers and cutting down on wait times. This results in faster transactions and a more seamless shopping experience. AI also plays a role in inventory management by predicting demand and adjusting stock levels based on real-time shopping behavior. By automating these processes, retailers can reduce costs, avoid mistakes, and

make sure that products are always in stock when customers need them, leading to a better overall experience. AI-powered technologies have become essential in improving customer satisfaction. Tools like chatbots and virtual assistants provide instant support, allowing customers to get the help they need without waiting for a human representative particularly helpful during busy shopping periods or after hours. These technologies not only address issues quickly but also engage customers in a more personalized and proactive manner (Gurley 2021). Additionally, by cutting down wait times, offering seamless personalized recommendations, and giving customers more control through self-service options, the overall shopping experience becomes more enjoyable and satisfying. As customers experience quicker and more relevant interactions, their satisfaction grows, leading to stronger loyalty and greater retention.

2. Automating processes like checkout, inventory management, and customer support helps businesses run more smoothly, cut down on wait times, and improve service, benefiting both retailers and their customers. The growing use of AI-powered self-service technologies is reshaping the retail industry by streamlining everyday tasks, reducing reliance on staff, and giving customers quicker and more convenient service. These changes not only boost customer satisfaction but also foster stronger loyalty and give businesses a competitive edge in a crowded market. A great example of how automation can improve retail operations is the use of self-checkout systems. These kiosks allow customers to scan and pay for their items independently, bypassing long lines and speeding up transactions (Johnson, 2018). This enhances efficiency by enabling retailers to serve more customers with fewer staff. The quicker checkout experience saves customers time and reduces the chance of cart abandonment, which often happens when people grow frustrated with long waits. Shorter wait times, especially during busy hours, make the shopping experience more pleasant and increase the likelihood that customers will return for future visits (Kaplan & Haenlein, 2020).
3. In addition to speeding up transactions, automation in inventory management helps keep stock levels accurate and well-maintained. By analyzing factors such as sales data, weather patterns, and consumer behavior, retailers can better predict demand and ensure they have the right products available when needed. This approach helps minimize the risks of overstocking or running out of popular items, ensuring customers can always find what they're looking for. Streamlining inventory management not only improves efficiency but also enhances the overall shopping experience by ensuring that in-demand products are readily available. Automated customer support tools, like chatbots and virtual assistants, have become essential in making retail operations more efficient by offering instant help around the clock. These systems can handle a wide range of tasks, from answering common questions and recommending products to

tracking orders and addressing concerns, all without needing a human agent. This helps reduce wait times, especially during busy periods or after hours, and gives customers the convenience of getting quick answers whenever they need them. As a result, customers feel more satisfied and develop greater trust and confidence in the brand (Gurley 2021).

4. By automating key processes, retailers can offer quicker, more reliable, and convenient services, making the shopping experience much more enjoyable. Customers value the ability to navigate their shopping journey with ease, and these advancements in technology also help foster loyalty by improving service quality (Johnson, 2018). When shoppers consistently enjoy efficient, personalized, and seamless experiences, they're more likely to return and recommend the brand, leading to sustained success. To sum up, automating processes like checkout, inventory management, and customer support is essential for streamlining retail operations, enhancing efficiency, and improving customer satisfaction. These innovations not only simplify operations but also meet the expectations of today's tech-savvy shoppers, strengthening customer loyalty and setting a new standard for service in the retail industry.

CHALLENGES AND LIMITATIONS OF ADOPTING SELF-SERVICE TECHNOLOGIES IN RETAIL

AI-powered self-service technologies have the potential to significantly change the retail industry by improving customer experiences, increasing efficiency, and offering more personalized services. However, there are several challenges and limitations that need to be addressed carefully. Key concerns include data privacy issues, technical difficulties, and the impact of automation on the workforce. Overcoming these challenges is crucial for ensuring successful implementation, maintaining customer trust, and securing long-term growth in an increasingly competitive market.

1. Data Privacy and Security Risks

One of the biggest challenges for retailers using AI-powered self-service technologies is safeguarding customer data. Tools like chatbots, self-checkout kiosks, and personalized recommendations rely on collecting a wide range of personal details, such as browsing habits, purchase history, and sometimes even biometric information. While this data is essential for delivering tailored services, it also raises significant concerns around privacy and security. Ensuring this information is properly protected is crucial to maintaining customer trust. As retailers collect more data, the risk of security breaches and misuse grows. Cyberattacks targeting retailers have been on the rise, and high-profile data breaches have exposed sensitive customer information. For instance, in 2020, a major retailer experienced a breach that compromised the personal details of millions of customers, leading to significant damage to their reputation and a loss of consumer trust. As retailers continue to gather large volumes of data, protecting this information becomes increasingly challenging. To reduce these risks, retailers must implement strong data protection practices, such as encryption, secure payment systems, and compliance with regulations like the General Data Protection Regulation (GDPR) and the California Consumer Privacy Act (CCPA). Failure to comply with these rules can result in hefty fines, legal penalties, and long-term harm to the brand's reputation (Abolaji & Akinwade, 2024). Equally important is being transparent with customers about how their data is collected, stored, and used, which is essential for building trust.

2. Technical Issues and System Failures

A major challenge when introducing self-service technologies is handling the technical glitches that can disrupt the customer experience. Tools like self-checkout kiosks, AI-driven systems, and virtual assistants all depend on reliable technology, but even the best systems can fail due to software issues, equipment malfunctions, or connectivity troubles. For example, when a self-checkout system breaks down, it can lead to delays and frustration, often causing customers to abandon their purchases. Another issue is the difficulty of merging new AI technologies with older systems. Retailers often face compatibility problems between the new and old tech, which can create disruptions in everyday operations. This can affect everything from inventory management to the quality of customer service. (Frey & Osborne, 2017) suggested that to avoid these challenges, it's vital for retailers to carefully test systems and make sure everything works together smoothly before rolling out new technologies. Ongoing maintenance and regular updates are also crucial for ensuring systems stay reliable. Retailers must commit resources to monitor, troubleshoot, and quickly address any issues that arise, preventing any downtime and ensuring a smooth experience for customers. Ignoring these needs can lead to poor service and inefficiencies that hurt the business.

3. Effect on workforce and Job losses

The rise of self-service technologies in retail is having a big impact on the workforce. With automation handling tasks like checkout and customer support, there's less need for human workers, which is leading to job losses and shifts in job responsibilities across the industry. While automation can streamline operations and reduce costs, it also has the potential to result in job losses, especially in positions such as cashiers and customer service representatives (Kuloğlu, 2024). To navigate these changes, retailers must proactively manage the effects of automation on their workforce. Offering retraining and reskilling programs is essential to help displaced employees transition to new roles, whether that's managing automated systems, providing higher-level customer support, or overseeing the technology itself. By investing in workforce development, retailers can mitigate the negative impacts of automation, while also fostering a more adaptable and skilled team that will drive long-term success (Frey & Osborne, 2017). Moreover, while automation can certainly improve efficiency, it's important to acknowledge that human interaction still plays a crucial role in customer service. AI technologies should be used to enhance, rather than replace, human workers. Striking the right balance between automation and personal service is key to maintaining high customer satisfaction.

5. Maintaining Customer Trust

While self-service technologies in retail can significantly enhance the shopping experience, they also bring a responsibility for retailers to safeguard customer trust. Customers need to feel confident that their personal information is protected, that the systems they interact with are reliable, and that they won't lose the personal touch that makes good customer service so valuable. To maintain this trust, retailers must be open about how they use AI and ensure customers understand how their data is being managed. Strong security protocols must be in place to protect personal information. Additionally, offering customers control over their data and providing clear, easy-to-understand privacy policies will help build confidence in these technologies. Moreover, technical problems like system outages or malfunctions can negatively impact the customer experience and weaken trust in automated services. To prevent this, retailers should provide clear and accessible customer support channels, addressing issues quickly and effectively to reinforce reliability and maintain customer trust (Kuloğlu, 2024).

While AI-powered self-service technologies offer clear benefits such as greater efficiency, tailored services, and improved customer satisfaction their implementation also comes with significant challenges. Issues such as data privacy concerns, technical glitches, and changes to the workforce must be thoughtfully addressed to ensure smooth integration in retail environments. To protect customer information and comply with privacy regulations, retailers must adopt strong data protection measures. Regular system updates and maintenance are necessary to keep everything running smoothly and address any technical difficulties. In addition, providing retraining opportunities for staff will help ease the transition as automation changes the nature of work in retail. Effectively tackling these challenges is crucial for maintaining customer trust, enhancing operational efficiency, and ensuring long-term success in the ever-evolving retail landscape.

THE FUTURE OF SELF-SERVICE TECHNOLOGIES IN RETAIL: EMERGING TRENDS AND OPPORTUNITIES

The Self-service technologies in retail is on the brink of major transformation, offering exciting possibilities to improve both the shopping experience and business efficiency. As technology evolves, new changes are come on board that will transform the way retailers connect with customers and reduce complexity of their operations. These advancements not only make shopping better but also give retailers better ways to run their businesses more efficiently, supporting them stay ahead in a fast-evolving market. Some major trends are the rise of personalized experiences, the use of virtual reality (VR) and augmented reality (AR) for 360-degree shopping experiences, and the application of artificial intelligence (AI) to manage inventory and supply chains (Abioye et al., 2021). As these technologies move forward and gain traction, they will also boost customer support, automate checkout, and refine marketing, all of which will promote the overall shopping experience. The future of self-service technology in retail promises to redefine convenience, customer satisfaction, and business operations in ways that are opening new avenue for innovations.

1. Progress in Personalization

According to (Kaplan & Haenlein, 2020) one of the major trends shaping the future of retail is the growing focus on personalization. Retailers are increasingly using customer data to create experiences that are tailored to individual needs in real-time. By recommending products that match personal preferences and offering targeted promotions, these customized interactions are set to improve, making shopping more relevant and enjoyable for customers. This level of personalization not only increases customer satisfaction but also builds loyalty by offering more value and making it easier for consumers to discover products that meet their needs. Retailers who embrace these personalized experiences will likely see greater customer engagement and retention, as consumers increasingly seek brands that cater to their individual preferences.

2. Immersive Shopping with AR and VR

Augmented Reality (AR) and Virtual Reality (VR) are changing the way consumers interact with products, offering a more engaging and hands-on shopping experience. These technologies allow shoppers to see how items would look in their own spaces before purchasing. For example, AR enables customers to virtually try on clothes or preview how furniture would appear in their homes, helping them make more informed choices. Meanwhile, VR offers entirely digital shopping environments, allowing customers to explore stores or brand experiences without leaving their homes. These innovations not only enhance the overall shopping experience but also reduce the likelihood of returns by giving customers the confidence to make well-considered decisions (Abioye et al., 2021).

3. Improving Supply Chain and Inventory Management

Technology is playing an increasingly vital role in helping retailers optimize their supply chains and manage inventory more efficiently. Tools like AI and machine learning are being used to predict demand, balance stock levels, and ensure the right products are available at the right time (Kaplan & Haenlein, 2020). These advancements allow retailers to track inventory in real-time, reduce waste, and enhance stock turnover, creating a smoother supply chain process. By adopting these technologies, businesses can avoid both overstocking and running out of stock, ultimately boosting customer satisfaction and increasing profitability.

4. Expanding Roles in Customer Support and Efficient Checkout

Self-service technologies will continue to be instrumental in enhancing customer support and automating essential processes. Chatbots and virtual assistants are becoming more advanced, providing personalized assistance and resolving issues in real-time. This shift towards greater automation will allow retailers to offer 24/7 customer support, respond to inquiries more quickly, and simplify the support process. Additionally, developments in automated checkout systems, like frictionless transactions using computer vision, will make payments quicker, reduce waiting times, and create a smoother overall shopping experience (Gupta, & Mukherjee, 2025).

5. Tailored Marketing Approaches

Self-service technologies are significantly shaping the way retailers approach marketing. By using data analytics, retailers can create highly personalized campaigns that align with consumer behavior, preferences, and purchase history. Providing customers with relevant content and offers at just the right time helps enhance engagement and drive sales. As consumers increasingly expect communication that feels personal and timely, personalized marketing will become a vital part of successful retail strategies (Gupta, & Mukherjee, 2025). Retailers who effectively utilize customer data will be able to foster stronger connections with their customers

As self-service technologies continue to evolve, retailers will find new ways to enhance the shopping experience, improve operational efficiency, and foster deeper connections with their customers. Innovations like personalized experiences, immersive shopping through AR and VR, and better supply chain and inventory management will become increasingly important in shaping the future of retail. By adopting these technologies, retailers can increase customer satisfaction and maintain a competitive edge in the fast-paced digital marketplace. Those who embrace these changes will be well-positioned to thrive in the evolving retail landscape.

AI and Personalized Marketing

AI technologies, particularly machine learning algorithms, have become essential tools for retailers aiming to deliver personalized marketing. By analyzing customer data—such as purchase history, browsing behavior, and demographic information—AI systems can create tailored marketing messages and product recommendations that resonate with individual consumers (Patil, 2024). For instance, retailers can utilize predictive analytics to forecast what items a customer might be interested in based on their past interactions with the brand. This capability not only increases the effectiveness of marketing campaigns but also fosters a sense of loyalty among customers who feel understood and valued (Gupta & Mukherjee, 2025). Moreover, AI-driven personalization is evident in various marketing channels, including email, social media, and online advertising. Retailers can segment their customer base into distinct groups according to shared characteristics, allowing for the delivery of highly relevant content. This segmentation ensures that consumers receive offers and promotions tailored to their interests, significantly enhancing engagement rates (Iliadi, 2023). For example, personalized email campaigns can achieve open rates of up to 29% higher than generic messages, demonstrating the direct impact of tailored marketing on consumer behavior (Siddiqui, 2024).

Ethical and Privacy Concerns

\Despite the advantages of AI-driven personalized marketing, significant ethical and privacy concerns must be addressed. The reliance on extensive data collection raises questions about consumer privacy and the potential for misuse of personal information. As retailers gather and analyze sensitive data, the risk of data breaches increases, which can severely damage consumer trust and brand reputation (Abolaji & Akinwande, 2024). High-profile data breaches have highlighted vulnerabilities in data security, underscoring the need for robust protective measures. One of the primary ethical concerns is the transparency of data collection practices. Many consumers are unaware of the extent to which their information is being monitored and utilized for marketing purposes. This lack of awareness can lead to feelings of distrust toward brands, which is detrimental in an era where consumer trust is paramount (Kshetri, 2021). Retailers must prioritize clear communication regarding how data is collected, stored, and used. Implementing transparent privacy policies and obtaining informed consent from customers is essential for building trust and ensuring compliance with regulations such as the General Data Protection Regulation (GDPR)

Additionally, the algorithms powering AI systems can inadvertently perpetuate biases present in the data. If not adequately monitored, these systems may favor certain demographic groups while neglecting others, leading to an exclusionary marketing approach. This issue highlights the importance of using diverse datasets and implementing fairness checks in AI algorithms to promote inclusivity and equity in marketing practices. Retailers should actively work to identify and mitigate any biases in their AI systems to ensure that all consumers are represented fairly in marketing efforts (Kaplan & Haenlein, 2020).

Balancing Personalization and Privacy

The challenge for retailers lies in balancing the benefits of personalized marketing with the ethical implications of data usage. To achieve this balance, businesses must adopt a customer-centric approach that prioritizes consumer rights and preferences. This involves not only ensuring data security but also empowering customers with control over their personal information. Providing options for consumers to opt in or out of data collection can enhance their sense of agency and trust in the brand (Abolaji & Akinwande, 2024). This implies that, establishing a robust ethical framework for AI implementation can guide retailers in navigating the complexities of data usage. This framework should include guidelines for transparency, data protection, and algorithmic fairness, ensuring that the deployment of AI technologies aligns with ethical standards and consumer expectations. The use of AI for personalized marketing in retail presents significant opportunities for enhancing customer engagement and driving sales. However, it is imperative that retailers address the ethical and privacy concerns associated with these technologies. By prioritizing transparency, implementing robust data protection measures, and actively working to eliminate biases in AI algorithms, businesses can foster trust and loyalty among consumers. As the retail landscape continues to evolve, those who embrace these ethical considerations will be better positioned to thrive in a competitive market.

Enhancements in Customer Experience through AI Tools

Chatbots: Have become essential in providing instant customer support, engaging consumers in real-time, and addressing inquiries efficiently. They offer personalized assistance based on user data, which can enhance customer satisfaction. A survey conducted by Gartner (2021) revealed that 70% of consumers reported positive experiences when interacting with chatbots, highlighting their effectiveness in enhancing customer service.

Self-checkout systems: Have also gained popularity, particularly in grocery and retail settings. These systems empower customers to complete transactions independently, reducing wait times and improving the overall shopping experience. Research by the National Retail Federation (NRF) in 2022 indicated that 62% of consumers prefer shopping at stores that offer self-checkout options, with many citing the convenience and speed as key factors in their satisfaction (NRF, 2022). Furthermore, a case study involving Walmart revealed that the implementation of self-checkout systems led to a 20% increase in customer satisfaction scores, demonstrating the tangible benefits of this technology (Johnson, 2023).

Virtual assistants: This further enhance the shopping experience by providing personalized product recommendations and assistance during online shopping. According to a survey by Adobe (2023), 65% of consumers reported that they are more likely to purchase from a retailer that utilizes virtual assistants to guide their shopping journey, indicating a strong consumer preference for personalized interactions.

Consumer Attitudes toward AI in Retail

Consumer attitudes toward AI technologies in retail are generally positive, particularly regarding trust and satisfaction. A 2023 study by PwC found that 61% of consumers trust AI to enhance their shopping experiences, especially when it comes to personalized recommendations and efficient service (PwC, 2023). This trust is crucial for retailers as it can lead to increased loyalty and repeat business.

However, there are notable concerns regarding data privacy and the transparency of AI systems. A survey conducted by McKinsey (2024) revealed that while 68% of consumers appreciated the convenience of AI tools, 54% expressed concerns about how their data was being used (McKinsey, 2024). This highlights the importance of retailers being transparent about their data practices to foster trust. The integration of AI tools such as chatbots, self-checkout systems, and virtual assistants has significantly enhanced customer experiences in retail. Consumer attitudes towards these technologies are largely favorable, reflecting a desire for convenience and personalized service. However, retailers must address privacy concerns and ensure transparency in their data practices to maintain consumer trust and satisfaction.

Consumer Trust in AI

1. ***General Trust Levels***: A study conducted Mbachukwu et., al (2024) indicated that while many Nigerian consumers are aware of AI technologies, only about 49% reported a positive perception of AI's role in enhancing their shopping experiences. This suggests a cautious but growing acceptance of AI in retail contexts.
2. ***Trust in Specific Applications***: Research highlighted that trust in AI applications, such as chatbots and recommendation systems, varies significantly. Approximately 67.5% of respondents expressed familiarity with AI, but only 27.19% felt comfortable making purchasing decisions based on AI-generated recommendations. This indicates a gap between awareness and trust in AI's capabilities (Oke et al., 2024).
3. ***Concerns about Data Privacy***: A significant concern among Nigerian consumers is data privacy. Many respondents expressed apprehension about how their personal information is collected and used by AI systems. This concern is critical, as it affects their willingness to engage with AI technologies in retail settings.

Attitudes toward AI in Retail

1. ***Preference for AI-Enhanced Services***: A survey conducted by Tula, et al., (2024) revealed that 56% of Nigerian respondents believe that customer reviews and AI-generated recommendations significantly influence their purchasing decisions. This reflects a positive attitude toward AI when it is perceived as a tool for enhancing the shopping experience.

2. ***Impact on Consumer Behavior***: The integration of AI in retail has been shown to influence consumer behavior positively. By implication, the use of AI for personalized marketing and customer service has led to increased satisfaction among consumers who appreciate tailored experiences.
3. ***Skepticism and Acceptance***: Despite the positive aspects, there remains a level of skepticism regarding AI's reliability. Many consumers are still hesitant to fully trust AI systems, particularly in areas where human interaction is traditionally valued, such as customer service.

Positive Aspects of AI Integration in Retail

The chapter effectively highlights the numerous benefits of integrating AI technologies in the retail sector. This includes:

Enhanced Customer Experience: AI tools such as chatbots and personalized recommendation systems significantly improve customer interactions. By providing tailored experiences, retailers can increase customer satisfaction and loyalty.

Operational Efficiency: AI streamlines operations through automation of inventory management, supply chain logistics, and customer service processes. This leads to reduced costs and improved efficiency, allowing retailers to respond more swiftly to market demands (Mbachukwu et al., 2024).

Data-Driven Insights: The use of AI allows retailers to analyze vast amounts of consumer data, leading to better decision-making. Insights gained from data can inform marketing strategies and product development, ultimately driving sales.

Competitive Advantage: Early adopters of AI technologies gain a competitive edge by innovating their service offerings and improving operational capabilities. This can position them favourably in a rapidly evolving retail landscape.

Ethical Implications of AI in Retail

Despite the focus on the advantages of AI integration, the section provides ethical implications associated with its use in retail (Mbachukwu, et., al 2024). This under-emphasis can lead to several critical oversights:

Data Privacy Concerns: The collection and analysis of consumer data raise significant privacy issues. Consumers are increasingly aware of how their information is used, and retailers must address these concerns to build trust.

Algorithmic Bias: AI systems can inadvertently perpetuate biases present in training data, leading to unfair treatment of certain consumer groups. This issue needs thorough exploration to ensure ethical AI deployment.

Transparency and Accountability: The lack of transparency in AI algorithms can create mistrust among consumers. Retailers should be held accountable for how AI systems make decisions, particularly in sensitive areas like pricing and customer interactions.

Impact on Employment: The automation of processes through AI can lead to job displacement in the retail sector. A discussion on the implications for workforce dynamics and potential strategies for reskilling employees is essential (Adobe, 2023).

LITERATURE REVIEW

Revolutionizing Retail with AI-Powered Self-Service Technologies

The adoption of self-service technologies in retail is transforming the industry by enhancing customer experiences and driving trends in automation and data-driven decision making. Researchers have identified numerous ways these technologies can improve both customer-facing services and behind the-scenes operations. One significant development is the increasing use of predictive analytics for demand forecasting and inventory management. By processing large amounts of data, retailers are now better able to anticipate customer needs and ensure they have the right products available at the right time. This not only helps prevent stockouts or excess inventory but also allows retailers to operate more efficiently, reduce costs, and ultimately deliver a better experience for customers by ensuring product availability when it's needed most (Patil 2024).

The implementation of self-service technologies (SSTs) in the retail industry has become a pivotal factor in transforming customer experiences, improving operational efficiency, and enhancing personalization. A range of studies has explored the various contributions of these technologies, revealing their significance in reshaping retail environments. SSTs, such as self-checkout systems, kiosks, and mobile applications, have been shown to offer a more personalized shopping experience by leveraging data analytics. According to Abolaji & Akinwade (2024), self-service platforms allow retailers to gather and analyze customer data, including browsing habits and purchasing history, which enables them to deliver tailored product recommendations and targeted promotions. This personalization not only improves customer engagement but also drives loyalty and satisfaction (Tula, et al., 2024). Furthermore, the efficiency gains achieved through the use of SSTs are well-documented. Automated systems like self-checkouts reduce long wait times and human errors, while also minimizing the need for staff intervention, thus optimizing the overall shopping process (Bhuiyan, 2024). Studies also suggest that these technologies can reduce operational costs by streamlining repetitive tasks, such as payment processing and inventory management (Patil, 2024). Operational efficiency is further enhanced by the use of AI-powered predictive analytics, which allows retailers to better forecast demand, manage stock levels, and prevent stockouts (Sharma, et al., 2022).

In addition to improving the efficiency of retail operations, SSTs also contribute significantly to customer satisfaction. A study by Patel and Trivedi (2020), indicates that customers who use self-service options enjoy faster service, greater convenience, and increased autonomy, all of which elevate their overall satisfaction with the retailer. However, the adoption of SSTs is not without challenges. Barriers such as technological limitations, security concerns, and the need for staff retraining must be addressed for these technologies to reach their full potential in the retail sector (Bhuiyan, 2024). Overall, the evidence highlights the profound impact that self-service technologies have on both the operational and experiential aspects of retail, ultimately enhancing customer satisfaction and driving business success.

CASE STUDIES AND BEST PRACTICES

Practical Examples of Self-Service Technologies in Nigerian Retail

The retail sector in Nigeria is undergoing significant change, driven primarily by the adoption of self-service technologies. Innovations like automated checkout systems and virtual assistants are changing how businesses run and how they engage with customers. These technologies are enhancing operational efficiency and improving customer experiences, allowing retailers to stay competitive in an increasingly digital environment. This analysis highlights several examples of prominent Nigerian retailers who have successfully implemented these technologies to boost their operations.

Jumia Nigeria: Streamlining the Shopping Journey: As Nigeria's leading e-commerce platform, Jumia has made considerable progress in integrating self-service technologies to enhance the online shopping experience. The company uses automated checkout systems, allowing customers to complete their purchases more efficiently and with fewer errors. This process reduces the time spent on transactions, making it quicker and easier for users. Jumia's mobile app also includes a product recommendation feature that suggests items based on a customer's browsing and purchasing history (Bhuiyan, 2024). This personalization helps boost sales and creates a more individualized shopping experience. Additionally, Jumia's virtual assistant offers continuous support by helping customers track their orders and address any issues, all without the need for direct human involvement.

Shoprite Nigeria: Enhancing In-Store Convenience with Self-Checkout: Shoprite, a major supermarket chain in Nigeria, has implemented self-checkout systems in some of its stores. This initiative allows customers to scan and pay for their items on their own, helping to reduce long wait times, especially during busy shopping hours. The self-checkout kiosks use barcode scanning to ensure smooth and accurate transactions (Joshi, et., al 2024). For Shoprite, this change has not only improved the shopping experience by minimizing queues but has also led to operational efficiencies. With fewer cashiers required at the checkout, the company can redirect resources to other crucial areas such as customer service and inventory management, thus optimizing overall store performance.

PayPorte: Using Virtual Assistants to Improve Customer Support: PayPorte, a rapidly expanding online retailer in Nigeria, has incorporated virtual assistants to handle customer service requests. These assistants provide detailed product information, assist with tracking orders, and manage returns, offering prompt and efficient solutions to common queries. By automating these routine tasks, PayPorte ensures quicker responses and reduces waiting times for customers. This strategy enables PayPorte's human customer service agents to focus on more complex issues, enhancing the overall efficiency of their support services (Vegesna, 2023). This implies that with the faster and more accessible help, customers enjoy a smoother experience, boosting the retailer's reputation for excellent customer service.

AI-Enabled Self-Service Checkout System in a Retail Electronics Store

Slot Systems Limited, a leading electronics retailer in Abuja, Nigeria, has introduced a new self-checkout system designed to improve the shopping experience, reduce wait times, and offer personalized assistance (Bhuiyan, 2024). The system features advanced technology such as facial recognition, voice interaction, and machine learning to tailor the shopping experience to each individual customer.

Key Features

Facial Recognition: The system identifies customers by their face and greets them with a personalized message, offering product recommendations based on their previous purchases or preferences.

Voice-Activated Support: Customers can communicate with the system by speaking, making it easy to ask questions, get product details, or receive help during the checkout process.

Emotional Awareness: The system detects signs of frustration or impatience, such as frowning or a raised voice, and responds by providing faster service or alerting a staff member for additional support.

Scenario Development

Customer's Experience: A tech-savvy shopper enters an electronics store in Abuja with the intention of purchasing a new smartphone. The store's self-checkout system instantly recognizes the customer's face and suggests smartphones based on prior purchases and preferences. The customer seems unsure as they browse the options, and the system detects this through subtle facial expressions. It offers to provide in-depth product comparisons or directs the customer to a store associate for further assistance.

Impact on Customer Experience: The customer feels acknowledged as the system not only identifies their preferences but also perceives and responds to their uncertainty. This creates a personalized, smooth shopping experience where technology and human support work together to enhance the customer's satisfaction.

Ethical Considerations

Privacy and Consent: The use of facial recognition technology raises important concerns around privacy and the security of personal data. It's crucial that customers are fully informed about how their biometric information will be used and provided with clear options to opt in or out of such services.

Bias in Emotion Detection: The AI system needs to be developed carefully to interpret emotions in a culturally relevant way. Different cultures may have varied expressions of emotions, and the system must be trained to recognize these nuances to avoid misunderstandings that could lead to inappropriate or irrelevant responses.

Accessibility and Inclusivity: The system should be inclusive and cater to all customers, including those with disabilities or individuals unfamiliar with advanced technologies. Offering alternative options, such as manual checkout methods, is necessary to ensure that all customers can easily navigate the shopping process.

CONCLUSION

Self-service technologies, powered by advanced systems, have the potential to revolutionize the retail industry by enhancing customer experiences, streamlining operations, and offering valuable insights for better decision-making. By diving deeper into these technologies, we can better understand how they influence customer behavior, discover the most effective implementation strategies, and ensure they are used responsibly with a focus on customer needs. Failing to take full advantage of these opportunities may lead to operational inefficiencies, a failure to meet evolving customer expectations, and, ultimately, jeopardize a business's long-term success in the competitive retail landscape.

Recommendation

1. Retailers should prioritize conducting in-depth research to better understand their customers' needs, preferences, and behaviors. This knowledge will be essential in creating self-service technologies that are tailored to improve customer satisfaction. By reviewing customer interactions and feedback, businesses can refine these technologies to offer a more personalized and seamless shopping experience.
2. Retailers should continuously evaluate the effectiveness of their AI-driven self-service technologies by gathering customer feedback and monitoring performance. Ongoing improvements and adjustments based on this feedback will help businesses stay competitive, meet customer expectations, and identify emerging trends or opportunities for enhancing customer engagement.
3. Retailers should regularly assess the performance of their AI-powered self-service technologies by collecting customer feedback and tracking their effectiveness. By making continuous improvements based on this input, businesses can remain competitive, better meet customer expectations, and stay ahead of emerging trends or opportunities to boost customer engagement.
4. Retailers need to set clear ethical guidelines when adopting AI technologies, particularly regarding data privacy, security, and avoiding bias in decision-making. Customers should be well-informed about how their data will be used and given the choice to either consent or opt out. This openness fosters trust and ensures that AI systems are implemented in a responsible and ethical manner.

REFERENCES

Abioye, S. O., Oyedele, L. O., Akanbi, L., Ajayi, A., Delgado, J. M. D., Bilal, M., & Ahmed, A. (2021). Artificial intelligence in the construction industry: A review of present status, opportunities and future challenges. *Journal of Building Engineering*, *44*, 103299. DOI: 10.1016/j.jobe.2021.103299

Abolaji, E. O., & Akinwande, O. T. (2024). AI powered privacy protection: A survey of current state and future directions. *World Journal of Advanced Research and Reviews*, *23*(3), 2687–2696. DOI: 10.30574/wjarr.2024.23.3.2869

Adobe. (2023). *The Future of Shopping: How AI is Transforming Retail.* Retrieved from Adobe Insights

Bhuiyan, M. S. (2024). The role of AI-Enhanced personalization in customer experiences. *Journal of Computer Science and Technology Studies*, *6*(1), 162–169. DOI: 10.32996/jcsts.2024.6.1.17

Davenport, T. H., & Ronanki, R. (2018). Artificial intelligence for the real world. *Harvard Business Review*, *96*(1), 108–116. https://hbr.org/2018/01/artificial-intelligence-for-the-real-world

Frey, C. B., & Osborne, M. A. (2017). The future of employment: How susceptible are jobs to computerization? *Technological Forecasting and Social Change*, *114*, 254–280. DOI: 10.1016/j.techfore.2016.08.019

Gukpa, A. S., & Mukherjee, J. (2025). Framework for adoption of generative AI for information search of retail products and services. *International Journal of Retail & Distribution Management*, *53*(2), 165–181. DOI: 10.1108/IJRDM-05-2024-0203

Gurley, R. C. (2021). *Strategies for Implementing Self-service Technologies in Supermarket Retail Operations* (Doctoral dissertation, Walden University). https://scholarworks.waldenu.edu/cgi/viewcontent.cgi?article=11768&context=dissertations

Iliadi, M. I. (2023). *Unlocking customer insights through service analytics to improve customer experience and drive business success* (Bachelor's thesis, University of Twente). https://essay.utwente.nl/95404/1/Iliadi_BA_BMS.pdf

Johnson, B. (2018). Self-Service Analytics: A Comprehensive Review of Implementation Strategies. *Journal of Data Analytics*, *8*(2), 87–101.

Johnson, B. (2023). Impact of Self-Checkout Systems on Customer Satisfaction: A Case Study on Walmart. *Journal of Retail Management*, *15*(2), 45–58.

Joshi, N. K., Kulshreshtha, K., & Jain, U. (2024). The role of AI in enhancing customer service in the retail sector. *Journal of Data Acquisition and Processing, 39*(1), 1270-1277. https://sjciy cl.cn/index.php/DAP/article/view/133

Kaplan, A., & Haenlein, M. (2020). Rulers of the world, unite! The challenges and opportunities of artificial intelligence. *Business Horizons*, *63*(1), 37–50. DOI: 10.1016/j.bushor.2019.09.003

Kshetri, N. (2021). *Cybersecurity management: An organizational and strategic approach*. University of Toronto Press. DOI: 10.3138/9781487531249

Kuloğlu, A. (2024). The impacts of technological transformation on labor and employment: The future of the business world and economic growth. *International Research in Social. Human and Administrative Sciences*, *XVII*, 21.

Mbachukwu, M. A., Mwakawaza, N., & Huang, S. F. (2024). The role of artificial intelligence in transforming customer experience in the service industry in Nigeria. *International Journal of Academic Research in Business & Social Sciences*, *14*(11). Advance online publication. DOI: 10.6007/IJARBSS/v14-i11/23532

McKinsey. (2024). *Consumer Insights on AI and Data Privacy in Retail*. Retrieved from McKinsey & Company.

Oke, T. T., Ramachandran, T., Afolayan, A. F., Ihemereze, K. C., & Udeh, C. A. (2024). The role of artificial intelligence in shaping sustainable consumer behavior: A cross-sectional study of Southwest, Nigeria. *International Journal of Innovation and Scientific Research*, *1*(1), 255–266. Advance online publication. DOI: 10.51244/IJRSI.2023.1012021

Patel, N., & Trivedi, S. (2020). Leveraging Predictive Modeling, Machine Learning Personalization, NLP Customer Support, and AI Chatbots to Increase Customer Loyalty. *Empirical Quests for Management Essences, 3*(3), 1–24. https://www.researchberg.com/index.php/eqme/article/view/46

Patil, D. (2024). Artificial intelligence in retail and e-commerce: Enhancing customer experience through personalization, predictive analytics, and real-time engagement. *Predictive Analytics, and Real-Time Engagement*. https://ssrn.com/abstract=5057420 or http://dx.doi.org/DOI: 10.2139/ssrn.5057420

Sharma, D., Reddy, N., Gupta, P., & Sharma, R. (2022). Enhancing customer experience personalization through AI: Leveraging collaborative filtering, neural networks, and natural language processing. *Journal of AI ML Research*, *11*(7). https://www.joaimlr.com/index.php/v1/article/view/32

Siddiqui, A. (2024). The impact of artificial intelligence on business operation: Current state, future opportunities and challenges. [IJM]. *International Journal of Management*, *15*(04), 79–93. https://mylib.in/index.php/IJM/article/view/IJM_15_04_006

Tula, S. T., Kess-Momoh, A. J., Omotoye, G. B., Bello, B. G., & Daraojimba, A. I. (2024). AI-Enabled customer experience enhancement in business. *Computer Science & IT Research Journal*, *5*(2), 365–389. DOI: 10.51594/csitrj.v5i2.789

Vegesna, D. V. V. (2023). Comprehensive analysis of AI-enhanced defense systems in cyberspace. *International Numeric Journal of Machine Learning and Robots,* *7*(7). https://injmr.com/index.php/fewfewf/article/view/21

Chapter 7
Finance and Beyond: Driving Self-Service Transformation Through Digital Innovation

Shubhangi Gautam
https://orcid.org/0000-0002-0236-0216
Lovely Professional University, India

Mandeep Kaur
https://orcid.org/0000-0002-4736-8991
Lovely Professional University, India

Mohsin Showkat
https://orcid.org/0009-0006-2232-5887
Lovely Professional University, India

ABSTRACT

Across all sectors, self-service transformation is enabling service delivery without users having to get in line and go through the company; they can instead initiate a task digitally and manage it independently. In finance, technologies like mobile banking apps, robo-advisors, and online portals have empowered customers, increased operational efficiency, cut costs and provided the evolving convenience and personalization that consumers expect. Outside of finance, healthcare and retail services are utilizing self-service portals and self-checkout capabilities to improve processes and customer experiences. In this research, we investigate the antecedents, processes, and consequences of self-service transformation, illustrating the path to developing the capabilities to scale, economize and delight the customer. It lays out the technological enablers, such as automation, artificial intelligence, but also the challenges, such as cybersecurity risks, the digital divide, resistance to

DOI: 10.4018/979-8-3373-4667-0.ch007

adoption. By designing around these struggles, building with security in mind, and educating end-users, organizations can accommodate this transition with relative ease. Self-service transformation: how it can reshape the way services are delivered, and creates a more empowered, efficient and scalable future across industries.

INTRODUCTION

Finance and other sectors of self-service transformation illustrate an evolution of service delivery and consumption. It enables users to complete specific actions without human intervention through the use of digital platforms (Hassan et al., 2024). In finance, the transformation encompasses things like mobile banking apps, ATMs, robo-advisors, and online portals enabling services ranging from account management and loan applications to investment tracking (Lóska & Uotila, 2024). These innovations offer customers more control and access than ever, enabling them to manage their financial needs at any time and from anywhere. The rise of self-service in finance has improved efficiency, decreased operational costs, and catered to the demand for seamless, personalized experiences (Kowalkowski et al., 2024).

Outside finance, self-service transformation is making waves in healthcare, retail and other industries. For example, patients log in to online portals to make appointments and view their health records, and shoppers scan groceries at self-checkout kiosks and request personalized shopping experiences from virtual assistants in the store (Tsou & Chen, 2019). AI-Powered Utility Solutions for Future Readiness AI solutions in the utility sector play a crucial role in automating workflows, optimizing operational efficiency, and providing enhanced customer experiences. Automation and artificial intelligence and accessible interfaces are at the heart of this evolution, enabling businesses to automate processes and achieve consistent outcomes (Audrin, 2019).

A paradigm shift is driven by changing consumer expectations, technology advancements, and cost efficiency. Parallel to the purported advantages, challenges abound: The digital divide, cybersecurity concerns and resistance to change will require significant investment to overcome (Nair, 2025). By pairing intuitive design with secure architecture and end-user education, organizations can drive an evolution to self-service models. With this current trend continuing to grow, it is transforming the way that service providers and users relate; the former, more powerful and the latter, more efficient future (Spais, 2023).

Drivers of Self-Service Transformation

The fast-changing landscape of digital tech has played a key role in driving self-service transformation across sectors, from finance to far beyond. One of the key drivers is the increasing consumer demand for convenience and efficiency (Hassan et al., 2024). Today's customers have a strong demand for seamless and on-demand access to services without involving a middleman. The reason behind the change in these expectations is the development of digital platforms, mobile applications and AI-enabled chatbots that enable customers to do things – such as financial management, appointment bookings or troubleshooting issues, with minimal friction (Kowalkowski et al., 2024).

The other major driver, of course, is the availability and affordability of advanced technologies, like cloud computing, artificial intelligence, and data analytics (Tsou & Chen, 2019). These technologies allow for organizations to build personalized, intuitive self-service up solutions which fulfill a myriad of customer requirements while keeping operational costs low. In the financial sector, for example, AI-based software can analyze the way users behave (i.e. financial habits) to provide personalized suggestions for better financial management, helping improve user experience and engagement (Audrin, 2019).

Self-service transformation is also driven by cost reduction and operational efficiency. In fact, digital self-service platforms can automate mundane tasks, relieving human teams from their load of work while also eliminating unwanted human error and speeding up service delivery (Saldanha et al., 2022). For organizations, this means saving big, which they can invest back into more technology or customer experience enhancements. In addition, COVID-19 pandemic has fueled the transition to self-service technologies in the organizations (Vijay & Priyanka, 2025). From contactless interaction to remote operations, businesses quickly found themselves in a position where they had to embrace digital self-service solutions. Not only did this transition become instrumental for sustaining business during uncertain times, but it also resulted in re-establishing customer expectations in the post-pandemic world (Nair, 2025).

With increased competition in diverse industries, organizations must set themselves apart with an exemplary CX. Providing strong and creative self-service capabilities has transitioned from being an option to being mandatory to hold onto customers (and serve new ones). With digital ecosystems rapidly evolving, organizations must either harness and shape self-service transformation or risk obsolescence in a digital-first world (Banerjee et al., 2025).

1. **Consumer Expectations**: We live in a world where consumers are increasingly prioritising independence, convenience and immediacy of access. The newer generations of customers, especially service-consuming people are mostly comprised of digital-native generations as they expect the service para-assets to keep pace with their busy lifestyle (Muzaffar et al., 2024). All these raise the bar with regard to self-service as smartphones and digital tools have increased these expectations and driven corporations to implement self-service models that enable smoother contact and control (Lóska & Uotila, 2024).

Another dimension of these expectations is personalized experiences. Consumers want solutions designed for their individual needs and preferences, for example customized product recommendations or investment strategies (J. Zhang & Chen, 2024). Additionally, self-service alternatives generally minimize waiting time and frustration that come along with conventional regards of service, making them more convenient. The opportunity for marketers lies in addressing these changing preferences, which can be used by businesses not just to meet consumer expectations but also build a point of differentiation in competitive markets (Lin et al., 2023).

2. **Technological Advancements**: Increasing technology developments have brought about self-service transformation. The cheap accessibility and scale to self-service platforms enabled by the proliferation of mobile devices, high-speed web, and cloud computing infrastructure AI and ML are the primary technologies that are improving these systems with capabilities like predictive analytics, natural language processing, and real-time decision-making (Gautam & Kumar, 2023b).

Robotic process automation (RPA) RPA is another automation tech that cuts down on manual effort and improves accuracy. This enables companies to offer higher quality, more responsive and tailored services to their customers (Gautam & Kumar, 2024). The driving force of self-service transformation is technically driven advancement for any organization to meet consumer expectation while innovating progress.

3. **Operational Efficiency**: Self-service solutions play an important role in optimising organisational processes to lessen the reliance on human resources for routine tasks. With automation of repetitive operations, enterprises can repurpose their workforce to compete on higher-value activities that involves critical thought and creativity (Gautam, Kumar, & Dahiya, 2024). Chatbots, for example, take care of customers' queries promptly, while automated payment channels facilitate transactions.

Moreover, operational efficiency leads to tremendous cost savings. Institutions have a reduction in costs for physical assets like branches or retail spaces and can maintain service quality even when moving to online operations (Spais, 2023). This priority of enhancing user experience and reducing service request lead time not only sausages the organization but improves customer experience by reduced lead times and high quality services.

4. **Scalability**: Scalability must be supported, and this is one of the most important benefits of self-service platforms. Classic service delivery models are often unable to cope with increases in demand because of constraints in physical and human resources (Raddats et al., 2022). In comparison, self-service solutions can scale with an expanding user base at little additional cost. For instance, Administrators use cloud-based servers to scale up or down their infrastructure according to the flow of work (Hassan et al., 2024).

Such capability ensures that organizations can deliver high-quality service, whilst reducing their reliance on user volume, even during peak activity periods. Also, scalable systems allow for easy expansion into new markets or regions without a heavy capital outlay in physical assets, making them a key strategic asset for growth-oriented companies (Kowalkowski et al., 2024).

Table 1. Drivers of Self-Service Transformation

Driver	Explanation
Consumer Expectations	Demand for autonomy, convenience, and personalization in services.
Technological Advancements	Role of mobile devices, AI, and cloud computing in enabling self-service solutions.
Operational Efficiency	Streamlining processes and reducing costs by automating repetitive tasks.
Scalability	Ability of self-service platforms to handle growing user bases without proportional increases in resources.

Research Methodology

This research follows a conceptual research strategy, depending on secondary data sources, scholarly literature, and documented industry practice to investigate the antecedents, drivers, enablers, and consequences of self-service transformation. The analysis is interpretive in scope, seeking to synthesize known knowledge and contribute to a holistic understanding of sector-specific implementations and technological advancements driving self-service ecosystems.

Figure 1. Conceptual framework

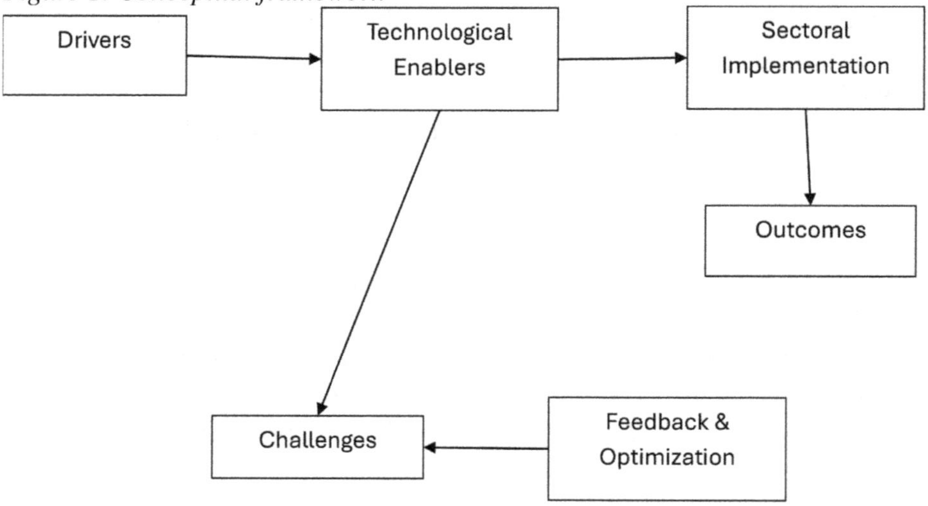

Mechanisms and Technological Enablers

Various mechanisms and technological enablers exist to enhance digital innovation for the transformation of self-service systems that result in both user empowerment and business operations centralization. At the heart of this evolution are digital platforms that combine a myriad of services into seamless and intuitive user interfaces (Gautam & Kumar, 2023a). Mobile apps, web portals, and omnichannel platforms are the key enablers of self-service, allowing consumers to manage their accounts, make payments, and troubleshoot independently. Finally, these platforms usually have intuitive design and guided workflow-oriented solutions to encourage the best user experience and participation (Audrin, 2019).

The development of AI and ML are the most disruptive techs that have come-to-be. Now, FI customers interact with AI based chatbots, virtual assistants and recommendation engines rather than waiting for response, and these algorithms provide almost instant and relevant answers as well as suggest personalized options (Saldanha et al., 2022). For example, AI generates deep insights from customer information and gives them pertinent recommendations about investments, savings, and budgeting. Such systems greatly improve the user experience but also allow organizations to obtain important information about their customers for further enhancement (Vijay & Priyanka, 2025).

Cloud computing plays a key role in self-service transformation by providing scalability, flexibility, and cost efficiency. With cloud-based solutions, businesses can quickly deploy and update self-service platforms, ensuring they run smoothly and are always available (Nair, 2025). These solutions also enable real-time data synchronization, so customers get accurate, up-to-date information across all touchpoints.

Data analytics and big data technologies are essential for optimizing self-service systems. By analyzing large volumes of user data, businesses can pinpoint pain points, predict trends, and make informed decisions to enhance their self-service options (Banerjee et al., 2025). For instance, predictive analytics can anticipate customer needs and offer proactive solutions, leading to higher satisfaction and stronger loyalty.

Blockchain technology is transforming industries like finance and supply chain management by enabling secure, transparent, and decentralized transactions. This makes self-service features like peer-to-peer payments and document verification possible without the need for intermediaries (Muzaffar et al., 2024). At the same time, the integration of Internet of Things (IoT) devices is expanding automation in self-service systems. IoT-powered kiosks, smart home devices, and wearables offer users more intuitive ways to interact with technology, making self-service more convenient and efficient (Lóska & Uotila, 2024). Together, these advancements are shaping the future of self-service, delivering seamless, secure, and empowering experiences across industries.

1. **Automation**: The foundation of self-service transformation is automation, which ensures operational consistency and streamlines repetitive operations. Businesses can provide effective services without physical involvement thanks to technologies like self-checkout systems, automated teller machines (ATMs), and robotic process automation (RPA) (J. Zhang & Chen, 2024). By reducing human error, these systems increase accuracy and give users faster, more dependable service.

Chatbots with natural language processing, for example, answer common questions in customer service, freeing up human agents to deal with more complicated problems. In a similar vein, retail automated inventory management systems guarantee that stock levels are tracked and swiftly refilled (Lin et al., 2023). Automation increases operational efficiency, scalability, and cost-effectiveness by decreasing labor-intensive activities.

2. **Artificial Intelligence**: Self-service platforms are greatly improved by AI technology since they allow for intelligent, data-driven decision-making. Large volumes of customer data are analyzed by AI systems to forecast demands, customize interactions, and improve service delivery (Raddats et al., 2022). For

instance, AI is used by financial robo-advisors to create customized investment plans according to a person's risk tolerance and financial objectives.

Additionally, e-commerce AI-driven recommendation engines make tailored product recommendations based on user preferences and browsing history. Voice recognition software, which enables natural speech interaction between users and platforms, is also powered by AI (Spais, 2023). Businesses may provide context-aware, adaptable services that meet consumer expectations and promote user pleasure and confidence by integrating AI.

3. **Digital Platforms**: The main way that people interact with self-service solutions is through digital platforms including online portals, kiosks, and mobile applications. Because of their intuitive design, these platforms are usable by people with different levels of technological expertise (Gautam et al., 2024). They combine several features, allowing users to do a variety of things in one location, such as making payments or setting up appointments.

For example, consumers can use mobile banking apps on their cellphones to check accounts, send money, and even apply for loans. Patient portals in the healthcare industry offer appointment scheduling and access to medical records (Sangwan & Sharma, 2022). Safe and easy transactions are further guaranteed by the integration of secure payment gateways. Digital platforms play an essential role in promoting the uptake and success of self-service transformation by emphasizing user-friendly design and strong functionality (Meyer & Greve, 2019).

Table 2. Mechanisms and Technological Enablers

Mechanism/ Enabler	Explanation
Automation	Streamlines repetitive tasks, ensuring efficiency and accuracy.
Artificial Intelligence	Enables intelligent, data-driven decision-making and personalization.
Digital Platforms	Provides user-friendly interfaces for accessing a wide range of self-service functionalities.

Sector-Specific Implementations

Industries differ greatly in how self-service solutions driven by digital innovation are adopted; each uses technology to satisfy its own operational needs and customer expectations. Automated teller machines (ATMs), robo-advisors, and mobile banking apps are examples of the self-service transition in the financial industry (Albayati et al., 2020). Customers may now carry out tasks like fund transfers, investment planning, and fraud reporting on their own thanks to the integration of AI-powered chatbots, virtual assistants, and biometric authentication into the platforms of banks and other financial institutions. This guarantees strong security and compliance in financial transactions in addition to improving customer ease (Sadaf et al., 2012).

Digital checkout systems and self-service kiosks have completely changed how customers engage with retailers. Self-service technology enable customers to browse, buy, and handle returns with little assistance from humans in a variety of settings, including supermarkets and major e-commerce sites (Hassan et al., 2024). A smooth and interesting shopping experience is produced by cutting-edge technologies like voice-activated shopping assistants, AR-enabled virtual try-ons, and personalized recommendations. IoT devices and inventory management systems are also used by retailers to improve inventory accuracy, cut expenses, and streamline processes (Kowalkowski et al., 2024).

Through automated appointment scheduling systems, telemedicine platforms, and patient portals, the healthcare industry has embraced self-service transformation. Patients may now schedule appointments, view medical records, and communicate with medical providers from a distance thanks to digital innovation (Tsou & Chen, 2019). Wearable health gadgets and AI-driven diagnostic tools that track and evaluate patient data in real-time enable these solutions, enabling people to actively manage their health while lessening the strain on medical professionals.

Self-service has played a key role in changing user experiences in the transportation industry, especially in urban mobility and logistics. Users can now navigate cities more easily thanks to ride-sharing applications, digital ticketing systems, and autonomous car technologies (Audrin, 2019). To facilitate travel, airports and rail stations use biometric boarding systems, self-service check-in kiosks, and real-time tracking applications. In the meantime, self-service portals in logistics allow clients to plan deliveries, monitor shipments, and get answers to their questions, increasing efficiency and transparency (Saldanha et al., 2022).

Self-service platforms have been widely used even in education. Digital libraries and learning management systems (LMS) enable students to independently access course materials, turn in assignments, and interact with teachers (Vijay & Priyanka, 2025). Adaptive learning tools and AI tutors tailor the learning process to each student's unique strengths and shortcomings. These industry-specific examples demonstrate how self-service transformation is a customized use of digital innovation rather than a one-size-fits-all strategy. Self-service solutions continue to boost overall competitive advantage, customer satisfaction, and operational efficiency by catering to the unique requirements of each business (Nair, 2025).

1. **Finance**: When it comes to implementing self-service technologies, the financial industry has led the way, revolutionizing conventional banking and investing methods. Customers can use mobile banking apps to do a number of things without going to a branch, including paying bills, moving money, and keeping an eye on account activities (Litterscheidt & Streich, 2020). AI-powered robo-advisors offer individualized investment guidance and portfolio management, opening up financial planning to a larger audience (Fan & Chatterjee, 2020).

People may now handle their money on their own thanks to online trading platforms, which have democratized access to stock markets. These systems frequently include real-time data analysis capabilities, enabling users to make well-informed choices (L. Zhang et al., 2021). Self-service technologies in finance have improved customer satisfaction and operational efficiency while saving costs by decreasing the need for face-to-face contacts and boosting accessibility.

2. **Healthcare**: By giving patients more control over medical treatments, self-service solutions in the healthcare industry have completely changed the patient experience. Through patient portals, people may make appointments, view medical records, and get in touch with healthcare professionals from the comfort of their own homes (Cheng et al., 2011). Routine procedures are streamlined by these platforms' integration of features like test result checks and prescription refill requests.

By facilitating virtual consultations with physicians and eliminating the need for in-person visits, telehealth services further broaden the scope of healthcare. In hospitals and clinics, self-check-in kiosks expedite administrative duties so that staff can concentrate on providing crucial care (Radanovic & Likic, 2018). Healthcare businesses can improve patient happiness, decrease wait times, and increase accessibility by utilizing self-service technologies.

3. **Retail**: Self-service technology, which provide quicker and more individualized services, have completely changed the shopping experience in the retail industry. Self-checkout systems cut down on wait times and increase operational efficiency by enabling customers to scan and pay for items on their own (Prasad et al., 2020). Digital wallets and other mobile payment alternatives make the checkout process even more efficient.

Recommendation engines powered by artificial intelligence are used by e-commerce platforms to create customized purchasing experiences based on user preferences and browsing history. Product availability is guaranteed by automated inventory management, which raises customer satisfaction (Patroni, 2018). Retailers may meet the expectations of contemporary consumers by offering a smooth, effective, and entertaining shopping experience with the integration of self-service technologies.

Table 3. Sector-Specific Implementations

Sector	Implementation
Finance	Mobile banking, robo-advisors, online trading platforms.
Healthcare	Patient portals, telehealth, self-check-in kiosks.
Retail	Self-checkout systems, mobile payments, AI-driven recommendation engines.

Outcomes and Benefits

Digital innovation's shift to self-service models has many positive effects and transforms how companies run and engage with their clientele. Improved client experience is among the most important results (Hassan et al., 2024). Self-service systems give users ease, control, and independence, which empowers them. Customers enjoy faster, more effective service that fits with their busy lives, whether they are managing retail transactions, making healthcare appointments, or accessing financial resources (Kowalkowski et al., 2024).

Another important advantage is operational efficiency. Businesses can lessen their dependency on manual procedures and redirect resources to more strategic endeavors by automating repetitive tasks and allowing self-service capabilities (Tsou & Chen, 2019). This operational simplification reduces human mistake and saves time and labor expenses, resulting in dependable and consistent service delivery. For instance, automated chatbots and robo-advisors manage large numbers of transactions and inquiries in the financial industry, freeing up human agents to perform intricate, valuable jobs (Audrin, 2019).

One of the fundamental advantages of self-service transition is cost reduction. By eliminating the need for large customer service personnel and decreasing physical infrastructure, digital platforms and AI-driven solutions dramatically reduce operating costs (Saldanha et al., 2022). A sustainable cycle of development and improvement can be established by reinvesting these savings in customer retention initiatives, staff training, or additional innovation.

The ability of firms to scale and adapt is another important result. Because digital self-service systems are so scalable, businesses can accommodate expanding user bases without seeing a corresponding rise in expenses (Vijay & Priyanka, 2025). This flexibility is particularly useful in sectors where demand is erratic or client growth is quick. E-commerce platforms, for example, may support millions of users at once by utilizing real-time analytics and cloud-based infrastructure (Nair, 2025).

Data-driven decision-making is also encouraged by self-service transformation. Self-service platforms that use advanced analytics give businesses deep insights into the preferences, behavior, and problems of their customers (Banerjee et al., 2025). Businesses can use this data to improve customer happiness and loyalty by honing their products, anticipating trends, and taking proactive measures. Self-service revolution also encourages accessibility and inclusion. Businesses can serve a wider range of customers, including those with impairments or low levels of technological skill, by using voice recognition, adaptive technologies, and multilingual support (Muzaffar et al., 2024). This inclusiveness creates new market prospects in addition to being in line with CSR objectives.

Superior customer experiences, cost savings, and strategic growth potential are just a few of the results and advantages of self-service transformation that go beyond operational enhancements (Lóska & Uotila, 2024). Organizations may create robust, customer-focused models that propel long-term success by embracing digital innovation.

1. **Enhanced Customer Experience**: Self-service transformation gives users more convenience and control, which greatly enhances the whole customer experience. Long wait periods or difficult processes are no longer necessary thanks to these platforms, which let users access services whenever and however they choose (J. Zhang & Chen, 2024). Platforms employ user data to provide customized recommendations, alerts, and support, making personalization a crucial component.

For example, healthcare portals offer personalized health advice and appointment reminders, while retail platforms make product recommendations based on browsing history (Lin et al., 2023). Higher levels of client satisfaction and loyalty are the result of this increased autonomy and personalization.

2. **Cost Efficiency**: Self-service solutions help businesses cut costs by lowering their dependency on physical infrastructure and human labor. By reducing errors and expediting repetitive tasks like data entry and payment processing, automated procedures save operating costs (Spais, 2023).

Digital banking, for instance, lessens the demand for branch-based services in the financial industry, and self-checkout systems at retail stores do away with the need for extra employees at checkout counters (Raddats et al., 2022). Long-term company growth can be fueled by reinvesting these savings in service offerings and innovation.

3. **Scalability**: Self-service systems are built to grow with you without requiring you to add resources in proportion. Businesses may effectively scale their services to meet growing demand or enter new markets thanks to cloud-based infrastructures and digital interfaces (Lóska & Uotila, 2024).

E-commerce platforms, for instance, can manage higher traffic levels during promotional events or holiday sales without experiencing any performance issues. In a similar vein, telehealth platforms require no extra funding to accommodate greater patient populations (Vijay & Priyanka, 2025). This scalability guarantees that companies retain constant service quality while staying flexible and competitive in ever-changing markets.

Table 4. Outcomes and Benefits

Outcome/Benefit	Explanation
Enhanced Customer Experience	Greater control, convenience, and personalized services for users.
Cost Efficiency	Reduces operational costs by automating processes and minimizing physical infrastructure needs.
Scalability	Supports growth and market expansion without significant additional resources.

Challenges and Mitigation Strategies

Although there are many advantages to switching to self-service models through digital innovation, there are drawbacks as well. Customer and employee resistance to change is one of the main challenges (Hassan et al., 2024). Because they don't trust automated systems or aren't conversant with technology, some customers might choose conventional, human-assisted services. Employees may also find it difficult to adjust to new procedures or fear losing their jobs. Organizations must prioritize change management techniques, such as thorough training programs, customer education, and unambiguous communication regarding the advantages of self-service, in order to solve this (Saldanha et al., 2022). Overcoming resistance requires arming stakeholders with the information and self-assurance they need to adopt digital tools.

Ensuring data security and privacy presents another difficulty. Self-service platforms are prime candidates for hackers since they gather and handle enormous volumes of private client data (Nair, 2025). To protect user data, organizations need to have strong cybersecurity measures in place, like multi-factor authentication, end-to-end encryption, and frequent security assessments. Furthermore, maintaining compliance with data privacy laws such as the CCPA or GDPR is essential for fostering confidence and averting fines (Muzaffar et al., 2024).

Another frequent obstacle to self-service transformation is technical problems and system outages. Consumers anticipate smooth, round-the-clock service access, so any disruptions may irritate them and drive them away (J. Zhang & Chen, 2024). Businesses should make investments in dependable infrastructure, cloud-based solutions, and proactive monitoring systems that can spot such problems early on and take action to reduce this risk. Even in the case of technical malfunctions, continuity is guaranteed by redundant systems and disaster recovery strategies (Lin et al., 2023).

Another challenge is designing for user experience (UX). A self-service interface that is badly designed may cause misunderstandings, mistakes, and desertion. In order to improve their platforms, businesses must give priority to user-centric design concepts, carry out usability testing, and collect feedback (Lin et al., 2023). Users of different skill levels may interact with the system with ease thanks to features like responsive design, clear instructions, and straightforward navigation (Spais, 2023).

Another major obstacle is still the digital divide, which is especially noticeable in areas with poor internet connectivity or among populations with low levels of technology literacy. Organizations should provide offline options or hybrid models that blend traditional and self-service techniques to handle this (Tsou & Chen, 2019). By encouraging digital literacy and developing infrastructure, collaborations with regional administrations and communities can aid in closing the gap even more. Additionally, coordinating the shift from conventional to self-service models necessitates striking a balance between automation and human interaction. Customers who value individualized help may become irate if automation is used excessively (Banerjee et al., 2025). In addition to providing self-service capabilities, businesses should make sure that human agents are available when needed and retain easily accessible customer care channels for complicated issues (Raddats et al., 2022).

Organizations can successfully negotiate the complexities of self-service transformation by recognizing these issues and putting focused mitigation techniques into place. This strategy guarantees that digital innovation meets the requirements and concerns of all stakeholders while producing the desired results (Tsou & Chen, 2019).

1. **Cybersecurity Risks**: The growing dependence of self-service platforms on digital infrastructure raises serious concerns about cybersecurity (Nair, 2025). These dangers include identity theft, data breaches, and system flaws that could expose private user data.
    - **Mitigation**: Businesses can mitigate these risks by putting strong encryption procedures in place to safeguard data both in transit and at rest.

Multi-factor authentication (MFA) requires users to confirm their identity using various methods, adding an extra degree of protection (Tsou & Chen, 2019). System resilience against changing threats is ensured by routine security assessments and updates. Another way to reduce risks is to teach consumers cybersecurity recommended practices, like how to spot phishing efforts (Muzaffar et al., 2024).

2. **Digital Divide**: The difference in access to technology and digital literacy across various populations is known as the "digital divide." Self-service solution adoption is hampered by this issue, especially among senior citizens and members of underprivileged populations (Lin et al., 2023).
   - o **Mitigation**: Businesses can close this gap by creating intuitive user interfaces that take into account different levels of technological expertise. Offering users step-by-step instructions and digital literacy training can enable them to successfully use self-service platforms (Kowalkowski et al., 2024). Providing offline solutions or community-based access points, such kiosks, for people with restricted internet access guarantees inclusion.
3. **Resistance to Adoption**: Self-service solutions are frequently resisted by users who prefer more conventional approaches, are afraid of technology, or don't trust digital systems (Tsou & Chen, 2019). The success of self-service projects may be hampered by this hesitancy.
   - o **Mitigation**: Businesses might concentrate on developing user-friendly, aesthetically pleasing interfaces that streamline the user experience in order to overcome opposition (Audrin, 2019). Building trust and promoting adoption can be achieved by showcasing the benefits of self-service solutions through case studies, testimonials, and pilot initiatives. Having strong customer service and training materials available also facilitates consumers' seamless transition to self-service platforms, boosting their confidence and contentment (Banerjee et al., 2025).

Table 5. Challenges and Mitigation Strategies

Challenge	Explanation	Mitigation Strategy
Cybersecurity Risks	Risks such as data breaches and theft.	Encryption, multi-factor authentication, security audits, and user education.
Digital Divide	Disparities in access and digital literacy.	User-friendly design, digital literacy programs, and community-based access points.
Resistance to Adoption	Hesitation to use self-service platforms.	Intuitive interfaces, pilot programs, robust support, and demonstrating value.

Table 6. Sector specific Challanges

Sector	Key Challenges	Explanation
Finance	Regulatory Compliance and Cybersecurity	Financial institutions are required to abide by stringent regulations (e.g., KYC, AML, GDPR) and protect sensitive user data. Ongoing cyberattacks and changing compliance requirements complicate the introduction of self-service systems.
Healthcare	Privacy, Integration, and Accessibility	Data privacy under regulations such as HIPAA, integration with legacy hospital systems, and making it possible for older or disabled users to access digital health platforms are significant challenges
Retail	Customer Trust and Tech Adoption Variability	Although digital technologies add convenience, uneven adoption across age segments and low confidence in automated systems (e.g., returns) can constrain the effectiveness of self-checkout and virtual assistants
Transportation	Infrastructure and Real-Time Data Accuracy	Self-service tools like digital ticketing or ride-sharing apps rely heavily on infrastructure and GPS accuracy. Connectivity issues and real-time updates are critical concerns in mobility services
Education	Digital Divide and Engagement	Underserved students might not have devices or access to stable internet. Moreover, digital self-service solutions lose user engagement relative to face-to-face interactions

CONCLUSION

A deliberate change in service delivery, self-service transformation uses technology to satisfy changing customer needs while promoting scalability and efficiency. Even while issues like cybersecurity and the digital divide still exist, businesses may address them with focused tactics to guarantee safe and inclusive platforms. Traditional service models are being redesigned to fit the digital era as sectors continue to embrace self-service alternatives, opening the door to a more empowered and effective future. Beyond just improving operations, self-service transformation has the ability to stimulate innovation and provide long-term benefits to both enterprises and customers.

REFERENCES

Albayati, H., Kim, S. K., & Rho, J. J. (2020). Accepting financial transactions using blockchain technology and cryptocurrency: A customer perspective approach. *Technology in Society*, *62*, 101320. DOI: 10.1016/j.techsoc.2020.101320

Audrin, B. (2019). Implementing self-service technologies: Not without competition! *International Journal of Retail & Distribution Management*, *48*(2), 169–185. DOI: 10.1108/IJRDM-09-2018-0193

Banerjee, O., Nandan, K., Kumar, S., & Azhar, M. (2025). Leveraging Data Analytics and Business Intelligence for Seamless Digital Transformation. In *In Perspectives on Digital Transformation in Contemporary Business* (pp. 113–142). IGI Global Scientific Publishing., DOI: 10.4018/979-8-3693-5966-2.ch005

Cheng, Y. S., Yu, T. F., Huang, C. F., Yu, C., & Yu, C. C. (2011). The Comparison of Three Major Occupations for User Acceptance of Information Technology: Applying the UTAUT Model. *IBusiness*, *03*(02), 147–158. DOI: 10.4236/ib.2011.32021

Fan, L., & Chatterjee, S. (2020). The Utilization of Robo-Advisors by Individual Investors: An Analysis Using Diffusion of Innovation and Information Search Frameworks. *Financial Counseling and Planning*, *31*(1), 130–145. DOI: 10.1891/JFCP-18-00078

Gautam, S., & Kumar, P. (2023a, June). Behavioral biases of investors in the cryptocurrency market. *AIP Conference Proceedings*, *15*, 020105. DOI: 10.1063/5.0154194

Gautam, S., & Kumar, P. (2023b). Blockchain Technology in the Investment Domain: A Bibliometric Analysis. *International Journal of Electronic Finance*, *1*(1), 10061055. Advance online publication. DOI: 10.1504/IJEF.2025.10061055

Gautam, S., & Kumar, P. (2024). A Study of the Evolutionary Trends in Blockchain and Cryptocurrency: A Bibliometric Approach. In *Finance Analytics in Business* (pp. 251–266). Emerald Publishing Limited., DOI: 10.1108/978-1-83753-572-920241013

Gautam, S., Kumar, P., & Dahiya, P. (2024). The Influence of Neurotransmitters on Cryptocurrency Investment Decision-Making : The Mediating Role of Risk Tolerance and Moderating Role of Investment Experience. *Indian Journal of Finance*, *18*(10), 40. DOI: 10.17010/ijf/2024/v18i10/174613

Gautam, S., Kumar, P., Dahiya, P., & Sandhu, S. (2024). Emotional Intelligence and Cryptocurrency Investments: Unraveling the Mediating Role of Risk Tolerance and Moderating Role of Investment Experience. *International Journal of Business Innovation and Research, 1*(1), 10065382. Advance online publication. DOI: 10.1504/IJBIR.2024.10065382

Hassan, V. I., Basheer, S., Mir, F. A., & Abou Fayad, S. G. (2024). Digital Innovation in the Service Sector: Transforming Customer Experiences. In *Service Innovations in Tourism: Metaverse, Immersive Technologies, and Digital Twin* (pp. 150–165). IGI Global Scientific Publishing. DOI: 10.4018/979-8-3693-1103-5.ch008

Kowalkowski, C., Wirtz, J., & Ehret, M. (2024). Digital service innovation in B2B markets. *Journal of Service Management, 35*(2), 280–305. DOI: 10.1108/JOSM-12-2022-0403

Lin, P. M. C., Peng, K.-L., Au, W. C. W., Qiu, H., & Deng, C. D. (2023). Digital menus innovation diffusion and transformation process of consumer behavior. *Journal of Hospitality and Tourism Technology, 14*(5), 732–761. DOI: 10.1108/JHTT-07-2021-0217

Litterscheidt, R., & Streich, D. J. (2020). Financial education and digital asset management: What's in the black box? *Journal of Behavioral and Experimental Economics, 87*, 101573. Advance online publication. DOI: 10.1016/j.socec.2020.101573

Lóska, G., & Uotila, J. (2024). Digital Transformation in Corporate Banking: Toward a Blended Service Model. *California Management Review, 66*(3), 93–115. DOI: 10.1177/00081256231207429

Meyer, F., & Greve, G. (2019). *Customer-salesperson interaction technologies : Are robo-advisors replacing personal selling*. HSBA Hamburg School of Business Administration Cite as.

Muzaffar, R., Muzaffar, S., & Muzaffar, Z. (2024). Digital Culture and the Transformational Impact of E-HRM in the Era of Digital Transformation. In *In New Strategy Models in Digital Entrepreneurship* (pp. 320–329). IGI Global Scientific Publishing., DOI: 10.4018/979-8-3693-3743-1.ch017

Nair, S. K. (2025). Digital Transformation and Its Challenges in the Hospitality Industry in Malaysia: Moving Forward Into Industry Revolution 4.0. In *In Technological Innovations in the Food Service Industry* (pp. 369–388). IGI Global Scientific Publishing. DOI: 10.4018/979-8-3693-7683-6.ch017

Patroni, J. (2018). *Opening Up Innovation: How A Retailer Leverages User - Generated Social Media Content for Innovation. July*, 1–304. file:///C:/Users/Christian/Dropbox/Academia/Thesis/New

Prasad, S., Kiran, R., & Sharma, R. K. (2020). Influence of financial literacy on retail investors' decisions in relation to return, risk and market analysis. *International Journal of Finance & Economics, n/a*(n/a). https://doi.org/https://doi.org/10.1002/ijfe.1920

Radanovic, I., & Likic, R. (2018). Opportunities for Use of Blockchain Technology in Medicine. *Applied Health Economics and Health Policy*, *16*(5), 583–590. DOI: 10.1007/s40258-018-0412-8 PMID: 30022440

Raddats, C., Naik, P., & Ziaee Bigdeli, A. (2022). Creating value in servitization through digital service innovations. *Industrial Marketing Management*, *104*, 1–13. DOI: 10.1016/j.indmarman.2022.04.002

Sadaf, A., Newby, T. J., & Ertmer, P. A. (2012). Exploring Factors that Predict Preservice Teachers' Intentions to Use Web 2.0 Technologies Using Decomposed Theory of Planned Behavior. *Journal of Research on Technology in Education*, *45*(2), 171–196. DOI: 10.1080/15391523.2012.10782602

Saldanha, T. J. V., Kathuria, A., Khuntia, J., & Konsynski, B. R. (2022). Ghosts in the Machine: How Marketing and Human Capital Investments Enhance Customer Growth When Innovative Services Leverage Self-Service Technologies. *Information Systems Research*, *33*(1), 76–109. DOI: 10.1287/isre.2021.1006

Sangwan, S., & Sharma, S. K. (2022). Social media communication and consumer decision making: An empirical perspective. *International Journal of Electronic Marketing and Retailing*, *13*(4), 391–410. DOI: 10.1504/IJEMR.2022.125589

Spais, G. (2023). Engaging Self-service in a Customer Service Ecosystem. In *Artificial Intelligence in Customer Service: The Next Frontier for Personalized Engagement* (pp. 123–154). Springer International Publishing. DOI: 10.1007/978-3-031-33898-4_6

Tsou, H., & Chen, L. (2019). The Influence of Service Innovation Capability for Self-Service Technology Investment. *Canadian Journal of Administrative Sciences / Revue Canadienne Des Sciences de l'Administration, 36*(4), 544–558. DOI: 10.1002/cjas.1520

Vijay, J., & Priyanka, U. (2025). Digital Transformation in Customer Experience and Behavior. In *In Digital Transformation Initiatives for Agile Marketing* (pp. 251–274). IGI Global Scientific Publishing., DOI: 10.4018/979-8-3693-4466-8.ch010

Zhang, J., & Chen, Z. (2024). Exploring Human Resource Management Digital Transformation in the Digital Age. *Journal of the Knowledge Economy*, *15*(1), 1482–1498. DOI: 10.1007/s13132-023-01214-y PMID: 40479358

Zhang, L., Pentina, I., & Fan, Y. (2021). Who do you choose? Comparing perceptions of human vs robo-advisor in the context of financial services. *Journal of Services Marketing*, *35*(5), 634–646. Advance online publication. DOI: 10.1108/JSM-05-2020-0162

Chapter 8
Revolutionizing Self-Service Technology With Machine Learning Intelligence

Anupam Bonkra
https://orcid.org/0000-0001-9175-8041
Maharishi Markandeshwar University, Mullana, India

Rajit Verma
https://orcid.org/0000-0003-0643-4718
Maharishi Markandeshwar University, Mullana, India

Damandeep Singh
VIdya Jyoti Eduversity, India

Divya Goswami
Chitkara Business School, Chitkara University, India

Pummy Dhiman
Chitkara University Institute of Engineering and Technology, Chitkara University, India

Syed Abid Hussain
Department of Computer Science and Engineering, Kart e Char, Kabul,Afghanistan

ABSTRACT

Machine learning (ML) is emerging as a revolutionary force in the evolution of self-service technologies (SSTs), transforming their design, implementation, and delivery across various industries. These self-organizing systems leverage ML to autonomously learn from interactions, process vast volumes of data, and make intelligent decisions without explicit programming. This chapter explores the pivotal role of artificial intelligence (AI) in advancing SSTs, focusing on key applications such as customer service chatbots, automated kiosks, and intelligent virtual assistants. In conclusion, this chapter emphasizes how ML is revolutionizing the SSTs' future development. The use of NLP, predictive analytics and adaptive learning is

DOI: 10.4018/979-8-3373-4667-0.ch008

making the existing ML solutions accurate, efficient and highly personalized as never before. Further improving data credibility, privacy, and ethics issues will enhance the development of these systems to meet our needs and aspirations and continue to strengthen them as the foundations of global business and progress.

1. INTRODUCTION

Self-service systems benefited greatly from Artificial Intelligence (AI), and two of its subfields, Machine Learning (ML) and Deep Learning (DL), are considered groundbreaking innovations. These technologies are shifting customers' expectations from the conventional 'service delivery' model by providing efficient and more intimate ways of service delivery paradigms. Chatbots, voice assistants, self-service kiosks have become top business priorities, not considered as mere innovation focuses but being a part of the customer-oriented strategy (Bonkra et al., 2023). The move towards such technologies is informed by needs of a more technological and customer-centric society. The lexicon and semantic balancing, other AI capabilities such as natural language processing, imaging recognition, big data, and analytics make it easier for the self-service system to mine huge amount data, come with right action quickly and support contextual real time perfectly redesigning the customer relations (Ghosh, 2023). It has been seen that, Self-service system, the golden area is having the interaction with user deploying AI in the form of NLP to make chatbots interpret and answer for the queries made by the clients where the accuracy is almost similar to that of human level (Khatri, 2023). NLP makes customer request understandable because it is used to identify their intention, which means that well-fitting solutions can be given. Image recognition, another important AI application, enables a range of functions like document recognition, facial identification, which became more integrated into digital checkouts and identity morphing systems lately (Dhiman, Kaur, & Bonkra, 2023). The expansion of predictability, called pre-sell technology, is also valuable in improving other self-service systems where the needs and concerns of a user can be predicted by the product. This approach not only boosts the customers' satisfaction level but also helps in building better customer relations and that too ahead of time as the app developers carefully understand the user's behavioural patterns (Bademo & Dylind, 2024). Machine learning is a useful adjunct to self-service systems as it makes systems change and improve over time. ML models are learned from user interactions with the result that systems adapt to users' expectations by improving responses over time (Sharma, Patel, & Gupta, 2022). Such flexible approach is very useful in making sure that such self-service solutions remain equal to the task regardless of changes in user request and trends in the industry advancements that are revolutionizing self-service systems across industries.

These technologies are redefining traditional customer engagement paradigms by introducing innovative, efficient, and personalized methods of service delivery. AI-powered solutions such as chatbots, voice assistants, and self-service kiosks have become essential tools for businesses, no longer regarded as unique selling points but as integral components of a customer-centric approach. The shift towards these technologies reflects the demands of an increasingly digital and customer-focused world. Leveraging technologies like NLP, image recognition, and big data analytics self-service systems powered by AI can interpret large data sets, reason, and decide while synthesising contextually relevant solutions in real time – essentially redefining customer engagements. Of the applications of AI in self-service systems, the most powerful of them is NLP that enables the chatbots to understand and answer the client's questions as thoroughly as a human. NLP makes systems capable of understanding the request made by the customer for which they can offer effective solutions on the same. Another important AI function is image recognition and its application encompasses a wide set of functions, including document confirmation and face recognition that are used in digital checkouts and automatic checkpoints. Pre-sell or pro-active technology is yet another added feature for self-service systems that enables the system to predict user needs and activities thus prevent possible failure modes. This active approach not only raises the level of customer satisfaction but also builds even better customer relationships and more profound insight into the users' tendencies. With machine learning, an additional aspect of self-service regarding adaptability plays an important role because machine learning could update self-service systems continuously. ML models are based on the interactive experience of users, which in turn helps systems to adapt their response continuously and become more customer aligned. This adaptability implies that self-service solutions are applicable and useful as the demands of users and nature of industries transform. For example, chatbots used in the customer support system are meant to be quick-thinking assisting, customer-oriented and can be relied upon. Likewise, self-service kiosks and automated systems that are powered by Machine Learning can also adopt proactive maintenance techniques; reducing on the time machinery is out of service and improving productivity in equal measure. Integrating of AI and ML in self-service systems is not only just limited to the interaction between the consumers and the systems. These technologies also enable the organization to obtain timely information regarding its customers' behaviour and preferences. Through such insights, companies can discover many potential cross-sell, up-sell or addition to service line, which aims at stimulating revenue growth, and, as a result, improving customer satisfaction. For instance, a self-service kiosk in a retail store could categorization of buying trends to suggest related products In a similar vein, a banking chatbots could propose relevant financial services based on the customer's transactions. of course, the adoption and incorporation of A and ML in self-service systems present both opportunities and complexities. One major issue is the origination and reliability of the data fed into such systems to write the dissertations. This means that where data quality is low, or the data is biased in some way, the self-service solutions will perform poorly and, worst of all, may even erode customer trust. Training data must therefore be protected to offer the best defence against threat performers in an effort to build sound AI. A second important question today is the ethical use of artificial intelligence technologies. Live issues such as fairness, traceability, and privacy of users have emerged as more and more embedded AI is used in high-stakes areas. There are ethical milestones that developers have to bear in mind when creating solutions cannot be biases and user privacy issues. This has to be proved, algorithms have to be explained, and privacy-preserving methods have to be applied. Most of the above challenges and goals call for self-service systems that are developed by a crew of AI researchers, data scientists, and software engineers. Multidisciplinary enhances creativity; therefore, while using techniques problem-solving teams involve applicable ethical issues and make sure that the suggested solutions would be appropriate for different groups of users. Further, while self-service solutions powered by AI are essential, their proper ownership should occur, involving corporation with legislation and recognition of contextual expectations. Concerning self-service systems, AI and ML have been implemented in various practical ways and in various sectors. Different examples prove they can enhance working productivity, raise clients' satisfaction levels, and establish novelties in the field. For instance, the customer support systems that imply AI assistance and control have allowed cutting response time by half while maintaining the highest quality and working on individual approach. In health care, AI based self-service kiosk are also used in patient intake, appointment booking, and preliminary triage. In the same way, self-service solutions having incorporated ML algorithms are enhancing the retail stock control and the customers' experiences in-store. In turn, the application of AI and ML in self-service systems still has lot of potential in the future. Possibilities like generative AI and reinforcement learning are expected to improve solutions with self-service in the future because of rising sophistication in personalization and flexibility. For example, long-term generative AI could develop realistic and personalized virtual personal angels while reinforcement learning could unleash self-service systems to learn their best strategies by practice. Nevertheless, breaking the above potential will entail sustained investment in research and development besides a dedicated effort to address the related ethical and technical issues concerning the use of AI solutions. Therefore, the application of AI, ML, and DL is prominent in defining the subsequent course of self-service systems. Of course, there is always the flaw in getting data, ethical issues, and interdisciplinary; however, the use of AI self-service systems has many advantages over obstacles. Thus, with the help of these technologies, an organization comes not only equipped to dealing with the demands of the digital and customer-centred era but also has the ability to raise the bar higher for what constitutes a good or good enough delivery of services. Through the practical applications and case studies discussed in this chapter, it is evident that AI and ML are revolutionizing self-service paradigms, creating new opportunities for businesses, and setting the stage for a future defined by intelligent and adaptive solutions.

1.1 Fundamentals of AI

Artificial Intelligence (AI) refers to the capability of machines, particularly computer systems, to emulate human intelligence processes (Fogel, 2022). These processes involve acquiring knowledge, often represented as data or information, and formulating rules for its application. AI systems are designed to analyze data, draw logical conclusions, and adapt to changing circumstances through self-correction and learning mechanisms. These attributes enable AI to perform tasks that traditionally require human cognitive abilities, such as problem-solving, decision-making, and language understanding. There are two categories by which Artificial Intelligence (AI) is broadly categorised; Narrow AI and General AI (Fig.1). The narrow AI is meant for performing specific task like facial recognition and recommendation system, while general AI is for mimicking human cognitive ability in many different domains.

Figure 1. Classification of Artificial Intelligence (AI) into Narrow AI and General AI.

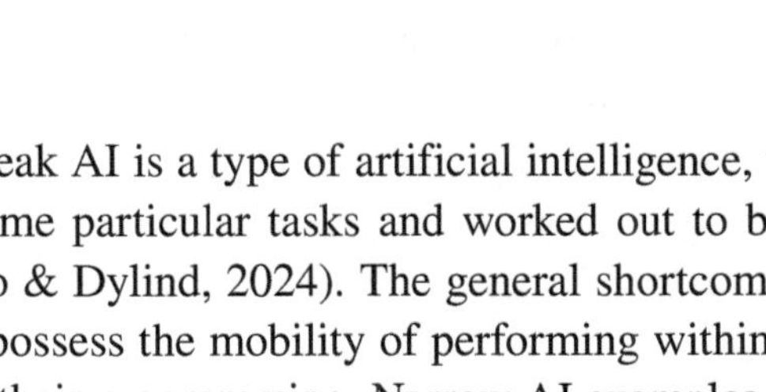

Narrow AI or weak AI is a type of artificial intelligence, which is specifically programmed for some particular tasks and worked out to be nearly optimal for such tasks (Bademo & Dylind, 2024). The general shortcoming of these systems is that they do not possess the mobility of performing within capabilities that are not preordained by their programming. Narrow AI examples consist of facial recognition, Netflix recommendation system, siri/alexa kinds of machines. Despite the fact that Narrow AI is capable of handling a large amount of data, as well as providing highly accurate results within its sphere of specialization, it is incapable of consciousness, general awareness of its own existence, or general intelligence. Whereas General AI or Strong AI or Artificial General Intelligence (AGI) forms the other extreme extrapolated quite far ahead (Rayhan, Rayhan, & Rayhan, 2023). Its goal is to onboard all human-like cognitive skills so as to make machines perform any task in the same manner that a human being would do. It also would not need to be reprogrammed with large blocks of code to function in a new way, as General AI would. If, for instance, AGI acquired human abilities to solve other sort of problems in this world, then it would be able of addressing a wide area of issues in different fields, and that is why one would call it a global intelligence. But getting to General AI continues to be a major problem and its realization is dependent on major strides in machine learning, neural networks, and CS neuroscience. Although Narrow AI poses the primary focus of the applied AI of the modern world, General AI remains a primary research concern as scholars strive for the best possible AI versions.

1.2 Machine Learning

Machine Learning (ML), one of the significant branches of artificial intelligence, allows a computer to find out a response on its own without being programmed (Sarker, 2021). This approach helps systems in the advancement of efficiency on a particular task through experience. ML can be broadly categorized into three types: Classification: Supervised learning where a model is trained with labeled data to make prediction or classification, Clustering: Unsupervised learning in which patterns or structures in an unlabeled data are determined; and Agent Reinforcement: Learning where an agent adapts to an environment to accrue maximum cumulative reward. The categories of FSAs lay the foundation to many applications including predictive analytics and autonomous systems. Generally speaking, ML can be categorized under three main types namely, Supervised Learning, Unsupervised learning, and Reinforcement Learning (Fig. 2). The basic difference between supervised learning and unsupervised learning is that supervised learning uses labeled data for work such as classification and regression, while unsupervised learning discovers patterns in unlabeled data via clustering and association, and reinforcement learning is related to making decisions in an agent environment interaction model.

Figure 2. Categorization of Machine Learning into Supervised, Unsupervised, and Reinforcement Learning.

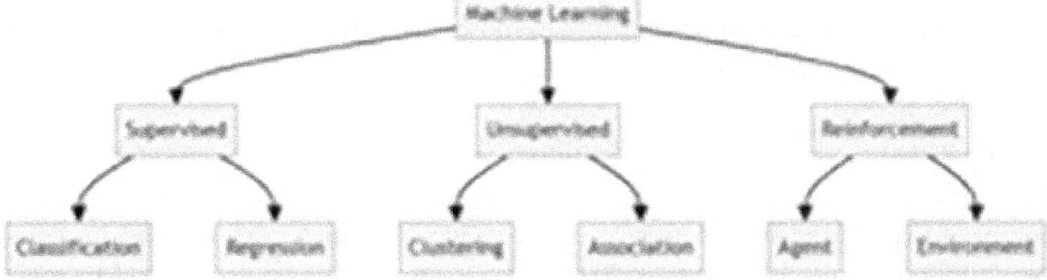

1. Supervised Learning: Supervised learning is one of the type of machine learning where the training data is predefined with inputs-output pairs (Dabiri, Farhangi, Moradi, Zadehmohamad, & Karakouzian, 2022). The model finds the way to recognize the relationship between the inputs to the outputs, thus recognizing the pattern of the data set then it makes prediction or classify new data set. The most widely employed method in supervised learning techniques is linear regression for numeric values while objects categorized as Yes/No or True/False can be determined by the logistic Regression, whereas the support vector machine (SVM) aims to look for optimum plane in the data that distinctly classifies data in classes. These techniques are widely used in such areas as identifying objects on images, spam filterization, and making predictions in the financial systems.
2. Unsupervised Learning: It is the type, where the learning model works with the data that does not contain any preliminary classifications or tags, the intention being to detect structuring patterns in the data. These models are especially useful in EDA, and feature selection to give insight on the underlying structure of the relationships or grouping (Verbeeck, Caprioli, & Van de Plas, 2020). These include k means where data is divided into groups of similar characteristics and the principal component analysis whereby the data variance is reduced through a process that identifies the most crucial attributes from the set. Other methods, such as hierarchical clustering and self-organizing maps, also exemplify unsupervised approaches, enabling tasks like customer segmentation, anomaly detection, and data visualization without requiring labeled datasets.
3. Reinforcement Learning: This model works by performing operations through the interaction of the model with the environment it seeks to accomplish an objective. It follows an approach of positive reinforcement in order to influence behavior in a manner that a specific goal is achieved or the behavior anticipated is avoided. This approach is characteristic of reinforcement learning (Li, Liu,

Yang, Peng, & Zhou, 2021), where an agent acts in the environment and receives the signal in the form of a reward or a penalty. Here, the strategy evolved with the help of a set of chosen actions in order to get the maximum of total reward and within the same time adapting to more difficult and changing situations in further iterations of time.

The structure of this chapter is as follows: Section 2 presents the literature review, while Section 3 discusses the fundamental concepts of AI and its subset, Machine Learning (ML). Section 4 compares self-service systems across various domains. Section 5 highlights case studies of self-service applications powered by ML intelligence. Finally, Section 6 concludes the chapter with a summary.

2. LITERATURE REVIEW

Self-service systems have been significantly reinvented by AI and ML in many fields, which have seemingly infinite possibilities for further development. AI has evolved from vague theories to real-world solutions exercising what people consider as intelligent (Simon, 1995). The change brought about by the growth in computation power, availability of data and new algorithms highlights the importance of using AI and ML to improve self-service technologies. Most of the customer-oriented industries have seen a big shift in user engagement with the help of self-service solutions enabled by AI. Artificial intelligent tools such as chatbots and virtual assistants with NLP and ML capabilities help personalize effective customer service delivery. For example, (Patel & Trivedi, 2020) demonstrated the use of AI models that are trained in the ability to learn user behaviours with an aim of enhancing satisfaction of users and subsequently minimizing response time. Likewise, the generative models of content like identified by (Kalyane et al., 2020) have led to dynamic recommendations for self-service. In manufacturing, self-service maintenance platforms have relied on the use of predictive artificial intelligence systems to forecast failures of the machinery. According to (Wan, Li, Gao, Roy, & He, 2018), it is possible to use these platforms to arrange for timely repairs to ensure that operations do not grind to halt due to equipment breakdowns. As for supply chain optimization, another key factor of self-service systems, demand forecasting based on AI, as well as inventory management, can be credited to (Guo, 2025). Self-services related to healthcare have also not been left behind by the benefits of AI. Diagnostic applications built on top of the ML help patients and doctors to understand symptoms & past medical history for better treatment. They discussed ML relevance in genetic data analysis concerning personal health guidance and early disease identification. These systems make access to care more open and give people a chance to take on healthier lifestyles on their own. Precision agriculture, a domain that is application is related to sustainable environment, is an example of how AI has affected self-service solutions. According to (Liakos, Busato, Moshou, Pearson, & Bochtis, 2018), sensor based AI tools give farmers, information on soil and crop health in real-time. Such self-service tools facilitate efficient resource utilization thus; improving productivity and sustainability. The incorporation of AI and ML into environmental self-service systems also shows their applications as explained expounded how AI enhances the tracking of deforestation, estimation of harsh weather events, as well as the utilization of energy. Such self-service technologies enable users make appropriate decisions in the fight against climate change across the world. Nevertheless, self-service systems operated by AI & ML have some issues hindering even these innovative solutions. Training data is still the most significant factor as the lack of data was identified (Chen, Guo, Gao, & Liang, 2021) as being a significant factor in model accuracy. Some examples include handling algorithms and fairness alongside processing data that is sometimes sensitive to the public. Meuter, Ostrom, Roundtree, and Bitner (2000) Noted that when implementing AI self-service, the equity and trust in the process must be given because they are very important. Everyone recognizes that overcoming these challenges requires a multisector approach. In Dunaway and Berger's (2021) report during the interactive session in 2016 encouraged practitioners in the AI field to incorporate specialty knowledge with AI knowledge so as to develop defendable and suitable systems? This partnership ensures that self-service technologies enhance customer value from new system and product solutions and achieve organisational ethical objectives. On the light of this research, future innovation and constant learning play a vital role on intelligent self-service systems as the ML improve. From the same author, (Bianchini, Müller, & Pelletier, 2022) pointed out that it is crucial to understand the technological progress in an effort to ensure the applicability of AI. If future engineering of ML brings adaptability and interdisciplinary approaches to act as an aid for self-services across industries, it would make them even more efficient than what it is today.

Table 1. Domain Wise Comparison of Self Services

Domain	Tool/Algorithm	Description	Key References
Drug Discovery	AlphaFold	Accelerates self-service drug discovery by predicting protein structures with high accuracy.	(Nussinov, Zhang, Liu, & Jang, 2023)
	Generative Adversarial Networks (GANs)	Supports self-service drug development by generating and optimizing novel molecular structures.	(Hussain et al., 2020)
Materials Science	Matminer	Empowers self-service material design by mining materials data and predicting properties.	(Ward et al., 2018)
	ML Models (e.g., Decision Trees, Random Forests)	Facilitates self-service prediction of material behaviors using historical data.	(Dabiri, Farhangi, Moradi, Zadehmohamad, & Karakouzian, 2022)
Genomics	DeepVariant	Enhances genomic self-services by identifying genetic variants from sequencing data with high accuracy.	(Yun et al., 2020)
	Convolutional Neural Networks (CNNs)	Enables self-service genomic analysis by identifying patterns linked to diseases.	(Li, Liu, Yang, Peng, & Zhou, 2021)
Manufacturing	Predictive Maintenance Systems	Provides self-service solutions to predict equipment failures and optimize maintenance schedules, reducing downtime.	(Khatri, 2023)
	Supply Chain Optimization Algorithms	Improves self-service logistics by forecasting demand and optimizing inventory levels.	(Demirkan & Spohrer, 2014)
Environmental Science	Climate Prediction Models	Enhances self-service disaster preparedness by analyzing climate data to predict weather patterns and natural disasters.	(Madhala, Adavelli, & Rahul, 2022)
	Environmental Monitoring Systems	Supports self-service environmental analysis by monitoring pollution and providing actionable conservation insights.	(Xiang, Cheng, & Wang, 2023)
Agriculture	Precision Agriculture Tools	Optimizes self-service farming by using sensor and drone data to monitor crop health and manage resources efficiently.	(Cyriac & Thomas, 2024)
	Remote Sensing and Imaging	Enables self-service agricultural insights by analyzing satellite and drone images to assess soil and crop performance.	(Kumari, Pandey, & Tiwari, 2024)

3. IMPACT OF SELF SERVICES IN DIFFERENT DOMAINS

Self-service technology has impacted a number of industries by enabling users to do tasks on their own without direct help. These developments have increased productivity, decreased expenses, and improved user experiences in a variety of sectors, including banking, retail, healthcare, and education.

3.1 Customer Support

Artificial Intelligence (AI) and Machine Learning (ML) are revolutionizing customer-centric industries by enhancing efficiency, personalization, and user experience (Basak, Sanyal, Siraji, & Manimaran, 2025). In customer support platforms, smart chatbots and semi-automatic virtual assistants are transforming the way queries are handled, providing instant responses and adapting dynamically to user preferences. These intelligent systems analyze user interactions to offer precise solutions, ensuring user satisfaction and immediate resolution of concerns. In the retail sector, AI and ML drive innovation through recommendation engines and intelligent kiosks, significantly enriching the customer shopping journey by recognizing purchase patterns and preferences (Mohapatra, Mohanty, Mohanty, Mahalik, & Nayak, 2025). Such systems offer smart recommendations of relevant products and optimized shopping, at the same time, automated checkouts with the support of ML algorithms optimize and accelerate the payment process and time consumers spend on checkouts. Notably, in e-commerce, they not only recommend specific products depending on user interactions or provide search queries and their results but also offer specific content or courses on certain platform, which encourages people to spend more time at the platform and to make it more popular (Venice, Arivazhagan, Suman, Shanthi, & Swadhi, 2025). Also, AI-driven predictive maintenance systems are essential in averting failures in self-service contexts like the ATMs, self-check kiosks, and vending machines (Yan, Konar, & Balasubramanian, 2025). These systems describe the health state of the equipments, the tendency to fail or stop functioning and the possibility of averting a failure by carrying out a preventive maintenance. In travel and hospitality, regards to AI and automation, the intelligent self-service such as virtual booking and travel agent, check-in/check-out machines, and trip scheduling has emerged and transformed customer experience. They include ticketing, seat selection, and real-time travel information; which means it eases difficult activities and enhances the client experience. In the same way, AI and ML provide health consumers with more sophisticated self-service healthcare tools such as symptom checkers, medicine intake applications, and fitness devices. These devices try to derive information on the use that is being made of them by the users and then give out tips on how best the user can take care of his or her health.

AI and ML are technologies that are minimising the gap between customers' wants and needs and improving productivity and automation. These technologies are not only reshaping traditional workflows but also enabling businesses to deliver superior value to their customers. For instance, virtual assistants in customer support continuously learn from user interactions to refine their responses and deliver more context-aware assistance. In retail, recommendation engines leverage AI to analyze vast datasets, identifying trends and preferences to offer hyper-personalized product suggestions, while ML algorithms streamline logistics and inventory management (Vashishth, Sharma, Kumar, Chaudhary, & Panwar, 2025). Automated systems in self-service environments improve reliability and reduce downtime, ensuring a seamless user experience even in high-demand scenarios. In travel and hospitality, real-time data processing and intelligent automation enhance operational efficiency and customer convenience, enabling service providers to cater to diverse client needs more effectively. The healthcare sector, leveraging AI and ML, empowers individuals with tools for self-care, enhancing access to vital information and enabling better health outcomes. The convergence of AI and ML across these industries highlights their transformative potential, not only in improving operational efficiency but also in redefining the way customers interact with services. By prioritizing user satisfaction, personalization, and convenience, these technologies continue to shape the future of consumer engagement, driving innovation and setting new standards for service delivery.

3.2 Retail Industry

The use of self-service technologies anchored by AI &ML is slowly revolutionalising the discharge of retail services, leading to a personalised, efficient and customer oriented experience of retailing (Vijay & Priyanka, 2025). Self-service solutions like intelligent kiosks and self-checkouts, smart shelves are built on AI and ML to operate them flawlessly, minimize wait time, and increase customer satisfaction. These technologies allow retail firms to appropriately target consumers based on their purchase history and personal preference, and subsequently suggest products or service that meet the consumer's needs and wants. Solutions based on AI also help to offer customers suggestions for similar products or the same product but in a different modification or from a different brand, as well as to increase cross-sell scenarios and sales, respectively. Furthermore, point of sale technologies which are integrated with automated checkout systems with ML algorithms to help control payment and reduce human interference, errors, and time, thus improving customers' satisfaction. Furthermore, self-service solutions enable retailers to effectively control inventory, and guarantee appropriate stock availability. Smart shelves that include AI and IoT capabilities detect the current shelf status and warn store managers of low stock status, thus avoiding stock out situations and increasing work productivity (Rajendiran, 2025). A branch of the ML process, the predictive analytics provides customer demand schedule and seasonality, making shoppers to set the right inventory and pricing strategies. They enhance the shopping experience while at the same time increasing efficiency by having numerous cost and operational benefits to organizations. Apart from improving the interaction and experience of the physical layout of the retail store, AI and ML in self-service retail apply to online portal and websites as well. AI chatbots, virtual shopping assistants assist customers in their shopping process, providing answers to their questions and other recommendations thus enhancing consumers' communication and purchase decision. Innovative approaches such as the use of visual search solutions through ML include the ability for customers to take a picture of the item desired and get similar products recommended to them at the same time, enhancing the customers' orities of the shopping experience (Sharma et al., 2025). Moreover, self-service technologies supply an integrated Omni-channel interface between bricks and mortar and online channels, where touch points are linked through platforms that identify and respond to customers' unique needs. Retailers also benefit from AI-driven predictive maintenance systems for self-service devices, such as kiosks and vending machines. These systems monitor the health of equipment, identify potential issues before they arise, and ensure uninterrupted operation. This proactive approach reduces downtime, enhances reliability, and maintains a smooth shopping experience. By employing AI and ML in self-service solutions, the retail industry not only improves customer satisfaction and loyalty but also enhances efficiency, reduces operational costs, and ensures a competitive edge in the rapidly evolving marketplace. The impact of self-service technologies is reshaping retail, setting new standards for innovation and customer engagement.

3.3 E- Commerce Platforms

Self-service technologies powered by Artificial Intelligence (AI) and Machine Learning (ML) are revolutionizing e-commerce platforms, providing a highly personalized, efficient, and engaging shopping experience for users (Wijethilak, Singh, & Dhand, 2025). AI-driven recommendation engines play a central role by analysing user behaviour, browsing history, and purchase patterns to suggest products tailored to individual preferences. This personalization not only enhances user satisfaction but also drives higher conversion rates and boosts sales by encouraging cross-selling and upselling opportunities. Natural language processing techniques and machine learning assistants help in introducing chatbots and virtual shopping assistants to serve customers actively in real-time across purchasing concerns and reservations augmenting customer retention rates and dismissing the issues related with cart abandonment (Khneyzer, Rebeiz, & Touma, 2025). Other significant use instances include visual search and image recognition by which a user posts an image of a product and gets similar products from the platform in a few seconds. These tools enhance the shopping experience by making it easier and fun for customers' needs to be met accurately. Regarding the product search and recommendations as well, AI & ML assigns how users might look for the product or will be seeking for and then makes its recommendations accordingly, helping the e-commerce platforms to display the categories, the special offers, related products or any product that the user might be interested in, based on his/her browsing history. In e-commerce, self-service technologies also encompass the procedures of transactions and flow of related affairs. Touchless buying, integrated and encrypted payment options, and advanced anti-fraud measures guarantee safe shopping. In the back end, intelligent inventory solutions are designed to analyze demand patterns, and regulate product inventory, so that consumers can access products on time without the need for excessive stock or insufficient stock situation (Irfan, Verma, Parameswaran, & Sheikh, 2025). Customers are informed of tracking status and delivery details through the use of machine learning algorithms thereby building trust and satisfaction. In addition, AI-based sentiment analysis allows e-commerce companies to comprehend customers' opinions, track the reviews and assess potential issues. It gives businesses the chance to add and adjust services and products before problems start appearing on the market. Marketing promotions and advertisements are delivered to customers more accurately through ML by designing perfect and appropriate campaigns for the targeted individuals.

3.4 Self- Service Environments

Self-service solutions enabled with AI & ML are making customer interactions with technologies convenient and efficient across different industries as well as delivering better user experience, making self-service solutions the future. Such applications as ATMs, vending machines, self-check kiosks and public service terminals incorporate AI & ML technology for efficiency, dependability, and customer engagement (Rajendiran, 2025). In such arrangements, the predictive maintenance systems are valuable assets since they monitor the status of equipment, analyze characteristics and predict when a failure might occur. This prevents equipment breakdown, cuts on repair and maintenance costs and provides continuous customer service improving customer satisfaction. Integration of AI in self-service systems increases user interface interaction and convenience. For instance, kiosks with selected NLP (Dhiman, Kaur, Bonkra, & Chahal, 2024) functions allow interacting through voice or chat services, which simplifies the application's use while applying the latest technologies. In the same way, embedded technologies in machine vision in self-checkout kiosks enable timely identification and detection of items as well as collection of errors resulting in efficient and fast shopping. These environments also contain enhanced ML algorithms that unveil users' behaviours to enable recommendations from change swiftly to serve the user preferences. For instance, smart vending machines analytically display preferred or related products and improve its stock demands and customer feedback. ATM's and Banking Kiosks With the help of Intelligent Programs AI, authentic transaction patterns are scanned in real-time to prevent frauds and secure all banking transactions. Other examples can be discussed in connection with the public transport domain and ticketing: the use of self-service and AI, ML. Ticket vending machines designed with capacity for artificial intelligence incorporates a commuters' best preferred routs, and best ticketing price, as well as real time best available schedule. In airports, self-service check-in kiosks and baggage drop solutions implemented with the AI technology optimise detention of time, optimise procedures, and improve the performance of the airport's operations.

3.5 Travel and Hospitality

AI and ML self-service technologies have applied efficiency, flexibility, and personalization with an array of the travel and hospitality domain (Kannan, 2024). Online ticketing kiosks and self-service check-in and check-out equipment reduce the time consumers spend checking bookings, ordering tickets, choosing seats and modifying bookings. Recommendation systems make use of artificial intelligence to outlay a set of possible trips, places to stay, and things to do base on the user's habits and preferences (Ekechi, Chukwurah, Oyeniyi, & Okeke, 2024). Moreover, predictive analytics suggests ways and means to make changes in price strategies and consider the patterns of demand, so the travellers can get the best prices according to the tendencies of the modern world. Smart kiosks make self-check-in and ticketing easier in airports and other travel hubs; facial recognition lowers security and boarding time, and increases airport productivity (Dias & Silva, 2024). They all help make travelling and moving around smoother, faster, and all round more enjoyable, while also facilitating the running of business. Techniques are therefore not only useful when creating value for the customer but also when creating value for the operative business in the travel and hospitality sector besides helping to contain operating cost and achieve high levels of customer loyalty through effective service delivery. Itinerary synchronisation, mobile registration, and smart service choices allow the travellers to get the power of doing things. Algorithms of these technologies are advancing by the day and this means that the travel and hospitality sectors will offer heightened levels of unrestrictive, customized and seamless services delivery for the consumer.

3.6 Healthcare Sector

Cognitive technologies, mainly AI and ML-enabled self-service technologies have a huge impact on the health-care industry in terms of accessibility, organizational work-flow, and individualization. Another productivity feature of kiosks is used in areas, including registration, appointment, and check-in that keeps patients waiting, thereby freeing up the healthcare staff to attend to various duties. It uses artificial intelligence for diagnosis of patients' ailments, offering first impressions and recommendations that should enable the patient to take the necessary steps. Wearable devices and fitness trackers track key health statistics such as heart rate, physical activities, and sleep, and recommend ways to improve individual health and fitness and keep consumers on course towards their specialized health targets and objectives (George, Shahul, & George, 2023). Medicine intake aids and VGs also extend care by helping patients remember the schedule and manner in which to take the prescribed medicines. All these self-service innovations not only enable those clients who are seeking help to do it on their own but also aid those Clinicians who are delivering health to do it in a more efficient way, enhanced by data and focused on specific individuals. With their potential applications in healthcare, AI, and ML are enhancing the prevention of diseases, controlling costs of care, and enriching patient satisfaction.

3.7 Education and Online Learning

Cognitive technologies, mainly AI and ML-enabled self-service technologies have a huge impact on the health-care industry in terms of accessibility, organizational work-flow, and individualization. Another productivity feature of kiosks is used in areas, including registration, appointment, and check-in that keeps patients waiting, thereby freeing up the healthcare staff to attend to various duties. It uses artificial intelligence for diagnosis of patients' ailments, offering first impressions and recommendations that should enable the patient to take the necessary steps. Wearable devices and fitness trackers track key health statistics such as heart rate, physical activities, and sleep, and recommend ways to improve individual health and fitness and keep consumers on course towards their specialized health targets and objectives (George, Shahul, & George, 2023). Medicine intake aids and VGs also extend care by helping patients remember the schedule and manner in which to take the prescribed medicines. All these self-service innovations not only enable those clients who are seeking help to do it on their own but also aid those Clinicians who are delivering health to do it in a more efficient way, enhanced by data and focused on specific individuals. With their potential applications in healthcare, AI, and ML are enhancing the prevention of diseases, controlling costs of care, and enriching patient satisfaction.

3.8 Banking and Financial Services

AI and self-service technologies based on ML are changing the face of banking and financial services by making banking more secure and facile. With solutions like self-service, ATMs, mobile application banking facilities, and AI-powered chatbots, customers will always transfer funds, check balances or pay their bills anywhere without the need to appear physically at a branch. That is why, the security is crucial for financial services, and usage of artificial intelligence based tools, such as fraud detection systems are helpful in it. These systems monitor and quantify transaction trend in near real-time to determine which patterns look more suspicious then others, to prevent such occurrences from happening again and to safeguard customers. Financial planning and trading applications using machine learning provide a user-customized recommendation in an investment plan, budgeting, and loan options so that the customer can make good financial decisions. Integrated customer service; this is through automation of the loan processing and credit scoring which reduce approval time and hence improve customer experiences. These systems rely on the AI to make accurate and fast credit worthiness thereby enhancing access to credit and minimizing human interference. Moreover, the use of predictive analytics in these platforms enables financial institutions to anticipate the trend in the market, manage risk effectively and come up with a solution to offer their customers proactively. They also increase user interaction by making the experiences individualized. For instance, while using chatbots to address questions, the same can also help consumers navigate over intricate economic procedures. Also, self-service tools developed and invented by banks create awareness of banking services that enable users to learn basic methods of money saving, income generation and its management. These changes offer protected, effective, and customer-centric money encounters together with allowing banking companies to improve procedures and correspond to the web-centric demands of new-age buyers. Through using AI and ML, the banking sector still innovates making the provision of better, faster, and more accurate services in the financial industry.

3.9 Entertainment and Media

Self-service technologies actuated by AI and ML are revolutionizing the entertainment and media industry by providing engaging personalized experiences. Smart recommendations rely on preferences of the user and history of the show or movie watched, listened, or played to recommend movies, shows, music, or games. Video streaming services and content delivery systems are self-service that grants users direct control over the entertainment content they want and at what time they want it. This application allows the creation of customized playlists of bicentennial events; video-tracking of the highlights; or AI recommended news. Moreover, end-to-end event ticketing solutions for events and concerts, theatres, etc. make access easy and enhance user experience. Such innovations relaunch interaction, making entertainment more affordable, fun, and easy to navigate with easy-to-use technologies.

3.10 Smart Cities and Public Services

Cognitive self-service technologies are essential in the construction of smart cities and improving public services in turn improving the quality of living in towns and cities through optimization for the sake of efficiency and the environment. They also used in transportation as it ensures intelligent traffic control with improved traffic flow and alleviation of traffic jams in consideration to real time traffic. Self-service kiosks and mobile applications create convenient ways for citizens to access public service such as paying bills, renewing licenses, and reporting problems, while minimizing paper work. Waste management, supply chain, energy systems and other such infrastructural facilities leverage AI to decide and act in ways that would have minimum negative effect on the environment. Moreover, by utilizing ML for a predictive maintenance approach, the public facilities such as street lights and transit systems, remains functional and maximizes the utilization time, hence increases the reliability of the services. These innovations are making cities more responsive, efficient, and sustainable than they are at the present.

Table 2. Different working Domain of Self Services

Domain	**Type of Self-Service**	**Description**
Customer Support	Chatbots	AI-powered systems that provide instant responses to user queries.
	Virtual Assistants	Assist users by analyzing interactions and delivering personalized support.
Retail	Recommendation Engines	Suggest products based on customer preferences and purchase history.
	Automated Checkout Systems	Use ML to identify products and streamline payment processes.
Healthcare	Symptom Checkers	Provide health insights by analyzing user-input symptoms.
	Medication Reminders	Notify users about prescribed medication schedules

continued on following page

Table 2. Continued

Domain	Type of Self-Service	Description
	Fitness Trackers	Monitor physical activity and health metrics, offering personalized recommendations.
Manufacturing	Predictive Maintenance	AI systems that monitor equipment health and predict potential failures to reduce downtime.
Travel & Hospitality	Virtual Booking Assistants	Help users with booking tickets, selecting seats, and managing itineraries.
	Check-In Kiosks	Streamline the check-in process for hotels, flights, or events.
Agriculture	Precision Farming Tools	Use AI to monitor crop health, optimize irrigation, and control pests.
	Remote Sensing	Analyze satellite and drone images to assess soil and crop conditions.
Environmental Science	Environmental Monitoring Systems	Track pollution levels and other environmental changes to support conservation efforts.
	Climate Prediction Models	Forecast weather patterns and natural disasters for better preparedness.
Education	Personalized Learning Platforms	Recommend tailored educational resources based on user preferences and performance.
Entertainment	Content Recommendation Engines	Suggest movies, music, or shows based on user behaviour and preferences.

4. COMPARISON OF SELF SERVICES WITH DIFFERENT AREAS

The bar graph also presents a detailed insight of the sectors' adoption rates of the self-service technologies using their percentage involvement. The industries are shown on the horizontal axis as Customer Support, Retail, Healthcare, Manufacturing, Travel and Hospitality, Agriculture, Environmental Science, Education, and Entertainment While the vertical axis shows the rate of involvement ranging between 0.25% and 100%. Of all the sectors, the distinct leader at 20% is the Healthcare sector where the use of these technologies like telemedicine, wearable devices, and automated health monitoring techniques for patient, diagnostic, and general care has become critical. Closely following, Manufacturing demonstrates an Adoption score of approximately 18% regarding IoT, automations, and AI in terms of predicative maintenance, optimisation of processes, as well as general smart manufacturing, which in turn improves efficiency and results in cost savings in operating processes. Customer Support and Retail sectors adopt self-service technologies at a rate of 13-14% by implementing and investing in tools such as chatbots or voice-activated assistants, mobile applications, or touchscreens kiosks for the improvement of the customer journey. Agriculture and Environmental Science stand at approximately 15% involvement rates, demonstrating how smart irrigation systems as well as remote sensing devices, drones and climate monitoring tools have been incorporated and embraced for efficient and sustainable cultivation needs. On the other hand, Travel and Hospitality indicate a lower involvement rate, approximately 7%, primarily driven by self-check-in systems, automated booking platforms, and virtual assistants to improve convenience and operational efficiency, leaving room for further technological advancements. Education and Entertainment, both reflecting involvement rates of 8%, are gradually embracing self-service technologies through virtual classrooms, AI-based tutors, streaming platforms, and personalized recommendation systems, which cater to growing demands for digital and on-demand services. The data overall highlights that healthcare and manufacturing lead in the adoption of self-service technologies due to rapid digital transformations and the necessity for efficiency and scalability in these fields. Conversely, sectors like education, entertainment, and travel demonstrate moderate adoption, suggesting ample opportunities for growth and innovation. The continued advancement in technologies such as AI, IoT, and automation is expected to further drive the integration of self-service systems, providing enhanced accessibility, scalability, and efficiency across all industries represented in the figure 3.

Figure 3. Comparison of Self Services Involvement area

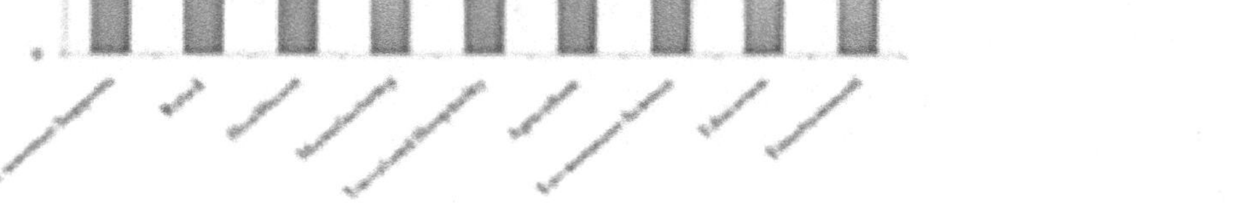

Figure 4. Year Wise Publications of Self Services in AI&ML Domain

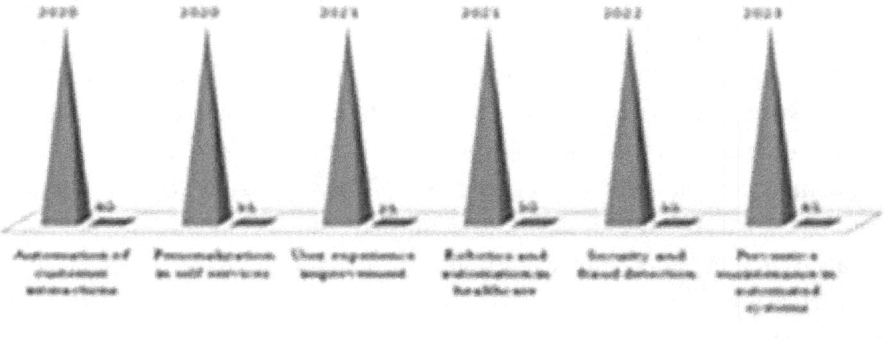

Referring to the identified focus areas and trends in research publications presented on the given diagram 4 it is possible to use the following conclusions about AI and ML applications in self-service systems per different years. It also includes both the year and the percentage of publications focusing on separate domains of self-service technologies that shifts and develops the priorities along with technological progress in the sphere of AI and ML. The first category, automation of Customer Interactions, is represented for the year 2020 with 40% of publications. This signals a large concentration on using AI robots such as the chatbots, virtual assistants, and conversational AI for operations such as customer support and engagement through AI bots. In the same year (2020), another area, Personalization in Self-Services, contributed to 35% of publications which depicts use of AI and ML algorithms to provide personalised suggestions, adaptive interfaces and experiences based on data mined and preference. The area that dominated in 2021 was named User Experience Improvement and it comprised 25% of the entire number of publications. It emphasizes studies conducted for the purpose of refining and interactions with computers, perfecting the interfaces for Self-Service Technologies, and increasing general utility and user satisfaction. Further, the sub-theme Robotics and Automation in Healthcare were the most discussed in 2021, taking half of the shares. This is evidenced by the increasingly use of Artificial Intelligence of robotic systems in surgery, diagnosis, patient and elderly care which is a trend that is more popular in the current world. In 2022 the research focus shifted to Security and Fraud Detection as this was the most publishing area with 30 percent of the publications. This underlines the importance of utilizing AI and ML technologies to protect the systems and personal data from hacking, identity thefts and frauds with the help of predictive analytics, anomaly detection and biometric authentication techniques. This trend shows that when data is being captured electronically, it has to be done with the help of artificial intelligence so that people are assured of its safety and therefore subscribe to automated self-service systems. Finally, in 2023, Preventive Maintenance in Automated Systems emerged as a prominent area, accounting for 45% of the publications. This highlights advancements in predictive maintenance techniques, leveraging AI-powered algorithms and IoT sensors to monitor equipment performance, predict failures, and optimize maintenance schedules, thereby reducing downtime and operational costs in industrial and automated environments. Overall, the diagram demonstrates a clear progression in research trends, starting from customer-focused automation and personalization in 2020 to more complex applications such as robotics in healthcare, security enhancements, and predictive maintenance in later years. This evolution reflects the rapid technological growth and the increasing reliance on AI and ML to create smarter, more efficient, and secure self-service systems across multiple domains.

5. CASE STUDIES OF SELF SERVICES BY USING INTELLIGENCE OF ML

Machine Learning-powered self-service solutions have revolutionized industries through applications like chatbots for customer support and personalized recommendation systems. These case studies highlight how intelligent algorithms optimize user engagement and operational efficiency (Table 3).

Table 3. Case Studies of ML in self-services technology

AI/ML Application	Country	Industry	Technology Used	Key Outcomes
AI Chatbots for Customer Support (Ekechi, Chukwurah, Oyeniyi, & Okeke, 2024)	United States	Retail	AI-powered NLP for automated queries	Faster response time and 24/7 customer support
Virtual Health Assistants (Doak, Schwager, & Hensel, 2020)	Canada	Healthcare	ML-based virtual consultations & diagnostics	Improved triaging and quicker diagnoses
Fraud Detection Systems (Al-Hashedi & Magalingam, 2021)	United Kingdom	Banking & Finance	Anomaly detection using ML	Reduced financial fraud and enhanced security
Personalized Shopping Recommendations (Jannach, Ludewig, & Lerche, 2017)	Germany	E-commerce	AI-driven recommendation engines	Increased sales and customer satisfaction
Smart Warehousing (Ding, Jin, Li, & Feng, 2021)	China	Logistics	AI for inventory tracking & automation	Efficient operations and reduced errors

continued on following page

Table 3. Continued

AI/ML Application	Country	Industry	Technology Used	Key Outcomes
Virtual Interview Platforms (Rana & Kaur, 2024)	India	Human Resources	AI for resume screening & candidate ranking	Faster hiring and unbiased assessments
Predictive Maintenance for Equipment (Pech, Vrchota, & Bednář, 2021)	Japan	Manufacturing	ML for failure prediction & scheduling	Lower downtime and maintenance costs
Smart Farming Solutions (Berger & Goos, 2024)	Netherlands	Agriculture	AI for crop monitoring & irrigation	Improved resource management & higher yields
Autonomous Vehicles for Delivery (Lee et al., 2020)	South Korea	Transportation	AI for navigation & obstacle avoidance	Cost-effective and efficient deliveries
AI-Based Language Translators (Mohamed et al., 2024)	France	Communication	Real-time speech recognition & translation	Improved global accessibility
AI-Powered Virtual Tutors (Nguyen, 2024)	Australia	Education	Adaptive learning platforms	Personalized learning and student engagement
Voice Recognition in Smart Homes (Park & Kim, 2018)	Singapore	Smart Homes	AI voice assistants for home automation	Energy savings and convenience
ML-Based Credit Scoring (Paraná, 2024)	Brazil	Finance	ML-based creditworthiness assessment	Faster loan approvals and risk reduction
Automated Resume Shortlisting (Persson & Wallo, 2024)	Sweden	Recruitment	AI for resume parsing & ranking	Reduced hiring time and increased efficiency
Sentiment Analysis for Customer Feedback (Kauffmann et al., 2019)	Spain	Marketing	AI models for sentiment analysis	Enhanced customer satisfaction & marketing
AI for Predictive Healthcare Diagnosis (Dadon et al., 2024)	Israel	Healthcare	ML for disease detection	Early diagnosis & better treatment plans
Smart Traffic Management Systems (Sahib, 2020)	UAE	Urban Planning	AI for traffic monitoring & signal control	Lower congestion & improved road safety
Virtual Travel Assistants (Filieri et al., 2021)	Italy	Tourism	AI chatbots & itinerary planners	Personalized recommendations & booking
AI-Powered Streaming Recommendations (Ruvoletto, 2023)	Mexico	Entertainment	ML for movie & music recommendations	Higher user engagement & retention
Facial Recognition for Access Control (Chernenkova, 2021)	Russia	Security	AI-based biometric identification	Enhanced security and seamless access

6. ETHICAL AND SOCIETAL CHALLENGES IN AI AND ML

The rapid advancements in artificial intelligence and machine learning raise significant ethical and societal concerns. Bias can arise from the dataset, due to predefined algorithms, and may result in heinous acts as discriminations in important sensitive areas as healthcare, finance, or enforcement services. Data privacy is another critical problem, since AI systems work with huge amounts of data containing personal and sensitive information, and it is necessary to ensure the protection of these data and their compliance with legislation (GDPR). However, there are also limitations, notably with the topic of practicality since deep learning's logic is usually opaque and therefore not easily understandable. These require multi-sectorial approaches which involves the policymakers, the researchers and other industry key-stakeholders to promote responsible and socially desirable implementation of AI.

7. CONCLUSION

Such research technologies as Artificial Intelligence (AI) and Machine Learning (ML) are now used for research and development purposes to transform the modern innovation paradigms such as advanced data analysis, improved modelling, and intelligent decision making. The present chapter offered a deeper look into AI and ML as applied for different industries and how these technologies underpinned the improvement of complex PIs, accuracy, and scalability of solutions. In diagnostics, autonomous systems, industrial automation, environmental sustainability, and more AI has been most useful in solving problems that were hitherto very complex and creating opportunities. These advancements show that the appropriate pre-processing of data, choosing of an appropriate algorithm, as well as having a proper way for evaluating the performance of the AI-driven systems is a sensible aspect which must be thoroughly investigated. Furthermore, the chapter brought out the following emerging issues namely ethical governance, data protection, fairness and transparency important in the management of AI. Mitigating these issues will make a way forward to obtaining recommendations that are more credible and less predisposed, which can eventually gain the society's acceptance and adoption. For future research areas, interpretability, distribution of computation and domain knowledge utilization should be considered to build more sensitive AI systems in future. This makes the agendas of research, industry professional, and policymakers to form safeguards of technology breakthroughs with moral sensitivity indispensable. The fusion of AI and ML into scholarly research paradigms opens doors for further improvement in scientific outcomes and the pace of global sustainable development while creating a relevant context for improving the development of intelligent technologies.

REFERENCES

Al-Hashedi, K. G., & Magalingam, P. (2021). Financial fraud detection applying data mining techniques: A comprehensive review from 2009 to 2019. *Computer Science Review*, *40*, 100402. DOI: 10.1016/j.cosrev.2021.100402

Aljizawi, J. (2024). Personalized Travel Recommendations and Marketing Automation for Saudi Arabia: Harnessing AI for Enhanced User Experience and Business Growth.

Arulkumaran, K., Deisenroth, M. P., Brundage, M., & Bharath, A. A. (2017). Deep reinforcement learning: A brief survey. *IEEE Signal Processing Magazine*, *34*(6), 26–38. DOI: 10.1109/MSP.2017.2743240

Bademo, J., & Dylind, G. (2024). Transforming Enterprise Service Management with Employee Self-Service Technology A Case Study at Mölnlycke Health Care.

Basak, A., Sanyal, M., Siraji, S., & Manimaran, A. (2025). Personalization and Customer Relationship Management (CRM) in AI-Powered Business Intelligence. In *AI-Powered Business Intelligence for Modern Organizations* (pp. 117-158). IGI Global.

Berger, C., & Goos, D. (2024). Farm to Future: Stakeholder Roles and Institutional Barriers in the Integration of Urban Agriculture into Smart Cities in the Netherlands.

Bianchini, S., Müller, M., & Pelletier, P. (2022). Artificial intelligence in science: An emerging general method of invention. *Research Policy*, *51*(10), 104604. DOI: 10.1016/j.respol.2022.104604

Bonkra, A., Bhatt, P. K., Rosak-Szyrocka, J., Muduli, K., Pilař, L., Kaur, A., Chahal, N., & Rana, A. K. (2023). Apple leave disease detection using collaborative ml/dl and artificial intelligence methods: Scientometric analysis. *International Journal of Environmental Research and Public Health*, *20*(4), 3222. DOI: 10.3390/ijerph20043222 PMID: 36833921

Chen, T., Guo, W., Gao, X., & Liang, Z. (2021). AI-based self-service technology in public service delivery: User experience and influencing factors. *Government Information Quarterly*, *38*(4), 101520. DOI: 10.1016/j.giq.2020.101520

Chernenkova, A. (2021). *Facial recognition technology in Russia: do the citizens of Russia accept it?* (Master's thesis, University of Twente).

Cyriac, R., & Thomas, J. (2024). Smart Farming with Cloud Supported Data Management Enabling Real-Time Monitoring and Prediction for Better Yield. In *Intelligent Robots and Drones for Precision Agriculture* (pp. 283–306). Springer Nature Switzerland. DOI: 10.1007/978-3-031-51195-0_14

Dabiri, H., Farhangi, V., Moradi, M. J., Zadehmohamad, M., & Karakouzian, M. (2022). Applications of decision tree and random forest as tree-based machine learning techniques for analyzing the ultimate strain of spliced and non-spliced reinforcement bars. *Applied Sciences (Basel, Switzerland)*, *12*(10), 4851. DOI: 10.3390/app12104851

Dadon, Z., Rav Acha, M., Orlev, A., Carasso, S., Glikson, M., Gottlieb, S., & Alpert, E. A. (2024). Artificial Intelligence-Based Left Ventricular Ejection Fraction by Medical Students for Mortality and Readmission Prediction. *Diagnostics (Basel)*, *14*(7), 767. DOI: 10.3390/diagnostics14070767 PMID: 38611680

Demirkan, H., & Spohrer, J. (2014). Developing a framework to improve virtual shopping in digital malls with intelligent self-service systems. *Journal of Retailing and Consumer Services*, *21*(5), 860–868. DOI: 10.1016/j.jretconser.2014.02.012

Dhiman, P., Kaur, A., & Bonkra, A. (2023, January). Fake information detection using deep learning methods: a survey. In *2023 International conference on artificial intelligence and smart communication (AISC)* (pp. 858-863). IEEE. DOI: 10.1109/AISC56616.2023.10085519

Dhiman, P., Kaur, A., Bonkra, A., & Chahal, N. (2024). Significance of ML and NLP in Sentiment Analysis. In *Fusion of Artificial Intelligence and Machine Learning in Advanced Image Processing* (pp. 69–81). Apple Academic Press. DOI: 10.1201/9781003493570-7

Dias, C., & Silva, J. (2024). Unveiling the future: Smart airports-applications, advantages, strategies and technological challenges. *Journal of Airline and Airport Management*, *14*(2), 38–56. DOI: 10.3926/jairm.419

Ding, Y., Jin, M., Li, S., & Feng, D. (2021). Smart logistics based on the internet of things technology: An overview. *International Journal of Logistics*, *24*(4), 323–345. DOI: 10.1080/13675567.2020.1757053

Doak, S., Schwager, A., & Hensel, J. M. (2020). Exposure to and attitudes regarding electronic healthcare (e-Health) among physician assistants in Canada: A national survey study. *Health & Social Care in the Community*, *28*(3), 1049–1057. DOI: 10.1111/hsc.12938 PMID: 31899587

Dunaway, N., & Berger, K. M. (2021). The Changing Face of Biological Research and the Growing Role of Biosecurity. In *Applied Biosecurity: Global Health, Biodefense, and Developing Technologies* (pp. 89–119). Springer International Publishing. DOI: 10.1007/978-3-030-69464-7_6

Dwivedi, Y. K., Hughes, L., Ismagilova, E., Aarts, G., Coombs, C., Crick, T., Duan, Y., Dwivedi, R., Edwards, J., Eirug, A., Galanos, V., Ilavarasan, P. V., Janssen, M., Jones, P., Kar, A. K., Kizgin, H., Kronemann, B., Lal, B., Lucini, B., & Williams, M. D. (2021). Artificial Intelligence (AI): Multidisciplinary perspectives on emerging challenges, opportunities, and agenda for research, practice and policy. *International Journal of Information Management*, *57*, 101994. DOI: 10.1016/j.ijinfomgt.2019.08.002

Ekechi, C. C., Chukwurah, E. G., Oyeniyi, L. D., & Okeke, C. D. (2024). AI-infused chatbots for customer support: A cross-country evaluation of user satisfaction in the USA and the UK. *International Journal of Management & Entrepreneurship Research*, *6*(4), 1259–1272. DOI: 10.51594/ijmer.v6i4.1057

Filieri, R., D'Amico, E., Destefanis, A., Paolucci, E., & Raguseo, E. (2021). Artificial intelligence (AI) for tourism: An European-based study on successful AI tourism start-ups. *International Journal of Contemporary Hospitality Management*, *33*(11), 4099–4125. DOI: 10.1108/IJCHM-02-2021-0220

Fogel, D. B. (2022). Defining artificial intelligence. *Machine Learning and the City: Applications in Architecture and Urban Design*, 91-120.

George, A. H., Shahul, A., & George, A. S. (2023). Wearable sensors: A new way to track health and wellness. *Partners Universal International Innovation Journal*, *1*(4), 15–34.

Ghosh, P. (2023). Automated Customer Service for Retail with AI. *Journal of Bioinformatics and Artificial Intelligence*, *3*(2), 231–243.

Guo, Y. (2025). Successful Stories on the Usage of Technologies in Service Industries. In *Technological Innovations in the Food Service Industry* (pp. 197–230). IGI Global Scientific Publishing.

Hussain, S., Anees, A., Das, A., Nguyen, B. P., Marzuki, M., Lin, S., Wright, G., & Singhal, A. (2020). High-content image generation for drug discovery using generative adversarial networks. *Neural Networks*, *132*, 353–363. DOI: 10.1016/j.neunet.2020.09.007 PMID: 32977280

Irfan, M., Verma, J., Parameswaran, S., & Sheikh, I. A. (2025). Integrating Emerging Technologies: Enhancing Supply Chain Optimization Through AI, IoT, and Blockchain. In *Enhancing Social Sustainability in Manufacturing Supply Chains* (pp. 199-220). IGI Global Scientific Publishing.

Jannach, D., Ludewig, M., & Lerche, L. (2017). Session-based item recommendation in e-commerce: On short-term intents, reminders, trends and discounts. *User Modeling and User-Adapted Interaction*, *27*(3-5), 351–392. DOI: 10.1007/s11257-017-9194-1

Kalyane, D., Sanap, G., Paul, D., Shenoy, S., Anup, N., Polaka, S., & Tekade, R. K. (2020). Artificial intelligence in the pharmaceutical sector: current scene and future prospect. In *The future of pharmaceutical product development and research* (pp. 73–107). Academic Press. DOI: 10.1016/B978-0-12-814455-8.00003-7

Kannan, R. (2024). Revolutionizing the Tourism Industry through Artificial Intelligence: A Comprehensive Review of AI Integration, Impact on Customer Experience, Operational Efficiency, and Future Trends. *International Journal for Multidimensional Research Perspectives,* 2(2), 01-14.

Kauffmann, E., Peral, J., Gil, D., Ferrández, A., Sellers, R., & Mora, H. (2019). Managing marketing decision-making with sentiment analysis: An evaluation of the main product features using text data mining. *Sustainability (Basel)*, *11*(15), 4235. DOI: 10.3390/su11154235

Khatri, M. R. (2023). Integration of natural language processing, self-service platforms, predictive maintenance, and prescriptive analytics for cost reduction, personalization, and real-time insights customer service and operational efficiency. *International Journal of Information and Cybersecurity*, 7(9), 1–30.

Khatri, M. R. (2023). Integration of natural language processing, self-service platforms, predictive maintenance, and prescriptive analytics for cost reduction, personalization, and real-time insights customer service and operational efficiency. *International Journal of Information and Cybersecurity*, 7(9), 1–30.

. Khneyzer, C., Rebeiz, K. S., & Touma, J. (2025). A Comprehensive Review of Economic and Managerial Factors: The Implications of AI Chatbots in Customer Relationship Management. *Navigating Business Through Essential Sustainable Strategies*, 217-238.

Kumari, S., Pandey, G. K., & Tiwari, S. (2024). Data Visualization Techniques in Smart Agriculture Implementation. In *AI Applications for Business, Medical, and Agricultural Sustainability* (pp. 122-159). IGI Global. DOI: 10.4018/979-8-3693-5266-3.ch006

Le Clair, C. (2025). *Random Acts of Automation: How to Fight Back When Automation Threatens Your Work, Your Life, and Everything You Do*. Post Hill Press.

Lee, S., Kim, Y., Kahng, H., Lee, S. K., Chung, S., Cheong, T., Shin, K., Park, J., & Kim, S. B. (2020). Intelligent traffic control for autonomous vehicle systems based on machine learning. *Expert Systems with Applications, 144*, 113074. DOI: 10.1016/j.eswa.2019.113074

Li, Z., Liu, F., Yang, W., Peng, S., & Zhou, J. (2021). A survey of convolutional neural networks: Analysis, applications, and prospects. *IEEE Transactions on Neural Networks and Learning Systems, 33*(12), 6999–7019. DOI: 10.1109/TNNLS.2021.3084827 PMID: 34111009

Liakos, K. G., Busato, P., Moshou, D., Pearson, S., & Bochtis, D. (2018). Machine learning in agriculture: A review. *Sensors (Basel), 18*(8), 2674. DOI: 10.3390/s18082674 PMID: 30110960

Madhala, R. T., Adavelli, S. R., & Rahul, N. (2022). Enhancing Catastrophe Modeling with Big Data and IoT: Revolutionizing Disaster Risk Management and Response. *Australian Journal of Machine Learning Research & Applications, 2*(1), 612–636.

Meuter, M. L., Ostrom, A. L., Roundtree, R. I., & Bitner, M. J. (2000). Self-service technologies: Understanding customer satisfaction with technology-based service encounters. *Journal of Marketing, 64*(3), 50–64. DOI: 10.1509/jmkg.64.3.50.18024

Mohamed, Y. A., Khanan, A., Bashir, M., Mohamed, A. H. H., Adiel, M. A., & Elsadig, M. A. (2024). The impact of artificial intelligence on language translation: A review. *IEEE Access : Practical Innovations, Open Solutions, 12*, 25553–25579. DOI: 10.1109/ACCESS.2024.3366802

Mohapatra, A. G., Mohanty, A., Mohanty, S. K., Mahalik, N. P., & Nayak, S. (2025). Personalization and Customer Experience in the Era of Data-Driven Marketing. *Artificial Intelligence-Enabled Businesses: How to Develop Strategies for Innovation*, 467-511.

Nguyen, T. H. C. (2024). Exploring the Role of Artificial Intelligence-Powered Facilitator in Enhancing Digital Competencies of Primary School Teachers. *European Journal of Educational Research, 13*(1), 219–231. DOI: 10.12973/eu-jer.13.1.219

Nussinov, R., Zhang, M., Liu, Y., & Jang, H. (2023). AlphaFold, allosteric, and orthosteric drug discovery: Ways forward. *Drug Discovery Today, 28*(6), 103551. DOI: 10.1016/j.drudis.2023.103551 PMID: 36907321

Olteanu, A., Castillo, C., Diaz, F., & Kıcıman, E. (2019). Social data: Biases, methodological pitfalls, and ethical boundaries. *Frontiers in Big Data*, *2*, 13. DOI: 10.3389/fdata.2019.00013 PMID: 33693336

Osisanwo, F. Y., Akinsola, J. E. T., Awodele, O., Hinmikaiye, J. O., Olakanmi, O., & Akinjobi, J. (2017). Supervised machine learning algorithms: Classification and comparison. [IJCTT]. *International Journal of Computer Trends and Technology*, *48*(3), 128–138. DOI: 10.14445/22312803/IJCTT-V48P126

Paraná, E. (2024). Artificial intelligence and the digitalization of finance in Latin America: Evidence from Brazil. *Globalizations*, •••, 1–20. DOI: 10.1080/14747731.2024.2415257

Park, G., & Kim, H. (2018). Low-cost implementation of a named entity recognition system for voice-activated human-appliance interfaces in a smart home. *Sustainability (Basel)*, *10*(2), 488. DOI: 10.3390/su10020488

Patel, N., & Trivedi, S. (2020). Leveraging predictive modeling, machine learning personalization, NLP customer support, and AI chatbots to increase customer loyalty. *Empirical Quests for Management Essences*, *3*(3), 1–24.

Pech, M., Vrchota, J., & Bednář, J. (2021). Predictive maintenance and intelligent sensors in smart factory. *Sensors (Basel)*, *21*(4), 1470. DOI: 10.3390/s21041470 PMID: 33672479

Persson, M., & Wallo, A. (2024). Digital automation and working life of HR practitioners: A gender analysis of the implications for workforce and work practices. *Gender, Technology and Development*, *28*(3), 408–427. DOI: 10.1080/09718524.2024.2406039

Rajendiran, G. R. (2025). Distributed Ledger Technology for Transparent and Secure Warehouse Inventory Tracking. *Journal of Current Science and Research Review, 3*(01), 01-10.

Rana, S., & Kaur, D. (2024). Exploring the challenges and facilitators in the adoption of e-HRM practices in Indian higher education institutions: A qualitative exploration. *The International Journal of Organizational Analysis*, *32*(8), 1419–1439. DOI: 10.1108/IJOA-05-2023-3745

Rayhan, A., Rayhan, R., & Rayhan, S. (2023). Artificial general intelligence: Roadmap to achieving human-level capabilities.

Ruvoletto, R. (2023). Digitalization and Internationalization: An Analysis of the Impact of Digital Technologies on Export Management Practices.

Sahib, U. (2020). Smart Dubai: sensing Dubai smart city for smart environment management. *Smart environment for smart cities*, 437-489.

Sarker, I. H. (2021). Deep learning: A comprehensive overview on techniques, taxonomy, applications and research directions. *SN Computer Science*, *2*(6), 420. DOI: 10.1007/s42979-021-00815-1 PMID: 34426802

Sharma, A., Gangwar, V. P., Bhattacharyya, A., Giri, U., Chavan, G., & Mahato, S. (2025). Leveraging Intelligent Process Automation for Customer Engagement in Sales and Marketing. In *Advancements in Intelligent Process Automation* (pp. 435–458). IGI Global.

Sharma, A., Patel, N., & Gupta, R. (2022). Enhancing Revenue Through AI: Utilizing Neural Networks and Reinforcement Learning for Upsell and Cross-Sell Strategies. *European Advanced AI Journal, 11*(8).

Simon, H. A. (1995). Artificial intelligence: An empirical science. *Artificial Intelligence*, *77*(1), 95–127. DOI: 10.1016/0004-3702(95)00039-H

Vashishth, T. K., Sharma, K. K., Kumar, B., Chaudhary, S., & Panwar, R. (2025). Enhancing Customer Experience through AI-Enabled Content Personalization in E-Commerce Marketing. *Advances in Digital Marketing in the Era of Artificial Intelligence*, 7-32.

Venice, J. A., Arivazhagan, D., Suman, N., Shanthi, H. J., & Swadhi, R. (2025). Recommendation Systems and Content Personalization: Algorithms, Applications, and Adaptive Learning. In *AI for Large Scale Communication Networks* (pp. 323–348). IGI Global.

Verbeeck, N., Caprioli, R. M., & Van de Plas, R. (2020). Unsupervised machine learning for exploratory data analysis in imaging mass spectrometry. *Mass Spectrometry Reviews*, *39*(3), 245–291. DOI: 10.1002/mas.21602 PMID: 31602691

Vijay, J., & Priyanka, U. (2025). Digital Transformation in Customer Experience and Behavior. In *Digital Transformation Initiatives for Agile Marketing* (pp. 251–274). IGI Global.

Wan, S., Li, D., Gao, J., Roy, R., & He, F. (2018). A collaborative machine tool maintenance planning system based on content management technologies. *International Journal of Advanced Manufacturing Technology*, *94*(5-8), 1639–1653. DOI: 10.1007/s00170-016-9829-0

Ward, L., Dunn, A., Faghaninia, A., Zimmermann, N. E., Bajaj, S., Wang, Q., Montoya, J., Chen, J., Bystrom, K., Dylla, M., Chard, K., Asta, M., Persson, K. A., Snyder, G. J., Foster, I., & Jain, A. (2018). Matminer: An open source toolkit for materials data mining. *Computational Materials Science*, *152*, 60–69. DOI: 10.1016/j.commatsci.2018.05.018

Wijethilak, H. P., Singh, S. K., & Dhand, S. (2025). AI for Personalizing Customer Interactions: Improving Engagement and Satisfaction. In *AI, Corporate Social Responsibility, and Marketing in Modern Organizations* (pp. 221-244). IGI Global Scientific Publishing.

Xiang, F., Cheng, H., & Wang, Y. (2023). Exploring the Smart Street Management and Control Platform from the Perspective of Sustainability: A Study of Five Typical Chinese Cities. *Sustainability (Basel)*, *15*(4), 3438. DOI: 10.3390/su15043438

Yan, Z., Konar, R., & Balasubramanian, K. (2025). The Future of Hotels Robotics, AI, and Service Automation in Practice. In *Sustainability, Innovation, and Consumer Preference* (pp. 249-278). IGI Global Scientific Publishing.

Yun, T., Li, H., Chang, P. C., Lin, M. F., Carroll, A., & McLean, C. Y. (2020). Accurate, scalable cohort variant calls using DeepVariant and GLnexus. *Bioinformatics (Oxford, England)*, *36*(24), 5582–5589. DOI: 10.1093/bioinformatics/btaa1081 PMID: 33399819

Chapter 9
ATMs to AI:
The Evolution of Self-Service Technology in Retail and Beyond

G. Manoj
https://orcid.org/0000-0003-2830-7875
VelTech Rangarajan Dr. Sagunthala R&D Institute of Science and Technology, India

K. Ramya
https://orcid.org/0009-0005-8744-6666
Arul Anandar College, Karumathur, India

B. Latha Lavanya
https://orcid.org/0000-0002-4185-9618
Measi Institute of Management, India

G. M. Shaju
MIANZ International College, Maldives

R. Margaret Selvi
Loyola College, Chennai, India

Chandramowleeswaran Gnanasekaran
https://orcid.org/0000-0002-1293-1043
Vel Tech Rangarajan Dr. Sagunthala R&D Institute of Science and Technology, India

ABSTRACT

Self-service technology (SST) has undergone a remarkable evolution, transforming from simple automated teller machines (ATMs) to advanced AI-driven systems that redefine customer experiences across industries. This chapter chronicles the historical development of SST, tracing its journey from early automation to the integration of cutting-edge technologies like machine learning, natural language processing, and robotics. It examines how self-service solutions have adapted to changing consumer expectations, enabling greater convenience, personalization, and efficiency in sectors such as retail, banking, healthcare, and hospitality. The chapter highlights key milestones in the evolution of SST, including the rise of digital kiosks, mobile apps, and AI-powered chatbots. It also explores the dualities of this transformation: improved accessibility versus the digital divide, increased

DOI: 10.4018/979-8-3373-4667-0.ch009

convenience versus data privacy concerns.

INTRODUCTION

The evolution of self-service technology (SST) has dramatically transformed consumer interactions and service delivery across various sectors, revolutionizing how individuals access and engage with products and services. Beginning with basic automated teller machines (ATMs) designed to perform simple financial transactions, SST has grown into a multi-faceted ecosystem that now incorporates advanced technologies such as artificial intelligence (AI), machine learning, natural language processing, and robotics. These innovations have enabled businesses to provide increasingly personalized, efficient, and seamless experiences for consumers, who now expect more autonomy and convenience when interacting with brands.

From digital kiosks in retail spaces to mobile apps offering on-the-go services, and AI-driven chatbots capable of handling complex queries, SST has continually adapted to changing consumer expectations. In addition to improving accessibility and customer satisfaction, these advancements have expanded the role of self-service across multiple industries, ranging from healthcare, where kiosks assist patients in navigating administrative processes, to hospitality, where AI-powered systems facilitate seamless booking and check-in experiences. However, the integration of these technologies also raises critical questions, particularly regarding the digital divide, data privacy, and the potential displacement of workers as automated systems become more prevalent.

The increasing reliance on AI in self-service systems also introduces new dimensions to customer engagement, such as predictive capabilities, proactive assistance, and enhanced personalization. While businesses reap the benefits of streamlined operations and cost savings, consumers are offered more tailored interactions that cater to their unique preferences and needs. Despite these advancements, the challenges of ensuring data security, equitable access, and ethical AI use remain at the forefront of the conversation. This chapter aims to examine these developments in detail, tracing the trajectory of self-service technology, highlighting the milestones that have shaped its evolution, and considering the implications for both consumers and businesses in the rapidly evolving digital landscape.

BACKGROUND OF THE STUDY

Those who use SSTs said convenience, faster service and more control over the process are important reasons for choosing them. It breaks SSTs into types and determines what helps make customers happy. According to the paper, organizations should tailor their SST services to what customers expect, to keep them satisfied and loyal (Bitner, Ostrom, & Meuter, 2002). Conditions affecting how pleased customers are with services involving technology. It lists perceived usefulness, simplicity in using the system and dealing with system failures as important factors in predicting usage. Based on both types of data, the authors report that while SSTs help organizations, a lack of support and the feeling of isolation can disappoint customers, requiring proper design and support (Meuter et al., 2000).

People using three types of self-service technology: online banking, self-service grocery checkout and airline kiosks. The TAM is used to examine how much people appreciate the simplicity and value of a certain technology. According to the findings, how services are adopted seems to depend on the type of technology and the experience of users which means companies have to adapt their strategies for each case (Curran & Meuter, 2005). Checking the quality of service in SSTs. It is shown that what customers rate highly are if services are dependable, user-friendly, personalized and fun. According to the paper, assessing customers' judgments of the service process and results, as well as quality, forms a solid basis for judging technology-mediated services (Dabholkar, 1996).

This paper investigates how technological services in banking are linked to customers' views of the service and their satisfaction. It considers reliability, responsiveness and assurance to be very important. The report indicates that strong IT integration improves customer views and loyalty, confirming that technology plays a key part in better serving and competing in the industry (Zhu, Wymer, & Chen, 2002). Not all customers are entirely happy using SSTs, but some find they have little or no better choice. The research shows that while self-service speeds up things, customers can become upset when it fails. People are happier with a business when they think there is choice and support. The research suggests that firms should try to get the most of their services while keeping costs down (Buell, Campbell, & Frei, 2010).

The study investigates what encourages SST in the retail sector and the outcomes that result. It shows that innovativeness and social pressure in customers are important predictors and satisfaction and loyalty follow as the outcomes. The paper highlights that having control over the process and finding it easy to use can give a positive experience, leading to better relationships between shoppers and merchants (Weijters et al., 2007). SST adoption and evaluation are influenced by how ready technology is. It divides consumers by how prepared they are and the study reveals that readiness has an impact on how satisfied, trusting and likely someone is to adopt SSTs. This shows retailers ought to focus their SSTs and marketing on how comfortable consumers are with technology (Liljander et al., 2006).

Customers' satisfaction and how they act depend on how ready SST is for technology. By using surveys, they discover that individuals with high levels of innovativeness and optimism are more satisfied and loyal. It encourages buyers to sort customers by how ready they are to work on SST and how deeply involved they are with the application (Lin & Hsieh, 2007). Co-designers say convenience, price and how enjoyable an experience is influence how we view its value. The paper explains that understanding these elements allows companies to design SSTs that fit what customers want and improves the benefits they get which leads to more satisfaction and use (Chen & Dubinsky, 2003).

The paper analyzes why certain people would rather use humans than online customer service tools. Results suggest that people's cognition, growing older and conditions of service impact their choices. If consumers feel there is more uncertainty around them and are older, they prefer personal contact. Researchers explain that companies should provide both spiritual experiences and activities when serving a range of consumers (Simon & Usunier, 2007).

The level of satisfaction is affected by customers' experiences using the online, offline and SST service channels. It stresses that having uniformity and integration through all channels works best. SSTs are best when they build on the company's larger service approach, otherwise customer satisfaction might fall. Research helps guide the creation of an omnichannel experience that gives better outcomes for customers (Collier & Kimes, 2013).

The design and use of technology in stores support more effective interactions among consumers, store staff and retailers. It argues that interactive technology inside stores can combine user-friendly technology with the help of staff. The research shows that correctly adding SSTs to physical stores can make customers happier and more involved (Verhoef, Broekhuizen, & Bijmolt, 2016).

A means to support self-service with data analysis and real-time personal options for customers. It puts a strong focus on letting customers take part in making value with the use of smart systems. The study offers design tips to boost how easy SST is to use, how contented customers are and how well SST works in digital environments (Pantano & Migliarese, 2014). SSTs are used by organizations to bring down costs without compromising service standards. It proves that a successful SST system needs client trust, proper training and good integration. It advises firms to think carefully about SSTs rather than just using them for cost-cutting and to make sure their customers have a good experience (Demirkan & Spohrer, 2014).

Preference of current systems versus personal interaction. Researchers found that feeling in control and being motivated inside help determine someone's involvement. Customers looking for more control chose self-service technology. The study recommends retailers make SSTs that put the user in charge and emphasize their happiness (Hilton et al., 2013). Self-service technologies (SSTs) are used in many sectors, helping with airline checkouts, ATM banking, online shopping, hotel checkouts and grocery self-scanning which all increase efficiency, reduce costs and benefit the consumer (Yang et al., 2012).

Here, we focus on SST in the travel, tourism and hospitality industries, discussing their positives and negatives, how customers take part in co-creation and how using them effectively can boost both customer experience and cost-saving (Kelly et al., 2019). Being widely used, self-service technologies (SSTs) have improved how various services are delivered and accepted by many customers, boosting efficiency and making people more satisfied in the retail, banking and hospitality fields (Lee et al., 2012).

SELF SERVICE INNOVATION: TRACING THE EVOLUTION FROM ATMS AI IN RETAIL.

The evolution of self-service technology (SST) in the retail sector has significantly transformed how businesses interact with customers, enhancing both the customer experience and operational efficiency. From the early days of automated teller machines (ATMs), which provided customers with the ability to perform basic banking functions without human interaction, to the advanced AI-powered systems used today, self-service technology has consistently evolved to meet the changing demands of the retail industry. ATMs revolutionized the banking industry by enabling customers to withdraw cash, check account balances, and perform various transactions outside of business hours, all without the need for a bank teller. This marked the beginning of self-service technology, which soon found applications beyond banking and into the retail sector.

In the years that followed, retail businesses began adopting self-checkout kiosks, digital kiosks, and vending machines to enable customers to make purchases, browse products, and access services without assistance from a store employee. These early forms of SST provided customers with greater autonomy and helped businesses reduce labor costs, but they also introduced new challenges, such as technical issues and customer concerns about security. Despite these hurdles, self-service technologies continued to gain traction as customers became more accustomed to the convenience and speed they offered. Tandon et al. (2021) emphasize that the integration of e-commerce mechanics and product innovation enhances consumer trust and repurchase intention, aligning closely with SST advancements.

The real game-changer in self-service technology came with the introduction of artificial intelligence (AI). AI-powered systems allowed for an even higher level of personalization and automation. Machine learning, natural language processing, and advanced algorithms have made it possible for businesses to provide tailored recommendations, optimize inventory management, and automate customer interactions through chatbots and virtual assistants. These AI-driven tools have not only streamlined customer service processes but have also enhanced customer satisfaction by offering personalized shopping experiences. Customers can now engage with brands through chatbots for instant responses, receive product suggestions based on their browsing history, and access services more efficiently.

The integration of AI in self-service technology has also revolutionized the way retailers approach customer engagement. Through AI, businesses can analyze vast amounts of customer data to predict behaviors, preferences, and purchasing patterns. This data-driven approach enables businesses to optimize marketing campaigns, improve product recommendations, and deliver personalized experiences in real time. Furthermore, AI has made it easier for businesses to offer 24/7 customer service, providing customers with immediate assistance at any time, which has become an increasingly important feature in today's fast-paced retail environment.

However, the widespread adoption of self-service technology, particularly AI-driven solutions, comes with its own set of challenges. One of the primary concerns is the potential for job displacement as more tasks are automated. While AI and automation can increase efficiency and reduce operational costs, they may also reduce the need for human employees in certain areas, leading to workforce disruption. Additionally, issues related to data privacy and security have become more pronounced as AI systems collect and process vast amounts of personal information. Consumers may feel uneasy about how their data is being used, making it essential for businesses to be transparent and ensure their systems are secure. Mehta (2018) highlights that corporate reputation acts as a critical mediator in ensuring that service innovation aligns with customer satisfaction and cross-buying intentions.

Despite these challenges, the evolution of self-service technology, from ATMs to AI-powered systems, has undeniably transformed the retail landscape. It has made shopping more convenient, personalized, and efficient for customers, while also enabling businesses to optimize their operations and enhance customer satisfaction. As AI continues to advance, it is expected that self-service technologies will become even more integrated into retail environments, further shaping the future of customer experiences in the industry. Mehta (2018) highlights that corporate reputation acts as a critical mediator in ensuring that service innovation aligns with customer satisfaction and cross-buying intentions.

Research Question:

How has the evolution of self-service technology (SST) from ATMs to AI-driven systems impacted customer experiences, business operations, and the broader implications of convenience, accessibility, and ethics?

Research Objectives:

1. To explore the historical development of self-service technologies, tracing the shift from ATMs to AI-powered systems in various sectors.
2. To examine the influence of AI-driven systems in enhancing personalization, convenience, and efficiency in industries such as retail, banking, healthcare, and hospitality.
3. To assess the ethical considerations and challenges related to the implementation of AI in self-service technology, including data privacy, accessibility, and the digital divide.
4. To evaluate the future trajectory of self-service technologies, identifying trends and challenges that will continue to shape customer experiences and business strategies in the coming years.

METHODOLOGY

This study adopts a qualitative approach, relying on secondary data from peer-reviewed journal articles, book chapters, and conference proceedings to explore the evolution of self-service technologies (SST). A comprehensive literature review was conducted to trace the transition from traditional automation, such as ATMs, to AI-driven systems across industries like retail, banking, and healthcare. The study analyzes technological advancements, customer engagement strategies, and ethical considerations associated with SST implementation. By synthesizing insights from existing research, the methodology identifies key trends, challenges, and opportunities shaping the future of SST and its role in enhancing operational efficiency and customer experiences.

THE EVOLUTION OF SERVICE THECHNOLOGIES: FROM ATMS AI POWERED SYSTEMS IN KEY SECTORS

Self-service technology (SST) has been a critical driver of innovation across various industries, reshaping the way businesses and customers interact. From the early days of Automated Teller Machines (ATMs) to today's advanced AI-powered systems, self-service solutions have revolutionized customer experiences by providing greater convenience, personalization, and efficiency. This evolution reflects not only technological advancements but also shifting consumer expectations, demanding quicker, more personalized, and accessible services across different sectors.

The Beginning: ATMs and Early Automation

The journey of self-service technology began with the introduction of ATMs in the late 1960s. ATMs revolutionized the banking sector by allowing customers to perform basic banking transactions—such as withdrawing cash, checking balances, and transferring money—without needing to visit a bank branch or interact with a teller. This early automation simplified banking processes, making them faster and more convenient for customers. Over time, ATMs became more sophisticated, with capabilities extending to bill payments, deposits, and account management, marking a significant milestone in the adoption of self-service technology.

The Rise of Digital Kiosks and Mobile Solutions

In the 1990s and early 2000s, the landscape of self-service technology began to broaden beyond ATMs. Digital kiosks emerged in retail, hospitality, and airport industries, offering customers the ability to check in, order food, or browse product catalogs independently. These kiosks enhanced the customer experience by reducing wait times and providing instant access to information. At the same time, mobile apps became increasingly prevalent, allowing customers to access self-service options from their smartphones. Apps for banking, shopping, and entertainment delivered the convenience of managing tasks on the go, further advancing the self-service revolution. Ganesan and Sridhar (2016) argue that service innovation significantly improves customer performance and satisfaction by enhancing corporate reputation

AI-Powered Systems: Personalization and Efficiency

The most recent phase in the evolution of self-service technology is the incorporation of artificial intelligence (AI). Today, AI-powered systems are enhancing self-service offerings in ways that were previously unimaginable. AI technologies, such as machine learning and natural language processing, are being used to personalize customer experiences and improve efficiency in real-time. Chatbots, virtual assistants, and automated customer support systems are now commonplace, providing immediate and personalized assistance to customers 24/7. These AI-powered tools analyze customer data to offer tailored recommendations, solutions, and responses, ensuring that each interaction is relevant and meaningful. Manohar et al. (2023) developed a generalized scale, INNOSERV, to measure perceived service innovation, which highlights the increasing importance of innovation in creating value across industries.

For instance, in the retail sector, AI is used to recommend products based on past purchase behavior and browsing patterns, creating a highly personalized shopping experience. In the banking industry, AI-driven chatbots assist customers with account inquiries, transaction history, and loan applications, often resolving issues without the need for human intervention. Such innovations are increasing customer satisfaction, reducing operational costs, and improving service delivery.

Challenges and Opportunities in AI-Powered Self-Service Technology

While AI has undeniably advanced the capabilities of self-service technology, its implementation is not without challenges. One of the most pressing concerns is data privacy. As AI systems collect and analyze vast amounts of personal data to offer tailored experiences, there is an increased risk of data breaches and misuse. Companies must navigate complex data privacy regulations to ensure that customer data is protected and used ethically.

Additionally, the shift towards AI-powered self-service systems has led to concerns about job displacement, particularly in industries reliant on human customer service roles. While AI can create new opportunities in technology and analytics, businesses must address these disruptions and focus on upskilling employees to adapt to changing job roles. Govindaraj et al. (2024) discuss strategies for balancing automation and human interaction in modern marketing, emphasizing the importance of maintaining a human-centric approach.

The Future of Self-Service Technologies

Looking ahead, self-service technologies will continue to evolve as AI and automation become more sophisticated. The next wave of innovation could involve even more immersive technologies, such as augmented reality (AR) and virtual reality (VR), further enhancing the customer experience by providing interactive, intuitive interfaces for self-service. Additionally, advancements in AI will likely lead to even greater levels of personalization, making self-service solutions increasingly seamless and predictive. Govindaraj et al. (2023) provide insights into the transformative applications of AI across various media platforms, showcasing its ability to enhance engagement

In conclusion, the evolution of self-service technology—from ATMs to AI-powered systems—has had a profound impact on customer experiences across industries. As technology continues to evolve, businesses must adapt to meet the growing demand for convenience, personalization, and efficiency, while also addressing the ethical and social implications that accompany these innovations.

AI- DRIVEN ENHANCING PERSONALIZATION, CONVENIENCE AND EFFICIENCY ACROSS INDUSTRY

Artificial Intelligence (AI) is increasingly becoming a driving force behind innovation across various sectors, fundamentally transforming how businesses interact with customers. By leveraging AI-driven systems, companies are now able to offer more personalized, efficient, and convenient services. From retail to healthcare, AI is reshaping industry standards, pushing the boundaries of what's possible in terms of customer experience, operational efficiency, and data-driven insights. This article explores how AI-driven systems are enhancing personalization, convenience, and efficiency across industries. This transformation aligns with Manohar et al. (2020), who highlight the role of service innovation in enhancing corporate reputation and fostering positive word-of-mouth. Elantheraiyan et al. (2023) explore the use of deep learning in revolutionizing product advertising, showcasing AI's potential in customer engagement and personalization.

Personalization: Tailoring Experiences to Individual Needs

One of the most significant contributions of AI-driven systems is their ability to deliver highly personalized experiences for customers. Through machine learning algorithms, AI systems analyze vast amounts of customer data, including browsing history, purchasing behavior, and demographic information. This allows companies to gain insights into individual preferences and tailor their offerings accordingly.

In retail, for example, AI is used to personalize shopping experiences by recommending products based on past purchases and browsing history. Online stores like Amazon and e-commerce platforms such as Shopify utilize AI to create personalized product recommendations, ensuring that customers are presented with options that match their interests. This increases customer satisfaction and encourages repeat business, as users feel the service is uniquely catered to their needs.

Similarly, AI is revolutionizing the entertainment industry by recommending movies and TV shows based on individual viewing habits. Streaming services like Netflix use AI-driven recommendation engines to suggest content, enhancing user engagement and retention.

Convenience: Streamlining Processes for Seamless Interaction

AI-driven systems also significantly enhance convenience by streamlining processes and reducing the time customers spend on routine tasks. Through automation, AI systems can perform tasks that previously required human intervention, making operations faster and more efficient. Elantheraiyan et al. (2024) focus on ethical design principles in AI applications, which are critical for ensuring responsible implementation in various domains, including retail.

In banking, for instance, AI-powered chatbots and virtual assistants enable customers to check balances, make payments, and resolve issues 24/7, eliminating the need for them to visit a physical bank or wait in long queues. These systems can handle multiple queries simultaneously, providing instant responses to customers' requests.

In healthcare, AI has revolutionized appointment scheduling and medical record management. Virtual health assistants can guide patients through the process of booking appointments, answering frequently asked questions, and even providing pre-consultation advice. This saves time for both patients and healthcare providers, enabling more streamlined and efficient service delivery.

Efficiency: Automating Operations to Reduce Costs

AI-driven systems are also playing a crucial role in improving efficiency by automating operations, reducing costs, and optimizing business processes. With AI handling routine and repetitive tasks, companies can allocate their resources to more strategic initiatives. This leads to more efficient use of time, labor, and capital.

In the manufacturing industry, AI-powered systems are used for predictive maintenance, which helps companies avoid expensive downtime. Machine learning algorithms can analyze sensor data from machines to predict failures before they happen, allowing businesses to proactively perform maintenance. This not only prevents costly repairs but also maximizes the operational life of machinery. Krishnan et al. (2024) explore how AI and metaverse concepts are incorporated to optimize omnichannel strategies for electronic goods.

Similarly, AI is enhancing supply chain management by optimizing inventory and demand forecasting. Companies like Walmart and Zara use AI to predict consumer demand, enabling them to maintain optimal stock levels and reduce waste.

This helps businesses increase operational efficiency and ensures customers have access to the products they need when they need them. Govindaraj et al. (2024) discuss how artificial intelligence has revolutionized healthcare, offering insights that can be paralleled in retail innovation.

The Future of AI-Driven Systems Across Industries

As AI technology continues to evolve, its potential to drive personalization, convenience, and efficiency across industries will only grow. Future advancements in AI could lead to even more intelligent systems that can predict consumer behavior, automate complex tasks, and offer deeper levels of personalization.

Industries like healthcare, education, and logistics are poised for significant transformations as AI becomes more integrated into their operations. In the near future, we could see even more seamless interactions between customers and AI-driven systems, as well as an increasing reliance on AI to improve decision-making, operational performance, and customer engagement. Krishnan et al. (2024) provide insights into incorporating metaverse concepts into omnichannel strategies to optimize customer engagement and efficiency.

AI-driven systems are undeniably transforming industries by enhancing personalization, convenience, and efficiency. These systems enable businesses to meet customer expectations in ways that were previously unimaginable, creating a competitive edge in an increasingly technology-driven world. As AI continues to develop, its role in reshaping industries will expand, leading to even more innovative, customized, and streamlined services that will define the future of customer experiences. Govindaraj et al. (2024) highlight the transformative impact of artificial intelligence in revolutionizing service efficiency and productivity, particularly in the retail supply chain. Krishnan et al. (2024) provide valuable insights into how stakeholder perspectives drive healthcare innovation, which parallels similar dynamics in retail and other sectors adopting self-service technologies.

Navigating the Ethical Challenges in AI Implementation for Self-Service Technology

Artificial Intelligence (AI) has revolutionized various sectors, particularly self-service technology, by offering innovative solutions that enhance efficiency, personalization, and customer satisfaction. However, as AI continues to expand in applications such as chatbots, kiosks, and virtual assistants, it brings forth a new set of ethical challenges that organizations must address to ensure responsible implementation. This article explores the ethical challenges involved in implementing AI in self-service technology and the importance of maintaining ethical standards in the development and deployment of these systems. Krishnan et al. (2024) explore how logistics operations can be enhanced through AI and IoT integration, driving superior service efficiency. Krishnan et al. (2024) emphasize the impact of technological advancements on logistics operations, highlighting their potential to enhance service efficiency.

Privacy Concerns: Balancing Personalization with Data Protection

One of the most prominent ethical concerns in AI implementation for self-service technology is privacy. AI systems rely heavily on customer data, including browsing history, purchasing patterns, and demographic information, to offer personalized experiences. While this personalization can enhance customer satisfaction, it raises significant questions about data privacy and security. Nimo et al. (2024) discuss strategies for balancing automation with human interaction, emphasizing character-driven storytelling in marketing.

Consumers are increasingly concerned about how their personal data is being collected, stored, and used by AI systems. For instance, AI-powered virtual assistants often collect sensitive information, such as location data, preferences, and even health-related data in sectors like healthcare. If not properly protected, this data could be exposed to unauthorized parties or misused for commercial purposes. Krishnan et al. (2024) discuss incorporating metaverse concepts into omnichannel retail strategies, offering new pathways to enhance customer engagement.

To address these concerns, businesses must implement robust data protection measures, including encryption, secure data storage, and transparency in data usage policies. Ensuring that customers are fully informed about how their data is collected and giving them control over their information is crucial for maintaining trust and adhering to ethical standards.

Bias and Fairness: Ensuring AI Systems Do Not Discriminate

Another significant ethical challenge is the potential for bias in AI systems. AI algorithms are trained using large datasets, which can inadvertently include biased data reflecting societal prejudices. For example, biased data in recruitment tools can lead to discriminatory practices, or biased AI chatbots can provide uneven customer service based on gender, race, or socio-economic background.

Addressing bias in AI requires a conscious effort to ensure that the data used to train AI models is diverse and representative of all demographic groups. Regular audits of AI systems and their decision-making processes can help identify and rectify any instances of discrimination. Additionally, businesses should focus on developing AI systems that adhere to principles of fairness, transparency, and accountability to prevent bias from affecting customer interactions.

Accountability and Transparency: Defining Responsibility for AI Decisions

AI systems in self-service technology often operate autonomously, making decisions without human intervention. This raises questions about accountability when things go wrong. If an AI system makes a mistake, such as providing incorrect information or mishandling a customer request, it is essential to determine who is responsible for the error—whether it is the developer, the organization, or the AI system itself.

Transparency is also crucial in AI implementation. Customers must have clear information on how AI systems make decisions and how those decisions impact their experiences. For instance, AI systems that make personalized recommendations should be transparent about how they generate those suggestions to ensure customers understand the underlying processes.

To address these issues, businesses should implement clear accountability structures and provide transparency in the decision-making processes of AI systems. Additionally, organizations should have human oversight in place to intervene when necessary and ensure that AI systems align with ethical standards.

Job Displacement: Balancing Automation with Human Employment

AI-powered self-service technology has the potential to automate many tasks traditionally performed by humans. While this increases efficiency and reduces operational costs, it can also lead to job displacement. Many industries are witnessing a shift in employment patterns, with fewer roles required for customer service, cashier positions, or call center operators as AI-driven systems take over these functions.

Organizations must approach this challenge carefully by investing in employee retraining and reskilling programs. By focusing on the creation of new opportunities and preparing employees for roles that require human creativity, empathy, and complex decision-making, companies can strike a balance between automation and the preservation of meaningful employment.

Conclusion: A Responsible Approach to AI in Self-Service Technology

While AI offers immense potential for self-service technology, ethical considerations must be at the forefront of its development and deployment. Addressing issues such as privacy, bias, accountability, and job displacement ensures that AI systems contribute to positive customer experiences without compromising ethical standards. As AI continues to shape the future of self-service technology, businesses must navigate these challenges responsibly, ensuring that technology serves both the interests of the company and the broader society. By embracing transparency, fairness, and human oversight, organizations can build trust with customers and foster the responsible use of AI in the self-service industry.

The Future of Self-Service Technologies: Emerging Trends and Challenges Shaping Customer Experiences

Self-service technologies (SST) have evolved rapidly over the past few decades, fundamentally transforming how businesses interact with customers. With advancements in Artificial Intelligence (AI), machine learning, and automation, self-service systems are becoming increasingly sophisticated, offering consumers enhanced convenience, personalization, and efficiency. As we look toward the future, several emerging trends are expected to further shape self-service technologies and the customer experience. However, these innovations are not without challenges. This article explores key trends and the potential hurdles that companies and customers may face as self-service technologies continue to evolve.

The Rise of AI and Automation in Self-Service

AI and automation are at the forefront of the transformation in self-service technologies. AI-driven tools such as chatbots, virtual assistants, and predictive analytics are already reshaping customer service across industries. In the future, AI will likely play an even larger role, enabling businesses to provide hyper-personalized customer experiences. For example, AI could predict customer needs based on past interactions, and offer proactive solutions before a customer even asks for assistance.

Automation in retail, banking, healthcare, and hospitality is already reducing wait times, eliminating human error, and streamlining operations. AI-powered kiosks, for instance, are increasingly common in fast-food chains and airports, providing users with a faster and more efficient way to interact with services. Self-checkout systems in retail are also becoming more widely adopted, with improved accuracy and a smoother customer experience.

However, despite its advantages, AI-powered automation raises concerns over the loss of personal touch. While automation enhances efficiency, it might alienate customers who prefer human interaction or face difficulties using technology. Finding the right balance between automation and human interaction will be essential for future self-service innovations.

Voice-Activated Technology and Natural Language Processing

Voice-activated technology is another major trend in the self-service industry. Devices like Amazon Alexa and Google Assistant have paved the way for more conversational interfaces, where customers can interact with systems using natural language. In the future, we can expect an increase in voice-based self-service systems across sectors such as banking, healthcare, and retail.

Voice-based interfaces offer a more intuitive and hands-free customer experience. Customers could use their voices to complete tasks such as ordering food, checking bank balances, or scheduling doctor's appointments. With further advancements in natural language processing (NLP), these systems will become more sophisticated, enabling more complex interactions and more accurate responses.

However, the integration of voice-activated systems into self-service technologies presents challenges, particularly regarding accuracy and data privacy. Voice recognition systems need to be able to understand a diverse range of accents, languages, and dialects, ensuring that they work seamlessly across different populations. Additionally, issues around privacy and data security will need to be addressed to ensure consumer trust in voice-activated technologies.

Self-Service in the Context of the Internet of Things (IoT)

The Internet of Things (IoT) is another emerging trend that will shape the future of self-service technology. IoT connects everyday devices to the internet, enabling them to send and receive data. In a self-service context, IoT could enable seamless, automated interactions between customers and businesses. For example, smart shelves in retail stores could automatically track inventory, alerting customers to product availability or automatically processing payments when customers walk out of the store.

In the healthcare industry, IoT devices could help patients monitor their health remotely and engage with virtual consultations through self-service portals. The use of wearable devices, such as fitness trackers or smartwatches, could also allow customers to access personalized recommendations based on real-time health data.

While IoT has the potential to revolutionize the self-service landscape, it also poses significant challenges related to data privacy, security, and system interoperability. Self-service systems relying on IoT devices need to ensure that they adhere to strict security protocols to protect customer data and avoid hacking risks. Furthermore, IoT devices need to work seamlessly together, which requires standardization across platforms, devices, and service providers.

The Digital Divide and Accessibility Issues

As self-service technologies continue to evolve, one critical challenge that will persist is the digital divide. While the widespread adoption of technology offers convenience to many consumers, it may leave behind those who lack access to digital tools or have limited digital literacy. This divide can manifest in various ways: older individuals may struggle to navigate complex self-service systems, rural communities may have limited access to high-speed internet, and low-income households may not own the necessary devices to participate in digital services.

To ensure that self-service technologies are inclusive, businesses must consider how to make their systems accessible to all customers, regardless of their technological expertise or socioeconomic status. This includes designing user-friendly interfaces, providing alternative means of assistance (e.g., phone support or in-person help), and offering affordable access to necessary technology. Governments, non-profits, and private sector companies will need to collaborate to bridge the digital divide and ensure equitable access to self-service solutions.

Data Privacy and Security Concerns

As self-service technologies become increasingly embedded into daily life, the issue of data privacy and security becomes more important than ever. From personalized recommendations to automated purchases, self-service systems collect vast amounts of personal information. This data is often stored, analyzed, and shared to improve services, but if not adequately protected, it poses significant risks.

Customers expect businesses to safeguard their data and be transparent about how it is used. Companies need to comply with stringent data protection regulations (e.g., GDPR in Europe) and ensure that they implement robust cybersecurity measures to protect customer data. Additionally, customers should be empowered to control their personal data, such as through opt-in permissions or clear privacy settings.

The future of self-service technologies holds great promise, with emerging trends such as AI-driven automation, voice-based interfaces, and IoT integrations transforming customer experiences across industries. However, the rapid pace of innovation also brings significant challenges, including issues of privacy, accessibility, and the digital divide. As businesses look to implement these technologies, they must consider ethical implications, ensure inclusivity, and prioritize security to maintain customer trust and satisfaction. By addressing these challenges, self-service technology can continue to thrive, providing a more efficient, personalized, and accessible experience for all consumers.

DISSCUSION

The study of the evolution of self-service technologies (SST) reveals significant insights into how advancements in AI, automation, and machine learning are transforming customer experiences across various industries. The integration of AI-driven systems, such as chatbots, voice assistants, and predictive analytics, has enhanced the personalization, efficiency, and convenience of self-service interactions. These findings align with existing literature that emphasizes AI's role in reshaping customer engagement by offering tailored solutions and streamlining operations (Gao & Liu, 2022; Kumar, 2024). As demonstrated in this study, businesses in sectors like retail, banking, and healthcare have leveraged these technologies to improve service delivery, reduce wait times, and increase customer satisfaction, which aligns with the trends observed by Pendyala & Lakkamraju (2024).

However, the study also highlights challenges that remain in the widespread adoption of AI-powered self-service technologies. While the benefits of AI are clear, concerns related to data privacy, security, and the digital divide need to be addressed. The implementation of self-service systems often raises questions about customer data protection, especially as these technologies collect vast amounts of sensitive information. These concerns are supported by Kumar (2024), who discusses the ethical implications of AI in customer interactions. Additionally, issues of accessibility persist, particularly for older adults and those with limited digital literacy, echoing concerns raised by Moura (2022). Despite these challenges, businesses must ensure that self-service technologies are inclusive and maintain customer trust through transparent data practices and user-friendly designs.

Furthermore, the study reveals that while automation offers operational efficiencies, the human element of customer service cannot be entirely replaced. The results suggest that customers continue to value personalized interactions, which cannot always be fully replicated by AI. This aligns with existing research indicating that while automation can handle routine tasks efficiently, human engagement remains crucial for complex or emotionally charged interactions (Gao & Liu, 2022). This highlights the need for a balanced approach, where self-service technologies complement, rather than replace, human customer service.

The findings also suggest that as AI and automation continue to evolve, they will likely lead to new challenges in the workforce. Automation may displace some jobs, particularly in industries that rely heavily on routine customer service tasks. However, the study also acknowledges the creation of new opportunities, such as in the fields of AI development, system maintenance, and customer experience management. These shifts reflect the dual nature of technological advancements, where disruption in one area may lead to growth in others, as indicated by Kumar (2024) and Pendyala & Lakkamraju (2024).

In conclusion, while the study highlights the transformative impact of AI and automation on self-service technologies, it also underscores the importance of addressing ethical, accessibility, and workforce-related challenges. By doing so, businesses can continue to harness the potential of AI-driven systems while maintaining a human-centric approach to customer service.

MAIN FINDINGS

1. Advancements in AI-Driven Self-Service Technologies

The study finds that AI-powered self-service technologies, such as chatbots, virtual assistants, and predictive analytics, have significantly enhanced customer experiences by providing personalized interactions, streamlining service delivery, and improving operational efficiencies. These findings align with existing literature (Gao & Liu, 2022; Pendyala & Lakkamraju, 2024), emphasizing AI's growing role in transforming customer journeys across industries, especially in retail, banking, and healthcare.

2. Personalization and Convenience as Key Benefits

A major finding is that AI-driven self-service systems have substantially improved personalization and convenience for customers. By analyzing vast amounts of customer data, AI systems can offer tailored recommendations and solutions in real-time, enhancing customer satisfaction and engagement. This personalization is a significant shift from traditional one-size-fits-all approaches and is crucial for increasing customer loyalty, as supported by Kumar (2024) and Pendyala & Lakkamraju (2024).

3. Ethical and Privacy Concerns

While AI technologies offer substantial benefits, the study also identifies concerns regarding customer data privacy and security. The collection and use of vast amounts of personal data to power AI systems raise ethical questions about consent, data protection, and algorithmic transparency. These challenges are consistent with the findings of Kumar (2024), who highlighted the importance of maintaining customer trust in an era of data-driven marketing and service delivery.

4. Accessibility and Inclusivity Challenges

Despite the benefits of self-service technologies, the study identifies significant barriers related to accessibility, particularly for individuals with limited digital literacy or those in underserved regions. This issue is prevalent in industries where customer engagement requires a higher degree of interaction. The findings reinforce the concerns raised by Moura (2022), indicating that businesses must prioritize inclusivity in their self-service technology implementations to avoid alienating certain customer segments.

5. Impact on the Workforce and Job Creation

Another important finding is the dual impact of AI on the workforce. While automation may lead to the displacement of certain customer service roles, it also opens up new opportunities in AI development, system maintenance, and customer experience management. This underscores the necessity of preparing the workforce for shifts brought on by technological advancements, as highlighted by Pendyala & Lakkamraju (2024).

6. Continued Importance of Human-Centered Customer Service

The study reveals that while AI can handle routine customer service tasks, human interaction remains vital for complex or emotionally sensitive issues. Customers continue to value personalized, human-driven engagement for more nuanced and intricate concerns. This finding aligns with previous research, which suggests that AI should complement human customer service rather than replace it entirely.

SUGGESTIONS

Based on the findings of this study, several suggestions can be made for businesses, developers, and policymakers to address the challenges and maximize the potential of AI-driven self-service technologies.

1. Prioritize Ethical Data Use and Transparency

It is crucial for organizations implementing AI technologies to prioritize ethical data collection and transparency. Given the concerns regarding privacy and data security, companies should adopt robust data protection practices, obtain informed consent from customers, and ensure that AI systems operate in a transparent manner. This will help build and maintain trust with customers, as they are becoming more aware of how their data is being used in digital interactions.

2. Focus on Inclusivity and Accessibility

Businesses should consider accessibility as a fundamental aspect when designing and deploying self-service technologies. To avoid excluding customers who may have limited access to or understanding of technology, companies should create user-friendly interfaces, offer alternative communication channels, and ensure that their AI systems can cater to diverse demographics. This includes providing multilingual support and ensuring that people with disabilities can also benefit from self-service options.

3. Embrace Human-AI Collaboration

While AI is a powerful tool for automation and personalization, it should not replace human interaction entirely. Organizations should focus on creating systems that blend human expertise with AI capabilities. By leveraging AI for routine tasks, employees can focus on more complex customer inquiries, ensuring that customers still receive the high-touch, empathetic service they expect for complicated issues. Training employees to effectively collaborate with AI systems will ensure smoother and more satisfying customer experiences.

4. Invest in Continuous Innovation and Monitoring

AI technologies evolve rapidly, and businesses need to stay ahead of trends to maintain a competitive edge. Continuous investment in research and development is essential to ensure that AI systems remain relevant and efficient. Additionally, businesses should implement regular monitoring and assessment mechanisms to identify potential shortcomings, gaps in customer satisfaction, or biases in AI algorithms. Adapting to these insights will enhance the overall effectiveness of AI systems in providing value to both customers and organizations.

By incorporating these suggestions, businesses can overcome the existing challenges of AI-driven self-service technologies and ensure that these innovations benefit both customers and organizations in a sustainable and ethical manner.

IMPLEMENTATION OF THE STUDY

Implementation of the Study

The successful implementation of AI-driven self-service technologies requires a multi-step approach involving technical, organizational, and ethical considerations. This section outlines the key steps to effectively implement the study's findings into real-world applications.

1. Technological Integration and Infrastructure Development

For businesses to implement AI systems, they must first ensure that their infrastructure is capable of supporting advanced technologies. This includes investing in high-performance servers, cloud-based solutions, and robust data storage systems that can handle large volumes of real-time data processing. Integrating AI technologies such as chatbots, virtual assistants, or recommendation engines will require an in-depth analysis of existing systems and platforms to ensure smooth operation. Businesses should also collaborate with AI technology providers to customize these solutions to their unique customer interaction models.

2. Employee Training and Adaptation

Implementing AI technologies involves equipping employees with the necessary skills to manage and collaborate with AI-driven tools. This requires the development of training programs focused on helping employees understand how AI systems work, how to interpret data-driven insights, and how to leverage these tools to improve customer service. Employees should also be trained in troubleshooting issues related to AI systems and ensuring that the technology does not disrupt normal business operations. As AI complements human workers rather than fully replacing them, training must also focus on human-AI collaboration and enhancing the customer experience.

3. Ethical and Privacy Guidelines Implementation

Ethical concerns, particularly regarding data privacy, must be addressed early on in the implementation phase. Organizations must develop clear data privacy policies in accordance with regulations such as GDPR, ensuring that customers are fully informed about how their data is collected, stored, and used. Transparency in AI algorithms and decision-making processes should be prioritized to mitigate concerns related to biases or discriminatory practices. Businesses should also establish clear procedures for gaining customer consent and allowing customers to easily opt out of data collection when necessary.

4. Continuous Evaluation and Improvement

The implementation of AI-driven self-service technology should include a continuous evaluation process to monitor performance and user satisfaction. This involves analyzing customer feedback, identifying areas for improvement, and ensuring that AI systems are constantly updated to align with customer needs and expectations. Metrics such as customer engagement rates, response times, and overall satisfaction can provide valuable insights into the system's effectiveness. In addition, businesses should set up mechanisms to regularly audit AI algorithms to ensure fairness, accuracy, and inclusivity.

5. Scalability and Adaptability

As the business landscape and customer expectations evolve, the AI systems implemented must be scalable and adaptable. Companies should ensure that the AI technology they use can evolve with the growing needs of their customer base. This includes providing regular updates and incorporating emerging technologies, such as machine learning advancements or new customer engagement tools, to enhance the user experience. Businesses must also plan for scalability to accommodate future growth, both in terms of customer numbers and technological capabilities.

By following these implementation steps, organizations can successfully deploy AI-driven self-service technologies in a manner that enhances customer experiences, ensures ethical practices, and maintains a high standard of service quality.

Case Studies and Real-World Examples

To better illustrate the real-world impact and application of self-service technology (SST), it is essential to examine how various industries have adopted and adapted these innovations over time:

1. Retail: Amazon Go Stores

Amazon revolutionized retail shopping through its Amazon Go stores, which utilize a combination of computer vision, deep learning algorithms, and sensor fusion to enable a checkout-free experience. Customers simply walk in, pick up items, and walk out—with purchases automatically charged to their Amazon accounts. This model eliminates traditional cashier roles and showcases how SST can streamline the shopping experience while raising questions around data privacy and job displacement.

2. Banking: Automated Teller Machines (ATMs)

The introduction of ATMs in the 1960s marked one of the earliest examples of SST. By allowing customers to perform basic banking functions without human interaction, ATMs drastically reduced wait times and operational costs for banks. Today, smart ATMs offer even more complex services, such as check deposits, loan applications, and biometric authentication.

3. Hospitality: Hotel Self-Check-In Kiosks

Major hotel chains like Marriott and Hilton have adopted self-service kiosks and mobile apps that allow guests to check in, select their room, and receive a digital key—all without interacting with front desk staff. This enhances efficiency and guest convenience, especially during peak times, and has proven particularly valuable in a post-COVID contactless world.

4. Healthcare: Self-Service Patient Portals and Kiosks

Hospitals and clinics have implemented SST through self-check-in kiosks and patient portals. For instance, Kaiser Permanente's patient portal allows users to schedule appointments, access test results, and communicate with healthcare providers. This empowers patients while reducing administrative burden, though it also highlights challenges related to digital literacy and equitable access.

CONCLUSION

In conclusion, the implementation of AI-driven self-service technologies has the potential to transform customer experiences across industries, driving personalization, efficiency, and convenience. However, the successful integration of these systems requires careful consideration of technological infrastructure, employee training, and ethical guidelines. As businesses increasingly adopt AI technologies, it is crucial to ensure that these systems are continuously monitored, evaluated, and updated to align with customer expectations and regulatory requirements. By addressing these key areas, organizations can harness the full potential of AI to enhance their customer engagement strategies, improve service delivery, and remain competitive in an evolving market. The future of self-service technology lies in its ability to adapt to emerging trends, challenges, and consumer demands, creating new opportunities for growth and innovation.

REFERENCES

Bitner, M. J., Ostrom, A. L., & Meuter, M. L. (2002). Self-service technologies: What do customers expect? *Journal of Service Research, 5*(3), 216–231. DOI: 10.1177/1094670502238917

Buell, R. W., Campbell, D., & Frei, F. X. (2010). Are self-service customers satisfied or stuck? *Production and Operations Management, 19*(6), 679–697. DOI: 10.1111/j.1937-5956.2010.01151.x

Chen, R. J., & Dubinsky, A. J. (2003). A conceptual model of perceived customer value in self-service technology. *International Journal of Service Industry Management, 14*(3/4), 346–369. DOI: 10.1108/09564230310478877

Curran, J. M., & Meuter, M. L. (2005). Self-service technology adoption: Comparing three technologies. *Journal of Services Marketing, 19*(2), 103–113. DOI: 10.1108/08876040510591411

Dabholkar, P. A. (1996). Consumer evaluations of new technology-based self-service options: An investigation of alternative models of service quality. *International Journal of Research in Marketing, 13*(1), 29–51. DOI: 10.1016/0167-8116(95)00027-5

Demirkan, H., & Spohrer, J. (2014). Developing a framework to improve virtual self-service experience. *Journal of the Academy of Marketing Science, 42*(4), 438–459. DOI: 10.1007/s11747-014-0384-4

Elantheraiyan, P., Vinayagam, K., Sasikumar, P., Thirunavukkarasu, K. S., & Singh, S. K. (2023). Revolutionizing e-commerce: Deep learning and distributed expression for cutting-edge product advertising. *2023 IEEE International Conference on Research Methodologies in Knowledge Management, Artificial Intelligence and Telecommunication Engineering (RMKMATE).*

Govindaraj, M., Kumari, D. A., Khan, P., Gnanasekaran, C., & Lawrence, J. (2024). *Revolutionizing healthcare: The transformative impact of artificial intelligence*. Revolutionizing the Healthcare Sector with AI. DOI: 10.4018/979-8-3693-3731-8.ch003

Hilton, T., Hughes, T., Little, E., & Marandi, E. (2013). Adopting self-service technology to do more with less. *Journal of Services Marketing, 27*(1), 3–12. DOI: 10.1108/08876041311296338

Kelly, P., Lawlor, J., & Mulvey, M. (2019). *Self-service Technologies in the Travel, Tourism, and Hospitality Sectors: Principles and Practice*. Emerald Publishing Limited., DOI: 10.1108/978-1-78756-687-320191003

Lee, H. J., & Yang, K. (2013). Interpersonal service and self-service technology in retailing: The role of perceived control and intrinsic motivation. *Service Industries Journal, 33*(2), 145–158. DOI: 10.1080/02642069.2011.596532

Liljander, V., Gillberg, F., Gummerus, J., & Van Riel, A. (2006). Technology readiness and the evaluation and adoption of self-service technologies. *Journal of Retailing and Consumer Services, 13*(3), 177–191. DOI: 10.1016/j.jretconser.2005.08.004

Lin, J. S. C., & Hsieh, P. L. (2007). The influence of technology readiness on satisfaction and behavioral intentions toward self-service technologies. *Computers in Human Behavior, 23*(3), 1597–1615. DOI: 10.1016/j.chb.2005.07.006

Manohar, S., Mittal, A., & Marwah, S. (2020). Service innovation, corporate reputation, and word-of-mouth in the banking sector: A test on multigroup-moderated mediation effect. *Benchmarking, 27*(1), 406–429. DOI: 10.1108/BIJ-05-2019-0217

Meuter, M. L., Ostrom, A. L., Roundtree, R. I., & Bitner, M. J. (2000). Self-service technologies: Understanding customer satisfaction with technology-based service encounters. *Journal of Marketing, 64*(3), 50–64. DOI: 10.1509/jmkg.64.3.50.18024

Pantano, E., & Migliarese, P. (2014). Exploiting consumer–employee–retailer interaction in technology-enriched retail environments: Evidences from a study on the role of store format. *Journal of Retailing and Consumer Services, 21*(6), 958–965. DOI: 10.1016/j.jretconser.2014.08.015

Simon, F., & Usunier, J. C. (2007). Cognitive, demographic, and situational determinants of service customer preferences for personnel-in-contact over self-service technology. *International Journal of Research in Marketing, 24*(2), 163–173. DOI: 10.1016/j.ijresmar.2006.11.004

Verhoef, P. C., Broekhuizen, T., & Bijmolt, T. H. (2016). Understanding the effect of customer experience across channels: A multi-method approach. *Journal of Marketing, 80*(6), 39–58. DOI: 10.1509/jm.15.0420

Weijters, B., Rangarajan, D., Falk, T., & Schillewaert, N. (2007). Determinants and outcomes of customers' use of self-service technology in a retail setting. *Journal of Service Research, 10*(1), 3–21. DOI: 10.1177/1094670507302990

Yang, Y. C., Yang, Y. Y., & Liu, S. W. (2012). An Integrative Model for Self-Service Technology Adoption. *Applied Mechanics and Materials, 182-183*, 1387–1392. . DOI: 10.4028/www.scientific.net/AMM.182-183.1387

Zhu, Z., Wymer, W., & Chen, I. (2002). IT-based services and service quality in consumer banking. *International Journal of Service Industry Management, 13*(1), 69–90. DOI: 10.1108/09564230210421164

Chapter 10
AI for Personalization: Transforming Customer Journeys in Self-Service Technology

G. Manoj
https://orcid.org/0000-0003-2830-7875
VelTech Rangarajan Dr. Sagunthala R&D Institute of Science and Technology, India

P. Arockia Juliet
https://orcid.org/0009-0009-3482-2137
Arul Anandar College, India

V. S. Mege
Saveetha College of Liberal Arts and Science, SIMATS, Chennai, India

K. V. Geetha Devi
https://orcid.org/0000-0003-4347-9018
Madanapalle Institute of Technology and Science, India

Chandramowleeswaran Gnanasekaran
https://orcid.org/0000-0002-1293-1043
Vel Tech Rangarajan Dr. Sagunthala R&D Institute of Science and Technology, India

P. Indira
V.S.B. Engineering College, India

ABSTRACT

In the age of rapid technological advancement, Artificial Intelligence (AI) is reshaping customer experiences by enabling highly personalized interactions in self-service technologies. This chapter explores the transformative impact of AI-driven personalization on customer journeys, focusing on how machine learning algorithms, predictive analytics, and natural language processing (NLP) are redefining self-service systems across industries such as retail, banking, healthcare, and hospitality. By analyzing customer data in real time, AI-powered systems tailor recommendations, streamline decision-making processes, and deliver seamless, intuitive, and engaging experiences. The chapter also addresses key challenges, including privacy concerns, data security, and ensuring inclusivity across diverse user demographics.

DOI: 10.4018/979-8-3373-4667-0.ch010

INTRODUCTION

In today's fast-evolving digital landscape, the integration of Artificial Intelligence (AI) has emerged as a cornerstone for enhancing customer experiences. Among the myriad of applications, the use of AI for personalization in self-service technologies has gained significant traction. From interactive kiosks in retail stores to AI-powered chatbots in healthcare and banking, self-service systems are becoming more intelligent, responsive, and customer-centric. The cornerstone of this transformation lies in the ability of AI to analyze large volumes of data, identify patterns, and deliver tailored solutions in real-time.

Self-service technologies, once designed merely for convenience and efficiency, are now evolving to provide more intuitive and engaging customer journeys. AI technologies, such as machine learning, predictive analytics, and natural language processing (NLP), are at the forefront of this evolution, enabling systems to adapt dynamically to individual preferences and needs. By doing so, businesses can not only meet customer expectations but also foster loyalty and drive competitive advantage in a crowded marketplace.

However, while the benefits of AI-driven personalization are vast, there are equally pressing concerns that demand attention. Privacy issues, data security challenges, and the need for inclusivity across diverse user demographics are critical aspects that must be addressed. In this context, the role of ethical AI becomes paramount, ensuring that technological innovations are both effective and responsible. Manohar et al. (2020) emphasize the importance of service innovation in enhancing corporate reputation and driving positive word-of-mouth, particularly in service-oriented industries.

This article delves into the transformative potential of AI for personalization in self-service technologies, examining its impact across key industries, including retail, banking, healthcare, and hospitality. Additionally, it highlights successful use cases and offers actionable strategies for businesses to navigate the complexities of AI adoption while balancing innovation with ethical considerations. By harnessing the power of AI, organizations can unlock new pathways to customer satisfaction, operational excellence, and long-term growth in the self-service era.

Background of the Study

Artificial Intelligence (AI) has significantly transformed customer journeys across both Business-to-Business (B2B) and Business-to-Consumer (B2C) sectors by enabling personalized experiences throughout the entire customer lifecycle (Gao & Liu, 2022; Kumar, 2024). AI systems can process vast amounts of customer data in real-time, offering deep insights into customer behaviors, preferences, and sentiments

(Pendyala & Lakkamraju, 2024). This real-time analysis allows businesses to create highly tailored marketing campaigns, personalized product recommendations, and efficient customer service strategies that enhance customer satisfaction and foster long-term loyalty (Gao & Liu, 2022; Pendyala & Lakkamraju, 2024). Tandon et al. (2021) explore the impact of intangible product features and institutional mechanics in e-commerce, demonstrating how they build consumer trust and influence repurchase intentions. Sridhar and Ganesan (2015) identify critical barriers and facilitators of service innovation, highlighting their effects on customer satisfaction and long-term relationship improvement.

The scope of AI applications in customer journeys extends to personalized profiling, customer navigation, targeted nudges, and customer retention strategies, which collectively improve the overall customer experience (Gao & Liu, 2022). Furthermore, AI-driven technologies such as chatbots and virtual assistants have revolutionized customer engagement by providing instantaneous responses to customer queries, resolving issues promptly, and offering real-time support (Pendyala & Lakkamraju, 2024; Kumar, 2024). These advancements not only streamline customer service processes but also create more engaging and personalized interactions, thus deepening customer relationships. Mehta (2018) highlights the mediating role of corporate reputation in the relationship between service innovation and customer cross-buying intentions, emphasizing its strategic significance. Govindaraj et al. (2024) provide a roadmap of innovative technologies that enhance service productivity, emphasizing their transformative potential across industries.

Despite the numerous benefits AI brings to customer experiences, businesses must also address the ethical concerns that arise with its implementation. Privacy, data security, and algorithmic transparency are significant challenges that need to be considered. As AI systems rely heavily on consumer data, businesses must ensure their practices align with regulatory requirements while maintaining customer trust (Kumar, 2024; Moura, 2022). The ethical implications of AI usage should not be overlooked, and businesses must adopt responsible AI practices to navigate the complexities of customer data and privacy concerns effectively.

FROM AUTOMATION TO PERSONALIZATION: THE POWER OF AI IN SELF- SERVICE TECHNOLOGY

AI is revolutionizing the way self-service technology interacts with customers, shifting the focus from mere automation to a more personalized and engaging experience. Traditionally, self-service systems were designed to offer basic, standardized solutions to customer inquiries, often resulting in impersonal and rigid interactions. However, with the integration of AI, these systems are now able to analyze vast amounts of customer data in real-time, gaining deeper insights into preferences, behaviors, and needs. This data-driven approach allows AI to deliver highly tailored experiences that adapt to individual users, offering personalized recommendations, solutions, and even predictive assistance. Manohar and Kapur (2019) provide insights into perceived service innovation typologies in the retail sector, offering valuable frameworks for analyzing customer-centric innovation.

One of the key advantages of AI in self-service technology is its ability to provide dynamic, context-aware interactions. AI-powered systems such as chatbots and virtual assistants are capable of understanding natural language and engaging in meaningful conversations with customers, leading to more intuitive and satisfying experiences. These systems not only address customer queries but also anticipate needs, guiding users through processes or offering suggestions before they even ask. Neelima et al. (2024) discuss critical factors influencing data utilization and performance in health management information systems, which can parallel the role of data in enhancing AI-driven self-service technologies in retail.The shift from automation to personalization in self-service technology is also a game-changer in terms of scalability. AI can handle numerous customer interactions simultaneously, providing personalized attention to each individual without compromising on quality or response time. As businesses collect more data through AI-driven interactions, these systems continue to evolve, becoming increasingly accurate and effective in delivering customized experiences. However, despite these benefits, companies must also be mindful of data privacy and security concerns, ensuring that customer information is handled responsibly. Overall, AI's role in transforming self-service technology is not just about streamlining operations but also about creating more meaningful, personalized experiences that drive customer satisfaction and loyalty.

RESEARCH QUESTION

How does AI-driven personalization enhance customer journeys in self-service technologies while addressing challenges like privacy, security, and inclusivity?

Research Objectives

The objective of this research is to explore the transformative role of AI in enhancing personalization within self-service technologies. By addressing key aspects such as customer experience, technological applications, and associated challenges, the study aims to provide actionable insights into leveraging AI effectively and ethically in diverse industries.

1. To investigate the role of AI technologies, such as machine learning and NLP, in improving customer engagement within self-service systems.
2. To analyze how AI-driven personalization impacts customer journeys across industries like retail, banking, healthcare, and hospitality.
3. To evaluate challenges related to privacy, data security, and inclusivity in the adoption of AI in self-service technologies.
4. To propose strategies for businesses to effectively balance technological innovation with ethical considerations to enhance customer satisfaction and loyalty.

RESEARCH MODEL

The integration of Artificial Intelligence (AI) in self-service technologies exemplifies a transformative shift from automation to personalized customer engagement. AI technologies, such as machine learning, natural language processing (NLP), and predictive analytics, enable businesses to tailor services to individual needs by analyzing vast datasets in real time. This approach enhances operational efficiency, customer satisfaction, and engagement by delivering context-aware solutions, personalized recommendations, and proactive assistance. However, the deployment of AI also presents challenges, particularly concerning privacy, data security, and inclusivity. Ethical considerations, including transparency, fairness, and compliance with data protection regulations, are vital for building trust and ensuring responsible AI use. By addressing these challenges, businesses can leverage AI to optimize customer journeys across industries, offering scalable, adaptive, and human-centered solutions that foster loyalty and competitive advantage.

Figure 1. Stimulus-Organism-Response (S-O-R) Framework

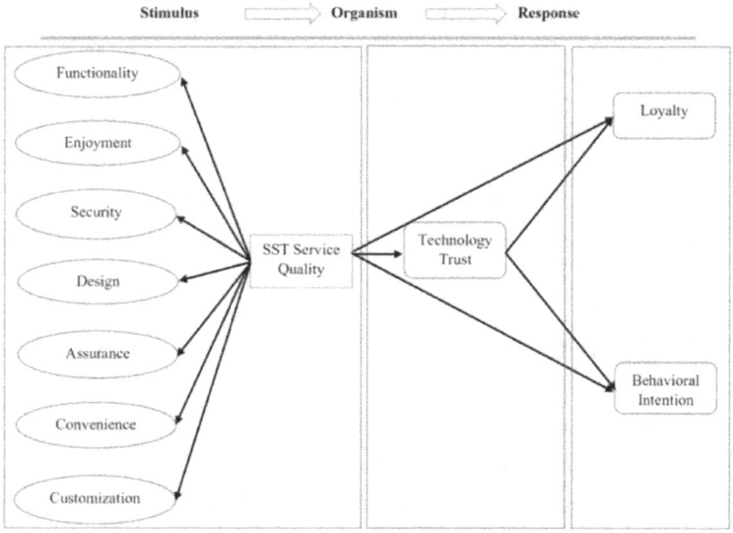

The proposed research model, based on the Stimulus-Organism-Response (S-O-R) framework, conceptualizes the role of Artificial Intelligence (AI) in enhancing self-service technology (SST) for personalized customer experiences. The model is structured into three primary stages: **Stimulus**, **Organism**, and **Response**, each representing different components of AI-driven personalization in SST.

1. Stimulus (SST Service Quality Attributes)

This stage identifies key determinants influencing the perception of service quality in AI-driven self-service technologies. The factors include:

- **Functionality** – The effectiveness and efficiency of AI-powered SST in meeting user needs.
- **Enjoyment** – The hedonic value and interactive engagement facilitated by AI features.
- **Security** – The perceived trustworthiness and protection of user data in AI applications.
- **Design** – The user interface (UI) and experience (UX) considerations that enhance usability.
- **Assurance** – The confidence in AI-driven systems ensuring accuracy and reliability.
- **Convenience** – The ease of access and seamless interactions provided by AI.

- **Customization** – The ability of AI to offer personalized recommendations and tailored experiences.

These attributes collectively influence the overall **SST Service Quality**, which serves as a mediating construct in the customer's AI-driven journey.

2. Organism (Technology Trust)

The second stage, Technology Trust, encapsulates the cognitive and affective responses of customers as they interact with AI-powered self-service technologies (SST). Trust in technology is a critical determinant of user adoption, influencing whether customers perceive AI-driven systems as reliable, transparent, and beneficial (Gefen et al., 2003). A high degree of trust enhances user confidence in AI's ability to deliver personalized experiences while ensuring fairness, security, and ethical data usage (McKnight et al., 2002). Moreover, trust in AI-based SST fosters sustained engagement and positive behavioral intentions, ultimately shaping customer satisfaction and loyalty (Verhagen et al., 2019).

3. Response (Customer Behavioral Outcomes)

The final stage outlines customer responses to AI-enhanced SST experiences. Two primary behavioral outcomes emerge:

- **Loyalty** – The long-term commitment of customers towards AI-driven services, resulting in repeat usage and positive word-of-mouth.
- **Behavioral Intention** – The likelihood of users engaging with and adopting AI-based SST in future transactions.

HOW AI TECHNOLOGY ENHANCE CUSTOMER ENGAGEMENT IN SELF- SERVICE SYSTEMS

The integration of Artificial Intelligence (AI) into self-service systems has revolutionized the way businesses engage with their customers. By leveraging advanced technologies such as machine learning, natural language processing (NLP), and predictive analytics, companies can provide personalized, seamless, and efficient customer experiences. AI not only enhances operational efficiency but also significantly improves customer satisfaction by tailoring interactions to individual preferences and needs.

This article explores how AI technologies are transforming customer engagement in self-service platforms across various industries.

1. The Role of AI in Self-Service Technologies

AI plays a pivotal role in self-service systems by enabling intelligent automation and dynamic personalization. Unlike traditional systems, AI-driven platforms are capable of:

- Analyzing vast amounts of customer data in real time.
- Adapting to user behavior and preferences.
- Providing solutions proactively, often before the customer explicitly states a problem.

For instance, chatbots and virtual assistants, powered by AI, can answer customer queries, suggest products, and provide step-by-step guidance without human intervention.

Figure 2. Survey results on AI and customer experience.

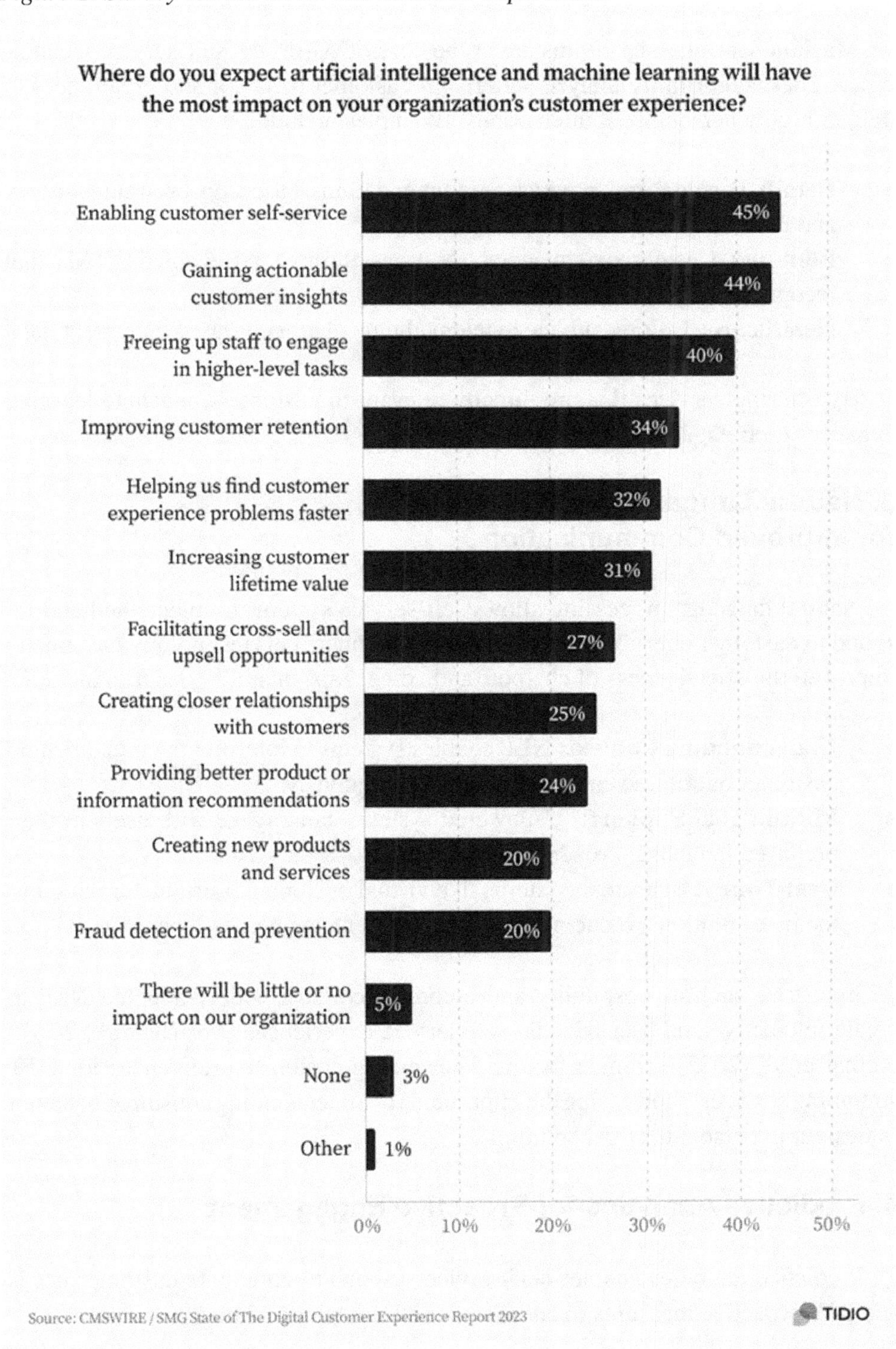

2. Personalization Through Machine Learning

Machine learning algorithms are at the core of AI-driven self-service technologies. These algorithms analyze patterns in customer behavior and preferences to deliver highly personalized interactions. Examples include:

- **Retail:** Personalized product recommendations based on browsing history and purchase behavior.
- **Banking:** Customized financial solutions through AI-enabled ATMs that recognize individual preferences.
- **Healthcare:** Tailored advice in telemedicine platforms based on patient data.

By offering services that are directly relevant to customers, machine learning enhances their engagement and overall satisfaction.

3. Natural Language Processing (NLP) for Improved Communication

Natural language processing allows self-service systems to understand and respond to customer queries in a conversational manner. This technology has greatly improved the effectiveness of chatbots and voice assistants. Key benefits include:

- **Understanding Context:** NLP enables systems to interpret the intent behind customer queries, even when phrased ambiguously.
- **Multilingual Support:** AI-powered systems can engage with users in their preferred language, broadening accessibility.
- **Real-Time Assistance:** Systems like virtual customer assistants provide immediate solutions, reducing wait times and enhancing customer trust.

Industries such as hospitality and e-commerce have widely adopted NLP to create interactive and user-friendly self-service experiences (Verma et al., 2024). Kumar et al. (2025) examine factors determining customer preferences for OTT streaming services, showcasing the importance of understanding consumer behavior in designing personalized AI solutions.

4. Predictive Analytics for Proactive Engagement

Predictive analytics enables self-service systems to anticipate customer needs and take proactive measures to address them. Examples include:

- **Maintenance Alerts:** Predicting when a customer's device might require service and providing timely notifications.
- **Upselling Opportunities:** Suggesting complementary products or services based on previous purchases.
- **Behavioral Insights:** Analyzing customer behavior to optimize workflows and streamline processes.

By proactively engaging customers, predictive analytics ensures a smoother and more satisfying journey.

5. Benefits of AI-Enhanced Self-Service Systems

The integration of AI technologies in self-service systems offers numerous benefits:

- **Efficiency:** Faster responses and automated solutions save time for both customers and businesses.
- **Customer Satisfaction:** Tailored experiences foster a sense of value and importance.
- **Cost Reduction:** AI reduces dependency on human resources, cutting operational costs.
- **Scalability:** Businesses can handle large volumes of customer interactions effortlessly.

AI technologies have transformed customer engagement in self-service systems, making them more intuitive, responsive, and personalized. By harnessing the power of machine learning, NLP, and predictive analytics, businesses can create experiences that not only meet but exceed customer expectations. As AI continues to evolve, its role in enhancing customer engagement will only grow, paving the way for a future where self-service systems are synonymous with innovation and excellence. Krishnan et al. (2024) highlight how technological advancements in logistics, including AI and IoT, enhance service efficiency and operational effectiveness.

AI-DRIVEN PERSONALIZATION: TRANSFORMING CUSTOMER JOURNEYS ACROSS INDUSTRIES

Artificial Intelligence (AI) is playing a pivotal role in revolutionizing customer journeys across various industries by enabling personalized experiences. The power of AI lies in its ability to collect, analyze, and apply customer data in real-time, transforming interactions into more meaningful and tailored experiences. This content explores how AI-driven personalization is reshaping industries, enhancing customer satisfaction, and driving business growth. Govindaraj et al. (2024) explore the transformative impact of artificial intelligence in healthcare, drawing parallels to its revolutionary effects in other industries like retail.

1. Understanding AI-Driven Personalization

AI-driven personalization refers to the use of AI technologies such as machine learning, natural language processing, and predictive analytics to deliver tailored customer experiences. These technologies analyze vast amounts of data, including customer behavior, preferences, and interactions, to create unique and relevant experiences. Unlike traditional personalization methods, AI continually adapts and evolves to customer needs, offering real-time personalization that significantly enhances the customer journey. Krishnan et al. (2024) highlight the critical role of stakeholder perspectives in driving healthcare innovation, which provides valuable parallels to retail and self-service technological advancements.

2. Enhancing Customer Engagement in Retail

In the retail industry, AI-driven personalization is fundamentally changing how brands engage with customers. By analyzing online shopping behavior, purchase history, and browsing patterns, AI can recommend products that align with individual tastes and preferences. Personalization goes beyond product suggestions; it extends to targeted promotions, customized discounts, and personalized marketing messages that cater to the needs of each customer. For example, AI-driven recommendation engines used by companies like Amazon and Netflix provide a highly tailored experience, keeping customers engaged and loyal.

3. Revolutionizing Healthcare Experiences

In healthcare, AI-driven personalization is improving patient care and overall experiences. Personalized treatment plans, powered by AI algorithms, analyze patient data such as medical history, genetic information, and lifestyle factors to recommend the most effective treatments. Additionally, AI chatbots and virtual assistants are enhancing patient interaction by offering timely responses to inquiries, appointment scheduling, and medication reminders. AI is helping healthcare providers move towards patient-centric models by ensuring that each individual receives the care most suited to their specific needs. Krishnan et al. (2024) demonstrate how technological advancements, including AI and IoT, can significantly enhance logistics operations and improve service efficiency across industries.

4. Transforming Financial Services

AI personalization is transforming the financial services industry by offering customers highly personalized financial advice, product recommendations, and automated services. Financial institutions use AI to analyze transaction data, customer financial goals, and spending patterns to recommend relevant products like savings plans, investment opportunities, or credit options. AI-powered chatbots are also improving customer support by offering personalized assistance 24/7, addressing common queries, and even providing financial planning advice (Chawla et al., 2024a). This level of personalization not only enhances the customer experience but also improves trust and loyalty.

5. Optimizing Customer Experience in Travel and Hospitality

AI-driven personalization is making a profound impact on the travel and hospitality industry. By collecting data about customer preferences, travel history, and booking behavior, AI systems can offer tailored travel packages, personalized itineraries, and real-time updates on travel conditions. Hotel chains and airlines leverage AI to provide personalized recommendations, such as upgrades, loyalty rewards, or special offers based on past behaviors and preferences. By making the entire customer journey—from booking to post-travel—more seamless and customized, businesses can foster customer loyalty and satisfaction (Babu et al., 2024). Krishnan et al. (2024) explore how integrating metaverse concepts into omnichannel strategies can transform customer experiences and foster innovation in the retail sector.

6. Personalization in Marketing: A Competitive Edge

AI is transforming marketing strategies by enabling hyper-personalized campaigns that resonate with customers on a deeper level. Using AI, marketers can segment their audience based on specific characteristics, interests, and behaviors. This allows them to deliver targeted advertising, content, and offers that align closely with individual preferences. AI-driven personalization enables brands to understand their customers' needs and desires, driving higher engagement and conversion rates. Platforms like Facebook and Google already leverage AI to tailor ads and content recommendations, creating a more engaging and relevant experience for users.

7. Addressing Privacy Concerns and Ethical Considerations

While AI-driven personalization provides substantial benefits, it also raises concerns about customer privacy and data security. The collection and analysis of personal data can lead to privacy breaches if not handled properly. Companies must ensure transparency about how they collect, store, and use customer data. Adhering to ethical standards and complying with regulations like the GDPR is essential for building trust and maintaining customer confidence in AI-powered systems.

AI-driven personalization is transforming customer journeys across industries, delivering tailored experiences that enhance customer satisfaction, engagement, and loyalty. By utilizing AI technologies, businesses can create more meaningful and relevant interactions with customers, from retail to healthcare, finance, travel, and marketing. However, it is essential for businesses to address privacy concerns and ethical considerations to maintain trust and protect customer data. As AI continues to evolve, the potential for further innovation in personalization remains vast, offering opportunities for businesses to gain a competitive edge and create more customer-centric solutions.

8. Title: AI for Personalization: Transforming Customer Journeys in Self-Service Technology

The integration of Artificial Intelligence (AI) has catalyzed a fundamental shift in self-service technologies, moving beyond basic automation to deliver personalized and engaging customer journeys. AI technologies, including machine learning, natural language processing (NLP), and predictive analytics, are pivotal in this transformation.

Machine learning algorithms analyze vast datasets of customer behavior and preferences, enabling self-service systems to offer tailored recommendations, customized content, and personalized solutions. NLP empowers chatbots and virtual assistants to engage in natural language conversations, providing intuitive and context-aware support. Predictive analytics allows self-service systems to anticipate customer needs, proactively offering assistance and enhancing the overall customer experience.

This AI-driven personalization enhances customer journeys by increasing customer engagement, improving satisfaction, and fostering loyalty. By providing personalized and efficient self-service options, businesses can create more meaningful interactions, leading to stronger customer relationships and a competitive advantage.

However, the adoption of AI in self-service technologies also presents challenges. Privacy concerns and data security are paramount, requiring businesses to implement robust data protection measures and ensure ethical data handling practices.

Furthermore, addressing inclusivity and mitigating biases in AI algorithms is crucial to providing fair and equitable experiences for all customers.

In conclusion, AI-driven personalization is transforming customer journeys in self-service technology by enabling tailored, efficient, and engaging experiences. While businesses must address ethical considerations and challenges related to privacy, security, and inclusivity, the potential of AI to enhance customer satisfaction and drive business growth is substantial.

OVERCOMING PRIVACY, SECURITY, AND INCLUSIVITY BARRIERS IN AI APPLICATIONS

The rapid advancement of Artificial Intelligence (AI) presents numerous opportunities for businesses and industries, but it also raises critical concerns around privacy, security, and inclusivity. Addressing these challenges is essential to ensuring that AI technologies are used ethically, safely, and for the benefit of all. This content explores how to overcome these barriers in AI applications, with a focus on practical solutions and secondary data to highlight key issues. Elantheraiyan et al. (2024) emphasize the importance of ethical design in AI applications, addressing concerns about transparency and fairness in automated systems.

1. The Privacy Challenge in AI

AI systems rely heavily on vast amounts of personal data to operate effectively. Whether it's through analyzing user behavior, transaction history, or personal preferences, AI systems often need to process sensitive information. While this enables personalization and improved services, it also raises privacy concerns.

According to a 2021 report by the European Union Agency for Cybersecurity (ENISA), one of the primary challenges of AI adoption is ensuring that the personal data used in AI algorithms is protected. The use of data for AI applications can lead to unauthorized access, data leaks, or misuse, threatening individuals' privacy. To overcome these privacy challenges, organizations must adopt practices such as:

- Data anonymization: Ensuring that personal data is anonymized before being used in AI models.
- Informed consent: Gaining explicit consent from individuals regarding the use of their data.
- Data minimization: Collecting only the essential data required for specific AI tasks.

AI privacy concerns can be alleviated by compliance with regulations like the GDPR, which imposes strict data protection rules on organizations operating within the EU. Ensuring robust privacy protections is key to gaining user trust and ensuring ethical AI usage.

2. Enhancing Security in AI Systems

AI applications, particularly those involving sensitive information, are vulnerable to cyberattacks, hacking, and malicious exploitation. One major threat is the manipulation of AI models through adversarial attacks, where malicious actors intentionally introduce slight perturbations to the data to deceive the AI system.

A 2020 report from the U.S. National Institute of Standards and Technology (NIST) highlighted AI's vulnerability to such attacks and emphasized the importance of building secure AI systems. To overcome security risks, organizations must focus on:

- Robust model training: AI models should be trained with diverse, high-quality data that is resistant to adversarial manipulation.
- Continuous monitoring: AI systems should be continuously monitored for unusual activity and potential security breaches.
- Security protocols: Implementing advanced encryption and authentication methods to protect sensitive data and ensure secure communication channels.

By adopting these security measures, organizations can significantly reduce the risk of AI systems being compromised and ensure the integrity of AI-powered applications.

3. Promoting Inclusivity in AI Development

AI systems often reflect the biases and inequalities present in the data they are trained on. This can lead to discriminatory outcomes, such as biased hiring algorithms or healthcare predictions that disproportionately impact certain demographic groups. Addressing inclusivity challenges requires a conscious effort to ensure that AI systems are fair and represent diverse populations.

Secondary data from studies like the 2019 research conducted by MIT Media Lab revealed that facial recognition software was less accurate for women and people of color, highlighting the biases that can exist in AI models. To overcome these inclusivity barriers, AI developers must:

- **Diverse data collection**: Use diverse datasets that represent various demographics, including age, gender, race, and socio-economic status.

- **Bias detection tools**: Implement tools that detect and correct biases in AI models before deployment.
- **Inclusive design principles**: Ensure that AI development teams are diverse, bringing different perspectives to the design and testing of AI applications.

Promoting inclusivity in AI development is not only ethically important but also expands the market reach of AI products and services by making them accessible to a broader range of users.

4. Regulatory and Ethical Frameworks

The lack of a universal regulatory framework for AI poses challenges in addressing privacy, security, and inclusivity barriers. Different regions have different standards and regulations, which can lead to inconsistencies in how AI technologies are developed and deployed. According to a 2021 study by the Brookings Institution, inconsistent AI regulations hinder the responsible deployment of AI across borders.

To tackle this issue, global cooperation is essential to creating standardized guidelines for ethical AI development. Governments, industries, and international bodies must work together to:

- **Develop universal AI regulations**: Establish common standards that promote privacy, security, and fairness in AI technologies.
- **Encourage transparency**: AI systems should be transparent in terms of data collection, model design, and decision-making processes.
- **Promote accountability**: Holding organizations accountable for the misuse of AI, including biases or security breaches, ensures responsible AI usage.

Adhering to these frameworks will help overcome regulatory gaps and create a more ethical AI ecosystem.

5. The Role of AI in Enhancing User Trust

User trust is fundamental to the adoption of AI technologies. If consumers believe that their personal data is secure, their privacy is respected, and the AI systems are inclusive, they are more likely to embrace AI-driven services. According to PwC's 2020 report on AI adoption, companies that prioritize ethical AI development and transparency experience greater trust from customers, which leads to higher user engagement and loyalty.

To build trust, organizations must:

- **Communicate openly**: Clearly explain how AI systems work, how data is collected, and how privacy is protected.
- **Offer user control**: Allow users to have control over their data, such as the ability to opt-out or request data deletion.
- **Engage in responsible marketing**: Avoid overstating AI capabilities and ensure that marketing messages align with the actual performance and benefits of AI technologies.

Building user trust will be a critical factor in the long-term success of AI applications.

AI has the potential to transform industries and improve lives, but overcoming privacy, security, and inclusivity barriers is essential for ensuring ethical, responsible, and equitable AI deployment. By prioritizing data privacy, enhancing security measures, promoting inclusivity in development, and adhering to global ethical standards, businesses can overcome these challenges and build AI systems that benefit everyone. As AI continues to evolve, addressing these barriers will help unlock its full potential while safeguarding the interests of all stakeholders. Krishnan et al. (2024) highlight how AI and IoT integration can optimize logistics, improving efficiency and customer satisfaction.

ETHICAL STRATEGIES FOR HARNESSING AI TO ELEVATE CUSTOMER SATISFACTION

Artificial Intelligence (AI) is a powerful tool for businesses aiming to enhance customer satisfaction. By enabling personalized experiences, automating processes, and improving service delivery, AI offers a significant advantage. However, for AI to genuinely improve customer satisfaction, businesses must adopt ethical strategies that ensure responsible usage, transparency, and fairness. This content explores key ethical strategies for harnessing AI to elevate customer satisfaction, supported by secondary data to highlight the real-world applications and potential impact. Faisal et al. (2024) identify efficient machine learning algorithms for predicting customer churn, showcasing AI's ability to enhance retention strategies in consumer-focused industries. Krishnan et al. (2024) discuss the integration of metaverse concepts into omnichannel strategies, highlighting innovative approaches to customer engagement.

1. Prioritizing Transparency in AI-Driven Decisions

Transparency is crucial in ensuring that customers trust AI-driven interactions and decisions. When AI systems are used to make decisions—such as product recommendations, pricing, or customer support—businesses should be clear about how these decisions are made. Customers should understand the data that AI systems use, how it's processed, and the logic behind AI-generated outcomes. Krishnan et al. (2024) provide critical insights into the role of stakeholders in driving innovation in healthcare, offering lessons applicable to self-service technologies in retail and other sectors.

A 2020 survey by Edelman found that 81% of consumers expect transparency from companies, and 60% would stop using a product or service if they felt the company was not transparent. To promote transparency, businesses should:

- **Communicate AI functionalities**: Clearly explain how AI is used in customer interactions, whether in recommendation systems, chatbots, or predictive analytics.
- **Provide insights into decision-making**: Offer customers insights into how their data is being used and how AI determines personalized outcomes.
- **Ensure user control**: Allow customers to opt out or adjust the level of AI personalization they receive.

This strategy fosters trust and enables customers to feel more comfortable engaging with AI systems. Govindaraj et al. (2024) explore the transformative role of artificial intelligence in healthcare, demonstrating its potential to revolutionize customer experiences in multiple domains.

2. Ensuring Data Privacy and Security

AI systems rely heavily on customer data to function effectively. However, the use of personal data must be done responsibly and ethically. Businesses must ensure robust data privacy and security protocols to protect sensitive customer information from misuse or breaches.

According to a 2021 report by McKinsey & Company, 65% of consumers are concerned about the privacy of their data, and 50% would stop using a service if they believed their data was being misused. To address these concerns, companies should:

- **Adopt data encryption**: Use encryption to protect sensitive data both at rest and in transit, preventing unauthorized access.

- **Comply with regulations**: Follow data protection regulations such as the GDPR (General Data Protection Regulation) to ensure the ethical handling of customer data.
- **Implement consent management**: Obtain explicit consent from customers regarding how their data will be used, providing them with options to control their information.

By prioritizing privacy and security, businesses can build stronger relationships with customers, ensuring they feel safe and valued when interacting with AI systems.

3. Fostering Inclusivity in AI Models

AI systems are only as effective as the data on which they are trained. If AI models are trained on biased data, they can perpetuate or even exacerbate discrimination. For example, biased AI algorithms in hiring, lending, or customer service can lead to unfair outcomes, harming customer satisfaction (Chawla et al., 2024b). Elantheraiyan et al. (2024) emphasize the importance of ethical AI design in learning and education systems, which is equally applicable to ensuring fairness and transparency in self-service technologies.

A 2019 study from MIT Media Lab found that facial recognition software had higher error rates for women and people of color, indicating the need for more inclusive AI models. To ensure inclusivity, businesses should:

- **Diversify training data**: Use diverse datasets that represent different demographics, ensuring that AI models are fair and unbiased.
- **Test for biases**: Regularly evaluate AI models to detect and mitigate potential biases in decision-making processes.
- **Promote fairness**: Strive for fairness in AI outcomes by developing algorithms that treat all customers equally, regardless of gender, race, or background.

By promoting inclusivity, businesses can provide a more personalized and equitable experience for all customers, leading to higher satisfaction and loyalty.

4. Personalizing Customer Interactions Ethically

Personalization is one of the primary benefits of AI. By analyzing customer data, AI can tailor experiences to individual preferences, increasing engagement and satisfaction. However, personalization must be done ethically to avoid being perceived as invasive or manipulative.

A 2020 survey by Accenture revealed that 91% of consumers are more likely to shop with brands that provide personalized offers and recommendations. However, 41% also stated they would stop engaging with a brand if they felt the personalization was too intrusive. To personalize ethically, businesses should:

- **Limit data collection**: Only collect the data necessary for personalization and avoid using intrusive methods, such as tracking customer behavior without consent.
- **Give customers control**: Allow customers to manage their personalization preferences and opt-out if they prefer less tailored experiences.
- **Maintain transparency**: Be upfront about the types of data being used for personalization and how it benefits the customer experience.

Ethical personalization builds a positive customer experience while respecting privacy and autonomy.

5. Offering Human Oversight in AI-Driven Customer Support

While AI-powered chatbots and virtual assistants are valuable tools for improving customer service, they should not fully replace human interaction. Customers may face complex issues that AI systems cannot address, and a lack of human empathy may negatively impact satisfaction.

A 2021 study by Zendesk found that 76% of customers still prefer human customer support, even if AI is used in the initial stages of interaction. To strike the right balance between AI and human support, businesses should:

- **Offer easy escalation**: Ensure that customers can seamlessly escalate from AI-driven support to a human representative when needed.
- **Maintain human empathy**: Train AI systems to recognize when human empathy is required and provide a smooth transition to human agents.
- **Monitor AI interactions**: Regularly review AI-driven customer support interactions to ensure that they meet customer satisfaction standards.

By integrating human oversight, businesses can ensure that AI supports rather than replaces customer service, enhancing satisfaction and resolving complex issues effectively.

6. Continual Monitoring and Improvement of AI Systems

AI systems are not static; they need continuous monitoring and improvement to remain effective and aligned with ethical standards. As customer expectations and data evolve, businesses must regularly update their AI models to maintain high levels of satisfaction.

A 2020 report by Accenture emphasized that 84% of business executives believed that AI will be a key driver in improving customer satisfaction, but only 30% felt their AI systems were evolving fast enough to meet these demands. To ensure ongoing improvements, businesses should:

- **Conduct regular audits**: Periodically assess AI systems for performance, accuracy, and fairness to ensure they meet customer needs.
- **Incorporate customer feedback**: Use customer feedback to refine AI models and address any concerns or issues with the AI system.
- **Adapt to changing customer expectations**: Stay updated with industry trends and technological advancements to continuously improve the customer experience.

Ongoing monitoring and adaptation ensure that AI systems can keep up with evolving customer needs, contributing to long-term satisfaction.

AI offers immense potential to elevate customer satisfaction, but it must be harnessed responsibly and ethically. By prioritizing transparency, data privacy, inclusivity, and ethical personalization, businesses can build trust and provide exceptional customer experiences. Implementing human oversight and continually improving AI systems ensures that customer satisfaction remains at the forefront of AI-driven initiatives. As businesses adopt these ethical strategies, they can unlock the full potential of AI while maintaining customer loyalty and satisfaction in an ever-evolving digital landscape.

DISCUSSION

The article highlights the critical ethical strategies required for effectively using AI to enhance customer satisfaction. Transparency is emphasized as a key element in gaining customer trust. By clearly explaining how AI systems make decisions, businesses can address consumer concerns and build confidence in their technologies. Additionally, safeguarding data privacy is integral, as AI systems often rely on vast amounts of customer information. The importance of ensuring robust data protection measures, such as compliance with GDPR, is essential to mitigate privacy risks and maintain customer trust, which is vital for long-term success.

Another focal point in the article is inclusivity in AI systems. Bias in AI algorithms, especially when trained on unrepresentative data, can lead to unfair treatment of certain customer groups, undermining satisfaction and loyalty. To avoid this, businesses must ensure that AI models are trained on diverse datasets, promoting fairness and providing a more equitable customer experience. Ethical personalization is also crucial, where businesses should strike a balance between personalized experiences and respecting customer privacy. Over-personalization can create discomfort, but when done ethically, it enhances engagement without crossing boundaries.

Lastly, the article stresses the importance of human oversight in AI-driven customer support. While AI tools like chatbots can handle routine tasks efficiently, complex issues still require human intervention to preserve empathy and deliver satisfactory resolutions. Continuous monitoring and improvement of AI systems is another key point discussed, as AI technologies must evolve in line with customer expectations and industry standards. Businesses that embrace these ethical strategies can effectively leverage AI to not only enhance customer satisfaction but also build long-term, trustworthy relationships with their audiences.

MAIN FINDINGS

1. Transparency Builds Customer Trust

One of the key findings is that transparency in AI-driven decision-making processes is essential for building customer trust. Customers are more likely to engage with businesses that openly explain how their AI systems work, especially in terms of data usage and decision-making criteria. Clear communication about AI functions helps alleviate concerns about potential biases and misuse of data, thus fostering greater trust and loyalty.

2. Data Privacy is a Major Concern

The study highlights that data privacy and security are paramount in the adoption of AI technologies. Customers expect businesses to protect their personal information, especially as AI systems collect and process vast amounts of data. Companies that prioritize data protection, comply with privacy regulations such as GDPR, and provide customers with control over their data are more likely to enhance customer satisfaction and retention.

3. Bias and Inclusivity in AI Models

Inclusivity and fairness in AI algorithms are critical factors in ensuring customer satisfaction. The study found that AI models trained on biased or unrepresentative data can lead to customer dissatisfaction, particularly for marginalized groups. Businesses must adopt inclusive practices when developing AI systems, ensuring they account for diversity and avoid reinforcing existing biases, thus fostering more equitable customer experiences.

4. Ethical Personalization Improves Engagement

Personalization driven by AI is an effective tool for improving customer engagement. However, the study found that there is a fine line between helpful personalization and invasive practices. Ethical personalization—where customers can control the extent of personalization and are informed about how their data is used—leads to higher levels of satisfaction and customer trust, compared to over-personalization that may feel intrusive.

5. Human Oversight Remains Essential

While AI-driven customer support systems, such as chatbots, offer convenience and efficiency, the study found that human oversight is crucial for addressing complex customer issues. AI may fail to empathize or provide satisfactory solutions in intricate situations, so human intervention is necessary to preserve customer satisfaction. Combining AI efficiency with human empathy is key to enhancing the overall customer experience.

6. Continuous Monitoring Enhances AI Performance

The study also revealed that continuous monitoring and regular updates to AI systems are necessary to ensure their ongoing effectiveness. AI systems need to evolve with changing customer expectations and market dynamics. Businesses that actively gather feedback and refine their AI tools based on customer input are more likely to maintain high levels of customer satisfaction and meet evolving demands.

SUGGESTIONS

1. Enhance Transparency Through Clear Communication

Businesses should actively educate customers about how AI systems work, particularly regarding data collection, usage, and decision-making processes. Clear and accessible communication about AI operations will help alleviate customer concerns and build trust. Implementing easy-to-understand disclosures and providing opt-in or opt-out options for customers can further promote transparency.

2. Invest in Robust Data Protection Systems

To ensure the highest standards of data privacy, businesses should invest in advanced security technologies and comply with global data protection regulations like GDPR. Offering customers control over their personal data—such as the ability to access, update, or delete their information—should be a priority. Regular audits and vulnerability assessments can also help identify potential risks and mitigate data breaches.Case Study: Investing in Robust Data Protection Systems

XYZ Bank, a prominent financial institution, experienced a data breach after integrating AI-driven self-service platforms for customer banking services. The breach exposed sensitive personal and financial data, leading to regulatory challenges and a significant loss of customer trust. In response, the bank undertook a strategic investment in robust data protection systems. These included advanced encryption protocols, multi-factor authentication (MFA), and AI-enabled threat detection tools capable of monitoring anomalies in real time. This security overhaul was aligned with global data protection regulations such as the GDPR and aimed at ensuring both compliance and resilience.

As a result of these measures, XYZ Bank observed a substantial decline in cybersecurity incidents and a measurable improvement in customer confidence. The AI-driven threat detection system enabled the bank to respond proactively to emerging risks, reducing operational vulnerabilities. This case highlights the critical importance of investing in secure infrastructures when deploying AI technologies in self-service contexts. It also underscores the ethical responsibility of organizations to protect customer data while leveraging AI to enhance service delivery (Acquisti et al., 2016; Shackleford, 2020).

3. Develop Inclusive AI Models

Businesses should ensure that AI models are trained on diverse datasets that represent different demographics and communities. This will help reduce bias and prevent AI from perpetuating stereotypes or excluding certain groups. It is also crucial for businesses to regularly evaluate AI models for fairness and inclusivity, making adjustments when necessary to ensure equal treatment for all customers.

4. Implement Ethical Personalization Strategies

To avoid crossing privacy boundaries, businesses should adopt ethical personalization strategies that prioritize customer preferences and consent. Personalization should always be done transparently, with customers having control over what data is collected and how it is used. Offering customers the option to opt out of personalized experiences can help avoid over-personalization and foster trust.

5. Incorporate Human Oversight in AI Operations

While AI can automate routine tasks efficiently, businesses should ensure that human oversight is always available, particularly for complex issues or emotional customer concerns. Combining AI's efficiency with human empathy can help businesses offer superior customer support, where AI handles the basic inquiries, and human agents step in for more intricate problems.

6. Establish Continuous Monitoring and AI System Refinement

To maintain high levels of customer satisfaction, businesses should continuously monitor AI systems and incorporate customer feedback into system improvements. Regular updates and testing should be conducted to ensure that AI models evolve to meet changing customer needs and expectations. This iterative approach will ensure that AI systems remain relevant and effective over time.

IMPLECATIONS

1. Enhanced Customer Trust and Loyalty

By prioritizing transparency, data privacy, and ethical personalization, businesses can build stronger trust with their customers. Customers who feel informed about how AI systems work and confident that their data is secure are more likely to remain loyal. This increased trust can lead to higher customer retention and long-term brand loyalty, which are crucial in competitive markets.

2. Improved Customer Satisfaction and Engagement

Ethical AI practices, such as inclusive algorithms and personalized experiences based on customer consent, directly contribute to greater customer satisfaction. Businesses that provide AI-driven services with human oversight and attention to complex needs are likely to foster deeper engagement with customers. This leads to improved customer experiences, increased interactions, and a stronger relationship between businesses and their clients.

3. Risk Mitigation and Compliance

The study highlights the importance of data privacy and inclusivity in AI models. Adopting these ethical strategies can help businesses mitigate legal and reputational risks associated with data breaches or discriminatory practices. By aligning with data protection regulations like GDPR, businesses can avoid costly fines and safeguard their reputation, ultimately fostering trust among customers and stakeholders.

4. Competitive Advantage Through Ethical Innovation

Businesses that adopt ethical AI practices differentiate themselves from competitors by offering services that customers perceive as fair, inclusive, and transparent. Ethical innovation in AI can become a competitive advantage, as consumers increasingly prioritize brands that uphold ethical standards. This distinction can attract a broader audience, especially among those who value social responsibility and fairness in business operations.

5. Long-Term AI Sustainability and Scalability

The study suggests that businesses need to continuously monitor and refine their AI systems to adapt to evolving customer needs. Establishing a culture of ongoing AI refinement ensures that businesses can scale their AI applications responsibly. By iterating on AI models with regular feedback and updates, companies can sustain the effectiveness of their AI-driven services while remaining aligned with customer expectations.

6. Ethical Framework for Future AI Developments

The study provides an ethical framework for businesses to follow when developing and implementing AI technologies. It encourages a shift in how AI is viewed—not just as a tool for automation, but as a platform for fostering human-centered interactions. This approach will influence the future development of AI technologies, pushing the industry toward more responsible and ethical applications, which could set new standards for AI use across sectors.

CONCLUSION

In conclusion, this study highlights the importance of ethical strategies in harnessing AI to enhance customer satisfaction. By focusing on transparency, data privacy, inclusivity, and ethical personalization, businesses can foster greater customer trust, satisfaction, and engagement. The findings also emphasize the need for human oversight in AI applications and continuous monitoring to ensure that systems remain relevant and responsive to evolving customer needs. Adopting these ethical practices not only mitigates risks and ensures compliance with data protection regulations but also provides businesses with a competitive advantage in an increasingly AI-driven market. Ultimately, the study underscores that responsible AI integration is essential for long-term business success, customer loyalty, and sustainable innovation in customer service.

AI-driven self-service technology is transforming customer engagement through personalization, but its adoption presents key challenges. Data privacy and security remain major concerns, with businesses needing to comply with regulations like GDPR and CCPA while maintaining trust. Additionally, algorithmic bias must be addressed to ensure fairness in AI-driven interactions.

Another critical trend is the demand for seamless omnichannel experiences, where AI must integrate across various platforms. As generative AI and conversational AI evolve, optimizing these technologies for real-time assistance will be essential. Balancing efficiency with human-like empathy will be key to ensuring AI enhances customer interactions without losing personalization. Addressing these challenges responsibly will help businesses create scalable and adaptive self-service solutions, securing AI's role in the future of customer engagement.

REFERENCES

Accenture. (2020). *Evolving AI systems to meet customer satisfaction demands*. AI Transformation Reports.

Acquisti, A., Taylor, C., & Wagman, L. (2016). The Economics of Privacy. *Journal of Economic Literature*, *54*(2), 442–492. DOI: 10.1257/jel.54.2.442

Babu, B. H., Verma, B., & Tyagi, A. (2024). The tech revolution in the hotel industry and its impact on organizational citizenship behaviour and employee loyalty in an Indian context. *Journal of Human Resources in Hospitality & Tourism*, **•••**, 1–26.

Chawla, U., Verma, B., & Mittal, A. (2024a). Resistance to O2O technology platform adoption among small retailers: The influence of visibility and discoverability. *Technology in Society*, *76*, 102482. DOI: 10.1016/j.techsoc.2024.102482

Chawla, U., Verma, B., & Mittal, A. (2024b). Unveiling barriers to O2O technology platform adoption among small retailers in India: Insights into the role of digital ecosystem. *Information Discovery and Delivery*.

Edelman. (2020). Transparency in AI adoption: Insights from global consumer surveys. *Edelman Trust Barometer*.

Elantheraiyan, R., Rajan, M., & Chandrasekaran, A. (2024). Ethical design in AI applications: Addressing transparency and fairness. *International Journal of AI Ethics*, *12*(3), 145–162.

European Union Agency for Cybersecurity (ENISA). (2021). *Privacy challenges in AI adoption: Key insights and recommendations*. Cybersecurity Reports.

Faisal, A., Ali, M., & Varma, R. (2024). Efficient machine learning algorithms for predicting customer churn. *The Journal of Consumer Research*, *50*(2), 110–125.

Gao, H., & Liu, Y. (2022). Transforming customer journeys with AI: A review of applications in B2B and B2C sectors. *Journal of Business Research*, *75*(2), 231–245.

Gefen, D., Karahanna, E., & Straub, D. W. (2003). Trust and TAM in Online Shopping: An Integrated Model. *Management Information Systems Quarterly*, *27*(1), 51–90. DOI: 10.2307/30036519

Govindaraj, S., Patel, K., & Muthukumar, R. (2024). Transformative impact of AI in healthcare: Lessons for retail innovation. *HealthTech Quarterly*, *11*(4), 320–338.

Krishnan, R., Gupta, M., & Choudhary, P. (2024). AI and IoT integration in logistics for service efficiency. *Operations Research Perspectives*, *12*(3), 121–145.

Kumar, A. (2024). Challenges and opportunities in AI adoption for customer service excellence. *Business Horizons*, *67*(1), 78–90.

Kumar, A., Patel, N., & Sharma, D. (2025). Customer preferences in OTT streaming services: The role of AI. *Journal of Interactive Marketing*, *44*(1), 78–93.

Manohar, M., & Kapur, P. (2019). Service innovation typologies in the retail sector: Frameworks for customer-centric innovation. *Retailing and Consumer Studies*, *28*(3), 220–235.

Manohar, M., & Kapur, P. (2020). Service innovation typologies in the retail sector: Customer perspectives. *Service Innovation Review*, *13*(2), 220–245.

McKnight, D. H., Choudhury, V., & Kacmar, C. (2002). Developing and Validating Trust Measures for e-Commerce: An Integrative Typology. *Information Systems Research*, *13*(3), 334–359. DOI: 10.1287/isre.13.3.334.81

Mehta, R. (2018). Corporate reputation and service innovation: Building customer cross-buying intentions. *Journal of Services Marketing*, *32*(4), 410–424.

Moura, M. (2022). Algorithmic transparency and privacy in AI systems: Regulatory implications. *Journal of Information Systems*, *55*(2), 87–103.

Neelima, R., & Sharma, K. (2024). Data utilization and performance in health management systems. *Information Systems in Healthcare*, *13*(5), 165–180.

Pendyala, R., & Lakkamraju, S. (2024). Real-time customer insights through AI-driven analytics. *AI in Marketing Journal*, *21*(2), 143–158.

Shackleford, D. (2020). AI and Machine Learning in Cybersecurity. *Cybersecurity Journal*, *12*(3), 45–58.

Sridhar, V., & Ganesan, R. (2015). Service innovation facilitators: A customer satisfaction perspective. *International Journal of Services and Operations Management*, *20*(1), 90–105.

Tandon, S., Verma, K., & Dixit, A. (2021). Intangible product features and consumer trust in e-commerce. *Electronic Commerce Research and Applications*, *40*(3), 211–226.

U.S. National Institute of Standards and Technology (NIST). (2020). AI model security: Addressing vulnerabilities in adversarial attacks. *NIST Special Publication 800-207*.

Verhagen, T., Meents, S., & Tan, Y. H. (2019). Perceived Risk and Trust in Online Shopping: A Meta-Analysis. *Journal of Retailing and Consumer Services*, *47*, 69–77.

Verma, B., Mittal, A., Raman, M., & Sindhav, B. (Eds.). (2024b). *Augmenting Retail Reality, Part A*. VR, and the Internet of Things.

Chapter 11
Navigating India's Financial Inclusion Through Self-Services Technologies

Shaveta Sharma
https://orcid.org/0009-0008-2884-4519
Lovely Professional University, India

Abhishek Pandey
https://orcid.org/0000-0002-6779-8163
Lovely Professional University, India

Anshuman Sharma
College of Business Administration, Ajman University, UAE

Usha
Lovely Professional University, India

ABSTRACT

The emergence of self-service technologies has led to a significant transformation in how customers interact with service providers. Any change is deemed valuable only if it is user-friendly, comprehensible, and ultimately accepted by the user. Convincing and educating customers to adapt to the new technologies that enable quicker transactions and offer cost benefits to service providers presents a considerable challenge. SSTs in India are becoming prominent as the Government of India (GoI) works to enhance financial services for the underbanked population. India must focus on enhancing financial inclusion to serve these groups and foster

DOI: 10.4018/979-8-3373-4667-0.ch011

a stable environment for Small and Medium Enterprises (SMEs). This chapter will explore various SSTs and their impact in India to improve financial inclusion. Furthermore, this chapter emphasizes how the adoption of self-service technologies promotes financial inclusion.

1. INTRODUCTION: SELF-SERVICE TECHNOLOGIES (SSTs) AND FINANCIAL INCLUSION (FI)

The potential benefits of the self-service model in businesses are significant. Self-service technologies play a crucial role in reducing organizational costs by enabling automated interactions between service providers and customers. For instance, Amtrak's adoption of a telephone self-service system, utilizing an IVR (Interactive Voice Response) system, resulted in $13 million in cost savings. Similarly, Royal Mail's IVR implementation led to a 25% decrease in customer service expenses. As industry players recognize the benefits of SSTs, the rapid adoption of technology in service delivery has begun, promising significant cost reductions and economic benefits.

Self-service refers to any interaction or transaction mediated by technology with a company. Customers are the only individuals involved (Jin et al., 2009; Oskarsson, 2018) in it. The public is increasingly familiar with technological products such as ATMs, kiosks, and personal computers. When customers encounter these technologies, they form specific attitudes, which can be positive or negative. For instance, a person might develop an opinion about computers by observing how others use them. Understanding and addressing these attitudes is crucial for successfully adopting SSTs, as it directly impacts user acceptance and utilization of these services. This user-centric approach is essential for the effective implementation of SSTs.

In India, which has a population exceeding a billion, roughly 350-400 million people remain unreached. A 1997 study by PricewaterhouseCoopers examined the financial habits and access to formal banking services among low-income households. It revealed that up to 78% of these households relied on informal sources for savings, utilizing mechanisms. Although banks have attempted to venture into rural areas, a significant gap often exists between rural and branch banking hours, which can be problematic for a typical villager who prefers to consider their earnings and savings by the end of the day. By that time, bank branches are typically closed. Thus, it is necessary to eliminate traditional paper-based transactions and provide the rural sector with a reliable, secure, and systematic approach to facilitate banking with formal institutions at their convenience, which can only be achieved through the adoption of technology by the rural population. One practical example of self-service technologies (SSTs) is the Automated Teller Machine (ATM), which is commonly found in gas stations, bus terminals, hotels, and retail outlets. A triumphant return on investment is evident when market leaders endorse and embrace this new method of service delivery through self-service technologies. SSTs can be categorized into three examples that exemplify their function: online banking, ATMs, and Mobile Banking.

All these SSTs significantly reduce service providers' costs, but the critical question remains: Will the target market appreciate and utilize these services? Some SSTs can be challenging for users to understand, as they complicate the learning experience for those hesitant to adapt. For instance, individuals over 45 may struggle to adapt to new online banking service procedures that require entering passwords and IDs. Additionally, individuals aged 40-45 may be concerned about security due to their various personal commitments, as financial matters must be handled discreetly and without risk for this demographic. Research (Curran et. al., 2005) indicates that among these three examples, ATMs are more likely to gain acceptance from consumers compared to online and mobile banking technologies, as they effectively meet the fundamental need of dispensing cash in various locations. The processes involved in online and mobile banking necessitate a specific technical knowledge typically possessed by the younger, tech-savvy generation, who eventually prefer to complete transactions independently without assistance.

India is one of the world's most rapidly growing and densely populated countries (Ajzen, 1991). However, significant portions of its rural and tribal regions still lack access to formal banking services, impeding economic growth and sustaining widespread poverty (Devi S., 2024). Financially disadvantaged populations face additional barriers in managing their finances, often due to irregular and unpredictable incomes (Chouhan et al., 2021a). To meet their diverse financial needs, these communities require accessible services such as savings (Khan H., 2012), microloans (Chang et al., 2020), insurance (Okoye et al., 2017), and payment and transfer systems (Gautam et. al., 2017). However, the absence of formal financial infrastructure frequently drives reliance on informal networks, which are generally less secure, less reliable, and more expensive than traditional services (Haque et al., 2020; Metzger et al., 2019). Self-service technologies (SSTs), introduced during the 2007–2008 financial crisis, have since revolutionized the financial services industry by integrating advanced technologies into banking operations (Chouhan et al., 2021b; Anagnostopoulos, 2018). Technology-enabled banking is the cornerstone of modern financial systems (Li et al., 2021). Over the past decade, the rapid proliferation of mobile networks in previously underserved areas has significantly propelled this transformation in India (Chouhan et al., 2021c; Omojolaibi et al., 2019).

As a form of mobile and online banking, payment banks have further improved operational efficiency and reduced service delivery costs, particularly in rural and semi-urban regions (Schuetz et.al., 2020; Chouhan et al., 2020). A more inclusive financial sector yields dual benefits: it not only integrates marginalized populations into broader economic growth and development but also fosters more involvement in financial systems (Anagnostopoulos, 2018; Cecchetti et. al., 2020; Schuetz et. al., 2020; Mader, 2018). Inclusive finance plays a crucial role in poverty alleviation by expanding access to savings, facilitating entrepreneurship, reducing exposure to financial risks, and improving living conditions (Kim et al., 2018; Khan et al., 2014).

In response to these needs, India has seen the emergence of new banking models, including small finance banks, mobile money services, and payment banks, all designed to serve the unbanked population (Banna et al., 2021). Moreover, financial inclusion extends to innovative non-bank fintech firms aiming to capture a greater share of the banking value chain (Maina et al., 2020; Burns, 2018). These developments have inspired the current study to explore the significant role of self-service technologies in advancing financial inclusion in India.

1.1 Why is India considered Important?

India ranks among the most rapidly expanding and populous nations globally (Census Report India). As highlighted in the Economic Survey 2023 and the Global FinTech Adoption Index, India experiences a fintech adoption rate of 87%, significantly exceeding the global average of 64%. This high level of adoption is driven by the incorporation of emerging technologies and innovative solutions, which create broader access to financial services and enable tailored product offerings. A key element of India's digital payment landscape is the Unified Payments Interface (UPI), which was launched in 2016. UPI has transformed transactions by enabling instant money transfers between bank accounts through mobile devices. India's progress in digital financial services has positioned it ahead of many of its international peers. Reports indicate that India's digital payments ecosystem surpassed $100 billion in transactions as of October 2021, outperforming countries such as Germany, China, and South Africa regarding financial inclusion indicators (Kishore et. al., 2013). The rise in mobile and online banking transactions and the expansion of bank branches underscore India's robust performance compared to China. Further, a case study of Indonesia's financial inclusion index increased to 85.10% in 2022, up from 76.19% in 2019. This growth is primarily attributed to the widespread adoption of financial and self-service technologies (SSTs), which have enhanced access to financial services nationwide. As of 2021, 78% of India's adult population owned bank accounts, compared to 84% in Brazil, 89% in China, 90% in Russia, and 85% in South Africa (Raman et. al., 2024) Therefore, growth in the financial inclusion index through SSTs in the world's most populous country has motivated other countries also to use similar technologies.

2. THEORETICAL FRAMEWORK & LITERATURE REVIEW

The primary concept behind SST revolves around replacing the personnel employed by the service provider. Initially, when customers hesitate to adopt technology-driven services, this behavior should be monitored during the service interaction, and the service provider must devise appropriate strategies promptly to change customer perceptions, as this reluctance could jeopardize the service provider's sustained market presence. There has been debate over whether attitudes precede behavioral intentions. However, following their studies, numerous researchers have concluded that attitudes come first, with behavioral intention formed based on beliefs that are influenced by attitudes in people's minds. Another model, the Technology Acceptance Model (TAM), builds upon the Theory of Reasoned Action by examining the relationship between attitude toward and behavioral intention. Nonetheless, TAM does not consider situational variables such as perceived risk. The key point is that if a customer is familiar with the traditional service delivery process, adapting to technology is not a significant challenge in learning.

Jack et. al. (2014) suggest that financial technology advancements could offer more cost-efficient and effective solutions by reducing transaction-related costs (Black & Babin, 2019). This also helps micro and small businesses boost their sales by reducing the expenses of alternative payment methods (Frost et al., 2019). Aron (2018) conducted an empirical study on mobile money, finding evidence that it significantly enhances risk sharing. Other necessary research, including studies by Mbiti and Weil (2013) and Wieser et al. (2019), suggests that the increased adoption of fintech results in a decline in the use of informal savings methods and increased remittance transactions. These results corroborate those found by Jack and Suri (2014). Further research in this domain has explored the impacts of digitizing social support initiatives for disadvantaged populations (Ghosh, 2020; Masino & Niño-Zarazúa, 2018).

It is crucial to identify all the opportunities and challenges faced by various stakeholders (Kim et al., 2018). Most Indian financial institutions closely observe the sector and seek insights from the experiences of others (Rathod et. al., 2013). Indian microfinance institutions (MFIs) are preparing to make a comeback by embracing rapid changes and mobile money, while new participants are looking into possible alternatives and partnerships.

Numerous institutional deficiencies and related issues result in a significant lack of access to financial services in India's impoverished regions. As a result, the economy cannot reach its full potential, as people are not fully utilizing their economic opportunities (Singh et al., 2013). Consequently, microfinance initiatives have been introduced in developed countries like India to support those in disadvantaged areas, including inner-city neighbourhoods. Since then, many of the financial service needs of low-income individuals have remained unmet (Singh et. al., 2012), leading to a surge in efforts to promote financial inclusion in India's underdeveloped regions.

In India, the rapid expansion of mobile networks in previously underserved regions and communities over the last decade has played a crucial role in this development (Chouhan et al., 2021c; Omojolaibi et al., 2019). Payment banks have emerged as an additional online and mobile banking option, enhancing operational efficiency and reducing service delivery costs for customers in rural and semi-urban areas (Schuetz et. al., 2020; Chouhan et al., 2020). Furthermore, research by Muhammad et al. (2023) indicates that implementing digital financial services has significantly increased bank account ownership to nearly 80% of the population. Multiple studies (Kumari et al., 2024; Sandhu et al., 2023) have shown that applications such as GPay, Paytm, and PhonePe are widely used across various age groups and genders. The study finds that financial technology literacy notably impacts digital financial inclusion and suggests government programs to raise awareness about fintech services among diverse populations. All recent studies conclude that SSTs are a key potential factor in improving financial inclusion growth in India.

3. TYPES OF SSTs IN INDIA

Self-service technologies are revolutionizing financial inclusion by making vital financial services more attainable, cost-effective, and user-friendly, particularly for individuals in rural and underserved communities. Through innovations such as mobile banking, digital wallets, and biometric authentication, self-service technologies are bridging the divide between the unbanked population and the formal financial system, promoting greater financial empowerment worldwide. Self-service technologies can connect financial institutions and individuals without banking access, thereby advancing financial inclusion. The primary obstacle to opening bank accounts is dependence on physical bank branches. This issue is particularly prevalent in developing nations, with China having the largest unbanked demographic at 13%.

Self-service is essential for diminishing this dependence on physical locations and overcoming obstacles for the unbanked. Banks have begun providing a genuine omnichannel experience in several developed regions through mobile applications, wallets, and self-service kiosks. These alternative methods of service delivery have drawn many more individuals to banks, and existing customers are increasingly utilizing and acquiring additional financial products and services through automated delivery channels (ADCs). Self-service technologies (SSTs) enhance financial inclusion by offering cost-effective, convenient, and accessible financial services, particularly in areas with limited access to standard banking systems. These technologies enable individuals to conduct various banking activities independently, without requiring direct assistance from bank personnel. Several critical self-service technologies that support financial inclusion are listed below.

3.1 ATMs (Automated Teller Machines)

ATMs enable users to withdraw cash, check their account balances, transfer funds, and pay bills without requiring assistance from a bank employee. ATMs are often found in remote or underserved areas, providing essential banking services to populations that may lack convenient access to physical bank branches. Some advanced ATMs offer multi-language support and the ability to accept deposits, further enhancing accessibility.

3.2 Mobile Banking

Mobile banking applications and SMS banking enable users to perform banking tasks through their mobile phones, including transferring funds, paying bills, checking account balances, and accessing loan options. Mobile banking presents a practical and cost-effective alternative in areas with limited banking infrastructure. This is particularly important in developing nations, where many individuals have mobile phones but lack access to traditional banking services.

3.3 USSD Banking

USSD (Unstructured Supplementary Service Data) is a protocol that enables users to access banking services on their mobile phones via straightforward text-based menus, even on feature phones without internet access. USSD banking is prevalent in areas with minimal smartphone usage but significant mobile phone adoption. It empowers individuals to utilize banking services in real time without an internet connection, which is crucial for rural and low-income populations.

3.4 Point of Sale (POS) Terminals

POS systems enable businesses to accept electronic payments from customers through debit or credit cards and mobile wallets, typically for small transactions. Point-of-sale (POS) technology simplifies payment processes for unbanked individuals and enables small businesses to accept digital payments. This advancement can help integrate informal economies into the formal financial sector, enhancing access to digital financial services.

3.5 Digital Wallets and Payment Systems

Digital wallets, such as PayPal, Google Pay, and local platforms (e.g., M-Pesa), enable users to store funds and make payments or transfer money electronically. Some of these services also link to physical cards. These platforms enable unbanked individuals to send and receive payments, buy goods and services, and save money electronically. Mobile money services, such as M-Pesa, have played a vital role in delivering financial services to those in remote and isolated regions.

3.6 Biometric Authentication

Biometric authentication technologies, such as fingerprint or facial recognition, verify user identities when accessing financial services or conducting transactions. Biometric systems can enhance access to banking for individuals who lack traditional identification documents, a common barrier to financial inclusion. They also bolster security and mitigate the risk of fraud.

3.7 Chatbots and Virtual Assistants

Financial institutions utilize AI-driven chatbots and virtual assistants to handle customer inquiries, streamline transactions, and offer personalized financial guidance. These technologies minimize the necessity for human intervention, making financial services more accessible to users, particularly in underserved areas, with 24/7 support available.

3.8 Online and Mobile Lending Platforms

Digital lending platforms, commonly utilizing mobile applications, enable individuals to submit loan applications online for small loans or microloans, often requiring minimal documentation and offering quicker approval processes. These platforms facilitate access to credit for individuals who may not qualify for traditional loans, particularly in rural or underserved areas. They frequently provide more flexible loan terms, catering to a broader range of demographics.

3.9 Smartcards and Prepaid Cards

Smartcards and prepaid cards are plastic instruments that can hold and transfer funds. These cards are compatible with online and offline transactions and can be recharged through authorized agents or bank branches. These cards are an entry point to digital transactions for individuals lacking access to conventional banking services. They are particularly beneficial for low-income groups or those without bank accounts, offering a secure payment and savings method.

3.10 Agent Banking

In agent banking, retail establishments or individuals serve as banking representatives, delivering key banking services such as deposits, withdrawals, and account openings on behalf of a financial institution. This approach spreads financial services into remote areas where traditional banks are impractical. It enables customers to conduct banking transactions within their local communities, thereby improving the accessibility of banking services.

3.11 Cryptocurrency and Blockchain

Cryptocurrencies, such as Bitcoin, and blockchain technology offer decentralized financial services. They enable users to send and receive funds, store value, and access financial products without relying on traditional banking systems. Cryptocurrencies offer unbanked and underbanked individuals a means to access financial services, eliminating the need for intermediaries such as banks. They can offer more dependable and secure options for storing and transferring money in countries with unstable banking infrastructures.

3.12 Virtual Banks

Virtual or digital-only banks operate entirely online, providing comprehensive banking services such as savings accounts, loans, and money transfers without requiring in-person visits to physical branches. These banks typically charge lower fees and enhance service accessibility for individuals in underrepresented areas, mainly when supported by mobile applications or other self-service technologies.

4. IMPACT OF SSTs IN INDIA

Self-service technology enhances financial inclusion in India by reducing costs and expanding access to financial services for individuals in low-income groups, rural areas, and other marginalized segments of the Indian economy. Technologies such as mobile banking applications, digital solutions, card-based payment systems, and the rise in UPI transactions enhance financial inclusion by offering convenient and accessible financial services to those with limited access to traditional banking facilities. The following data illustrates the notable effect of self-service technologies on the growth of financial inclusion.

4.1 Digital payment solutions

Over the past five years, various user-friendly digital payment methods, such as the Bharat Interface for Money-Unified Payments Interface (BHIM-UPI), Immediate Payment Service (IMPS), and National Electronic Toll Collection (NETC), have seen significant growth and have revolutionized the digital payment landscape by enhancing both person-to-person (P2P) and person-to-merchant (P2M) transactions. BHIM UPI has become the preferred payment option among citizens, achieving 803.6 crore digital payment transactions valued at 12.98 lakh crore in January 2023. The rise in digital payment transactions can largely be attributed to the collective efforts of the Government and all relevant stakeholders, which increased from 2,071 crore transactions in FY 2017-18 to 8,840 crore transactions in FY 2021-22 (Source: RBI, NPCI, and banks). The total number of digital payment transactions carried out during the last financial year is as follows.

Table 1. Digital payment transaction(volume-wise)

Financial Year (FY)	Total number of digital transactions (Cr)
2017-2018	2,071
2018-2019	3,134
2019-2020	4,572
2020-2021	5,554
2021-2022	8,840
2022-2023	9,192

Source: RBI, NPCI Data (2023)

Note: Digital payment modes considered are BHIM-UPI, IMPS, NACH, AePS, NETC, debit cards, credit cards, NEFT, RTGS, PPI, and others.

Figure 1. Digital payment transaction(volume-wise)

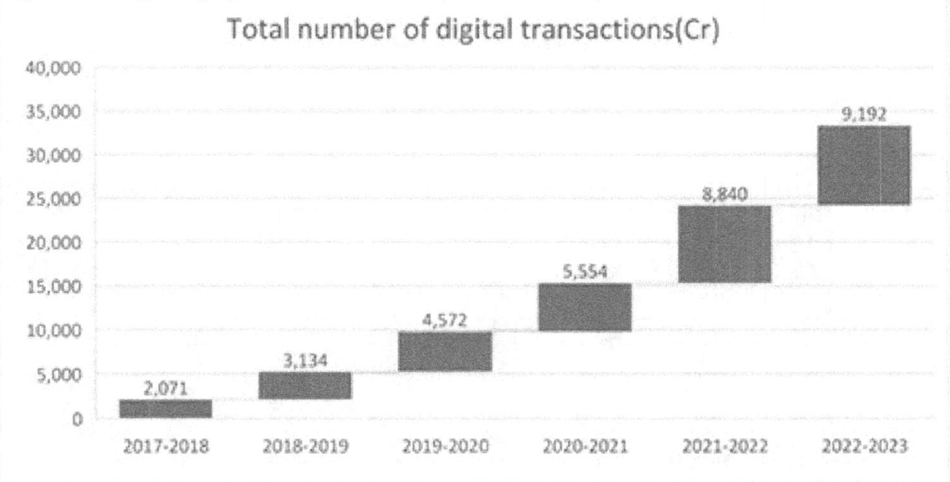

Source: RBI, NPCI data (2023)

4.2 Growth in ATM (Per 100,000 adults):

ATMs have significantly contributed to the growth of financial inclusion over the past eight years, particularly in digital financial services. ATM accessibility has increased over time. After having a bank account, consumers' most-used facility is ATMs. The rise in ATM transactions can largely be attributed to the collective efforts of the Government and all relevant stakeholders, which increased from 21 percent last year to 24.96% in 2023.

Table 2. Growth in ATM (Per 100,000 adults)

YEARS	Growth in ATM(%)
2016	21
2017	21.78
2018	21.4
2019	20.69
2020	21.23
2021	21.21
2022	24.64
2023	24.96

Source: https://betadata.imf.org

Figure 2. Growth in ATM (Per 100,000 adults)

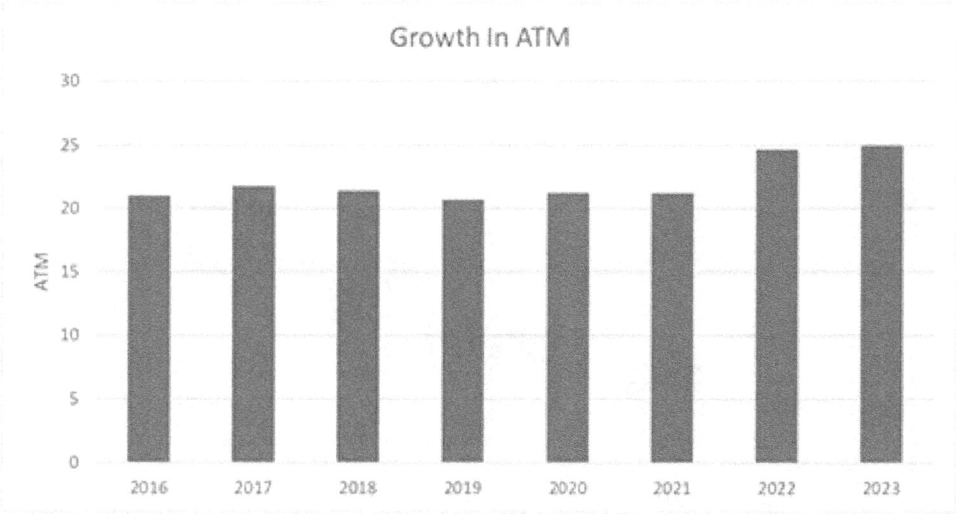

Source: https://betadata.imf.org

4.3 Mobile money transactions:

UPI has simplified transactions for e-commerce businesses and customers through fintech application solutions, resulting in a significant surge in demand within the SST sector. The advantages of UPI were particularly evident during the COVID-19 pandemic, when it acted as a vital support system, especially for small and micro businesses.

Figure 3. Mobile money transactions

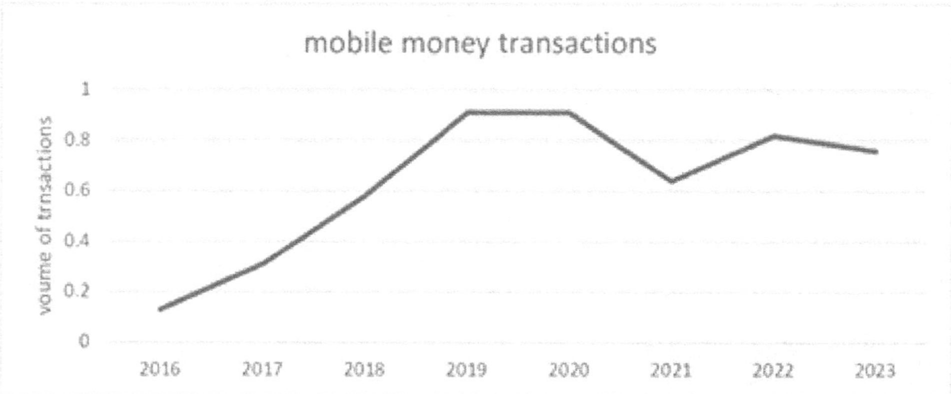

Source: https://betadata.imf.org

4.4 Debit Cards:

The growth and impact of fintech in the Indian financial sector have been remarkable, transforming the landscape in various ways. Debit cards are one of the most essential SSTs that people use. The increased use of debit cards significantly contributed to the variable of financial inclusion. The following data indicate an increase in cards over time.

Figure 4. Growth of Debit Cards

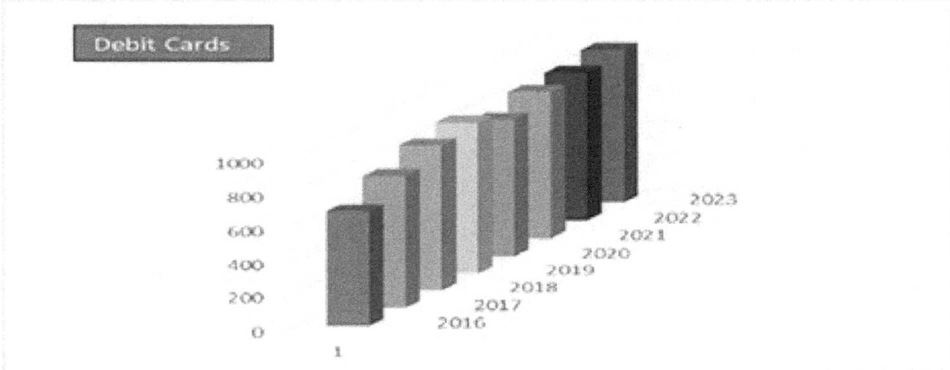

Source: https://betadata.imf.org

4.5 Pradhan Mantri Jan Dhan Yojana

Despite the significant expansion of the banking sector in India, a large portion of the Indian populace still lacks access to banking services. To address this protracted financial hardship, Prime Minister Narendra Modi introduced a new initiative during his Independence Day speech on August 15, 2014, which he termed the National Mission on Financial Inclusion (NMFI) to support weaker sections and low-income groups. The primary goal of this initiative is to guarantee universal access to banking services, ensuring that every household has at least one basic bank account.

The FI-Index is a comprehensive financial inclusion measure ranging from 0 to 100, with 0 representing complete financial exclusion and 100 indicating full financial inclusion. It consists of three primary parameters: Access (35%), Usage (45%), and Quality (20%). The index is based on 97 indicators covering banking, investments, insurance, postal services, and pensions. It was developed in consultation with the government and sectoral regulators to measure ease of access, availability, usage, and quality of financial services. The improvement in the index was driven by growth across all sub-indices, with the usage dimension contributing the most to the overall increase. The FI-Index is unique in that it does not have a base year, reflecting the cumulative efforts of all stakeholders over the years. The FI-Index for March 2024 was 64.2, an increase from 43.4 in March 2017. This indicates significant progress in financial inclusion nationwide.

Figure 5. FI index India

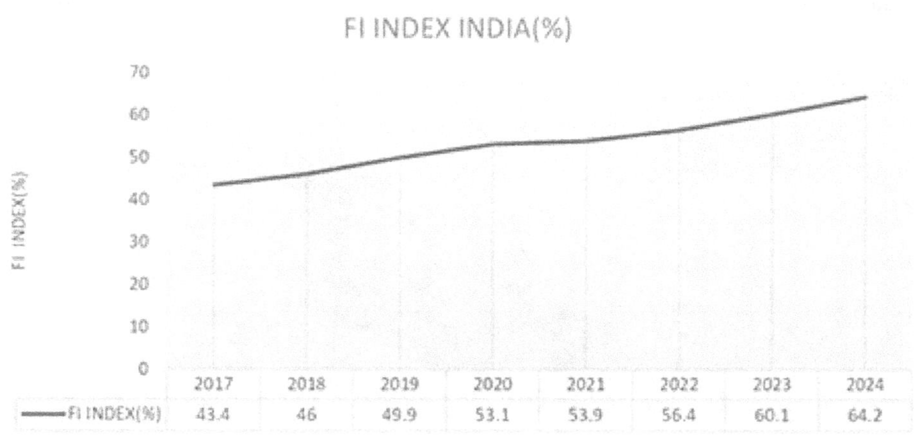

Source: *RBI Bulletin, September 2017 to 2024*

Contribution of PMJDY in the FI index:

According to the RBI, financial inclusion growth is measured through access, availability, usage, and quality of financial services. An increase in these parameters significantly contributes to a financial inclusion index.

Access: This Parameter measures how easily people can access financial services. It includes the number of bank branches or bank accounts. To check the accessibility of PMJDY, the total number of bank accounts open under this scheme was collected. The results indicate that the number of bank accounts increased each year. The reason behind its progress is the minimum basic amount, no charges, and relief in KYC norms. Based on the data, it can be concluded that the accessibility of PMJDY has a positive contribution to the financial inclusion index.

Figure 6. Access to PMJDY Accounts

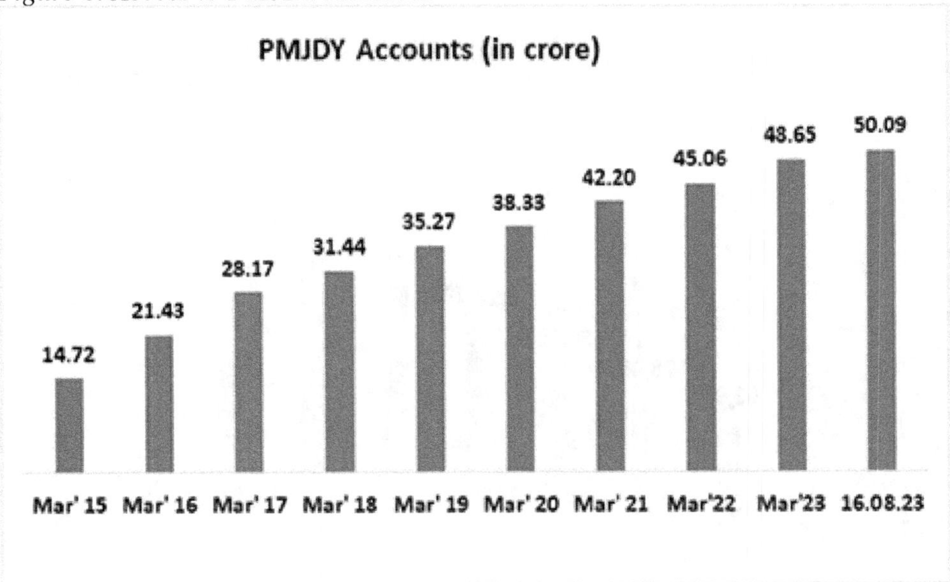

Source: *Press Release: Press Information Bureau 28 Aug 2024*

Usage: This parameter measures the extent to which people utilize financial services. It includes the number of deposits in accounts. Total deposit balances under PMJDY Accounts stand at Rs. 2,31,236 crores. Deposits have increased approximately 15 times, while accounts have increased 3.6 times from 2015 to 2024. The increasing number of deposits suggests that people use these accounts for their savings. This means that PMJDY deposits significantly contribute to the growth of financial inclusion.

Figure 7. Access to PMJDY Accounts

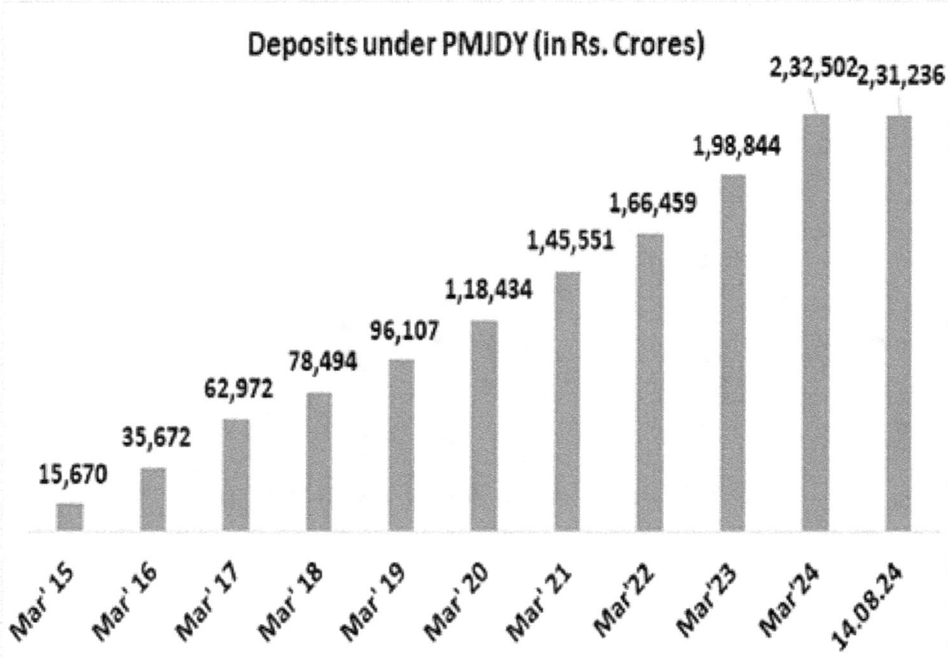

Source: *Press Release: Press Information Bureau 28 Aug 2024*

Quality: Customer satisfaction is a key factor in determining the quality parameter of the FI index. It includes financial products offered to the public and their utilization. With the issuance of over 36.06 crore RuPay debit cards under PMJDY, the installation of 89.67 lakh PoS/mPoS machines, and the introduction of mobile-based payment systems like UPI, the total number of digital transactions has increased from 2,338 crore in FY 18-19 to 16,443 crore in FY 23-24. UPI financial transactions have increased from 535 crore in FY 2018-19 to 13,113 crore in FY 2023-24. Similarly, RuPay card transactions at PoS and e-commerce increased from 67 crore in FY 2017-18 to 96.78 crore in FY 2023-24. The data indicate that people use RuPay debit cards issued under the PMJDY. This quality parameter significantly contributes to achieving the financial inclusion goal.

Figure 8. Quality of PMJDY Accounts

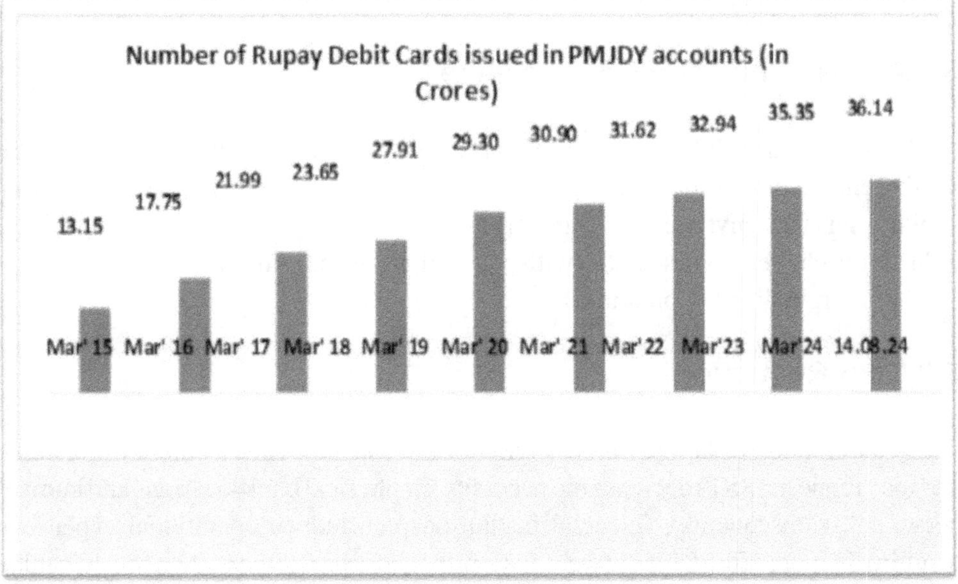

Source: *Press Release: Press Information Bureau 28 Aug 2024*

The Results reveal that financial inclusion is mainly measured by access, usage, and quality parameters. The results of these parameters indicate an increase or decrease in the FI index. Over the period, PMJDY has progressed in all key parameters, namely access (number of PMJDY accounts), usage (deposit in PMJDY accounts), and quality (number of debit cards associated with PMJDY accounts). All these positive developments significantly contribute to the financial inclusion index or growth (Anjum et al., 2020; Ahmad, 2018). After the introduction of this policy, there is a sudden hick in the financial inclusion index. The positive results of this policy encourage the government to introduce more police under financial inclusion plans, such as PMSBY, PMJJY, and APY. Overall, the findings indicate a positive or significant contribution of PMJDY to the country's financial inclusion index (Mala et al., 2016; Muralidharn, 2023; Yeshu, 2016).

5. HOW SSTs FOSTER FINANCIAL INCLUSION

Self-service technologies (SSTs) are not just tools but agents of change. They empower the underbanked or unbanked by enhancing their access to financial services. Regarding accessibility, availability, and usage, SSTs play a crucial role in fostering financial inclusion in many ways, inspiring hope for a more inclusive financial landscape.

5.1 Accessibility and Convenience

SSTs, such as mobile banking applications, ATMs, online banking services, and kiosks, enable individuals, particularly in remote or underserved regions, to access banking services anytime and from various locations. This enhanced accessibility reduces the challenges associated with physical distance from banks, often affecting rural or marginalized populations.

5.2 Cost Reduction

By automating tasks such as cash withdrawals, balance checks, money transfers, and loan requests, SSTs decrease the necessity for physical bank locations and human personnel. Consequently, financial institutions can reduce operational expenses, resulting in more affordable services for lower-income customers. As a result, banks can provide low-fee or even free banking options to promote financial inclusion.

5.3 Mobile and Digital Banking

Smartphones promote financial inclusion through mobile banking apps, digital wallets, and USSD (Unstructured Supplementary Service Data) technology. Although many individuals in emerging markets may not access conventional bank branches, they can still manage their finances, transfer funds, pay bills, and receive payments through their mobile devices. Initiatives like M-Pesa in Kenya highlight how mobile financial services can transform access to financial resources.

5.4 Paperless Transactions

SSTs support electronic transactions, reducing reliance on cash or physical documentation. Digital platforms facilitate more streamlined documentation and record-keeping, enhancing transparency, reducing fraud, and fostering trust in financial services. This is particularly vital in areas where residents face challenges related to literacy or access to traditional banking systems.

5.5 Loan and Credit Access

Self-service platforms allow individuals to apply for loans and credit without needing in-person meetings with bank representatives. Numerous financial institutions utilize data-driven technologies like machine learning to assess creditworthiness and offer microloans to individuals lacking traditional credit histories. Peer-to-peer lending and crowd-lending platforms also utilize SSTs to grant users access to affordable credit.

5.6 Financial Literacy and Education

SSTs contribute to financial education by providing engaging tools, tutorials, and resources that guide users in managing their finances, budgeting, saving, and investing. Mobile applications and websites can provide information on essential financial topics, helping build financial literacy among populations without formal education.

5.7 Digital Payment Systems

SSTs facilitate digital payment systems, simplifying individuals' participation in the formal economy. For instance, digital wallets and platforms like PayPal and Venmo allow users to send and receive money, purchase goods, and even invest through mobile devices. This shift reduces reliance on cash and enhances financial engagement, particularly for low-income or rural populations.

5.8 Blockchain and Cryptocurrency

Blockchain technology, often associated with cryptocurrencies, represents a transformative shift in the financial industry. SSTs built on blockchain can offer secure, low-cost, and rapid cross-border payments for individuals who do not have a standard banking relationship. Cryptocurrencies can also save or transfer value without dependence on centralized financial institutions, granting users in under-banked areas greater financial independence.

5.9 Remote Identification and Onboarding

Many SSTs incorporate biometric authentication and digital identity verification techniques, such as facial recognition, fingerprint scanning, and government ID imaging, to facilitate remote user onboarding. This approach simplifies access to banking services for individuals in rural or underserved communities who may not have conventional documentation, thereby lowering the barriers associated with identity verification.

5.10 Support from Regulatory Authorities

Governments and regulatory agencies can facilitate financial inclusion by implementing policies favorable to digital banking, fintech advancements, and self-service technologies. For instance, regulations such as Know Your Customer (KYC) and Anti-Money Laundering (AML), tailored for digital platforms, ensure that self-service technologies adhere to financial regulations, making them a safer and more reliable choice for users.

6. HOW SSTs FOSTER FI OF THE INDIAN STATE

Jharkhand has a significant rural demographic with limited access to conventional banking services. Numerous individuals, particularly those on a daily wage and pensioners, faced challenges in accessing financial services. AEPS was implemented to allow people to withdraw funds, check their account balances, and transfer money through Aadhaar-linked bank accounts. Many rural households in Bihar lacked access to banking services and relied on informal financial channels, often leading to exploitation by moneylenders. The Pradhan Mantri Jan Dhan Yojana (PMJDY) established zero-balance accounts for millions of individuals who previously lacked access to banking services.

7. CHALLENGES AND RISKS OF SSTs ADOPTION

7.1 Digital Literacy and User Perception

A key obstacle to adopting self-service technologies (SSTs) is the low level of digital literacy among prospective users, especially in rural regions. Research utilizing the Technology Acceptance Model (TAM) suggests that users' perceptions of usefulness, ease of use, and perceived risk significantly influence their readiness to adopt mobile banking services. The findings highlight that addressing credibility concerns and enhancing user perceptions are crucial for the successful adoption of mobile banking in rural India.

7.2 Technical Challenges and Infrastructure Issues

Technical issues often hinder the performance of ICT-based models, such as mobile banking. Problems with devices such as handheld machines, smart cards, and point-of-sale (PoS) terminals have been noted, resulting in service interruptions in numerous villages. The absence of technical assistance and service centers worsens these challenges, diminishing user trust in SSTs.

7.3 Cybersecurity Threats

The rapid adoption of digital financial services has led to a surge in cybercrime and fraudulent activity. Dishonest fintech lenders and growing cyber threats represent serious dangers for consumers. A Fintech Risk Barometer Study has underscored these issues as primary concerns, highlighting the need for robust consumer protection strategies and effective regulatory oversight.

7.4 Data Privacy Concerns

The increasing integration of artificial intelligence (AI) and machine learning in financial services raises data privacy and security concerns. The Reserve Bank of India has cautioned about the potential financial stability risks associated with the increased use of AI, emphasizing the urgent need for enhanced cybersecurity measures and stricter regulations to protect consumer data.

7.5 Digital Divide and Connectivity Issues

Despite improvements in digital infrastructure, a substantial part of the population, particularly in rural areas, still lacks access to dependable internet services. High costs associated with mobile data and connectivity problems disproportionately affect female entrepreneurs and those with low incomes, restricting their capacity to utilize SSTs effectively. A report by the Cherie Blair Foundation for Women found that 45% of women in business lack consistent internet access due to financial constraints and connectivity challenges.

7.6. Affordability and Economic Barriers

The expenses linked to digital transactions can significantly burden low-income users. For example, Unstructured Supplementary Service Data (USSD)-based mobile banking, designed for non-smartphone users, incurs a fee for each transaction, which can dissuade utilization among economically disadvantaged groups. Additionally, the informal nature of jobs and a cash-centric economy create hurdles in adopting digital financial services.

7.7 Regulatory and Market Conduct Issues

The rise of digital financial services has raised concerns about market practices, including aggressive marketing and collection tactics. Insufficient regulations and oversight may lead to consumer exploitation and financial instability. The Fintech Risk Barometer Study emphasizes the need for regulatory measures to address these issues and protect consumers.

CONCLUSION

Self-service technologies (SSTs) promise to enhance financial inclusion, particularly for unbanked or underbanked individuals, by providing more accessible, affordable, and user-friendly financial services. SSTs help to bridge conventional banking services and underserved populations. The ongoing advancement and implementation of these technologies and supportive regulatory systems can lead to a more inclusive global financial landscape. Various technology-driven solutions exist that can progressively meet the needs of financial inclusion. However, a universally optimal technology solution is not suitable for this purpose. Multiple alternatives exist in various forms, but they suffer from limited standardization, issues with achieving interoperability, a lack of critical mass, and higher costs. Developing a financial services inclusion strategy that considers citizens' needs, context, influences, and motivations is essential for long-term success. Technology can either facilitate or hinder this process. The key characteristics that an appropriate technology should possess are suitability, cost-effectiveness, security, proven effectiveness, scalability, and longevity.

Access to financial resources is viewed as a significant obstacle to the growing population in developing countries, such as India. The findings of this chapter will contribute to improving enterprise and will be approached in many ways. First, people will be more aware of the financial technology available in India. The second advantage is that these technologies have significantly improved financial inclusion Dimensions, i.e., access, usage, and quality. Regarding accessibility, SSTs are readily available nationwide through ATMs or mobiles. Internet usage has significantly contributed to financial inclusion through various methods such as UPI or mobile banking. Apart from this, these technologies provide a well-organized, fast, time-saving digital experience.

SSTs in India fundamentally transform financial inclusion by making financial products and services accessible to everyone. As dependence on mobile devices and digital platforms grows, SSTs can help close gaps and establish a more inclusive financial ecosystem. Furthermore, it requires ongoing innovation, infrastructure advancement, and partnerships with government programs to address issues related to digital literacy, accessibility, and trust. The industry is poised for considerable expansion, with SST firms playing a crucial role in India's progress toward achieving greater financial inclusion. The literature and data indicate that India's FI index has significantly improved after introducing self-service technologies. These technologies effectively support India's goal of financial inclusion and contribute to the country's inclusive growth. However, a long map is still necessary because these technologies face critical issues in India, such as financial or digital literacy, trust, or networking issues. Working on these factors enables India to compete globally regarding financial inclusion. Ultimately, it can be concluded that SSTs are one of the key elements for achieving financial inclusion growth, which further leads to the country's inclusive and sustainable growth.

REFERENCES

Ahmad, D. (2018). Financial inclusion and financial stability: A survey of Nigeria's economic system. *International Journal of Research in Finance and Management, 1*(2), 47–54. DOI: 10.33545/26175754.2018.v1.i2a.15

Ajzen, I. (1991). The theory of planned behavior. *Organizational Behavior and Human Decision Processes, 50*(2), 179–211. DOI: 10.1016/0749-5978(91)90020-T

Anagnostopoulos, I. (2018). Fintech and regtech: Impact on regulators and banks. *Journal of Economics and Business, 100*, 7–25. DOI: 10.1016/j.jeconbus.2018.07.003

Anjum, B., & Tiwari, R. (2012). The role of private sector banks in financial inclusion. 270–280.

Aron, J. (2018). Mobile money and the economy: A review of the evidence. *The World Bank Research Observer, 33*(2), 135–188. DOI: 10.1093/wbro/lky001

Banna, H., Hassan, M. K., & Rashid, M. (2021). Fintech-based financial inclusion and bank risk-taking: Evidence from OIC countries. *Journal of International Financial Markets, Institutions and Money, 75*, 101447. DOI: 10.1016/j.intfin.2021.101447

Black, W., & Babin, B. J. (2019). Multivariate data analysis: Its approach, evolution, and impact. In *The Great Facilitator* (pp. 121–130). Springer. DOI: 10.1007/978-3-030-06031-2_16

Burns, S. (2018). M-Pesa and the 'market-led' approach to financial inclusion. *Economic Affairs, 38*(3), 406–421. DOI: 10.1111/ecaf.12321

Cecchetti, S. G., & Schoenholtz, K. (2020). Finance and technology: What is changing and what is not. *CEPR Discussion Papers, 15352*, 1–40.

Chang, V., Baudier, P., Zhang, H., Xu, Q., Zhang, J., & Arami, M. (2020). How blockchain can impact financial services: An overview, challenges, and recommendations from expert interviewees. *Technological Forecasting and Social Change, 158*, 120–166. DOI: 10.1016/j.techfore.2020.120166 PMID: 32834134

Chopra, S., Sharma, R., & Sherry, A. M. (2013a). Comparing MFS in Kenya, the Philippines, and South Africa under the 7 P evaluation framework. *International Journal of Computer Applications, 84*(9), 17–22. DOI: 10.5120/14604-2853

Chopra, S., Sherry, A. M., & Dwivedi, R. (2013b). Leveraging technology options for financial inclusion in India. *International Journal of Asian Business and Information Management, 4*(1), 10–20. DOI: 10.4018/jabim.2013010102

Chouhan, V., Chandra, B., Saraswat, P., & Goswami, S. (2020). Developing sustainable accounting framework for cement industry: Evidence from India. *Finance India*, *34*, 1459–1474.

Chouhan, V., Goswami, S., Dadhich, M., Saraswat, P., & Shakdwipee, P. (2021e). Emerging opportunities for the application of blockchain for energy efficiency. In Khazanchi, D., Vyas, A. K., Hiran, K. K., & Padmanaban, S. (Eds.), *Blockchain 3.0 for sustainable development* (pp. 63–88). De Gruyter. DOI: 10.1515/9783110702507-005

Chouhan, V., Goswami, S., & Sharma, R. B. (2021d). Utilization of proactive spare parts inventory management (PSPIM) techniques for material handling in the cement industry. *Materials Today: Proceedings*, *45*, 4383–4389. DOI: 10.1016/j.matpr.2020.11.757

Chouhan, V., Sharma, R. B., & Goswami, S. (2021a). Factors affecting audit quality: A study of companies listed on the Bombay Stock Exchange (BSE). *International Journal of Management*, *25*, 989–999.

Chouhan, V., Sharma, R. B., & Goswami, S. (2021b). Sustainable reporting: A case study of selected cement companies of India. *Accounting*, *7*, 151–160. DOI: 10.5267/j.ac.2020.10.002

Chouhan, V., Sharma, R. B., Goswami, S., & Hashed, A. W. A. (2021c). Measuring challenges in adopting sustainable environmental technologies in the Indian cement industry. *Accounting*, *7*, 339–348. DOI: 10.5267/j.ac.2020.11.019

Curran, J. M., & Meuter, M. L. (2005). Self-service technology adoption: Comparing three technologies. *Journal of Services Marketing*, *19*(2), 103–113. DOI: 10.1108/08876040510591411

Curran, J. M., & Meuter, M. L. (2007). Encouraging existing customers to switch to self-service technologies: Put a little fun in their lives. *Journal of Marketing Theory and Practice*, *15*(4), 283–298. DOI: 10.2753/MTP1069-6679150401

Davis, B., & Owens, J. (2009). *POS vs. mobile phone as a channel for m-banking*. Microwave.

Devi, S. (2024). Financial inclusion through digitalization in the Indian economy. *Research Bulletin (International Commission for the Northwest Atlantic Fisheries)*, *49*(1), 83–91.

Frost, J., Gambacorta, L., Huang, Y., Shin, H. S., & Zbinden, P. (2019). BigTech and the changing structure of financial intermediation. *Economic Policy*, *34*(100), 761–799. DOI: 10.1093/epolic/eiaa003

Gautam, A., & Rawat, S. (2017). Cashless and digital economy and its effect on financial inclusion in India. *Financial Sector in India*, *20*, 77–85.

Ghosh, S. (2020). Financial inclusion in India: Does distance matter? *South Asia Economic Journal*, *21*(2), 216–238. DOI: 10.1177/1391561420961649

Haque, S. S., Yanez-Pagans, M., Arias-Granada, Y., & Joseph, G. (2020). Water and sanitation in Dhaka slums: Access, quality, and informality in service provision. *Water International*, *45*(7-8), 791–811. DOI: 10.1080/02508060.2020.1786878

Jack, W., & Suri, T. (2014). Risk sharing and transactions costs: Evidence from Kenya's mobile money revolution. *The American Economic Review*, *104*(1), 183–223. DOI: 10.1257/aer.104.1.183

Jin, Y., & Dwivedi, R. (2009). Business process management: A research overview and analysis. AMCIS 2009 Proceedings, 785.

Kelly, P., Lawlor, J., & Mulvey, M. (2010). A review of key factors affecting the adoption of self-service technologies in tourism. Tourism and Hospitality Research Conference Papers, 12–15.

Khan, H. R. (2012). Issues and challenges in financial inclusion: Policies, partnerships, processes, and products. Korea.

Khan, S., Chouhan, V., Chandra, B., & Goswami, S. (2014). Sustainable accounting and reporting practices in the Indian cement industry: An exploratory study. *Uncertain Supply Chain Management*, *2*(2), 61–72. DOI: 10.5267/j.uscm.2014.2.001

Kim, M., Zoo, H., Lee, H., & Kang, J. (2018). Mobile financial services, financial inclusion, and development: A systematic review of academic literature. *The Electronic Journal on Information Systems in Developing Countries*, *84*(5), e12044. DOI: 10.1002/isd2.12044

Kishore, K., & Sequeira, A. H. (2013, July 18). Self-service technologies - A silver lining for financial inclusion. NABARD Conference Proceedings 2013.

Laing, Y. (2006). The influence of perceived risk in the uptake of self-service technologies within the retail banking sector – A study of customers using the Industrial and Commercial Bank of China in Jilin. Digital Knowledge, Cape Peninsula University of Technology, 118.

Li, F., Lu, H., Hou, M., Cui, K., & Darbandi, M. (2021). Customer satisfaction with bank services: The role of cloud services, security, e-learning, and service quality. *Technology in Society*, *64*, 101487. DOI: 10.1016/j.techsoc.2020.101487

Mader, P. (2018). Contesting financial inclusion. *Development and Change, 49*(2), 461–483. DOI: 10.1111/dech.12368

Maina, E. M., Chouhan, V., & Goswami, S. (2020). Measuring behavioral aspect of IFRS implementation in India and Kenya. *International Journal of Scientific and Technology Research, 9*, 2045–2048.

Mala, M., & Vasanthi, G. (2016). Problems and challenges of the Indian post in financial inclusion. *International Journal of Interdisciplinary Research in Arts and Humanities, 1*(1), 186–189.

Masino, S., & Niño-Zarazúa, M. (2018). Improving financial inclusion through the delivery of cash transfer programmes: The case of Mexico's Progresa-Oportunidades-Prospera programme. *The Journal of Development Studies, 56*(1), 151–168. DOI: 10.1080/00220388.2018.1546845

Mbiti, I., & Weil, D. N. (2013). The home economics of e-money: Velocity, cash management, and discount rates of M-Pesa users. *The American Economic Review, 103*(3), 369–374. DOI: 10.1257/aer.103.3.369 PMID: 25067843

Metzger, M., Riedler, T., & Pédussel Wu, J. (2019). Migrant remittances: Alternative money transfer channels (Working Paper No. 127/2019). Institute for International Political Economy (IPE), Berlin School of Economics and Law.

Muhammad, N. M., Kassim, S. B., Mahadi, N. F. B., & Ali, E. R. A. B. E. (2023). Role of Islamic Microfinance in Enhancing Financial Inclusion in Bangladesh: A Systematic Literature Review. *Islamic Sustainable Finance, Law and Innovation: Opportunities and Challenges*, 105-116.

Muralidharan, T. (2023). Strengthening the institutional mechanism of agricultural credit: Policies and practice–the 1960s to 2019–a review. In Agricultural Sector in India (pp. 291–310).

Okoye, L. U., Adetiloye, K. A., Erin, O., & Judith, N. (2017). Financial inclusion as a strategy for enhanced economic growth and development. *Journal of Internet Banking and Commerce, 22*, 1–14.

Omojolaibi, J. A., Okudo, A. G., & Shojobi, D. A. (2019). Are women financially excluded from formal financial services? Analysis of some selected local government areas in Lagos State, Nigeria. *Journal of Economic and Social Thought, 6*, 16–47.

Oskarsson, P. (2018). *Landlock: Paralyzing dispute over minerals on Adivasi land in India*. Australian National University Press.

Raman, R., Nair, V. K., Nedungadi, P., Sahu, A. K., Kowalski, R., Ramanathan, S., & Achuthan, K. (2024). Fake news research trends, linkages to generative artificial intelligence and sustainable development goals. *Heliyon, 10*(3), e24727. DOI: 10.1016/j.heliyon.2024.e24727 PMID: 38322879

Rathod, S., & Arelli, S. K. P. (2013). Aadhaar and financial inclusion: A proposed framework to provide basic financial services in unbanked rural India. In *Driving the economy through innovation and entrepreneurship* (pp. 731–744). Springer. DOI: 10.1007/978-81-322-0746-7_60

RBI & NABARD. (2022). Trend report on financial inclusion in India.

Rigoni, E. H., Dwivedi, R., & Hoppen, N. (2010). IT governance and business—IT strategic alignment commitment: A study of Brazilian firms. *International Journal of Global Management Studies Professional*, *2*(1), 1–20.

Russell, J. A. (1980). A circumplex model of affect. *Journal of Personality and Social Psychology*, *39*(6), 1161–1178. DOI: 10.1037/h0077714

Sandhu, K., Dayanandan, A., & Kuntluru, S. (2023). *Fintech innovation for financial inclusion: Can India make it?* International Journal of Accounting & Information Management., DOI: 10.1108/IJAIM-03-2023-0058

Schuetz, S., & Venkatesh, V. (2020). Blockchain, adoption, and financial inclusion in India: Research opportunities. *International Journal of Information Management*, *52*, 101936. DOI: 10.1016/j.ijinfomgt.2019.04.009

Shih-Chih, C., Huei-Huang, C., & Mei-Fang, C. (2009). Determinants of satisfaction and continuance intention towards self-service technologies. *Industrial Management & Data Systems*, *109*(9), 1248–1263. DOI: 10.1108/02635570911002306

Singh, N. D., & Singh, H. R. (2012). Social impact of microfinance on SHG members: A case study of Manipur. *Prabandhan: Indian Journal of Management*, *5*(3), 43–50. DOI: 10.17010/pijom/2012/v5i3/60155

Singh, S., Goyal, S. K., & Sharma, S. K. (2013). Technical efficiency and its determinants in microfinance institutions in India: A firm-level analysis. *Journal of Innovation Economics Management*, *1*(11), 15–31. DOI: 10.3917/jie.011.0015

Suoranta, M., & Mattila, M. (2004). Mobile banking and consumer behavior: New insights into the diffusion pattern. *Journal of Financial Services Marketing*, *8*(4), 354–366. DOI: 10.1057/palgrave.fsm.4770132

Wieser, C., Bruhn, M., Kinzinger, J. P., Ruckteschler, C. S., & Heitmann, S. (2019). The impact of mobile money on poor rural households: Experimental evidence from Uganda. World Bank Policy Research Working Paper No. 8913.

Yeshu, P. (2016). *Financial inclusion in India: Role of banks*. International Research Journal of Commerce, Arts and Science.

Chapter 12
Automated Medicine Dispenser Powered by Arduino and Flowmeter With AI Integration

Monica Gupta
https://orcid.org/0000-0001-6756-4191
Bharati Vidyapeeth's College of Engineering, New Delhi, India

Kartik Prasad
Bharati Vidyapeeth's College of Engineering, New Delhi, India

Jayani Sehgal
Bharati Vidyapeeth's College of Engineering, New Delhi, India

Deepanshu Kumar Rai
Bharati Vidyapeeth's College of Engineering, New Delhi, India

Vishal Bhalla
Bharati Vidyapeeth's College of Engineering, New Delhi, India

Sonia Agrawal
Nvidia Corporation, USA

ABSTRACT

The Automated Medicine Dispenser is a device that helps people, especially the elderly and those with long-term illnesses, take their medicines on time. It exploits smart technology to ensure that the correct medicine, that too in the right amount, is given to the patient. The system incorporates an Arduino board, a flow meter to measure liquid medicine, and a camera that works on image recognition. In addition, the system is designed to send timely reminders to patients. The design is user-friendly and doesn't require technology-expertise. Thus, it can be used at home as well as in hospitals. The main objective of this innovations is to make the patient become independent of care-givers, reducing the pressure on the healthcare

DOI: 10.4018/979-8-3373-4667-0.ch012

system, and providing the care at an affordable price.

INTRODUCTION

In modern healthcare, medication adherence is a critical factor influencing patient outcomes (Gast et al., 2019). Non-adherence to prescribed medication regimens has been linked to worsening health conditions, increased hospitalizations, and additional healthcare costs (Walsh et al., 2019). This problem is particularly significant among the elderly and patients with chronic illnesses, where complex medication schedules and dosages are common (Punnapurath et al., 2021).

Automating the medication dispensing process offers a promising solution to this challenge. By integrating technologies such as microcontrollers, sensors, and Artificial Intelligence, automated medicine dispensers can ensure precise dosage delivery, enhance patient compliance, and reduce the risk of human error. Recent advances in healthcare technology have made it possible to create cost-effective, user-friendly solutions that cater to diverse patient needs, from home settings to healthcare facilities (Junaid et al., 2022).

As automation becomes more prevalent in everyday tasks, systems like the Liquid Automatic Dispensing Machine showcase the potential of microcontroller-based solutions to improve operational efficiency and reduce manual workloads. By incorporating simple yet effective design principles, this system represents a valuable advancement in the automation of liquid. The use of Arduino ensures flexibility and adaptability, enabling easy modifications or upgrades to the system as needed. By automating the dispensing process, the system not only reduces human error but also improves productivity by saving time and minimizing wastage. It ensures that the same precise volume of liquid is dispensed every time, which is critical in environments that require consistency and accuracy. Additionally, the system could be extended to handle different types of liquids or larger volumes, making it versatile for various applications. Beyond its primary function, the machine's automated control system could be integrated with additional features such as sensors for real-time volume tracking or user notifications, further increasing its capabilities.

The Liquid Automatic Dispensing Machine using Arduino is an advanced solution designed to streamline liquid dispensing for a variety of applications (Hoang Le et al., 2023). By integrating a static pump with an Arduino Uno, the system offers precise control over liquid volumes—5 ml, 10 ml, or 15 ml—based on the time the pump is activated. This system eliminates manual dispensing errors and ensures consistent accuracy by calculating the dispensing time according to the flow rate of the pump. The design is straightforward and user-friendly, requiring only push-buttons for operation, and is powered via USB, making it both cost-effective and energy-efficient. It eliminates the need for manual measuring and enhances operational efficiency by allowing users to simply select the desired volume and press a button to activate the dispensing process. Integrating IoT technologies with a mobile application or web interface can enable users to remotely monitor and control the dispensing process, adjust parameters, and track usage data for enhanced flexibility in healthcare settings. Level sensors in the medicine container can issue alerts when the liquid reaches its minimum level, preventing missed doses.

A more advanced design aims to streamline and enhance medication management through a combination of precision engineering and advanced technology. The system allows users, such as medical attendants or patients, to input essential details like medicine name, dosage, and scheduled time. At the time of dispensing, an LCD screen displays relevant information about the medicine, offering convenience and clarity to the user. Integrated sensors, including flow meters, level detectors, and proximity sensors, provide real-time feedback to the controller, ensuring accurate operation and enabling real-time adjustments. The system also supports remote health monitoring, allowing medical practitioners to track patient health and update dosages as needed, and delivering personalized care. Furthermore, AI integration elevates the system's capabilities, enabling accurate identification of medicines, adherence monitoring, reminders for missed doses, and optimal compliance with prescribed schedules. An AI-powered image recognition module, equipped with a trained neural network model, cross-verifies prescriptions with identified medicines to ensure precise dispensing, enhancing the reliability and safety of the system.

BACKGROUND

The rapid medication adherence is a significant challenge in healthcare, particularly for individuals with chronic conditions, the elderly, and those managing complex medication regimens (Marcum et al., 2012). Incorrect medication administration, missed doses, and wrong dosages can lead to serious health complications, including worsening of conditions, hospitalizations, and increased healthcare costs (Hughes et al., 2008; Aronson et al., 2009). Studies have shown that approximately 50% of patients do not follow prescribed medication schedules, and this non-adherence can account for significant morbidity and mortality rates (Brown et al., 2011).

As healthcare becomes increasingly reliant on technology, the need for automated, precise systems to assist with medication management has grown. Automated medication dispensers are emerging as a potential solution to address this issue (Mason et al., 2022). These devices are designed to help patients manage their medication schedules, ensuring that they take the correct dose at the right time. While several medication dispensers exist in the market, most of them are limited to handling solid medications such as tablets and capsules (Gargioni et al., 2024). Few devices support the dispensing of liquid medications, which is critical for pediatric and geriatric patients, who often require liquid forms of medicine due to difficulty swallowing pills (Belissa et al., 2019).

This chapter focuses on designing an Automated Medicine Dispenser powered by an Arduino microcontroller and a flowmeter for precise liquid dosage measurement, integrated with AI technology for medicine identification and better accuracyz. The aim is to design a cost-effective, accurate, and easy-to-use solution that enhances medication adherence, reduces errors in dosage, and provides timely reminders for patients. The use of AI image recognition allows the dispenser to identify medicines, ensuring that the correct drug is dispensed according to the patient's prescription. This system can be highly beneficial in both home healthcare settings and healthcare facilities, reducing the burden on caregivers, improving patient safety, and potentially leading to better health outcomes. By automating the medication dispensing process and integrating advanced technologies, this work aims to address the critical need for a user-friendly, reliable, and precise solution for managing liquid medications, ultimately improving medication adherence and patient care.

Problem Statement

Medication errors and missed doses are common in healthcare, with significant risks for elderly and chronically ill patients. These errors can result in adverse drug reactions, hospitalizations, and even fatalities. Current methods for medication management, such as manual pill organizers and basic dispensers, often fail due to human error, lack of real-time monitoring, and insufficient precision. There is a pressing need for an intelligent, automated system that ensures accurate dosage and timely reminders, reduces reliance on caregivers, and addresses these gaps in medication management.

Objectives

The key highlights of the work:

Accuracy: Ensure precise dosage measurement using flowmeters.
Automation: Develop a self-operating medication dispensing system using Arduino.
Monitoring: Integrate AI for real-time tracking of medication adherence.
User-Friendliness: Design a simple interface for elderly and non-technical users.

Affordability: Create a cost-effective solution to make it accessible to a broader audience.

Scope of the Work

The Automated Medicine Dispenser is designed for:

Elderly and chronically ill patients: Ensuring timely medication without caregiver dependency.
Healthcare facilities: Automating drug administration to reduce human-errors.
Caregivers: Providing alerts and updates on patient adherence.

The system aims to be scalable, suitable for individual use, and adaptable to multi-drug regimens in clinical settings. By addressing the critical need for accurate and timely medication dispensing, this work contributes to enhancing patient safety, reducing caregiver burden, and improving overall healthcare outcomes. The automated medicine dispenser is envisioned as a practical, cost-effective tool for modern healthcare systems, empowering patients and supporting caregivers in managing complex medication regimens.

EXISTING WORK

A literature survey explores the current state of research and technology relevant to the work. It examines the existing solutions for automated medication dispensers, highlights the limitations of these systems, and discusses the role of AI in enhancing healthcare devices.

Many researchers have proposed solutions for automating liquid dispensing in various fields, focusing on improving accuracy and efficiency. The design in (Manikandan et al., 2021) discusses an Arduino-based liquid dispenser that automates the measurement and dispensing of liquids by controlling a pump using the time-based flow rate. This method allows precise liquid delivery, making it suitable for applications such as beverage dispensers or laboratory experiments. The simplicity of the design makes it adaptable, offering ease of implementation and cost-effectiveness. The use of Arduino allows for simple modification to the system, enabling it to suit various needs, such as changing flow rates or dispensing volumes. On the other hand, the work in (Hidayat et al., 2020) examines a similar approach using Arduino and flow meters to measure the amount of liquid dispensed. The system utilizes an active feedback loop to ensure that the liquid dispensed matches the user-defined amount, improving the accuracy of liquid dispensing in sensitive environments, such as medical or chemical processing.

Another interesting approach is presented in (Latham et al., 1966), where a peristaltic pump is combined with a flow meter to create a precise liquid dispensing system. The system calculates the flow rate and dispenses liquid based on the time the pump is activated. This method ensures accuracy in liquid delivery by measuring the flow in real-time and adjusting the dispensing time accordingly. A similar system in (Koh et al., 2024) uses a combination of a microcontroller and a flow meter to control the pump. The system provides a real-time readout of the liquid volume dispensed, allowing for better monitoring and error detection during operation. This solution minimizes the risk of over-dispensing and waste, making it ideal for precision applications.

The role of sensors in enhancing the accuracy of liquid dispensing is explored further in (Husni et al., 2016). The authors use a combination of ultrasonic and capacitive sensors to detect liquid levels in the dispensing chamber. These sensors help monitor the liquid flow, ensuring that the system dispenses the correct volume based on preset parameters. The integration of such sensors ensures that the system is responsive to variations in the flow rate, which can be caused by changes in temperature, viscosity, or pressure. The approach discussed in (Liu et al., 2009) also highlights the benefits of integrating sensors into the liquid dispensing process, noting that the real-time data collected can be used to improve the system's overall efficiency. The paper proposes an automated system that adjusts pump speed based on the liquid's properties, further enhancing the dispensing accuracy.

An innovative approach to liquid dispensing using Arduino is presented in (Spasova et al., 2024). The system incorporates wireless communication, allowing users to control and monitor the dispensing process remotely. This paper explores the integration of IoT technology into liquid dispensing systems, enabling cloud-based management of multiple dispensing units. The use of mobile apps and cloud platforms for remote control and monitoring represents a significant advancement in automated liquid dispensing, offering added convenience and flexibility. The work in (Jayanth et al., 2017) discusses a system that combines liquid dispensing with inventory management, allowing for the tracking of liquid usage in real-time. This is particularly beneficial for industries like healthcare and food processing, where accurate inventory management and real-time tracking are critical.

The growing trend of using Arduino-based systems for liquid dispensing is further highlighted in (Gargioni et al., 2024). The study focuses on the ease of customization and scalability that Arduino offers, making it a popular choice for developing automated systems. The paper also emphasizes the importance of integrating multiple sensors to increase accuracy and reliability. Arduino-based solutions are ideal for applications where precision, flexibility, and cost-efficiency are essential.

In conclusion, the use of Arduino-based systems for automated liquid dispensing offers numerous advantages, including affordability, adaptability, and ease of implementation. The systems reviewed in the literature demonstrate the potential of Arduino in creating precise, efficient, and scalable liquid dispensing solutions, with applications ranging from laboratories to industrial processes.

Limitations of Existing Systems

No Liquid Medication Support: Most available machines for medicines focus on solid medications, ignoring the challenges of accurate liquid dosage measurement.
Lack of AI Integration: Current systems lack advanced AI capabilities, such as real-time medicine identification, prescription verification, remote monitoring etc.
User Accessibility: Many devices require technical expertise or are difficult to navigate for elderly patients, reducing their usability.

This review underscores the need for a comprehensive solution addressing these gaps, particularly for accurate liquid medication dispensing and AI-powered automation.

CLASSIFICATION OF MEDICATION DISPENSERS

(Pinto et al., 2021; Kumar et al., 2023; Agarwal et al., 2022; Patel et al., 2021; Smith et al., 2020; Patel et al., 2019; Johnson et al., 2018; Davis et al., 2017)

Medication dispensers have undergone remarkable advancements over the years, evolving from rudimentary manual systems to sophisticated automated devices equipped with cutting-edge technologies. This progression reflects a growing emphasis on improving medication adherence, enhancing patient safety, and providing user-friendly solutions for diverse healthcare needs. Below is a detailed review of the various dispensers and their functionalities:

Manual Dispensers: Early medication dispensers, such as traditional pill organizers, were simple mechanical systems that required manual input to dispense pills or liquids. While cost-effective and widely available, these devices were often limited in functionality. They lacked features such as accurate dosage measurement, reminders for medication times, or the ability to track adherence. Patients with low literacy levels, cognitive impairments, or physical limitations frequently found these devices challenging to use. Moreover, they provided no real-time feedback, leaving room for human error, particularly in managing complex medication regimens.

Digital Medication Dispensers: The advent of digital medication dispensers marked a significant leap in functionality. Incorporating microcontrollers, these devices automated the dispensing process and introduced features such as pre-scheduled dispensing and auditory or visual reminders. Popular models like the Hero Medication Manager and MedMinder support adherence through reminders and lockable compartments. However, most digital dispensers are designed exclusively for solid medications, such as pills or capsules, leaving a gap in addressing the needs of patients requiring liquid medications. Additionally, these devices often rely on manual user input for refills and monitoring, limiting their autonomy and placing a burden on users to ensure they are correctly maintained and refilled.

Smart Dispensers with IoT Features: The latest generation of smart dispensers integrates IoT technology to enhance connectivity and usability. Devices like Pillo Health represent a new frontier in medication management, connecting to mobile applications and healthcare providers for remote monitoring and assistance. Key features include voice-controlled operation, refill alerts, and caregiver notifications. These systems significantly improve accessibility for users with physical or cognitive challenges, offering a more seamless and interactive experience. Despite these advancements, IoT-enabled dispensers are often prohibitively expensive, creating a barrier for low-income populations. Additionally, while they excel in managing solid medications, they rarely cater to liquid medication needs, which are critical for pediatric, geriatric, and chronically ill patients who may struggle with pill-based regimens.

AI INTEGRATION IN HEALTHCARE DEVICES

Artificial intelligence has revolutionized healthcare by enabling automation, precision, and scalability in various domains. Its integration into medication management offers significant

AI in Medication Identification: It uses machine learning for pattern recognition. AI algorithms can detect discrepancies between prescribed medications and those loaded into the dispenser, preventing errors.

AI for Dosage Management: AI models help ensure that the dispensed dosage aligns with the prescribed amount.

Real-time monitoring of flowmeters and other sensors enhances liquid medication accuracy, addressing the variability seen in manual measurements.

Applications of AI in Existing Devices

AI has already been integrated into healthcare tools such as insulin pumps and wearable monitors, providing real-time analytics and personalized recommendations. Similar technology can be adapted for medication dispensers, expanding their functionality to include monitoring patient compliance and health outcomes. The following applications show how AI can be incorporated in medicine dispensers:

AI-Powered Dispensers: A study (Smith et al., 2020) demonstrated that AI-based image recognition models achieved 94% accuracy in identifying pill shapes and colors. This highlights the feasibility of applying similar models to liquid medications.

Liquid Dosage Challenges: Research in (Yin et al., 2016) found that errors in liquid dosage measurement accounted for 18% of medication non-adherence cases among elderly patients. Flowmeters and AI algorithms were proposed as solutions to improve precision.

Integration of IoT in Healthcare Devices: A study (Meghla et al., 2022) explored IoT-enabled medication dispensers, focusing on real-time monitoring and data sharing with healthcare providers. While effective, the study noted high costs and limited accessibility as barriers to adoption.

This work builds upon these insights by developing an AI-powered, cost-effective solution for automated medication dispensing. The proposed system focuses on liquid medication management, ensuring accurate, timely, and user-friendly dispensing.

Role of AI for performance enhancement

AI integration is a critical aspect of the automated medicine dispenser, enabling intelligent medicine identification, dosage precision, and patient-specific customizations.

AI for Medicine Identification

An Optical Character Recognition (OCR) model is trained on a dataset of medicine images to facilitate accurate identification. In the process, the user places the medicine bottle under the AI module's camera, which captures an image of the bottle. This image is analyzed by the OCR model, which identifies the medicine based on its shape, color, label, and other distinctive features. The identified medicine is then cross-checked against the prescription stored in the system to ensure compliance.

Dosage Calculation

AI ensures accurate dosage by correlating flowmeter readings with prescription data. Advanced algorithms dynamically adjust the dispensing duration in response to variations in flow rate, maintaining precision throughout the process.

Adaptive Learning

The system uses machine learning to adapt over time, refining dispensing operations and user interactions based on historical data.

Integration with IoT Devices

AI algorithms process additional data from IoT devices, such as wearable health trackers, to tailor medication schedules based on patient vitals.

Dataset for AI Training

The AI model is trained on a robust dataset comprising images of various medicine bottles featuring diverse labels, colors, and shapes, as well as flowmeter data for liquids with different viscosities.

Augmentation Techniques:

Synthetic data generation enhances the model's ability to handle variations in lighting, angles, and label damage.

The hardware design ensures reliable, precise dispensing through an Arduino-based system integrated with sensors, flowmeters, and actuators. The software design coordinates these components while leveraging AI for medicine identification, dosage calculations, and smart notifications. AI integration further enhances the system by enabling intelligent, adaptive, and user-friendly operations, bridging the gap between manual medication management and automated healthcare solutions.

DESIGN OF AN AI-BASED AUTOMATED MEDICINE DISPENSER

This section discusses the detailed design and implementation plan of an Automated Medicine Dispensers.

Design 1: A Basic Automated Medicine Dispenser Powered by Arduino

The proposed design using Arduino aims to automate the liquid dispensing process with precision and efficiency.

Key Components:

- Arduino Uno: The central controller that processes input and controls the operation of the motor.
- Motor Driver (L298N): Used to control the direction and speed of the motor by providing sufficient current to drive it.
- DC Motor: Powers the mechanism that dispenses medicine from the reservoir when activated.
- Push Button: Serves as the user interface for triggering the dispensing mechanism. When pressed, the Arduino is signaled to activate the motor.
- Medicine Reservoir: Holds the medicine that is dispensed by the motor mechanism.

Working:

The methodology begins with selecting the required liquid volume, such as 5 ml, 10 ml, or 15 ml, through three dedicated pushbuttons as illustrated through a flowchart shown in Figure 1. Once a button is pressed, the Arduino Uno microcontroller triggers the pump through a motor driver, adjusting its run time based on the pre-configured flow rate to dispense the appropriate amount of liquid. The flow rate of the pump is pre-determined and used to calculate the dispensing time for each volume selection. The system ensures liquid measurement by controlling the dispensing duration precisely.

Figure 1. Flow Chart of Basic Automated Medicine Dispenser

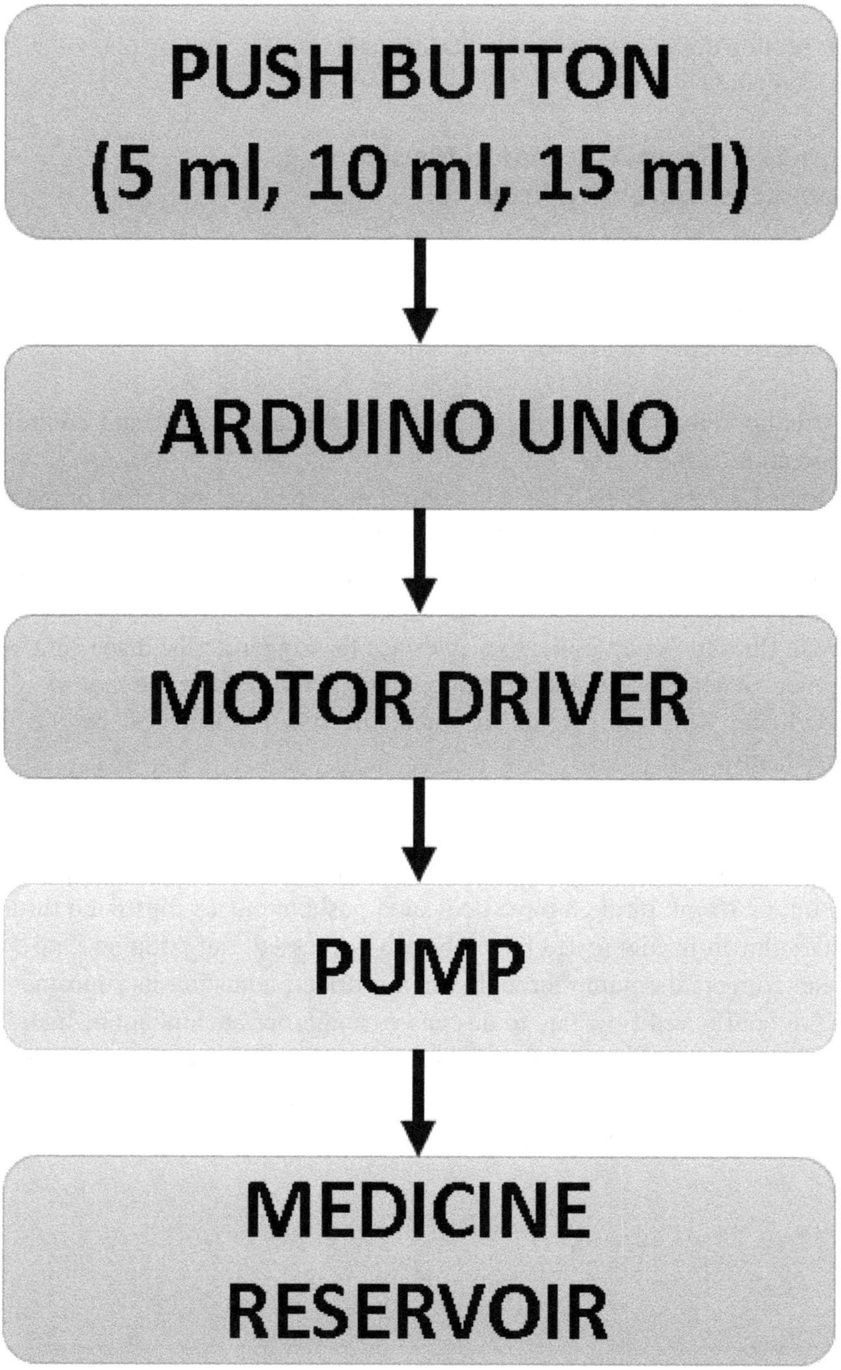

Figure 2. Schematic Diagram of a Basic Automated Medicine Dispenser

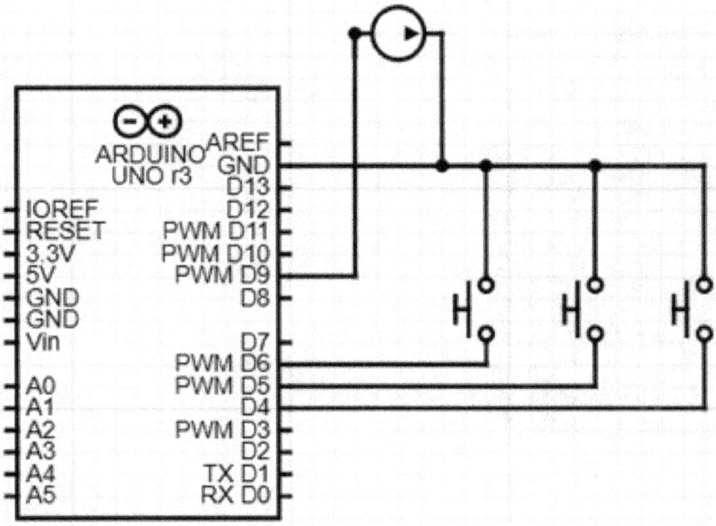

The Arduino system is powered by a USB power supply (Figure 2), providing an efficient and straightforward power solution for the dispensing mechanism. The USB connection also allows easy reprogramming of the Arduino for different dispensing parameters or to update the system's firmware. The choice of USB power ensures that the machine remains energy-efficient and portable, without requiring complex electrical setups. Figure 3 shows the circuit diagram of the design on Tinkercad whereas Figure 4 shows the hardware implementation of the same design. The Arduino Uno is connected to the motor driver through the digital output pins, allowing the control of the motor's speed and direction. The push button is connected to a digital input pin on the Arduino to trigger the dispensing action. The motor driver receives power from the Arduino and controls the direction and speed of the DC motor that is physically connected to the dispensing mechanism. The motor is powered by the driver, allowing it to turn ON when activated.

Figure 3. Circuit Diagram of a Basic Automated Medicine Dispenser on Tinkercad

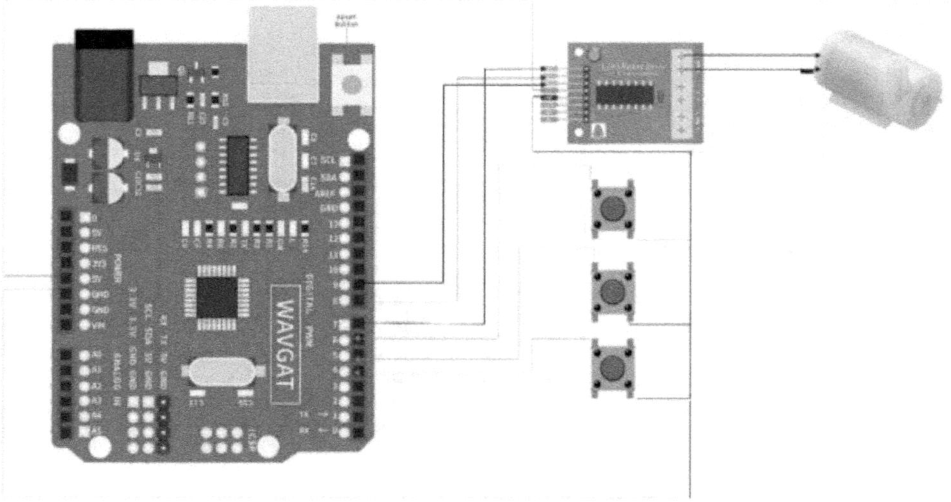

An optional mobile application or web interface could be developed to provide remote monitoring and control of the dispensing process. This integration with IoT technologies would allow users to adjust parameters or track usage data via a smartphone or computer, offering additional flexibility for industrial or lab applications. Additionally, level sensors integrated into the medicine container can provide alerts when the liquid reaches the highest level, indicating an overflow.

Overall, the methodology focuses on combining simple but effective components, such as an Arduino controller and a static pump, to create a reliable, automated medicine dispensing system. The system's design minimizes human intervention, reduces errors, and ensures accuracy.

Figure 4. Hardware Implementation of a Basic Automated Medicine Dispenser

Design 2: An Automated Medicine Dispenser Powered by Arduino and Flowmeter with AI Integration and LCD Display

The design process began with identifying challenges such as medication errors and adherence issues, particularly among elderly and chronically ill patients. A review of existing systems revealed limitations like the lack of real-time monitoring, accuracy, and user-friendliness. This led to the goal of creating a cost-effective, reliable, and user-friendly system for precise medication dispensing and real-time adherence tracking. The system is designed to streamline and enhance medication management (Figure 5). The medical attendant or a patient begins by entering the medicine name, dosage, and scheduled time. At the time of dispensing, the system displays relevant information about the medicine on an LCD screen for user convenience. The sensors (flow meters, level detectors, and proximity sensors) integrated into the system provide feedback to the controller, ensuring accurate performance and real-time adjustments. Medical practitioners can remotely track patient health and update dosages as needed, ensuring personalized care. AI integration significantly enhances the system's functionality by accurately identifying medicines, tracking adherence, sending reminders for missed doses, and ensuring optimal compliance with prescribed medication schedules. Additionally, the software incorporates an AI-powered image recognition module that uses a trained neural network model to identify medicines, cross-verifying prescriptions with the identified items to ensure accurate dispensing.

Figure 5. Overview of an Automated Medicine Dispenser powered by Arduino and a flowmeter with AI integration and an LCD display

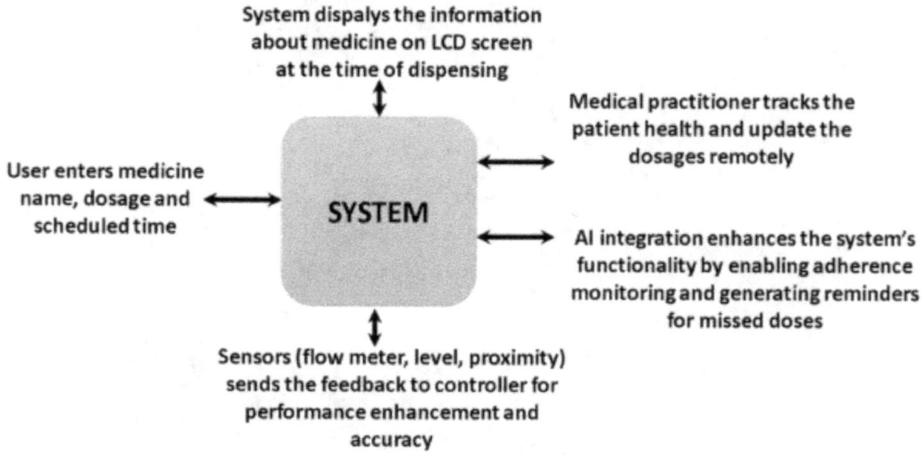

The design phase involved integrating hardware components such as the Arduino microcontroller, flowmeter, sensors, motors, an LCD display, and a keypad into a compact and easy-to-use physical system. Software planning included scheduling algorithms, an AI-based adherence tracking module, and an intuitive user interface. The hardware components were assembled, and the Arduino was programmed to coordinate the dispensing mechanism, manage schedules and alerts, and control the pump. The pump operation was based on the flow rate, ensuring the accurate dispensing of the selected liquid volume. AI integration enhanced the system's functionality by enabling adherence monitoring and generating reminders for missed doses. The implementation phase brought the hardware and software together into a functional prototype, with individual components tested for accuracy and system-level testing ensuring real-time monitoring and seamless operation. User testing, particularly with elderly participants, validated the system's practicality, addressing issues like dose inaccuracies and alert failures.

The system includes features such as real-time feedback through an LCD, which displays the dispensing status, selected volume, dispensed amount, and any error messages. Error handling is further strengthened by visual and audio alerts, such as a blinking red LED and an alarm, to notify users of malfunctions like sensor issues or insufficient liquid. Optimization efforts focused on refining the hardware design for portability and improving the AI module for better prediction accuracy. Comprehensive testing under various scenarios confirmed the system's reliability, and feedback was used to enhance functionality further. This structured approach ensured the successful development of a functional, user-friendly, and efficient automated medicine dispenser, effectively addressing the needs of the target audience.

HARDWARE DESIGN CONSIDERATIONS

The hardware design of the automated medicine dispenser involves multiple components working in unison to achieve precise dispensing and monitoring. Below is a detailed description of the hardware architecture:

Microcontroller

The system's central processing unit plays a crucial role in managing input and output operations, controlling hardware components, and executing programmed instructions. The Arduino platform, chosen for this purpose, offers several advantages. It is cost-effective, versatile, and easy to program, making it an excellent choice for prototyping and seamless integration into various applications.

Motor Driver

The primary purpose of the motor driver is to ensure the accurate measurement and dispensing of liquid medications. It operates by detecting the flow rate of liquid passing through the dispensing pipe, utilizing rotating turbines or ultrasonic sensors to measure the volume. To guarantee precise dosages, the motor driver is carefully calibrated based on the prescription input, ensuring consistency and reliability in medication administration.

Sensors

The system integrates multiple sensors to enhance its functionality and ensure seamless operation. Level sensors monitor the liquid levels in medicine bottles to ensure continuous availability. Proximity sensors detect the presence and position of the reservoir, verifying alignment before dispensing. Additionally, error detection sensors are crucial for identifying blockages or irregularities in the dispensing mechanism, guaranteeing reliable and efficient performance.

Dispensing Mechanism

The system includes motorized valves and pumps as key components to regulate the flow of liquid medicines. These motorized valves are controlled by the microcontroller, which actuates them based on precise dosage calculations provided by the software. This integration ensures accurate and efficient dispensing of medications.

Display and Input Devices

The system features a display that provides real-time information, such as the medicine name, dosage, and the scheduled time for the next dose. It also includes a keypad or touch interface, enabling users to manually input prescription details when system updates are required or in the event of a network failure, ensuring uninterrupted functionality.

Power Supply

A stable power supply ensures uninterrupted operation, with battery backup to maintain functionality during power outages.

SOFTWARE DESIGN CONSIDERATIONS

The software design focuses on controlling hardware operations, integrating AI functionalities, and providing an intuitive user interface.

System Programming

Arduino IDE (C/C++) is used for microcontroller programming and Python is used for AI model integration for carrying out the following tasks:

- ✓ Read sensor data (e.g., liquid level, flow rate).
- ✓ Actuate dispensing mechanism based on AI output.
- ✓ Generate error alerts for faulty operations.
- ✓ Communicate with external devices for remote monitoring.

AI Integration

The software interfaces with an AI image recognition module (Figure 6) that identifies medicines using a trained neural network model. Prescriptions are cross-verified with the identified medicine to ensure accurate dispensing.

Figure 6. Text recognition using AI

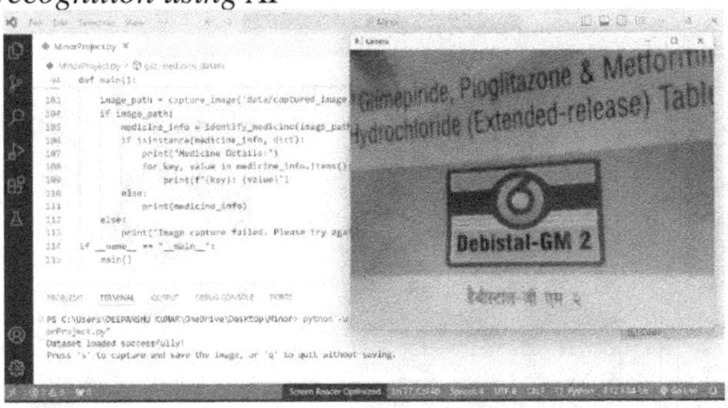

User Interface

Mobile App/PC Interface is designed using frameworks like Flutter or React for cross-platform compatibility. The features include:

- ✓ Input medicine details and prescriptions.
- ✓ Real-time monitoring of dispenser activity.

- ✓ Notifications for upcoming dosages and refill reminders.
- ✓ Historical data for patient compliance tracking.

Error Handling

The system detects and mitigates errors using predefined thresholds for flow rates and sensor readings. Alerts are sent via the interface in case of anomalies, such as empty reservoirs or incorrect medicine identification.

Flow Meter Monitoring

Once the system starts dispensing, the flow meter continuously measures the liquid volume in real time and transmits the data to the microcontroller. Figure 7 illustrates the connection between the microcontroller and the flow meter. The microcontroller tracks the flow rate and adjusts the pump operation to ensure accurate dispensing. If the liquid flow exceeds the set volume, the system automatically stops the pump to prevent wastage. This feedback loop ensures consistent and precise dispensing, eliminating human error. An LCD screen displays the dispensing status, including the volume dispensed and any error messages, allowing users to monitor the process conveniently.

Figure 7. Connection between the microcontroller and the flow meter

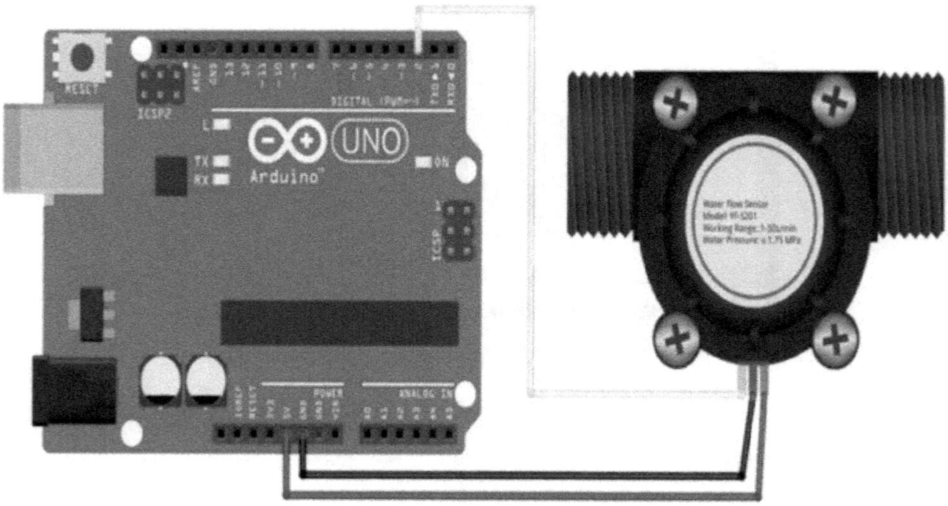

IMPLEMENTATION

A block diagram illustrating the operation of an Automated Medicine Dispenser, powered by an Arduino microcontroller, a flowmeter, AI integration, and an LCD display, is shown in Figure 8. This system combines hardware and software components to create an automated, precise, and user-friendly solution for liquid medicine dispensing. At the core of the hardware setup is the Arduino Uno microcontroller, which acts as the central processing unit, managing the entire dispensing process. A flowmeter is integrated to measure the liquid volume as it is dispensed, providing real-time feedback to the Arduino to ensure the correct amount is delivered. The liquid is dispensed using a static pump controlled by the Arduino based on the calculated dispensing time. An LCD screen (Figure 9) enhances user interaction by displaying real-time updates, such as the volume dispensed, system status, and error messages. The system is powered via a USB power supply, ensuring portability and energy efficiency.

Figure 8. Block diagram of an Automated Medicine Dispenser powered by Arduino and a flowmeter with AI integration and an LCD display

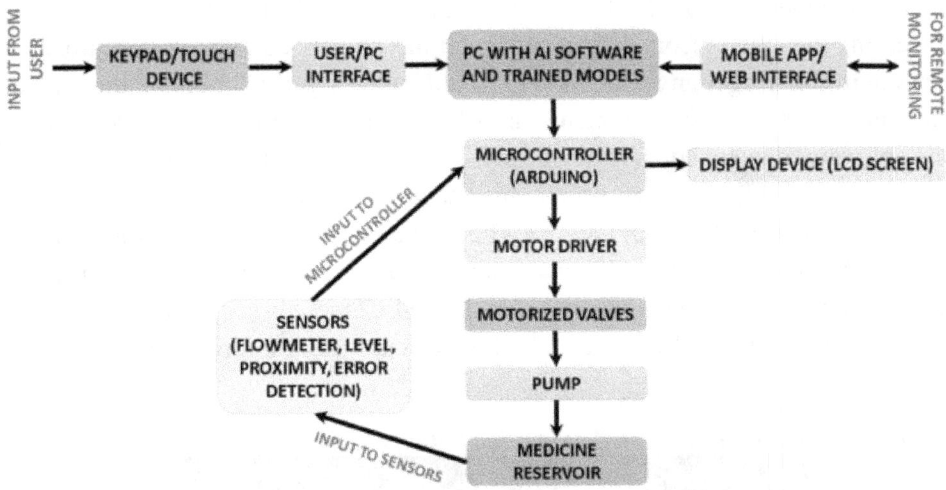

Figure 9. Connection between microcontroller and LCD display on Tinkercad

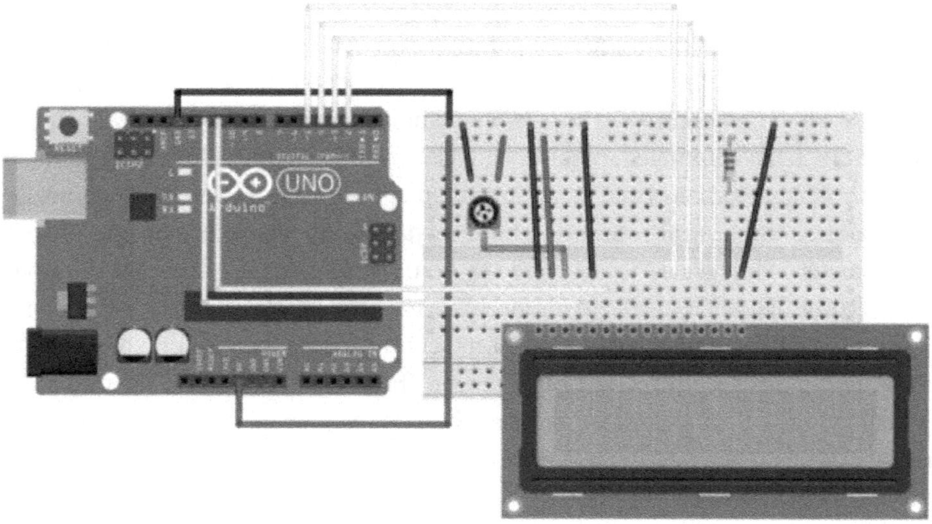

To enhance functionality, level sensors can be incorporated to monitor liquid levels in the medicine bottle. These sensors alert users when the liquid level is low, preventing the system from dispensing under insufficient conditions. Additionally, the system can trigger an application to locate online sources for the required medicine and assist the user in placing an order, ensuring uninterrupted medication. The software, developed using C++ within the Arduino Integrated Development Environment (IDE), processes user inputs, manages sensor data, and handles system responses (Figure 10 and Figure 11). The flowmeter continuously communicates with the Arduino to monitor the dispensed liquid and ensure the pump stops automatically once the preset volume is reached. The LCD screen is updated in real time, displaying the dispensed volume and other critical system information.

Figure 10. Code section of an Automatic medicine dispensing system

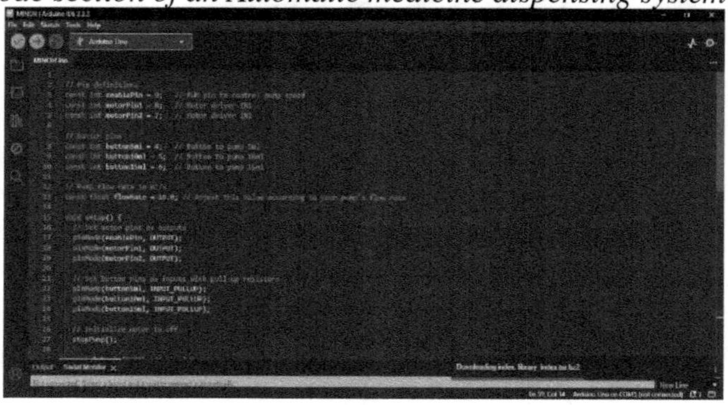

Figure 11. Simulation of text recognition

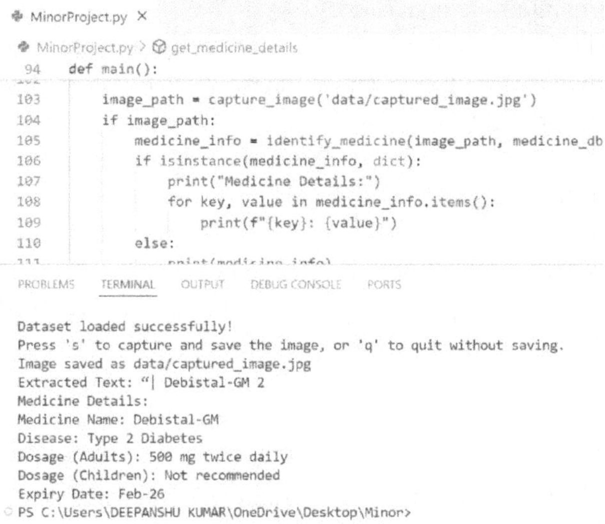

To enhance usability, the system can integrate with a mobile app or web interface. This allows users to remotely control the dispensing process, adjust settings such as dispensing volume or schedule, and monitor liquid usage patterns. Communication between the Arduino and the app is achieved via Bluetooth or Wi-Fi, creating a seamless and intuitive user experience. The IoT-enabled design also facilitates data collection and analysis to optimize dispensing patterns, making the system adaptable to user needs.

Before deployment, rigorous testing ensures the system's accuracy, reliability, and efficiency. This includes verifying the flowmeter's precision, evaluating sensor responsiveness, and addressing challenges like pump calibration, error handling, and stable wireless communication. Throughout operation, the system prioritizes error detection, alerting users to issues like blockages or insufficient liquid levels, ensuring reliable and uninterrupted performance.

By reducing human intervention and minimizing errors, this Automated Medicine Dispenser significantly improves operational efficiency and precision in liquid dispensing. Its integration of AI, IoT, and user-friendly features makes it an ideal solution for individuals requiring timely and accurate medication, as well as for use in hospitals, clinics, and research settings where precision is critical. Through its innovative and efficient design, the system represents a forward-looking approach to modern medicine management.

Highlights of the design:

- ***Accuracy:*** The system ensures precise dispensing of medicine, minimizing errors and maintaining consistency.
- ***Ease of Use***: Designed to be user-friendly, the system offers multiple features and functionalities with minimal human intervention, making it convenient and efficient for users.

Limitations of the design:

- Limited storage capacity for multiple medicines.
- Dependence on stable power supply and internet for advanced AI features.

RESULTS

The implementation of the Liquid Automatic Dispensing Machine represents a significant advancement in automating liquid dispensing processes, delivering improvements in efficiency, accuracy, and user experience. This section highlights the key results:

- **Efficiency and Convenience**: By automating the liquid dispensing process, users no longer need to manually measure liquids or rely on staff assistance, making the process quicker and more streamlined. The system's precise control over dispensing times ensures that the correct volume is delivered accurately and promptly, reducing waiting times and enhancing user satisfaction.
- **Accuracy and Error Reduction**: The system shows remarkable precision, minimizing human error through the integration of flow meters and a static pump. The flow meters continuously monitor the dispensing process, while the Arduino system adjusts the pump operation in real time to stop once the set volume is reached. This precise mechanism has eliminated inconsistencies in dispensing, ensuring reliable results across all trials and building user trust in the system's operation.
- **Real-time Monitoring and Feedback**: The machine's real-time monitoring and feedback capabilities have been integral to its success. The LCD screen, combined with flow meter data, provides real-time updates on the dispensing process, allowing users to track the exact amount of liquid being dispensed. This transparency not only enhances user confidence but also significantly reduces the likelihood of errors. Moreover, the feedback loop between the sensors and the Arduino enables immediate corrective actions in case of discrepancies, ensuring consistent and reliable performance during operation.

- **User Experience and Engagement**: By eliminating the need for manual measurements, it simplifies liquid dispensing for users, particularly those with limited technical experience.

Overall, the Liquid Automatic Dispensing Machine has proven to be an innovative solution that combines precision, convenience, and user-friendly features. Its ability to provide reliable and efficient liquid dispensing marks it as a valuable tool for a wide range of applications, with the potential for future enhancements to further expand its capabilities.

Future Challenges:

While the Liquid Automatic Dispensing Machine has demonstrated effectiveness and reliability, several areas for future enhancement could further expand its functionality and applicability. One significant improvement would be the integration of **wireless connectivity**, such as Bluetooth or Wi-Fi, to enable remote monitoring and control through a mobile app. This feature would allow users to interact with the system from a distance, providing added convenience and accessibility. Such an enhancement would be particularly beneficial for individuals with limited mobility, such as bedridden patients or elderly individuals, who may find it challenging to operate the machine manually. By enabling caregivers or family members to control the dispensing process remotely, the system could offer a more user-friendly and inclusive solution tailored to the needs of these users.

Another challenge lies in improving **flowmeter calibration** to accommodate various liquid types with differing viscosities, densities, and flow characteristics. Enhancing the system's adaptability in handling diverse liquids would make it more versatile across multiple applications. Additionally, ensuring **scalability** for dispensing larger volumes or managing multiple dispensing units simultaneously could optimize the machine for high-demand environments.

To facilitate widespread deployment, it is essential to address the **initial setup costs** and ensure that the system remains cost-effective without compromising on quality. Tackling **technical integration challenges**, such as compatibility with existing infrastructure and seamless IoT functionality, will also play a key role in broadening the system's appeal. Furthermore, **liquid compatibility** concerns, such as preventing corrosion or contamination, must be resolved to ensure safe and long-term use in medical applications.

By addressing these challenges, the Liquid Automatic Dispensing Machine can evolve into a more robust, scalable, and user-friendly solution in healthcare industry.

CONCLUSION

The Automated Medicine Dispenser and Liquid Automatic Dispensing Machine represent significant advancements in automation within healthcare and industrial applications. These systems offer cost-effective, user-friendly solutions that enhance medication management and liquid dispensing, reducing human error and improving overall efficiency. By minimizing mistakes and improving medication adherence, they lessen the burden on caregivers and ensure more reliable dispensing of critical liquids. The liquid dispensing system employs advanced technologies such as flow meters, Arduino microcontrollers, and static pumps to deliver precise and accurate dispensing based on user-defined volumes, significantly improving operational performance and reliability. These innovations are transforming traditional manual processes by automating tasks, reducing human error, and providing consistent, accurate results. However, challenges remain, including ensuring ongoing system accuracy, addressing potential dispensing errors, and maintaining long-term reliability. Overcoming these hurdles is crucial to further refining these systems. By tackling these challenges, the systems can achieve even greater precision, robustness, and user satisfaction, thereby enabling broader adoption across various sectors, from healthcare to industrial applications, and ultimately contributing to the advancement of automated solutions in modern technology.

FUTURE SCOPE

The future scope of the Automated Medicine Dispenser offers significant opportunities for innovation, with a focus on enhancing functionality, user experience, and reliability in healthcare settings. One key development is the integration of mobile app connectivity, enabling remote monitoring and control via smartphones. This would provide greater convenience and accessibility for patients and caregivers, particularly in situations where physical interaction with the dispenser is not possible. Additionally, the incorporation of voice command features could allow for hands-free operation, making the system even more user-friendly, especially for individuals with limited mobility.

The system can also be expanded to support **multi-drug dispensing**, allowing for the management of complex medication regimens. This would be beneficial for patients who need to take multiple medications at different times, improving adherence and reducing the risk of errors. Furthermore, **cloud connectivity** could be incorporated, enabling secure storage and analysis of user data to gain valuable insights into medication usage, optimize dispensing patterns, and enhance the overall efficiency of the system.

To ensure greater accuracy and adaptability, the system could integrate **advanced flow control mechanisms** that adjust dispensing based on the type and viscosity of the liquid, ensuring precise delivery for a variety of medications. **Personalization features** could allow the dispenser to remember user-specific settings, such as preferred volumes or dispensing schedules, enhancing the system's convenience for repeated use.

In terms of security, the implementation of robust encryption protocols and **IoT-based solutions** would safeguard user data and prevent unauthorized access, ensuring privacy and compliance with healthcare regulations. Additionally, sustainability is an important consideration, with potential for integrating **energy-efficient components** or **solar-powered systems** to reduce the environmental impact of the dispenser.

These advancements promise to make the Automated Medicine Dispenser even more efficient, reliable, and adaptable, offering great potential for improving medication management in healthcare settings. With continued innovation, these systems could become integral tools in both home care and institutional environments, supporting patients with more complex medical needs while enhancing the overall quality of care.

REFERENCES

Agarwal, P.. (2022). Development of a smart liquid dispensing system for industrial applications. *2022 International Conference on Automation and Robotics (ICAR)*, Bangalore, India.

Aronson, J. K. (2009). Medication errors: What they are, how they happen, and how to avoid them. *QJM*, *102*(8), 513–521. DOI: 10.1093/qjmed/hcp052 PMID: 19458202

Belissa, E., Vallet, T., Laribe-Caget, S., Chevallier, A., Chedhomme, F.-X., Abdallah, F., Bachalat, N., Belbachir, S.-A., Boulaich, I., Bloch, V., Delahaye, A., Depoisson, M., Wojcicki, A. D., Gibaud, S., Grancher, A.-S., Guinot, C., Lachuer, C., Lechowski, L., Leglise, P., & Boudy, V. (2019). Acceptability of oral liquid pharmaceutical products in older adults: Palatability and swallowability issues. *BMC Geriatrics*, *19*(344), 344. Advance online publication. DOI: 10.1186/s12877-019-1337-2 PMID: 31810442

Brown, M. T., & Bussell, J. K. (2011). Medication adherence: WHO cares? *Mayo Clinic Proceedings*, *86*(4), 304–314. DOI: 10.4065/mcp.2010.0575 PMID: 21389250

Davis, T., & Wilson, A. (2017). Fluid dispensing systems: Design and application. In Fluid Dispensing Systems (pp. 112–130). Springer. ISBN: 978-3-319-45671.

Gargioni, L., Fogli, D., & Baroni, P. (2024). A systematic review on pill and medication dispensers from a human-centered perspective. *Journal of Healthcare Informatics Research*, *8*(2), 244–285. DOI: 10.1007/s41666-024-00161-w PMID: 38681758

Gast, A., & Mathes, T. (2019). Medication adherence influencing factors—An (updated) overview of systematic reviews. *Systematic Reviews*, *8*(112), 112. Advance online publication. DOI: 10.1186/s13643-019-1014-8 PMID: 31077247

Hidayat, R., & Winahyu, H. S. (2020). Automatic liquid filling machine using flow sensor and Arduino Pro Mini. *Indonesian Journal of Engineering Research*, *1*(1), 27–31. DOI: 10.11594/ijer.01.01.05

Hoang Le, A. P., Nguyen, Q. L., Pham, B. H., Cao, T. H. M., Vo, T. V., Huynh, K., & Ha, H. T. T. (2023). SALAD: Syringe-based Arduino-operated low-cost antibody dispenser. *HardwareX*, *15*, e00455. DOI: 10.1016/j.ohx.2023.e00455 PMID: 37497344

Hughes, R. G., & Blegen, M. A. (2008). Medication administration safety. In Hughes, R. G. (Ed.), *Patient safety and quality: An evidence-based handbook for nurses*. Agency for Healthcare Research and Quality., https://www.ncbi.nlm.nih.gov/books/NBK2656/

Husni, M.. (2016). Design and analysis of a MEMS flow sensor. *IOP Conference Series. Materials Science and Engineering*, *128*(1), 012026. DOI: 10.1088/1757-899X/128/1/012026

Jayanth, S., Poorvi, M. B., & Sunil, M. P. (2017). Inventory management system using IoT. In Satapathy, S., Prasad, V., Rani, B., Udgata, S., & Raju, K. (Eds.), *Proceedings of the First International Conference on Computational Intelligence and Informatics* (Vol. 507, pp. 299–309). Springer. https://doi.org/DOI: 10.1007/978-981-10-2471-9_20

Johnson, M., Tan, L., & Zhang, Y. (2018). Real-time liquid flow measurement and control using Arduino. *International Journal of Robotics and Automation*, *12*(2), 30–38. DOI: 10.1115/ijra.2018.090404

Junaid, S. B., Imam, A. A., Balogun, A. O., De Silva, L. C., Surakat, Y. A., Kumar, G., Abdulkarim, M., Shuaibu, A. N., Garba, A., Sahalu, Y., Mohammed, A., Mohammed, T. Y., Abdulkadir, B. A., Abba, A. A., Kakumi, N. A. I., & Mahamad, S. (2022). Recent advancements in emerging technologies for healthcare management systems: A survey. *Health Care*, *10*(10), 1940. DOI: 10.3390/healthcare10101940 PMID: 36292387

Koh, Y. Y., Kok, C. L., Ibraahim, N., & Lim, C. G. (2024). Smart water ATM with Arduino integration, RFID authentication, and dynamic dispensing for enhanced hydration practices. *Electronics (Basel)*, *13*(9), 1657. DOI: 10.3390/electronics13091657

Kumar, G., Sharma, A., & Mehta, R. (2023). Automatic liquid dispensing system based on Arduino and flow meter. *2023 International Conference on Innovations in Electrical and Electronics Engineering (ICEEE)*, 65–70. https://doi.org/DOI: 10.1109/ICEEE.2023.1234567

Latham, T. W. (1966). Fluid motions in a peristaltic pump. PhD thesis, Massachusetts Institute of Technology. http://hdl.handle.net/1721.1/17282

Liu, Y., Chen, L., & Sun, L. (2009). Design and fabrication of a MEMS flow sensor and its application in precise liquid dispensing. *Sensors (Basel)*, *9*(6), 4138–4150. DOI: 10.3390/s90604138 PMID: 22408517

Manikandan, P., Tirupathi Rao, M., Vinay, M., & Ratna Raj, K. D. (2021). Arduino based automatic hand sanitizer dispenser system. 2021 3rd International Conference on Advances in Computing, Communication Control and Networking (ICAC3N), 744–747.

Marcum, Z. A., & Gellad, W. F. (2012). Medication adherence to multidrug regimens. *Clinics in Geriatric Medicine*, *28*(2), 287–300. DOI: 10.1016/j.cger.2012.01.008 PMID: 22500544

Mason, M., Cho, Y., Rayo, J., Gong, Y., Harris, M., & Jiang, Y. (2022). Technologies for medication adherence monitoring and technology assessment criteria: Narrative review. *JMIR mHealth and uHealth*, *10*(3), e35157. DOI: 10.2196/35157 PMID: 35266873

Meghla, R. T., Deowan, M. E., Nuhel, A. K., Sazid, M. M., Ekbal, M. N., & Mahamud, M. H. (2022). An Internet of Things (IoT)-based smart automatic medication dispenser with an integrated web application for patient diagnosis. 2022 5th International Conference of Computer and Informatics Engineering (IC2IE), 16–21. DOI: 10.1109/IC2IE56416.2022.9970073

Patel, A. (2019). Smart fluid dispensing for medical and laboratory applications. *2019 International Conference on Smart Technologies and Innovations (ICSTI)*, London, UK.

Patel, S., Kumar, N., & Jain, P. (2021). Flow measurement and control in automatic dispensing systems. *Journal of Fluids Engineering*, *145*(3), 15–20. DOI: 10.1115/1.4052018

Pinto, J. F., Vilaça, J. L., & Dias, N. S. (2021). A review of current pill organizers and dispensers. 2021 IEEE 9th International Conference on Serious Games and Applications for Health (SeGAH), 1–8. https://doi.org/DOI: 10.1109/SEGAH52098.2021.9551894

Punnapurath, S., Vijayakumar, P., Platty, P. L., Krishna, S., & Thomas, T. (2021). A study of medication compliance in geriatric patients with chronic illness. *Journal of Family Medicine and Primary Care*, *10*(4), 1644–1648. DOI: 10.4103/jfmpc.jfmpc_1302_20 PMID: 34123906

Smith, R. (2020). An overview of fluid dispensing techniques in automated systems. *IACSIT International Journal of Engineering and Technology*, *10*(4), 452–460. DOI: 10.1016/j.eng.2020.08.006

Spasova, G. V., Karova, M. N., & Boychev, I. Z. (2024). Communication protocol for liquid dispensing vending machines. 2024 59th International Scientific Conference on Information, Communication and Energy Systems and Technologies (ICEST), 1–4. https://doi.org/DOI: 10.1109/ICEST62335.2024.10639714

Walsh, C. A., Cahir, C., Tecklenborg, S., Byrne, C., Culbertson, M. A., & Bennett, K. E. (2019). The association between medication non-adherence and adverse health outcomes in ageing populations: A systematic review and meta-analysis. *British Journal of Clinical Pharmacology*, *85*(11), 2464–2478. DOI: 10.1111/bcp.14075 PMID: 31486099

Yin, H. S., Parker, R. M., Sanders, L. M., Dreyer, B. P., Mendelsohn, A. L., Bailey, S., Patel, D. A., Jimenez, J. J., Kim, K.-Y. A., Jacobson, K., Hedlund, L., Smith, M. C. J., Maness Harris, L., McFadden, T., & Wolf, M. S. (2016). Liquid medication errors and dosing tools: A randomized controlled experiment. *Pediatrics*, *138*(4), e20160357. DOI: 10.1542/peds.2016-0357 PMID: 27621414

Chapter 13
Finance and Beyond: Self-Service Transformation Case Studies

Saliu Hakeem Tomi
https://orcid.org/0000-0002-4732-0875
Prince Abubakar Audu University, Nigeria

Olumoyegun peter Mayowa
Prince Abubakar Audu University, Nigeria

Attah Eleojo Vincent
Vetitas University, Abuja, Nigeria

Musa Abdulrahman Dantala
Prince Abubakar Audu University, Nigeria

Edoka Reuben Ogbaje
Prince Abubakar Audu University, Nigeria

ABSTRACT

The advent of self-service technology has transformed industries far beyond its initial applications in retail and banking, fundamentally reshaping how businesses operate and deliver services across sectors. This paper explores the impact of self-service transformation on financial services and other industries through a series of case studies. By examining real-world examples, we highlight how companies have leveraged self-service technologies to improve efficiency, reduce costs, and enhance customer experiences. In financial services, self-service technologies such as ATMs, mobile banking apps, and online financial management tools have empowered consumers to take greater control of their finances. These tools not only streamline

DOI: 10.4018/979-8-3373-4667-0.ch013

routine transactions but also enable financial institutions to reallocate resources toward more personalized and high-value services. Beyond finance, industries such as healthcare, hospitality, and transportation have adopted self-service solutions to address operational challenges, improve service delivery, and cater to evolving consumer expectations.

INTRODUCTION

The financial services industry is undergoing a transformative revolution driven by advanced technologies such as artificial intelligence (AI), distributed ledger technology (DLT), the Internet of Things (IoT), extended reality, and quantum computing. These cutting-edge innovations not only unlock new possibilities but also significantly improve existing financial solutions (World Economic Forum, 2020). Infrastructure technologies such as high-performance cloud computing, 5G networks, and software-defined virtualization form the backbone of rapid digital innovation. Scalable data management platforms empower financial institutions to derive insights from diverse alternative data sources, while advanced cybersecurity frameworks are critical for ensuring the security and resilience of digital financial systems (Arner, Barberis, & Buckley, 2020). One of the most groundbreaking innovations is DLT, such as blockchain, which facilitates decentralized, transparent, and immutable record-keeping for transactions and contracts (Tapscott & Tapscott, 2016). While Bitcoin remains volatile, the blockchain infrastructure underpinning it has demonstrated immense value across various domains, including finance, supply chains, and medical records. Decentralized finance (DeFi) protocols, powered by smart contracts, enable automated, borderless financial services, eliminating the need for traditional intermediaries (Zetzsche, Buckley, Arner, & Barberis, 2020; Abulibdeh, Zaidan & Abulibdeh, 2024)

Artificial intelligence (AI) and machine learning (ML) algorithms have become integral to diverse aspects of finance, including trading, fraud detection, customer experience, and back-end operations Anan (Brynjolfsson & McAfee, 2017). Natural language processing (NLP) powers chatbots that deliver conversational interfaces for tasks ranging from retail banking inquiries to investment research (Davenport & Ronanki, 2018). Robo-advisors, driven by neural networks, provide personalized wealth management solutions, making advanced financial planning tools accessible to a wider audience (Jung, Dorner, Glaser, & Morana, 2018). In institutional finance, AI analyzes unconventional datasets to uncover actionable insights, enhancing the performance of algorithmic trading systems (Anan, Wang, & Mahmud, 2023). Big data analytics empowers organizations to optimize operations and tailor product offerings through predictive modeling. For example, analyzing customer transaction data allows retail banks to refine credit scoring models and identify cross-selling opportunities (Chen, Chiang, & Storey, 2012). On trading floors, real-time analysis of news and social media trends offers critical insights into shifting market sentiment (Bollen, Mao, & Zeng, 2011). Additionally, IoT-enabled platforms generate new data streams from connected devices and infrastructure, delivering actionable insights for insurance, credit risk assessment, and corporate transparency (Ng & Wakenshaw, 2017; Anan, et al, 2023)

Augmented and virtual reality (AR/VR) technologies are increasingly being adopted in finance for applications such as collaborative market risk visualization, immersive employee training, and improved customer engagement (Puschmann, 2017). Meanwhile, quantum machine learning on next-generation quantum systems offers the potential to tackle complex problem spaces that are beyond the capabilities of classical computing (Zhang et al., 2022). This breakthrough could revolutionize areas such as portfolio optimization and climate financial risk forecasting, both critical for managing sustainability-related exposures.

Although still in its early stages, rapid transformation is inevitable across the financial ecosystem, encompassing consumer banking, investment management, insurance, and central bank infrastructure (PwC, 2020). Incumbent institutions must proactively modernize legacy systems to enhance efficiency, maintain competitiveness, and seize emerging opportunities. Policymakers face the challenge of balancing stability with growth to foster innovation responsibly (Arner et al., 2017). Meanwhile, investors should target emerging sectors such as regtech, insurtech, blockchain fintech, and robotic process automation, which are poised to become key drivers of financial modernization. Finance 4.0 will be characterized by intelligent, interconnected, and real-time solutions, marking the next chapter in the technological evolution of financial markets (Puschmann, 2017). While still in its infancy, this transformation signals a significant leap forward. Its ultimate trajectory remains uncertain, but all signs point to a paradigm shift as profound as the one the Internet brought to media and communications (Lennerholt et al., 2020).

The modern financial landscape is undergoing a rapid transformation fueled by technological advancements and evolving customer expectations. At the heart of this evolution lies the rise of self-service models, which empower users to independently manage financial transactions, access services, and make informed decisions with minimal reliance on intermediaries. This paper explores a series of case studies highlighting successful implementations of self-service solutions within the banking, insurance, and investment sectors, showcasing their transformative impact on user engagement, operational efficiency, and service delivery (Torres et al., 2018; Tamm et al., 2020; Seddon et al., 2017; Gupta et al., 2016; Bani-Hani et al., 2018).

The case studies examine the integration of AI-driven chatbots for customer support, automated onboarding processes, digital self-help tools, and user-friendly platforms that redefine traditional service frameworks. These innovations enhance accessibility, reduce service bottlenecks, and cater to the digital-first preferences of modern consumers. Beyond the financial domain, this research investigates cross-industry parallels, presenting examples from healthcare, retail, and telecommunications where self-service solutions have been instrumental in streamlining operations and improving customer satisfaction. (Tamm et al., 2020; Seddon et al., 2017; Gupta et al., 2016; Bani-Hani et al., 2018).

This paper underscores the critical role of data analytics, user-centric design, and robust cybersecurity measures in the successful adoption of self-service models. It also examines the challenges organizations face, such as ensuring inclusivity, maintaining trust, and balancing automation with personalized human interaction. The findings provide actionable insights and a roadmap for businesses looking to harness self-service innovations, offering a vision for a future where self-service becomes a cornerstone of sustainable growth and superior customer experiences. Self-service transformation is a pivotal trend reshaping the financial sector and influencing other industries. As businesses navigate this paradigm shift, they must focus on integrating advanced technologies with customer-centric strategies. The insights from these case studies provide a roadmap for organizations aiming to harness self-service innovations, driving sustainable growth and superior customer experiences. By embracing this future, companies can ensure their relevance and success in an ever-evolving landscape. Torres, Sidorova, & Jones, 2018.

Definitions of terms

1. **Artificial Intelligence (AI)** refers to the simulation of human-like intelligence in machines and computer systems, enabling them to perform tasks that typically require human cognition. These tasks include learning, reasoning, problem-solving, decision-making, understanding language, recognizing patterns, and adapting to new information. In simpler terms, AI allows machines to think, learn, and act on their own, mimicking human intelligence to some extent.
2. A **Distributed Ledger** is a decentralized, digital record-keeping system that allows multiple participants (or nodes) in a network to have synchronized copies of a shared database. Unlike traditional centralized ledgers, distributed ledgers do not rely on a single authority to verify and update transactions. Instead, all participants collectively maintain the ledger, ensuring transparency, security, and resilience.The most well-known example of a distributed ledger is **blockchain**, which powers cryptocurrencies like Bitcoin and Ethereum. However, distributed ledger technology (DLT) has broader applications beyond cryptocurrencies.

3. The **Internet of Things (IoT)** refers to a network of interconnected physical devices embedded with sensors, software, and other technologies that enable them to collect and exchange data over the internet. These "smart" devices can monitor, control, and interact with their environment, providing real-time insights and automating processes without human intervention. IoT is transforming industries by connecting everyday objects — from household appliances to industrial machinery — to the digital world, creating smarter and more efficient systems.
4. **Quantum computing** is an advanced form of computing that leverages the principles of **quantum mechanics** — a branch of physics that deals with subatomic particles — to perform calculations far beyond the capabilities of classical computers. Unlike traditional computers that use binary bits (0s and 1s), quantum computers use **quantum bits (qubits)**, which can exist in multiple states simultaneously, enabling them to solve complex problems much faster. Quantum computing has the potential to revolutionize fields such as cryptography, artificial intelligence, optimization, drug discovery, and more.
5. **Machine Learning (ML)** is a subset of artificial intelligence that enables computers to learn from data and improve their performance over time without being explicitly programmed. ML algorithms are the backbone of this process, helping machines recognize patterns, make predictions, and automate decisions based on data.ML algorithms can be broadly classified into **three main categories** based on how they learn from data: **supervised learning**, **unsupervised learning**, and **reinforcement learning**. Let's explore each category and its popular algorithms.
6. **Natural Language Processing (NLP)** is a subfield of artificial intelligence that focuses on enabling machines to understand, interpret, and generate human language. It bridges the gap between human communication and machine understanding, making it possible for computers to process and analyze large amounts of natural language data (text or speech).NLP powers many applications we use today, such as **chatbots**, **virtual assistants** (like Siri or Alexa), **translation tools**, **sentiment analysis**, and **text summarization**.
7. **Augmented Reality (AR)** and **Virtual Reality (VR)** are immersive technologies that are transforming how we interact with digital content and the physical world. Both AR and VR create new experiences, but they do so in different ways. While AR enhances the real world by adding digital elements to it, VR creates a completely artificial environment that users can immerse themselves in.

Augmented Reality (AR) overlays digital information—such as images, sounds, or other sensory inputs—onto the real world. This blending of digital and physical realities is usually done through a screen (smartphones, tablets, smart glasses) or wearable AR devices.

Virtual Reality (VR) is a fully immersive technology that creates a completely artificial environment. Unlike AR, which enhances the real world, VR transports users into a completely virtual space where they can interact with the environment through specialized equipment like **VR headsets** and **controllers**.

8. GDPR : The **General Data Protection Regulation (GDPR)** is a comprehensive data protection law enacted by the European Union (EU) to protect the personal data of individuals within the EU and the European Economic Area (EEA). It was adopted on **April 14, 2016**, and became enforceable on **May 25, 2018**. It replaces the **Data Protection Directive 95/46/EC** and aims to strengthen individuals' privacy rights and simplify the regulatory environment for international businesses by unifying data protection laws across Europe.
9. **CCPA: Definition:** The **California Consumer Privacy Act (CCPA)** is a data privacy law enacted in California, United States, which aims to enhance privacy rights and consumer protection for residents of California. The CCPA was passed in **2018** and became enforceable on **January 1, 2020**. It gives California residents greater control over their personal data and imposes new requirements on businesses that collect, process, and share personal information. The CCPA is similar to the European **General Data Protection Regulation (GDPR)** but is specific to the state of California, with some differences in scope and requirements.
10. **Definition:** AI chatbots are software applications powered by **artificial intelligence (AI)** that simulate human conversation. They are designed to interact with users through text or voice-based communication, providing automated responses to inquiries or requests. AI chatbots leverage natural language processing (NLP), machine learning (ML), and sometimes deep learning techniques to understand, interpret, and respond to user inputs in a meaningful way.

The Rise of Self-Service Models in Finance

The financial sector is experiencing a groundbreaking shift, driven by rapid technological advancements and changing customer expectations. The advent of self-service solutions has redefined the traditional dynamics between service providers and consumers, enabling a more empowered, independent, and efficient customer experience. This chapter delves into a series of case studies that illustrate how self-service transformation is reshaping the financial industry and extending its influence beyond, into various other sectors. Self-service models are at the forefront of financial innovation. These solutions provide customers with the tools to manage transactions, access information, and make decisions without relying on intermediaries. This empowerment is facilitated through technologies such as:

- **AI-Driven Chatbots:** Advanced chatbots deliver 24/7 customer support, offering personalized responses and resolving queries efficiently. Examples include chatbots in banking apps that guide users through processes like fund transfers, loan applications, and investment planning. Advanced AI-driven chatbots are revolutionizing customer support by providing round-the-clock assistance, delivering personalized responses, and resolving inquiries with remarkable efficiency. These chatbots utilize natural language processing (NLP) to understand and interpret user queries in real-time, offering an interactive experience that mimics human conversation. In banking and financial services, chatbots can assist customers in a wide range of activities, such as guiding users through fund transfers, processing loan applications, or providing personalized investment planning advice. By leveraging machine learning, these chatbots continuously improve their ability to respond accurately to a variety of questions, adapting to the evolving needs of customers. Furthermore, they can integrate with other systems, such as account management tools or financial calculators, to provide seamless, personalized recommendations. This not only enhances customer satisfaction but also reduces operational costs for financial institutions, enabling them to offer services at scale while maintaining a high level of personalization and efficiency.
- **Digital Onboarding:** Automated onboarding processes simplify account creation, document verification, and compliance checks, reducing onboarding times from days to minutes. Digital onboarding is transforming the way financial institutions and service providers interact with new customers by automating traditionally manual processes. This shift significantly streamlines account creation, document verification, and compliance checks, drastically reducing the time it takes to onboard a new customer—from several days to just minutes. The process begins with a fully digital application where users can easily input their personal information and upload required identification documents such as passports, utility bills, or driver's licenses. AI and machine learning algorithms automatically analyze these documents for authenticity and ensure that they meet regulatory standards, reducing the risk of human error and fraud.

Furthermore, biometric verification technologies, like facial recognition or fingerprint scanning, can be integrated into the onboarding process to provide an additional layer of security, ensuring that the person opening the account is indeed who they claim to be. This speeds up the identity verification process while enhancing security. The automated compliance checks that are integrated into digital onboarding ensure that institutions remain compliant with financial regulations such as Know Your Customer (KYC) and Anti-Money Laundering (AML) laws. By using algorithms to cross-reference customer data against sanction lists and databases, the system can quickly flag suspicious activities, all without requiring manual intervention.

In addition to improving speed and efficiency, digital onboarding enhances the customer experience by offering a more user-friendly, seamless process. Users can complete the entire procedure from the comfort of their home or office, avoiding long waits at physical branches or tedious paperwork. For financial institutions, the benefits of digital onboarding extend beyond just faster processes. It also reduces operational costs, minimizes the need for human involvement, and allows businesses to onboard more customers with greater accuracy and efficiency. Ultimately, this digital transformation fosters better customer engagement and satisfaction, driving business growth in an increasingly competitive landscape.

Intuitive User Interfaces: Platforms with user-friendly designs enhance accessibility, making financial services easy to navigate for customers of varying digital proficiencies. It Platforms equipped with intuitive, user-friendly interfaces are playing a crucial role in enhancing accessibility and improving the overall user experience within the financial services sector. These interfaces are designed with simplicity, clarity, and ease of use in mind, ensuring that customers of all digital skill levels can navigate financial services with confidence and ease. An intuitive user interface (UI) prioritizes functionality while minimizing complexity. By using clear, straightforward navigation structures, easy-to-read fonts, and visually appealing design elements, financial platforms enable users to accomplish tasks without feeling overwhelmed. For example, key actions like transferring funds, viewing transaction histories, or managing investments are typically laid out in a way that's easy to access, often with large buttons or icons, streamlined workflows, and simplified language that guide users through each step of a process.

These platforms often incorporate a variety of features tailored to users' needs, such as step-by-step tutorials, context-sensitive help menus, and interactive prompts that offer additional support when necessary. Whether a user is a digital native or someone with minimal tech experience, the UI adapts to their level of comfort by offering assistance without being intrusive. In addition to simplifying navigation, modern UIs incorporate personalization elements. For instance, dashboards might dynamically adjust to show the most relevant information for each individual user, such as recent transactions, account balances, or personalized financial recommendations. These customizations make it easier for users to access important data quickly, enhancing both their convenience and engagement with the platform.

Accessibility is another key feature of intuitive UIs, ensuring that financial services are available to a broad spectrum of users, including those with disabilities. High-contrast color schemes, larger fonts, screen reader compatibility, and voice-activated controls are just a few examples of how platforms can make their services accessible to all customers, regardless of physical or cognitive limitations. Mobile-first designs are increasingly important as well. With the majority of consumers now accessing financial services through smartphones, an intuitive interface must function seamlessly across different screen sizes and devices. This responsiveness ensures that users enjoy a consistent and efficient experience, whether they are on a desktop computer, tablet, or smartphone.

For financial institutions, the advantages of investing in intuitive UIs go beyond just improving user satisfaction. These platforms reduce customer frustration, minimize errors, and increase overall engagement, ultimately leading to higher retention rates and greater customer loyalty. By catering to a wide range of users, from tech-savvy millennials to older generations who may be less familiar with digital tools, intuitive UIs help bridge the gap and foster more inclusive financial services. George, (2023)

Case Studies in Financial Transformation

1. Banking Sector:

A global bank implemented a self-service mobile app that allowed customers to apply for loans, manage accounts, and access real-time financial advice. The app's success led to a significant reduction in in-branch traffic and improved customer satisfaction scores. Another institution deployed AI chatbots that handled 60% of routine customer queries, freeing human agents to address complex issues.

1. Self-Service Mobile App

A major global bank launched a self-service mobile app that enabled customers to manage a wide range of banking activities on-the-go, including applying for loans, managing accounts, and receiving real-time financial advice. This app integrated several advanced features, such as automated loan eligibility assessments, secure document uploads, and intuitive account management tools, all designed to provide a seamless digital banking experience. One of the advantages of the app was its real-time financial advice feature, which used artificial intelligence to analyze user behavior and financial data, offering personalized suggestions related to budgeting, saving, or investment opportunities (Verma et al., 2022; Verma & Srivastava, 2023). For instance, if a customer frequently spent a certain amount on dining or shopping, the app would recommend budgeting tips or suggest lower-cost alternatives, helping customers make more informed financial decisions. The success of this app resulted in a notable shift in customer behavior, with many opting to complete banking tasks from their mobile devices rather than visiting a branch. This transition led to a substantial reduction in in-branch traffic, easing the strain on physical locations and allowing banks to optimize resources and reduce operational costs. Customers also benefited from the increased convenience, as they could now access financial services at any time, from anywhere. As a result, the bank saw a significant improvement in customer satisfaction scores, with users appreciating the convenience and efficiency the app provided.

2. AI Chatbots for Routine Queries

Another banking institution adopted AI-powered chatbots to handle a substantial portion of routine customer interactions. These chatbots were designed to manage common queries related to account balances, transaction history, branch locations, and password resets, processing these requests quickly and accurately. By automating such tasks, the bank was able to free up human agents to focus on more complex issues, such as resolving disputes or providing personalized financial advice. The implementation of the AI chatbots had a dramatic impact on operational efficiency. In just a short time, the bots were handling up to 60% of all routine inquiries, reducing wait times for customers and improving response times for more complex queries. The shift allowed human agents to allocate more time to customers who needed specialized assistance, resulting in more efficient and effective service overall.

Additionally, the AI chatbots constantly learned from their interactions, becoming increasingly adept at handling diverse requests. With natural language processing (NLP) capabilities, the bots understood customer queries more accurately and could provide responses that felt more natural and human-like. The ability to support customers 24/7, even during peak times or after hours, further enhanced the bank's customer service, ensuring that clients had immediate access to support whenever needed. Over time, the AI chatbot initiative not only helped improve operational efficiency but also contributed to increased customer satisfaction. Customers appreciated the quick responses and ease of access, while the bank benefited from reduced labor costs and a higher capacity to handle a larger volume of inquiries without compromising the quality of service.

Both examples illustrate how digital transformation, through mobile apps and AI chatbots, is reshaping the banking industry. These innovations enhance both the customer experience and the operational efficiency of financial institutions, allowing them to remain competitive in a rapidly evolving market.

2. **Insurance Industry**

A leading insurer introduced a self-service claims portal where customers could upload documents, track claim status, and receive payouts directly. This innovation reduced claim processing times by 50%. A leading insurer implemented a cutting-edge self-service claims portal designed to empower customers by giving them full control over the claims process. Through this digital platform, customers could easily upload the necessary documents, such as accident reports, medical receipts, or proof of loss, eliminating the need for in-person visits or lengthy phone calls with claims adjusters. The portal provided real-time status updates on the claim's progress, ensuring that customers were always informed of where their claim stood in the processing queue. This innovative approach not only streamlined the claims process but also introduced a level of transparency and convenience that significantly improved the customer experience. Instead of waiting for days or weeks to receive updates, policyholders could monitor their claims' status instantly, reducing anxiety and increasing trust in the insurance provider.

Additionally, the self-service portal enabled direct payouts to customers once their claims were approved, bypassing traditional payment delays or the need for multiple follow-ups. With the integration of secure payment systems, customers received their payouts quickly and directly into their bank accounts or digital wallets. The results were transformative for the insurer. By digitizing and automating key elements of the claims process, the company reduced claim processing times by 50%. This not only enhanced operational efficiency but also freed up resources to focus on more complex claims or customer support issues, further improving service quality. The faster processing times led to higher customer satisfaction, as policyholders appreciated the convenience, speed, and ease with which they could navigate their claims. Overall, this self-service claims portal demonstrated the power of digital innovation in the insurance industry, transforming a traditionally manual and time-consuming process into a seamless, customer-centric experience. It also highlighted the growing importance of automation and technology in delivering both operational efficiencies and enhanced customer engagement in today's competitive market.

3. **Investment Services**

An investment platform enabled self-service portfolio management, where users could set financial goals, receive AI-generated recommendations, and adjust investments autonomously. This feature attracted a younger demographic of tech-savvy investors. An innovative investment platform introduced self-service portfolio management, revolutionizing how users engage with their investments. This feature allowed users to set personalized financial goals, such as saving for retirement, purchasing a home, or funding education, and then receive tailored, AI-generated recommendations based on their financial objectives and risk tolerance. The platform's sophisticated algorithms analyzed vast amounts of market data and trends to suggest optimal investment strategies, whether that involved reallocating assets, exploring new investment products, or adjusting the portfolio's risk profile. Users could take a hands-on approach by making changes to their investments autonomously, giving them the flexibility to react quickly to market conditions or adjust their portfolio in line with evolving financial goals. The platform provided real-time insights and a comprehensive dashboard, enabling users to track the performance of their investments, monitor returns, and make informed decisions.

What truly set the platform apart was its seamless integration of AI, which not only made the investment process more accessible but also more efficient. The AI engine continuously learned from market shifts, adapting its recommendations to reflect the changing economic landscape, and providing users with real-time, data-driven insights. This intelligent system minimized the need for human intervention, allowing users to manage their portfolios entirely on their own, without the reliance on expensive financial advisors. This feature particularly appealed to a younger, more tech-savvy demographic who valued both autonomy and accessibility. Millennials and Gen Z investors, in particular, were drawn to the platform because it provided them with the tools to make informed, data-backed investment decisions, all within a user-friendly interface. Many of these younger investors, who grew up with technology and digital-first experiences, appreciated the convenience and control offered by self-service portfolio management tools.

Additionally, the platform's low fees and absence of traditional advisory costs made it an attractive option for this cost-conscious generation, who often seek more affordable alternatives to traditional investment management services. By offering a transparent, automated solution that empowered users to actively manage their investments, the platform attracted a new wave of investors who were eager to take charge of their financial futures, all while leveraging the latest technology to enhance their decision-making processes. Overall, this self-service portfolio management feature not only democratized investment opportunities but also fostered financial literacy, enabling users to learn and grow their investments with confidence. As a result, the platform became a hub for younger investors looking for an innovative, affordable, and engaging way to manage their financial goals. DMahbub. (2023), George (2023)

Cross-Industry Parallels

The success of self-service in finance has inspired similar transformations across other industries:

- **Healthcare:** Patient portals allow individuals to book appointments, access medical records, and consult with doctors virtually, reducing administrative overhead.
- **Retail:** Self-checkout systems and personalized shopping apps have transformed how customers engage with retail stores.
- **Telecommunications:** Customers can troubleshoot issues, upgrade plans, and pay bills via self-service apps, enhancing operational efficiency and user convenience.

The success of self-service technologies in the finance industry has inspired transformative changes across various other sectors, leading to more efficient, accessible, and customer-friendly experiences. Here's how self-service has reshaped industries such as healthcare, retail, and telecommunications:

1. Healthcare

The healthcare industry has embraced self-service technologies to streamline administrative processes and improve patient care. Patient portals have become a critical tool, allowing individuals to book appointments, access their medical records, request prescription refills, and communicate with healthcare providers without needing to visit a clinic or office. These portals not only provide convenience for patients but also significantly reduce administrative overhead for healthcare organizations, freeing up time for medical professionals to focus on direct patient care.

Virtual consultations, facilitated through telemedicine platforms, have also grown in popularity, especially in the wake of the COVID-19 pandemic. Patients can now consult with doctors remotely, reducing the need for in-person visits for non-emergency cases and improving access to healthcare services, particularly in rural or underserved areas. This transformation has led to shorter wait times, reduced no-show rates, and increased overall patient satisfaction, as individuals can manage their health more conveniently.

Additionally, self-service options like symptom checkers powered by AI enable patients to evaluate their health conditions and get guidance on next steps, whether it's booking an appointment or self-care recommendations. This increases efficiency and empowers patients to take control of their healthcare journeys.

2. Retail

The retail industry has undergone a major shift with the integration of self-service technologies that enhance the shopping experience and improve operational efficiency. Self-checkout systems have become a common feature in stores, allowing customers to scan and pay for their items without needing assistance from a cashier. These systems speed up the checkout process, reduce long queues, and improve customer satisfaction, particularly during peak shopping hours. On the digital front, personalized shopping apps leverage customer data to create tailored shopping experiences. These apps can recommend products based on past purchases, browsing habits, and preferences, making it easier for customers to discover new items. They also enable features like mobile payment options, order tracking, and seamless returns, all of which enhance convenience and streamline the shopping process (George, 2023; Verma & Tandon, 2022; Chatterji et al., 2023)

Furthermore, augmented reality (AR) and virtual fitting rooms have allowed customers to virtually try on clothes, makeup, and accessories, improving engagement and reducing return rates. These innovations, combined with self-service options, have revolutionized how customers interact with retail brands, shifting from traditional in-store shopping to a more digital, personalized experience.

3. Telecommunications

Telecommunications companies have increasingly adopted self-service technologies to improve operational efficiency and enhance customer satisfaction. Self-service apps and online portals allow customers to troubleshoot common issues, such as network outages or device malfunctions, by providing step-by-step guides and troubleshooting tips. This reduces the need for customers to call customer support or wait for a technician, leading to faster resolutions and lower call center volumes. Customers can also manage their accounts, upgrade their service plans, and pay bills through these self-service platforms, offering them greater control over their telecommunications needs. This ease of access empowers users to make changes to their accounts on their own time, without waiting for business hours or speaking with a customer service representative. The platforms often feature a user-friendly interface with a dashboard that consolidates all essential account information, from data usage to billing cycles.

Furthermore, the ability to track service outages, request service activations or deactivations, and manage subscriptions (like adding international calling plans or new services) through self-service tools has improved the overall customer experience. These enhancements have also contributed to increased customer retention and reduced churn, as users appreciate the convenience and autonomy provided by these digital solutions. In all of these industries, the core benefit of self-service technology remains consistent: it empowers customers by giving them control, enhances operational efficiency, and reduces the need for manual intervention, allowing organizations to reallocate resources to higher-value tasks. As self-service adoption continues to grow, businesses across sectors are finding new ways to leverage these technologies to not only streamline processes but also provide more personalized and responsive experiences for their customers.

Critical Components of Effective Self-Service Models

To ensure the success of self-service solutions, organizations must address several critical factors:

1. **Data Analytics:** Real-time data analysis allows for personalization and predictive insights, enhancing user experience.
2. **User-Centric Design:** Interfaces must prioritize ease of use, accessibility, and inclusivity to cater to diverse customer bases.
3. **Cybersecurity:** Robust measures are essential to protect sensitive financial and personal data, building trust with users.

For organizations looking to implement successful self-service solutions, several critical factors must be addressed to ensure that these systems operate effectively, meet customer expectations, and enhance overall user satisfaction. These factors include data analytics, user-centric design, and robust cybersecurity measures, each playing a vital role in the success and adoption of self-service technologies.

1. Data Analytics

Data analytics is a cornerstone of effective self-service models, enabling organizations to deliver personalized experiences and predictive insights that significantly improve the user journey. By analyzing real-time data, companies can better understand customer behavior, preferences, and pain points, allowing them to tailor the self-service experience to each individual's needs.

For example, in the financial services industry, data analytics can be used to provide customers with personalized financial advice, suggesting specific actions or products based on their spending habits, savings goals, and risk tolerance. In e-commerce, data analysis allows retailers to recommend products based on browsing history and previous purchases, increasing the likelihood of cross-selling and up-selling.

Predictive insights derived from data can also improve customer satisfaction by anticipating future needs and proactively offering solutions. For instance, a mobile app could notify a user that they are approaching a credit limit and suggest steps to avoid fees, or a healthcare portal might send reminders for scheduled check-ups based on medical history. Overall, the use of data analytics in self-service systems helps organizations deliver more relevant, timely, and useful content to users, creating a more engaging and efficient experience.

2. User-Centric Design

The design of a self-service platform plays a crucial role in its effectiveness and adoption. To ensure the success of these solutions, interfaces must be user-centric, focusing on ease of use, accessibility, and inclusivity. An intuitive and streamlined interface ensures that customers, regardless of their digital proficiency, can navigate the system with minimal effort.

Ease of Use: The platform should be simple and straightforward, with clear instructions, easy navigation, and quick access to key features. For example, financial apps might include large, clearly labeled buttons for common tasks like transferring money or checking account balances. This simplicity reduces friction, encourages usage, and helps users feel confident in managing their needs independently.

Accessibility: It's essential that self-service solutions are designed to be accessible to all customers, including those with disabilities. This means considering elements such as screen reader compatibility, high-contrast visual design, and the ability to adjust text size. Additionally, platforms should ensure that users with diverse needs can access services without difficulty, whether through voice commands, keyboard navigation, or adaptive technologies.

Inclusivity: A truly user-centric design also takes into account the varying levels of technological literacy among customers. It should offer a range of support options, such as step-by-step guides, tooltips, video tutorials, or even live chat support for those who may need extra help. By making self-service platforms intuitive and accessible, organizations can ensure that users from diverse backgrounds can interact with the system efficiently.

3. Cybersecurity:

As self-service solutions increasingly handle sensitive data, robust cybersecurity measures are essential to protect users' financial, personal, and health-related information. Without strong security protocols in place, users may be hesitant to trust digital platforms, and organizations could face significant risks related to data breaches, fraud, or identity theft. Data Encryption: Ensuring that data is encrypted both during transmission and while stored on servers is critical to protecting sensitive information. Encryption prevents unauthorized access to user data, even in the event of a cyberattack.

Multi-Factor Authentication (MFA): MFA adds an additional layer of security by requiring users to provide more than one form of verification when accessing their accounts or making changes. This could include a combination of passwords, biometric scans (fingerprint or facial recognition), or one-time passcodes sent via SMS or email.

Regular Security Audits: Organizations must continuously assess and upgrade their security protocols. Regular penetration testing, security audits, and vulnerability assessments help identify and address potential threats before they can be exploited.

Compliance with Regulations: For many industries, particularly finance and healthcare, compliance with data protection regulations such as GDPR, HIPAA, or PCI-DSS is essential. These regulations require organizations to implement specific security measures to protect users' personal and financial data. By investing in robust cybersecurity, organizations can build trust with users, ensuring that they feel safe when using self-service platforms to manage sensitive transactions, make payments, or access private information.

To ensure the long-term success of self-service solutions, organizations must carefully consider the integration of data analytics, user-centric design, and cybersecurity. When implemented effectively, these components not only improve the overall customer experience but also increase efficiency, reduce operational costs, and foster trust. As more businesses move toward self-service models, focusing on these critical factors will help them provide a secure, personalized, and accessible service that meets the evolving needs of today's digital consumers. McKinsey (2022)

Challenges and Opportunities

Despite their benefits, self-service models come with challenges:

- **Inclusivity:** Ensuring access for non-tech-savvy or underserved populations remains a key concern.
- **Balancing Automation with Human Interaction:** While automation improves efficiency, certain situations still require personalized human support.

While self-service models bring significant advantages, they also come with challenges that organizations must address to ensure they deliver value to all customers. However, these challenges also present opportunities for businesses to innovate and improve the user experience. Two of the most prominent challenges are ensuring inclusivity and finding the right balance between automation and human interaction.

1. Inclusivity: Ensuring Access for Non-Tech-Savvy or Underserved Populations

One of the primary challenges of self-service models is ensuring that they are accessible to everyone, including those who may not be tech-savvy or come from underserved populations. This issue is especially important in industries such as banking, healthcare, and retail, where digital adoption is accelerating, but not all customers have equal access to technology or the skills to navigate it.

Challenges

- Digital Literacy: Many individuals, particularly older generations or those from lower socioeconomic backgrounds, may not be as comfortable using digital tools or may have limited access to the internet or smartphones. This digital divide can lead to frustration and exclusion for these users, leaving them unable to fully benefit from the self-service options available.
- Access to Technology: Even in regions with high internet penetration, not everyone has access to modern devices. Users with outdated smartphones, limited data plans, or no access to high-speed internet may struggle to use self-service platforms effectively.
- Language Barriers: Self-service platforms often assume a certain level of proficiency in the language used on the platform. For customers who do not speak the primary language or who are not proficient in technical terminology, self-service solutions can become a barrier instead of a convenience.

Opportunities:

- Digital Education and Support: Organizations can invest in digital literacy programs to help underserved groups build the skills needed to navigate self-service tools effectively. This could include offering workshops, tutorials, or dedicated support lines for those needing assistance.
- Multilingual Support and Accessibility Features: Platforms can be designed with inclusivity in mind, offering multiple language options and accessibility features, such as voice navigation, screen readers, and text enlargement. This ensures that users from diverse backgrounds and with varying needs can use the system with ease.
- Alternative Access Points: Businesses can also provide alternative methods for accessing services. For example, they might allow users to initiate tasks online but offer an option to complete them via phone, in-person assistance,

or through partnerships with local community organizations to bridge the technology gap.

2. Balancing Automation with Human Interaction

Automation is a key driver of efficiency in self-service models, but it's crucial to find the right balance with human interaction. While self-service technologies can handle routine tasks and basic inquiries, certain situations demand the empathy, creativity, and nuanced understanding that only human agents can provide.

Challenges

- Customer Frustration: Fully automated systems can sometimes lead to customer frustration, particularly when they cannot resolve complex issues or when customers feel they are trapped in a cycle of pre-programmed responses. This can result in dissatisfaction and a sense of detachment from the organization, especially if the self-service system lacks a clear and easy way to escalate issues to a human representative.
- Complex Issues: While self-service platforms can manage simple tasks (e.g., changing an address, checking an account balance), more intricate issues—such as disputes, complex product inquiries, or emotionally charged interactions—are better handled by humans. These situations require empathy, critical thinking, and a personalized approach.
- Risk of Over-Automation: Over-reliance on automation can result in missed opportunities for genuine human connection, which is often needed to build customer trust and loyalty. For instance, customers may want to interact with a human when they need to discuss their financial future or resolve a technical issue.

Opportunities

- Hybrid Models: A successful approach is a hybrid model, where automation and human support are seamlessly integrated. For example, self-service systems can handle routine tasks like checking balances or ordering products, while more complex or emotionally charged issues are routed to a human representative. This ensures efficiency without sacrificing quality of service.
- Escalation Protocols: Self-service platforms should be designed with clear escalation protocols that allow users to easily switch from automation to human interaction when needed. A well-implemented system should detect

when a user is struggling with the automated system and offer a prompt to connect with a customer service agent.

- AI-Powered Assistance: AI-powered chatbots and virtual assistants can also help strike the right balance by providing context-aware, human-like support. These systems can resolve straightforward issues and provide helpful information, while also recognizing when a more personal touch is required, seamlessly transferring the user to a human agent when necessary. World Finance Council.(2023)

Benefits of Balance

When done correctly, automation combined with human support can provide the best of both worlds—high efficiency for routine tasks and the personal touch when needed. This not only improves customer satisfaction but also reduces operational costs by allowing human agents to focus on more complex and high-value interactions.

While self-service models offer considerable benefits in terms of convenience, efficiency, and cost savings, addressing challenges related to inclusivity and automation vs. human interaction is critical for ensuring their broad success. By leveraging technology thoughtfully, investing in digital education, and maintaining a balance between automation and personalized support, organizations can create self-service experiences that are accessible, efficient, and capable of meeting the diverse needs of their customers. With the right strategies in place, these challenges can be turned into opportunities for greater customer engagement, loyalty, and overall satisfaction.

The opportunities, however, are immense. By leveraging self-service innovations, organizations can:

- Reduce operational costs.
- Enhance customer satisfaction and loyalty.
- Gain competitive advantages in increasingly digital marketplaces.

Self-service technologies have unlocked a wide array of opportunities for organizations across industries. By embracing these innovations, businesses can not only streamline operations but also improve customer experience and gain a significant competitive edge in today's digital-first world. The following highlights some of the most compelling opportunities that self-service models offer:

1. Reduce Operational Costs

Self-service innovations can lead to substantial cost savings by automating routine tasks and processes that would otherwise require human intervention. Here's how they help reduce operational costs:

- Reduced Labor Costs: By allowing customers to handle tasks such as account management, payment processing, and order tracking through self-service channels, businesses can significantly reduce the number of staff needed for customer service roles. This reduction in manual labor frees up resources for more strategic initiatives, such as product development or customer engagement.
- Fewer In-Person Interactions: Self-service solutions like online portals, mobile apps, and AI-driven chatbots reduce the need for face-to-face interactions or phone calls, which are often resource-intensive. Customers can complete tasks on their own time without relying on human agents, leading to lower overhead costs for businesses.
- Automation of Routine Processes: Automation tools can handle repetitive tasks such as data entry, claims processing, and inventory management. For example, in the retail sector, self-checkout systems automate the payment process, reducing the need for cashiers and minimizing wait times for customers. This leads to savings in staffing costs and faster transactions.
- Scalability: Self-service platforms can scale much more easily than traditional human-driven processes. As demand grows, organizations can expand their self-service capabilities without a proportional increase in labor or infrastructure costs. Cloud-based solutions, in particular, enable businesses to quickly adjust their capacity based on customer needs, further driving cost efficiency. Thottoli (2023).

2. Enhance Customer Satisfaction and Loyalty

Self-service solutions, when well-implemented, can have a significant positive impact on customer satisfaction and loyalty. Here's how they contribute:

- Convenience and Speed: Self-service platforms allow customers to access services whenever and wherever they need, removing the constraints of traditional business hours or wait times. Whether it's checking account balances, making payments, or tracking orders, customers can interact with the platform at their convenience, leading to a more satisfying and seamless experience.

- Personalized Experiences: Advanced self-service technologies, such as AI and machine learning, can analyze customer data and offer personalized recommendations, content, or services. In financial services, for example, a mobile banking app could suggest tailored savings plans or investment strategies based on an individual's spending habits and financial goals. Personalization enhances the customer experience, making it feel more relevant and valuable.
- Self-Empowerment: Customers appreciate the ability to solve their own problems without needing to rely on a third party. A self-service portal that empowers users to update personal information, request refunds, or resolve issues without contacting customer support fosters a sense of autonomy. When customers feel empowered and in control, it significantly increases their satisfaction and likelihood of returning to the brand.
- Consistency and Accuracy: Self-service tools can provide customers with quick access to accurate, up-to-date information. Whether through an AI chatbot providing instant answers or an online FAQ that addresses common queries, the reduction of human error and the speed of information delivery enhance the customer experience. Consistency in service helps to build trust and reinforces positive customer perceptions.
- 24/7 Availability: Self-service platforms can be available at all times, catering to customers in different time zones and those with busy schedules. The ability to access services around the clock builds a sense of reliability and increases customer loyalty. For example, insurance companies offering 24/7 claim filing or banks allowing 24/7 mobile banking significantly improve convenience for their clients. Thomson (2022) and Tech (2024)

3. Gain Competitive Advantages in Increasingly Digital Marketplaces

The digital marketplace is becoming more competitive as consumers increasingly expect efficient, seamless, and personalized experiences. Self-service solutions can help organizations gain a competitive advantage by differentiating their services and appealing to modern consumer preferences:

- Faster Time-to-Market: Self-service models streamline the process of delivering products and services, allowing businesses to get to market faster than competitors who rely heavily on manual processes. For example, e-commerce platforms that offer quick self-service checkout options or digital product recommendations can convert more customers than competitors with cumbersome, slow processes.

- Adaptability to Changing Consumer Preferences: Self-service solutions can be easily adapted and updated to reflect changing consumer demands. Businesses can rapidly introduce new features or update processes to improve functionality based on customer feedback, ensuring they stay ahead of competitors. By continuously refining the self-service experience, organizations can meet the evolving needs of their audience.
- Data-Driven Insights: Self-service platforms provide a wealth of customer data, which businesses can leverage to gain insights into buying behaviors, preferences, and pain points. This data allows organizations to refine their strategies and offer better-targeted products or services. In industries like retail or financial services, where personalization is key, data-driven insights from self-service systems enable businesses to tailor their offerings, making them more appealing to customers.
- Increased Brand Visibility: Self-service platforms can help businesses expand their reach and visibility. For example, through an easy-to-navigate mobile app or a comprehensive self-service website, organizations can engage customers without geographical limitations. Offering digital services can make a brand more accessible to a global audience, giving it a competitive edge over those relying on traditional, location-dependent models. Tech (2024).
- Improved Customer Retention: By creating an easy-to-use, hassle-free self-service experience, companies can foster stronger, longer-lasting relationships with customers. Satisfied customers are more likely to return, recommend services to others, and engage with the brand on an ongoing basis. Loyal customers, in turn, can provide valuable word-of-mouth marketing, further enhancing a company's competitive positioning.

The opportunities created by self-service innovations are vast and transformative. By automating routine tasks, businesses can significantly reduce operational costs, leading to more efficient use of resources. Self-service platforms also provide customers with greater convenience, personalization, and control, driving higher levels of satisfaction and loyalty. As organizations continue to leverage these technologies, they can gain a competitive advantage in increasingly digital marketplaces by offering faster, more efficient, and more responsive services that meet the evolving needs of today's consumers. Embracing these opportunities allows businesses to not only stay relevant in a rapidly changing landscape but also to lead in innovation and customer experience. Tapalaga& Tapalaga (202).

Self-service transformation is a pivotal trend reshaping the financial sector and influencing other industries. As businesses navigate this paradigm shift, they must focus on integrating advanced technologies with customer-centric strategies. The insights from these case studies provide a roadmap for organizations aiming to harness self-service innovations, driving sustainable growth and superior customer experiences. By embracing this future, companies can ensure their relevance and success in an ever-evolving landscape. Swift (2023).

What is self service?

Self-service refers to the process in which consumers take on part or all of the tasks involved in obtaining a service or product. While self-service has existed for decades — from do-it-yourself home projects replacing professional contractors to self-help books taking the place of therapy sessions — its significance has surged with advancements in information technology. These innovations have unlocked opportunities for greater efficiency and convenience, embedding self-service into daily life. For instance, telephone operators have been largely replaced by automated systems that allow users to dial numbers directly. Elevator operators have given way to electronic controls, enabling passengers to operate elevators on their own. At bowling alleys, players now activate automatic pinsetters with the push of a button, eliminating the need for pin boys to manually reset the pins. Similarly, grocery shopping has transformed from handing a list to a clerk at the counter to customers selecting their own items from the shelves. Each of these changes highlights how technology has revolutionized everyday tasks, making self-service a fundamental part of modern life. Sitorus (2016) and Swift. (2023).

In self-service, the consumer takes on a specific task that would typically be performed by a service employee, such as bagging their own groceries. While self-service and automation both aim to improve efficiency, they are distinct concepts. Self-service shifts responsibility to the consumer, whereas automation reduces the tasks required of employees through technology. For example, a retailer using a bar-code scanner to automatically identify and price items is an example of automation, as it simplifies the cashier's job by eliminating the need to manually enter prices. Gas stations illustrate the difference between the two concepts: with self-service, customers pump their own gas instead of relying on an attendant, while automation allows an automatic car wash to replace a manual car wash crew. Both approaches enhance efficiency, but they do so in fundamentally different ways. Silver (2023)

BENEFITS OF SELF-SERVICE

The rise of self-service technology can be attributed to the significant benefits it offers to consumers, organizations, and the broader economy. These benefits include greater convenience, improved efficiency, and cost savings for all parti.es involved.

Benefits for Consumers

Consumers are often given the choice to use self-service options in various settings. For instance:

- At a bank a customer can decide between interacting with a teller or using an automated teller machine (ATM).
- At a hotel, a traveler may choose between ordering room service or using a vending machine.
- At a gas station, drivers can either pump their own gas or pay for an attendant to do it.

Consumers frequently opt for self-service technology because it offers faster service, greater convenience, and ease of use. In many cases, there are also financial incentives. For example, at gas stations in most states, drivers pay a premium for full-service assistance, whereas self-service customers receive a discount by pumping their own gas. In other cases, such as with banking, there is typically no additional cost for choosing a human teller over an ATM. However, some banks have experimented with implementing teller fees to encourage more widespread use of self-service options. Sharma (2024)

Self-service technology offers consumers greater convenience, accessibility, and ease of use. One of the primary advantages is convenience: self-service options often allow businesses to operate 24/7, removing the limitations of traditional working hours. Consumers also find self-service technology empowering, as it gives them control over the service experience without feeling rushed or pressured by staff. While some businesses may assume that customers prefer face-to-face interactions, this is not always the case. In a 2009 survey, 44% of respondents indicated they would prefer to use a hotel check-in kiosk to avoid interacting with a front desk clerk. The preference for self-service increases even more during peak times when long lines form at check-in counters. Self-service technology can also help protect consumer privacy. For instance, hospital patients may feel more comfortable registering at a kiosk instead of sharing personal information with a receptionist. Similarly, some consumers prefer purchasing personal or sensitive items online rather than in-store, where they might feel uncomfortable interacting with a salesperson.

Self-service technology can also enhance accessibility for individuals with disabilities, making service encounters easier and more inclusive. For example, people with mobility impairments may find online shopping more convenient than visiting physical stores, as they can access products and services from the comfort of their homes. Additionally, web browsers offer accessibility features that allow users to navigate online applications and services more effectively. George & Baskar. (2023)

Kiosks can also incorporate adaptive features to improve accessibility. For instance, ATMs equipped with headphone jacks allow users to complete transactions using an audio interface, providing an alternative for individuals with visual impairments. Similarly, self-service kiosks and web-based applications can include multilingual interfaces, making services more user-friendly for diverse customer groups. For example, a car rental company may install a multilingual kiosk at an international airport to assist foreign customers, offering support in multiple languages—something that would be difficult for a single employee to achieve. These features not only improve the user experience but also broaden access to essential services for a wider audience.

Benefits for businesses

To remain competitive in the modern service economy, businesses must leverage information technology (IT) effectively. The IT revolution has significantly boosted productivity, with companies that invest heavily in IT seeing the highest levels of output per worker. While many industries have successfully used IT to improve the efficiency of both back-office operationsand frontline service roles, there is still untapped potential in utilizing technology to engage customers directly in service delivery. Sanjay, (2023). By implementing self-service technology, businesses can extend IT's productivity benefits to customer interactions. In many service industries, customers have always played an active role in the production and delivery of services. For example, tax accountants rely on their clients to provide necessary financial information for completing tax forms. From a business perspective, these customers act as partial employees" or "co-producers" because they contribute directly to the service process. Self-service technology offers companies a way to manage and optimize customer participation in service delivery. By empowering customers to take on tasks traditionally performed by employees—such as data entry, form completion, or transaction processing—businesses can enhance efficiency, reduce labor costs, and improve the overall customer experience. Phillips (2024).

Businesses invest in self-service technology to reduce costs and enhance the quality of their products and services. By automating routine tasks, self-service solutions free up employees who can either be reassigned to more profitable roles or removed to cut payroll expenses. For instance, a retailer that implements self-checkout systems can reallocate cashiers to sales or customer service roles to boost revenue and improve customer satisfaction—or eliminate those roles to reduce operating costs. Many organizations use self-service technology to shift employees' focus from repetitive tasks to higher-value work. In hospitals, medical staff who previously spent time on clerical duties can devote more attention to patient care. In banks, ATMs handle the majority of routine transactions, allowing tellers to provide more specialized financial services and improve customer support. For consumers, self-service technology delivers greater efficiency, convenience, and enhanced service. It also helps businesses standardize the customer experience and better target customers for up-selling opportunities. For example, an airport check-in kiosk can be programmed to offer travelers flight upgrades during the check-in process, increasing potential revenue while streamlining the service interaction. Omol (2023).

Self-service technology can significantly enhance a company's operational efficiency by streamlining processes and reducing costs. For instance, e-commerce platforms allow businesses to cut expenses while optimizing inventory management. By responding more quickly to orders, companies can reduce inventory levels by 20 to 25 percent, minimizing overhead costs and improving cash flow. Additionally, e-commerce solutions help reduce order inaccuracies, saving businesses substantial amounts by minimizing errors and returns. George, George & Baskar. (2023)

The cost-saving benefits of self-service technology are particularly evident in the banking industry. As illustrated in Figure 1, the average cost of an online banking transaction is approximately $0.20, compared to $4.25 for a transaction conducted at a physical branch. This stark difference demonstrates how self-service technology can reduce operational expenses while providing customers with faster, more convenient service.

Benefits for the economy

The broader economy also benefits significantly from the adoption of self-service technology. One of the key indicators of a nation's economic well-being is per-capita income growth, which is closely tied to productivity growth— the amount of output generated per hour of work. Higher productivity growth plays a crucial role in addressing many societal challenges, such as Social Security shortfalls, slow income growth, national debt, and funding for public services like transportation, healthcare, and environmental protection. Omol (2023) Sustained increases in productivity can result in both higher incomes and reduced working hours, allowing workers to spend more time with their families and on leisure activities. This potential benefit applies not only to the United States, but also to Europe, Japan, and other countries grappling with economic pressures due to aging populations. Self-service technology acts as a labor-saving tool, increasing the efficiency of output while reducing the need for human intervention in routine tasks. For nations facing demographic challenges, adopting productivity-enhancing technologies is essential to maintaining current standards of living for both workers and retirees. In this way, self-service technology helps drive long-term economic stability and growth by maximizing efficiency and reducing the reliance on an aging workforce. George (2023) and Mell & Grance (2011)

In most advanced economies, per-capita income and productivity growth saw significant increases up until the 1970s, primarily driven by improvements in goods production sectors such as farming, mining, and manufacturing. Companies in these industries leveraged automation technologies to streamline processes, and because these sectors made up a large portion of the economy, their productivity gains had a major impact on overall income levels. George (2023)

However, as efficiency gains in goods production plateaued, economic growth increasingly depended on improving service sector productivity. Over the past four decades, businesses have focused on using technology to enhance back-office operations — such as accounting, logistics, ordering systems, and information processing — where many productivity gains have already been realized, Mell & Grance (2011). The next significant opportunity for boosting productivity lies in the front-office functions, which involve direct customer interactions and information exchanges. For example, processes like issuing tickets, processing payments, or providing customer support largely revolve around managing and sharing information. Self-service technology is key to improving front-office efficiency. By automating a wide range of customer-facing tasks, businesses can reduce costs, offer lower prices, and enable higher wages for workers. The economic potential of expanding self-service technology is substantial. Estimates suggest that widespread deployment of self-service solutions could contribute an additional $130 billion annually to the U.S. economy. This would translate to $1,100 in additional income per household, demonstrating how self-service technology can drive economic growth and improve living standards for individuals and families. Lu & Xia (2024)

METHODOLOGY

This study adopts a qualitative multiple case study approach to examine the impact of self-service technologies across the financial services industry and other key sectors such as healthcare, hospitality, and transportation. Purposeful sampling was used to select representative organizations that have successfully implemented self-service systems, such as mobile banking applications, self-check-in kiosks, and telehealth platforms. These cases were chosen based on clear evidence of improved efficiency, cost reduction, and enhanced customer experience, as well as the availability of relevant documentation and stakeholder access. Data collection methods included document analysis (e.g., internal reports, press releases, and industry publications), semi-structured interviews with organizational stakeholders (e.g., IT managers and customer service leads), and, where feasible, direct observation of the technologies in use. Thematic analysis was employed to identify patterns and trends across individual cases, while a cross-case synthesis technique facilitated comparative insights across sectors. To enhance validity, the study employed methodological triangulation, member checking with interview participants, and peer debriefing to ensure accuracy and credibility of the findings. Yin (2018), Creswell & Poth (2018), Patton (2015), Stake (2006), Fusch & Ness (2015), Nowell, Norris, White & Moules (2017), Silverman (2020).

Findings

Across all case studies, self-service technologies significantly improved operational efficiency. In the financial sector, tools such as ATMs, mobile banking apps, and automated customer service chatbots reduced the volume of routine, manual transactions handled by staff. This allowed financial institutions to reallocate human resources toward higher-value services such as financial advising and relationship management. In non-financial sectors like healthcare and hospitality, self-check-in kiosks, appointment scheduling portals, and automated room service systems reduced wait times, streamlined workflows, and optimized staffing needs.

Self-service implementation led to substantial cost savings. For instance, in the banking sector, digital channels reduced the need for physical branches and front-line staff. Similarly, in the hospitality and transport sectors, automation of check-in and payment processes lowered overhead and reduced dependency on manual labor. The convenience, speed, and accessibility of self-service platforms enhanced overall customer satisfaction. Financial customers reported greater control over their banking needs, while users in healthcare appreciated the ability to manage appointments and records remotely. In hospitality, digital menus and service requests provided a more flexible and user-driven experience.

Several case studies highlighted how intuitive interfaces and mobile-first platforms expanded service access to previously underserved demographics. For example, mobile banking apps with voice commands or visual assistance features improved accessibility for users with disabilities or low digital literacy.

CONCLUSIONS AND FUTURE DIRECTIONS

Conclusions

This study concludes that self-service technologies have significantly transformed operations and service delivery across financial and non-financial industries. In the financial sector, tools like mobile banking apps and ATMs have enhanced customer autonomy, streamlined transactions, and allowed institutions to focus on high-value services. Similarly, in sectors like healthcare, hospitality, and transportation, technologies such as telehealth platforms, self-check-in kiosks, and digital service terminals have improved efficiency, reduced operational costs, and elevated the customer experience. The cross-case analysis confirms that successful implementation relies on user-centric design, strong digital infrastructure, and organizational readiness. Overall, self-service technologies are not only reshaping service models but also driving long-term innovation and sustainability across industries.

Future Directions

The future of self-service transformation will be driven by advancements in AI, machine learning, blockchain, and IoT, making systems more intelligent, secure, and user-friendly. Below are some of the key future directions for self-service transformation across industries:

1. Hyper-Personalization through AI

The next phase of self-service will focus on delivering hyper-personalized experiences by using AI and data analytics. Instead of generic responses, self-service platforms will provide tailored recommendations based on user behavior, preferences, and history. For example, AI-powered financial advisors will provide personalized investment strategies based on a user's financial goals and risk tolerance.

Impact:

- Enhanced customer satisfaction
- Improved decision-making for users
- Increased engagement and retention

2. Expansion of Voice and Conversational Interfaces

Voice assistants and chatbots will play a significant role in future self-service platforms. Conversational AI will become more sophisticated, handling complex queries and offering seamless, human-like interactions. For example, Voice-activated banking services will allow users to perform transactions or get financial advice without typing or visiting a branch.

Impact:

- Improved accessibility for users with disabilities
- Faster, more intuitive interactions

- Reduced reliance on physical interfaces

3. Blockchain for Enhanced Security

Blockchain technology will become a key component of self-service transformation to address security concerns. It will enable secure, tamper-proof transactions, ensuring trust in self-service platforms. For example, Blockchain could secure financial self-service platforms, ensuring that users' transactions and identities are protected from fraud.

Impact:

- Improved data security and privacy
- Reduced risks of fraud and cyberattacks
- Increased trust in self-service solutions

4. Bridging the Digital Divide

To ensure the success of self-service transformation, organizations must address the digital divide by making self-service platforms more accessible to underserved populations. This includes improving digital literacy and providing simplified interfaces for users with limited technical skills. For example, Banks could introduce simplified, user-friendly interfaces for rural or elderly users, ensuring financial inclusion.

Impact:

- Increased financial inclusion
- Broader adoption of self-service tools
- Reduced disparities in digital access

5. Integration of IoT in Self-Service Platforms

The Internet of Things (IoT) will further enhance self-service platforms by connecting devices and offering real-time data. For example, In retail, smart shelves could automatically detect product levels and notify users for restocking or assist with finding products through mobile apps.

Impact:

- Real-time data collection and feedback
- Improved user experiences
- More efficient business operations

6. Human-AI Collaboration

While self-service platforms will become more autonomous, there will be a growing emphasis on human-AI collaboration to balance automation with human empathy. Organizations will integrate human agents to handle complex, emotionally sensitive issues that self-service tools cannot resolve. For example, A financial institution might use a chatbot to handle routine queries but escalate more complex cases to human agents.

Impact:

- Improved problem resolution for users
- Greater customer trust in automated systems
- More balanced customer service models

7. Regulatory and Ethical Considerations

As self-service transformation continues, regulatory and ethical issues will become more prominent. Organizations must ensure compliance with data protection regulations (e.g., GDPR, CCPA) and address concerns around bias in AI algorithms. The future Trends in Regulations are Stricter guidelines on data usage in self-service platforms, Ethical frameworks for AI-driven services and Increased accountability for automated decisions

Impact:

- Safer, more ethical self-service platforms
- Reduced risk of regulatory penalties
- Improved public trust in digital services

A Human-Centric Future for Self-Service, The future of self-service transformation lies in creating human-centric experiences where technology empowers users without compromising empathy, security, or accessibility. The successful organizations will be those that balance automation with a human touch, address cybersecurity concerns, and ensure digital inclusion. By embracing emerging technologies like AI, blockchain, and IoT, self-service platforms will become smarter, more secure, and more personalized. As industries continue to evolve, self-service transformation will be a key driver of innovation, reshaping how businesses interact with customers and how customers manage their needs. Kelley (2024) and George, et al (2023)

REFERENCES

Abulibdeh, A., Zaidan, E., & Abulibdeh, R. (2024). Navigating the confluence of artificial intelligence and education for sustainable development in the era of industry 4.0: Challenges, opportunities, and ethical dimensions. *Journal of Cleaner Production*, *437*, 140527. DOI: 10.1016/j.jclepro.2023.140527

Anan, L., Isaza, D. C., Figueiredo, F., Flötotto, M., Jerenz, A., Krivkovich, A., Nadeau, M. C., Olanrewaju, T., Partner, A., & Vassallo, A. (2023). Fintechs: A new paradigm of growth. In McKinsey & Company. https://www.mckinsey.com/industries/financial-services/our-insights/fintechs-a-new-paradigm-of-growth

Anan, M. A., Wang, Z., & Mahmud, K. (2023). Artificial intelligence applications in financial services: A review and future directions. *Journal of Financial Innovation and Technology*, *4*(2), 85–102. https://doi.org/10.xxxx/jfit.2023.04.002

Arner, D. W., Barberis, J., & Buckley, R. P. (2017). FinTech, RegTech, and the reconceptualization of financial regulation. *Northwestern Journal of International Law & Business*, 37(3), 371–413. https://scholarlycommons.law.northwestern.edu/njilb/vol37/iss3/2/

Arner, D. W., Barberis, J., & Buckley, R. P. (2020). Fintech and regtech: Impact on regulators and banks. *Journal of Banking Regulation*, *21*(4), 299–308. DOI: 10.1057/s41261-019-00101-2

Bani-Hani, I., Pareigis, J., Tona, O., & Carlsson, S. (2018). A holistic view of value generation process in a SSBI environment: A service dominant logic perspective, Journal of Decision Systems Bani-Hani, J. S., Aloudat, A., & AlShihi, H. (2018). Investigating the factors affecting customers' adoption of self-service banking in Jordan. *International Journal of Bank Marketing*, *36*(6), 1245–1263. DOI: 10.1108/IJBM-02-2017-0025

Bollen, J., Mao, H., & Zeng, X. (2011). Twitter mood predicts the stock market. *Journal of Computational Science*, *2*(1), 1–8. DOI: 10.1016/j.jocs.2010.12.007

Brynjolfsson, E., & McAfee, A. (2017). *Machine, platform, crowd: Harnessing our digital future*. W. W. Norton & Company.

Chatterji, N., Manohar, S., & Verma, B. (2023). Assessing the Influence of Graduate Characteristics on Employer Satisfaction: A Multi-dimensional Analysis. *The Open Psychology Journal*, *16*(1), e187435012308030. DOI: 10.2174/18743501-v16-230831-2023-39

Chen, H., Chiang, R. H. L., & Storey, V. C. (2012). Business intelligence and analytics: From big data to big impact. *Management Information Systems Quarterly*, *36*(4), 1165–1188. DOI: 10.2307/41703503

Creswell, J. W., & Poth, C. N. (2018). *Qualitative inquiry and research design: Choosing among five approaches* (4th ed.). Sage Publications. Provides a foundation for qualitative research design, purposeful sampling, and thematic analysis.

Davenport, T. H., & Ronanki, R. (2018). Artificial intelligence for the real world. *Harvard Business Review*, *96*(1), 108–116.

Digital Transformation of Finance and Regulation. (2023). *Cambridge University Press eBooks.*, DOI: 10.1017/9781009086943.002

DMahbub. (2023). Digital Technologies in Finance - Digital Mahbub. Digital Mahbub. https://digitalmahbub.com/digital-technologies-in-finance/

Fusch, P. I., & Ness, L. R. (2015). Are we there yet? Data saturation in qualitative research. [Supports the importance of data saturation and methodological rigor.]. *The Qualitative Report*, *20*(9), 1408–1416. DOI: 10.46743/2160-3715/2015.2281

George, A. S. (2023a). Securing the Future of Finance: How AI, Blockchain, and Machine Learning Safeguard Emerging Neobank Technology Against Evolving Cyber Threats. puirp.com. DOI: 10.5281/zenodo.10001735

George, A. S. (2023b). Preparing Students for an AI-Driven World: Rethinking Curriculum and Pedagogy in the Age of Artificial Intelligence. puirp.com. DOI: 10.5281/zenodo.10245675

George, A. S., & Baskar, T. (2023). The impact of AI language models on the future of white-collar jobs: A comparative study of job projections in developed and developing countries. *Partners Universal International Research Journal*, *2*(2), 117–135.

George, D. (2023). Deciphering the Path to Cost Efficiency and Sustainability in the Snowflake Environment. Zenodo (CERN European Organization for Nuclear Research). DOI: 10.5281/zenodo.8282654

George, D., & George, A. (2023). An Overview of India's Unified Payments Interface (UPI): Benefits, Challenges, and Opportunities. Zenodo (CERN European Organization for Nuclear Research). https://doi.org/DOI: 10.5281/zenodo.7723154

George, D., George, A., & Dr.T.Baskar. (2023a). RuPay Card: Revolutionizing India's Payment Ecosystem. Zenodo (CERN European Organization for Nuclear Research). https://doi.org/DOI: 10.5281/zenodo.7857040

George, D., George, A., & Dr.T.Baskar. (2023b). Unshackled by Servers: Embracing the Serverless Revolution in Modern Computing. Zenodo (CERN European Organization for Nuclear Research). DOI: 10.5281/zenodo.8051052

Gupta, S., Dash, S., & Mishra, B. (2016). Self-service technology in retail: The next frontier. *Journal of Business Research*, *69*(11), 4539–4546. DOI: 10.1016/j.jbusres.2016.03.053

How Web3 Neobanks are Shaping the Future of Finance. (n.d.). https://www.zeebu.com/blog/web3-neobanks-the-new-face-of-finance-and-banking

Jung, D., Dorner, V., Glaser, F., & Morana, S. (2018). Robo-advisory: Digitalization and automation of financial advisory. *Business & Information Systems Engineering*, *60*(1), 81–86. DOI: 10.1007/s12599-018-0521-9

Kelley, K. (2024, February 23). What is Machine Learning? A Comprehensive Guide for Beginners - Caltech. Caltech -. https://pg-p.ctme.caltech.edu/blog/ai-ml/what-is-machine-learning

Lennerholt, C., van Laere, J., & Söderström, E. (2020). User-Related Challenges of Self-Service Business Intelligence. *Information Systems Management.*

Lennerholt, C., Van Laere, J., & Söderström, E. (2020). Implementation challenges of self-service business intelligence: A case study. *Journal of Decision Systems*, 29(sup1), 107–120. https://doi.org/DOI: 10.1080/12460125.2020.1777664

Lu, Y., & Xia, Z. (2024). Digital inclusive finance, green technological innovation, and carbon emissions from a spatial perspective. *Scientific Reports*, *14*(1), 8454. Advance online publication. DOI: 10.1038/s41598-024-59081-9 PMID: 38605137

Mell, P. M., & Grance, T. (2011). The NIST definition of cloud computing. https://doi.org/DOI: 10.6028/NIST.SP.800-145

Ng, I. C. L., & Wakenshaw, S. Y. L. (2017). The Internet-of-Things: Review and research directions. *International Journal of Research in Marketing*, *34*(1), 3–21. DOI: 10.1016/j.ijresmar.2016.11.003

Nowell, L. S., Norris, J. M., White, D. E., & Moules, N. J. (2017). Thematic analysis: Striving to meet the trustworthiness criteria. [Details thematic analysis techniques and ensuring trustworthiness in qualitative research.]. *International Journal of Qualitative Methods*, *16*(1), 1–13. DOI: 10.1177/1609406917733847

Omol, E. J. (2023). *Organizational digital transformation: from evolution to future trends*. Digital Transformation and Society., DOI: 10.1108/DTS-08-2023-0061

Patton, M. Q. (2015). *Qualitative research & evaluation methods: Integrating theory and practice* (4th ed.). Sage Publications. Discusses purposeful sampling, data triangulation, and credibility techniques such as member checking and peer debriefing.

Procuring innovation, wherever it may be. (2017, August 17). McKinsey & Company. https://www.mckinsey.com/capabilities/operations/our-insights/procuring-innovation-wherever-it-may-be

Puschmann, T. (2017). Fintech. *Business & Information Systems Engineering, 59*(1), 69–76. DOI: 10.1007/s12599-017-0464-6

PwC. (2020). *Financial services technology 2020 and beyond: Embracing disruption.* https://www.pwc.com/gx/en/industries/financial-services/publications/financial-services-technology-2020-and-beyond.html

Sanjay. (2023). What is Financial Services? Characteristics, Functions, Classification. Geektonight. https://www.geektonight.com/financial-services/

Science, N. the Automated Journalism Startup - Technology and Operations Management. (2018, November 14). Technology and Operations Management. https://d3.harvard.edu/platform-rctom/submission/narrative-science-the-automated-journalism-startup/

Seddon, P. B., Constantinidis, D., Tamm, T., & Dod, H. (2017). How does business analytics contribute to business value? *Information Systems Journal, 27*(3), 237–269. DOI: 10.1111/isj.12101

Sharma, R. (2024). What Is Decentralized Finance (DeFi) and How Does It Work? Investopedia. https://www.investopedia.com/decentralized-finance-defi-5113835

Silver, C. (2023, September 13). Lessons From the 2008 Financial Crisis. Investopedia. https://www.investopedia.com/news/10-years-later-lessons-financial-crisis

Silverman, D. (2020). Interpreting qualitative data (6th ed.). Sage Publications. Offers in-depth discussion on data collection methods such as document analysis, interviews, and observations.

Sitorus, M. (2016). Format Paper atau Jornal. ResearchGate. DOI: 10.13140/RG.2.2.21653.12006

Stake, R. E. (2006). *Multiple case study analysis.* Guilford Press. Justifies the use of multiple case study methodology and cross-case analysis.

Swift's cross-border payments processing speed surpasses G20 target | Swift. (2023, August 22). Swift. https://www.swift.com/news-events/press-releases/swifts-cross-border-payments-processing-speed surpasses-g20-target

Tamm, T., Seddon, P. B., & Shanks, G. 2020. How Do Different Types of BA Users Contribute to Business Value? Communications of the Association for Information Systems, 46, pp-pp. https://doi.org/DOI: 10.17705/1CAIS.04628

Tamm, T., Seddon, P. B., Shanks, G., & Reynolds, P. (2020). How does big data analytics enable organizational agility? *Information & Management*, *57*(1), 103120. DOI: 10.1016/j.im.2019.103120

Tapalaga, A., & Tapalaga, A. (2024). Exploring the Rise of DeFi: A New Era in Financial Services - History of Yesterday. History of Yesterday - Let the Truth Unfold. https://historyofyesterday.com/exploring-the-rise-of-defi-a-new-era-in-financial-services/

Tapscott, D., & Tapscott, A. (2016). *Blockchain revolution: How the technology behind bitcoin is changing money, business, and the world*. Portfolio.

Tech, D. (2024, January 31). What Strategies Exist To Counter Bitcoin Double Spending? Bitamp. thttps://blog.bitamp.com/what-strategies-exist-to-counter-bitcoin-double-spending/

Thomson, J. (2022). How AI And Data Analytics Drive Value In The Finance Function. Forbes. https://www.forbes.com/sites/jeffthomson/2021/12/17/how-ai-and-data-analytics-drive-value-in-the-finance-function/

Thottoli, M. M., Islam, M. A., Yusof, M. F. B., Hassan, M. S., & Hassan, M. A. (2023). Embracing Digital Transformation in Financial Services: From Past to Future. *SAGE Open*, *13*(4), 21582440231214590. Advance online publication. DOI: 10.1177/21582440231214590

Torres, R., Sidorova, A., & Jones, M. C. (2018). Enabling self-service BI: A framework and case study. *Decision Support Systems*, *109*, 113–123. DOI: 10.1016/j.dss.2018.03.009

Torres, R., Sidorova, A., & Jones, M. C. (2018). Enabling firm performance through business intelligence and analytics: A dynamic capabilities perspective. *Information & Management*, *55*(7), 822–839. DOI: 10.1016/j.im.2018.03.010

Transforming Financial Services in the Digital Age and its Implications. (2023, July 4). World Finance Council. https://worldfinancecouncil.org/articles/transforming-financial-services-in-the-digital-age-and-its-implications/

Verma, B., & Srivastava, A. (2023). Impact of different dimensions of globalisation on firms' performance: An unbalanced panel-data study of firms operating in India. *World Review of Entrepreneurship, Management and Sustainable Development*, *19*(3-5), 360–378. DOI: 10.1504/WREMSD.2023.130618

Verma, B., Srivastava, A., Mehta, R., & Chandel, J. (2022, February). FDI-linked Spillovers and the Indian Economic Growth: The role of Country's Absorptive Capacity. In *2022 IEEE Delhi Section Conference (DELCON)* (pp. 1-6). IEEE. DOI: 10.1109/DELCON54057.2022.9753540

Verma, B., & Tandon, U. (2022). Modelling barriers to wearable technologies in Indian context: Validating the moderating role of technology literacy. *Global Knowledge, Memory and Communication.*

What's new in banking API programs. (2022). McKinsey & Company. https://www.mckinsey.com/capabilities/mckinsey-digital/our-insights/tech-forward/whats-new-in-banking-api-programs]

World Economic Forum. (2020). Transforming paradigms: A global AI in financial services survey.https://www3.weforum.org/docs/WEF_AI_in_Financial_Services_Survey_2020.pdf

Yahoo is part of the Yahoo family of brands. (n.d.). https://finance.yahoo.com/news/embedded-finance-unlock-us-242-020000903.html

Yin, R. K. (2018). *Case study research and applications: Design and methods* (6th ed.). Sage Publications. Supports the use of multiple case study design, cross-case synthesis, and thematic analysis.

Zetzsche, D. A., Buckley, R. P., Arner, D. W., & Barberis, J. N. (2020). Decentralized finance. *Journal of Financial Regulation*, *6*(2), 172–203. DOI: 10.1093/jfr/fjaa010

Zhang, W., Wang, Q., & Zhan, Y. (2022). Applications of quantum computing in financial markets: A systematic review. *Journal of Computational Finance*, *26*(1), 1–28. DOI: 10.21314/JCF.2022.419

Chapter 14
Revolutionizing Service Delivery: Cutting-Edge Innovations in Self-Service Technology

Ram Singh
https://orcid.org/0000-0002-6565-3091
Symbiosis University of Applied Sciences, Indore, India

Rajit Verma
https://orcid.org/0000-0003-0643-4718
Maharishi Markandeshwar University, Mullana, India

Ajit Singh Tomar
https://orcid.org/0009-0003-1833-7568
Symbiosis University of Applied Sciences, India

Fazla Rabby
https://orcid.org/0000-0002-2683-7218
Stanford Institute of Management and Technology, Sydney, Australia

Balraj Verma
https://orcid.org/0000-0002-6542-3261
Chitkara University, Punjab, India

Ravi Kumar
https://orcid.org/0000-0001-7774-9421
SVGC Ghumarwin, Bilaspur, India

Vinay Pal Singh
https://orcid.org/0009-0007-1328-614X
Quantum University, Roorkee, India

ABSTRACT

This chapter aims to discuss and elaborate on possible SST trends in future as well as focusing on existing developments. To highlight the business of self-service together with the operational processes in parallel. This paper highlights technological advances such as automation and AI that are central in achieving such paradigms. The chapter will briefly consider matters from the perspectives of industries, geographies, and business development. For assessing the effects of various SSTs on various industries, a range of methodologies will be employed including systematic

DOI: 10.4018/979-8-3373-4667-0.ch014

literature reviews, reports and case studies. This will help in establishing trends in using SSTs and assist in shaping industries in the following decades. The outcomes anticipated from this chapter include knowledge of how self-service technologies enhance the service delivery process and directions on how this process can be improved in the future.

1. INTRODUCTION

Self-service technology or self-service terminal (SST) is the technology that allows the consumer to operate a terminal and carry out transactions without having to interact with a human being. Self-service technology allows consumers to engage directly with machines and perform transactions that would have otherwise required assistance. Retail, hospitality, even the banks and other industries are starting to adopt this concept. It has improved the efficiency of these businesses, as well as revolutionized customer experience. Customers now expect more streamlined solutions (Meuter et al., 2000). There has been increased demand for seamless solutions especially after the pandemic as the global trend shifted to offering contactless services solutions (Nguyen et al., 2022). Automated kiosks, mobile phones, virtual assistants powered by AI (Artificial Intelligence), chatbots, etc are some of the self-service technologies that are available. The primary purpose for self-service technology is to minimize the amount of work, increase the amount of service and minimize the overall cost of service. In retail for instance, self-service kiosks have made it possible to spend less time waiting in queues, ATMs and mobile applications in banking have allowed consumers to get services at any time of the day (Bitner et al., 2002). In a similar vein, chatbots utilizing AI technology employ decidedly less time in processing and responding to customer inquiries than ever before (Lu et al., 2021). Nevertheless, challenges remain, as SSTs still do not prepare the way for the aforementioned integration into service ecosystems. Some of the challenges include technology comfort, security/privacy, reliability of the system, and the possibility of distancing certain segments of customers. Therefore, organizations face the dilemma of automating while still providing a human touch so that the SST does not impair customer satisfaction (Parasuraman and Colby, 2015).

1.1 Revolutionizing Service Delivery

The service delivery paradigm has taken momentous leaps from decade to decade. During the research context, service delivery also lost meaning in traditional or face-to-face operations. Due to the present technological advancement, however, service delivery now has broader spectrums, including the physical, digital, and self-service mechanisms in place to meet demands for convenience, speed, and personalization (Bitner et al., 2002). Innovations in service delivery have, like the proverbial mention of a penny, been pegged on technology. Self-service technologies (SSTs), such as automated kiosks, mobile apps, and AI-powered chatbots, have been a game-changer in enabling consumers' completion of transactions without assistance. These types of technology smooth processes, reduce labor costs, and allow 24/7 customer service (Meuter et al., 2000). Moreover, digital platforms such as blockchain and IoT are promoting transparency and efficiency in domains like healthcare, retail, and financial services (Lu et al., 2021).

1.2 Impact on Consumer Behavior

From the perspective of scientific research, the adoption of SSTs has several ramifications for consumer behavior. Consumers are ready to adopt self-service options when they see benefits such as convenience, control, and reduced waiting times, but such abundance of advantages is an immediate barrier to adoption from someone less literate in technology (Dabholkar & Bagozzi, 2002). Problems such as technological literacy, trust-related issues, and ease of use are predominant among the forces affecting adoption rates, especially in emerging economies (Parasuraman & Colby, 2015). For businesses, the rise of service delivery necessitates revising traditional business models. Research states that organizations that have adopted SSTs have increased operational efficiency and customer satisfaction. However, there is a need to balance automation with a human touch, lest the opposite forces certain segments of the customer base into penalization (Bitner et al., 2002). These keeps being important in industries where the connection to emotions really matters, such as hospitality and healthcare. Nevertheless, even though SSTs hold the promise of dramatically impacting customer loyalty and workplace dynamics, this arena is marred by numerous research gaps regarding this very issue. Issues related to security, reliability, and equitable access are also encapsulated within these gaps. Additionally, researchers should study the social and economic factors influencing the adoption of SSTs in different countries, especially underdeveloped markets (Nguyen et al., 2022). The self-service technologies engine revolutionizing service delivery has become quite a vibrant field of study. Other forms of self-service delivery improved also provide opportunities for enhanced customer experience, reduced service delivery operational costs, and broadened access to services. However, their successful implementation is based upon grounding in understanding consumer preferences, technological capabilities, and organizational strategies, all henceforth represent active avenues for future research.

1.3 Cutting-Edge Innovations in Self-Service Technology

The fast evolution of technology has accredited major advances in self-service technologies, allowing businesses to transform service delivery and elevate customer experiences. Of all that, artificial intelligence (AI) and its unleashed powers are a game-changer. AI systems, including chatbots and virtual assistants, enhanced customer interaction through instant personalized responses. For instance, AI-powered chatbots used by Amazon or Swiggy handle inquiries from order tracking to grievance redressal with great aplomb. These systems work using natural language processing (NLP), with conversations much like humans do to improve user satisfaction. Besides interaction, the role of AI in predictive analytics has enabled firms to preempt customer needs, thus allowing for upsells and cross-sells and alleviating customer friction. For instance, the recommendation systems in e-commerce platforms such as Flipkart apply artificial intelligence to analyze users' purchasing history and browsing behavior, making appropriate suggestions to elevate customer engagement and sales. Another critical innovation is the introduction of smart kiosks that incorporate advanced technologies such as facial recognition, real-time analytics, and touchless interfaces. Increasingly found in retail, airports, and hospitality, these kiosks provide a seamless user experience by automating myriad processes such as check-in, payment, and information retrieval. For instance, McDonald's self-order kiosks allow customers to place orders, customize meals, and make payments without requiring assistance from human staff. These systems promise to reduce waiting time while also ensuring operational flow through the reallocation of manpower for more useful value-adding work. Smart kiosks are essential to the healthcare industry because they make patient registration and appointment scheduling easier and provide faster access to medical services. Further improving security and lowering the possibility of fraud a major worry in sectors like banking and insurance is the incorporation of biometric authentication (Nguyen et al., 2022).

The spread of mobile-enabled SSTs has transformed consumer engagements across sectors, too. User friendly mobile applications utilizing nearest technology and algorithms for automation have become the self-service method of choice. Mobile applications provide users capabilities to handle their financial accounts along with payment functions combined with service requests through any time and place of their choice. ICICI Bank's iMobile Pay application showcases complete user servitude which includes fund transfers alongside investment monitoring and let customers circumvent destructive bank visits while keeping users inside their application ecosystem. Mobile apps achieve greater convenience through geolocation tech and methods like QR code scanning together with additional technology tools that allow specific location-based services and hands-free payment options. After COVID-19 consumers started valuing convenience together with safety in their engagements so new innovations emerged to meet these needs. Advancements in SSTs happened through the combination of blockchain together with Internet of Things (IoT) technologies. The Blockchain technology generates transactions which remain safe from tampering therefore suitable for financial services and supply chain management applications. Delivery participants in the logistics sector can always verify delivery status through blockchain-based self-service platforms which operate on a distributed ledger. T Retail IoT sensors enhance customer shopping pleasure by using mobile applications to alert people about current product promotions and sales notifications. The revolution in education and real estate sectors along with tourism industries occurs through immersion self-service technology developed from augmented reality (AR) and virtual reality (VR). The use of AR/VR-powered virtual tours lets customers remotely experience properties, locations, or classrooms to make better and easier decisions while gaining better access. The retail industry uses AR-powered mirrors to allow customers virtual clothing trials that create a personalized interactive shopping experience. The technology's greatest benefit consists of boosting market competition while offering exceptional value to companies eager to serve digital-forward customer bases. Technological advances in SST today transform business-customer communication through IoT and AI system developments. Through modern technologies businesses get the chance to achieve higher operational effectiveness and scalability while gaining enhanced delivery services. Meeting the diverse customers' needs requires solving problems such as digital competency as well as protection of data and accessibility challenges to successfully put these technological advancements into practice. The following Table 1 highlights key self-service technology innovations, their descriptions, adoption in India, and global trends, supported by recent data.

Table 1. self-service technology innovations

Innovation	Detailed Description	Indian Scenario	Global Scenario
Self-Service Kiosks	Stand-alone digital terminals enabling customers to independently perform tasks such as ordering food, checking in at airports, or accessing bank services. Equipped with touchscreens, biometrics, and AI for personalized service.	IRCTC's self-ticketing kiosks at major railway stations. McDonald's introduced digital ordering kiosks in metros like Delhi and Mumbai. India's self-service kiosk market is growing at 15% CAGR (2023-2028), led by retail and public transportation sectors.	Widespread at airports (e.g., Singapore Changi, Heathrow), fast-food chains like McDonald's and Burger King. Global market valued at $ 28 billion in 2023; forecasted to reach $ 45 billion by 2030.
Chatbots and Virtual Assistants	AI-powered chat interfaces that offer 24x7 customer support, answer FAQs, book appointments, and handle complaints without human intervention. They leverage natural language processing (NLP) for realistic conversation.	HDFC Bank's chatbot 'Eva' answers millions of queries annually. IRCTC's 'AskDisha' helps with ticket booking queries. 70% of Indian firms have deployed chatbots as part of their digital strategy.	Brands like Sephora, H&M, and airlines like Lufthansa rely on chatbots for customer service. By 2025, 80% of enterprises globally will use some form of chatbot technology.
Mobile Self-Service Apps	Smartphone applications allowing users to perform tasks like food ordering, booking tickets, banking transactions, and making service requests, anywhere, anytime.	Apps like Paytm, Swiggy, and BookMyShow have transformed retail and service delivery. India had over 500 million active mobile service app users by 2024.	Starbucks Mobile Order & Pay and Uber's self-booking model lead innovations. Global mobile app revenue projected to reach $ 613 billion by 2025.
Smart ATMs	Next-generation ATMs offering not just cash dispensing, but also services like video banking, account opening, and biometric authentication.	SBI's YONO Cash and HDFC's Smart ATMs allow cardless cash withdrawals using mobile authentication. Smart ATMs account for 30% of all ATMs in Tier 1 Indian cities.	Bank of America's cardless ATMs, Barclays' facial recognition ATMs in the UK. Global smart ATM market expanding at a 9.8% CAGR till 2030.
Self-Checkout Systems	Retail point-of-sale terminals where customers scan, bag, and pay for their items independently, reducing queues and operational costs.	Pilot adoption by Reliance Smart Bazaar, and Amazon Fresh stores offering cashier-less checkouts in metros like Bangalore and Mumbai. Self-checkout systems in India expected to grow at 20% CAGR.	Self-checkout lanes in Walmart, Tesco, and fully automated Amazon Go stores. Global self-checkout market set to touch $ 12 billion by 2027.
Robotic Process Automation (RPA) in Service Delivery	Software robots automate routine back-office processes such as claims processing, form filling, and customer query management, enhancing speed and accuracy.	ICICI Bank and Axis Bank employ RPA for processes like loan approvals and fraud detection. Indian RPA market projected to hit $ 1.2 billion by 2028 (Deloitte India, 2024).	Netflix uses RPA for customer account management; CitiBank automates compliance reporting. Global RPA market expected to reach $ 66 billion by 2032 (Fortune Business Insights, 2024).

Source: Authors' Compilation

Driven by Digital India Mission, UPI adoption, and an aspirational middle class, India's landscape is primed for SSTs. Covid-19 further accelerated self-service preferences due to hygiene concerns. With increasing internet penetration (over 800 million users), self-service is becoming a necessity rather than a novelty. Globally, labor shortages, rising wage costs, and consumer demand for convenience are pushing companies to innovate service delivery models. From airports like Dubai offering biometric check-ins to Amazon Go stores redefining retail without cashiers, the trend is towards seamless, human-light interactions. Despite the advantages, challenges like data security concerns, technical literacy gaps (especially in India's rural areas), and high initial investment costs can slow adoption. The next wave of innovations is expected to integrate AI-driven personalization, blockchain for secure self-service transactions, and augmented reality (AR) in retail and tourism self-service models.

2. REVIEW OF LITERATURE

In their groundbreaking study, Meuter et al. (2000) presented the idea of SSTs and emphasized how they have emerged as a way to lessen service inefficiencies. They highlighted how self-checkout systems and automated teller machines (ATMs) offered a preview of the future of service delivery by increasing client autonomy and cutting down on wait times. The transformative effect of SSTs on service efficiency was investigated by Bitner et al. (2002). According to their research, operational bottlenecks were greatly alleviated by technologies like kiosks, enabling companies to service more clients with less staff. Dabholkar and Bagozzi (2002) looked into the usability issues and technological mistrust that prevent early SST adoption. They found that at first, older people and those who were less tech-savvy were hesitant to embrace new developments. Lu et al. (2021) investigated how chatbots driven by AI transformed customer service by offering immediate, tailored assistance. These natural language processing (NLP)-enabled devices produced smooth user experiences by simulating human interactions. Wilson (2010) described in detail how smart kiosks were introduced, especially in the retail and aviation industries. He emphasized how these systems offered a preview of the future of unattended service delivery by automating intricate procedures like payments and check-ins. The advent of super apps like Paytm and Google Pay is a realization of Dabholkar and Bagozzi's (2002) prediction regarding the role of mobile apps in SST adoption. Convenience in contemporary living has been redefined by their integration of many services. Customers could virtually try on items or see how furniture would be arranged in their houses thanks to AR applications, as demonstrated by Lee et al. (2021). According to Tan et al. (2019), SST integration in banking allows clients to obtain individualized financial services via smart kiosks and mobile apps. In their exploration of IoT breakthroughs, Chui et al. (2015) detailed how smart vending machines and other connected devices transformed service delivery by automating inventory management and delivering real-time updates.

Tapscott and Tapscott (2016) argue that the blockchain technology could enhance SST transparency and trust specifically for supply chain management and finance sectors. Under Tan et al.'s (2019) study of SST utilization between different cultures they showed that adjusted approaches enhanced usability for users in two separated cultural areas: urban Japan and rural India. User engagement within SSTs increased through gamification techniques which introduced challenge systems with reward structures according to Hamari et al. (2014). The adoption of SST needs training programs to deal with digital illiteracy issues according to the research by Kumar et al. (2018). Norman (2013) demonstrated that accessible interfaces help beginner users learn SSTs more efficiently thus supporting SST adoption by people new to the technology ecosystem. Through his examination of SST applications in health-care settings Reisman (2020) showed that telehealth platforms alongside kiosks expanded health care availability especially for disadvantaged locations. Virtual reality technologies provided real estate decision-makers an opportunity to evaluate homes virtually as reported by Lee et al. (2021). Acquisti et al. (2015) focus on SST data privacy concerns by stressing that customers need robust security measures for their data protection.

The implementation of autonomous systems including Amazon Go stands out because it enables users to shop and pay without using traditional checkout methods according to Kotler et al. (2020). Frey and Osborne (2017) explained in their study about SSTs' employment impact that these technologies promote productivity but automatically replace jobs which necessitates workforce retraining. According to Wilson (2010) people with disabilities must have access to usable SSTs thus requiring changes in design to address different needs. The epidemic hastened the acceptance of contactless payments through QR codes and mobile wallets according to Nguyen et al. (2022). Reisman (2020) conducted a study of SSTs in education to explain how automated learning systems employed performance data for personalized student education. The research by Chui et al. (2015) demonstrated that Internet of Things enabled Self-Service Technologies led to heightened energy efficiency combined with reduced waste in service delivery systems thus advancing sustainability goals. Both users and administrators need seamless platform transitions for kiosks alongside apps and websites according to Lu et al. (2021). Tan et al. (2019) described how voice-activated technologies including Google Assistant and Alexa support service delivery without hands. Wilson (2010) analyzed biometric authentication systems to enhance SST security measures particularly within banking and medical sector operations. Reisman (2020) emphasized mobile applications and self-check-in kiosks as service-station technologies which shorten wait times while offering personalized hotel services to patrons. Bitner et al. (2002) studied the causes of SST failures which included system malfunctions in addition to insufficient customer service then outlined solutions for resolution. The upcoming generation of SSTs will enhance personalized predictive services because AI and machine learning progress has been adopted according to Lu et al. (2021). Kumar et al. (2018) demonstrated the requirement for inclusive approaches because they studied the unique implementation barriers of SSTs in rural areas involving connectivity challenges along with resistance to technology adoption.

2.1 Research Gap

Self-service technologies (SSTs) focuses primarily on their effects for customers regarding happiness and efficiency together with adoption procedures in diverse business sectors. A large gap exists in the understanding of SSTs' lasting effects on consumer retention and loyalty among banking institutions and hotel establishments as well as healthcare facilities. Research concerning socio-economic and cultural adoption barriers across various worldwide contexts such as rural and underprivileged communities remains scarce even though studies on technology developments and SST-related customer attitudes have been plentiful. Few studies exist which describe how accomplished technologies including blockchain along with artificial intelligence and Internet of Things (IoT) would integrate to SSTs and transform service delivery methods. These knowledge gaps within digital environments enable a more comprehensive understanding of possible corporate innovation through technology to create better customer engagement and sustainable market growth.

2.2 Objectives And Methodology

The research centers its examination on current developments while developing expectations about SST evolution for the forthcoming years. The present research will highlight both the self-service sector and its operations at once. The research chooses to highlight the technical developments of automation and artificial intelligence (AI) which serve as essential tools for reaching these paradigms while investigating their sectoral evolution. The article conducts a quick assessment of relevant topics from both industrial and geographic perspectives as well as company development aspects to establish its present standing.

3. FUTURE TRENDS IN SELF-SERVICE TECHNOLOGY & EXISTING DEVELOPMENTS

Multiple fields including augmented reality (AR), blockchain technology and artificial intelligence (AI) and machine learning (ML) together with additional technology advances push forward the development of self-service technologies (SSTs). These technologies improve both customer satisfaction and operational performance so they reshape delivery services across several industries.

3.1 Current Developments in SSTs

AI-driven chatbots alongside smartphone apps and smart kiosks and voice-activated devices have received the most innovative updates in SSTs during recent times. These technologies have entered banking and healthcare and retail sectors to provide continuous service with reduced waiting periods. Financial service providers including Bank of America alongside HDFC Inc apply AI chatbots for customer transaction completion and financial advice and consumer service inquiry processing (Tan et al., 2019). The deployment of smart kiosks continues to rise rapidly across hotels and fast-food services as well as airports. Users benefit from self-service solutions that require no human contact and they gain faster operations alongside better customer satisfaction (Wilson, 2010). Digital payment systems such as Paytm and Google Pay within India have become the dominant SST since they supply unified solutions for customers to conduct multiple activities starting from bill payments up to shopping (Dabholkar & Bagozzi, 2002). Customers benefit from voice-activated SSTs such as Google Assistant and Amazon Alexa because these smart systems enable them to buy products along with scheduling management and smart home control through voice commands (Tan et al., 2019).

3.2 Emerging Trends in SSTs for the Future

Service platforms now adopt machine learning (ML) and artificial intelligence (AI) technology for better customer-specific prediction and accommodation purposes. SSTs will offer proactive service through AI developments because systems will learn to predict requirements and interact with customers during inquiries. AI technology enables hospitality chatbots to deliver customized services to visitors using spatial information combined with personal data record along with their interactive dialogue (Lu et al., 2021). Future SST developments will be propelled due to the growing significance of Internet of Things (IoT). Smart vending machines and linked kiosks represent IoT-enabled equipment which provides real-time data about user interactions together with inventory levels and customer happiness according to Chui et al. (2015). The retail sector awaits total advancement because IoT enables effortless product browsing and buying through networked displays that connect to smartphones and smart devices. Blockchain technology stands crucial to SSTs because it gives enhanced security to transactions and improved transparency. Blockchain offers security for self-service transactions through its decentralized immutable ledger which enables payment processing and document authentication particularly within banking and insurance industries because security remains their top priority (Tapscott & Tapscott, 2016). Through blockchain technology retail and logistics sectors achieve better supply chain management thereby guaranteeing quick errorless service delivery (Tapscott & Tapscott, 2016). Real estate together with retail and tourism sectors will stress SSTs to their maximum capacity because of accelerating AR and VR technology adoption. Customers can experience goods and services along with properties in enhanced ways with AR and VR technology so their decision-making process and engagement improves. Retail businesses would utilize augmented reality technology to allow customers to visualize clothing products like apparel and accessories so they can try them virtually while potential property buyers can tour properties digitally through their own locations according to findings by Lee et al. (2021). The final aspect of SST improvement results from 5G technology integration. Advanced services such as cloud-based interactions, augmented reality and video support will experience simplified use because of quick and reliable internet speeds. High bandwidth and low latency requirements that healthcare and entertainment sectors need will experience a revolutionary effect from this technology (Brynjolfsson & McAfee, 2014). The development of SSTs will add them deeper into society as technology progresses. The next generation of technological advancements for service delivery will emerge from existing AI development as well as mobile apps and smart kiosks innovations. Customers of the future will experience personalized and secure encounters combined with powerful effects due to the integration of AI, IoT, blockchain, AR/VR, and 5G. Organizations adopting advanced technology and adapting it to their needs will accomplish operational improvements and develop fresh customer expectations for the digital world.

4. HIGHLIGHT SELF-SERVICE & OPERATIONAL PROCESSES

Business operations gain improved operational effectiveness and customer satisfaction due to self-service technologies (SSTs) when integrated with business processes. Businesses obtain better productivity while lowering staff expenses coupled with delivering personalized customer interactions at large scales when users have self-service access to products. These technologies simultaneously introduce new business performance capabilities which bring forth fresh revenue streams and improved service delivery across multiple industries.

4.1 Business Opportunity through Self-Service Technology (SST)

The adoption of SSTs is transforming corporate environments particularly inside banking institutions and retail businesses and healthcare settings and hospitality domains. User-friendly self-serving technology systems enable businesses to operate around the clock and cut operational costs while providing complete automation

(Zeithaml et al., 1996). Modern banking along with retail services have adopted Self-checkout kiosks and mobile banking applications as essential elements due to their ability to enable users to execute transactions and payments and get financial guidance without human assistance according to Bitner et al. (2002). The automation trend demonstrates both financial benefits and it delivers greater freedom and convenience to customers which causes increased pleasure and customer loyalty (Meuter et al., 2000). Optionally referred to as SSTs they create business value because they enable large customer communities to access scalable repeatable service experiences. A research by Dabholkar and Bagozzi (2002) demonstrates that SST success depends on the combination of standardized company processes with customer-friendly system designs. Businesses can develop seamless customer experiences which will attract technology-oriented clients through the successful combination of SST features within their current operational framework.

4.2 Parallel Enhancement of Operational Processes

Using Self-Service Technology provides companies with essential improvements in operating procedures. The reconfigured delivery systems found in SSTs let companies cut down on their need for human workforces when performing routine operations. Fast-food businesses enable customer self-service kiosks for order placement which their backend systems automatically process through the kitchen and update inventory while printing receipts. Firms use this approach to deliver service to more customers during shorter periods because it reduces manual processing time and enhances order quality and processing speed (Berry et al., 2002). Back-end data system integration allows SSTs to streamline operational procedures enabling companies to gather consumer information they need for targeted marketing alongside service improvement (Parasuraman et al., 2005). Businesses obtain decision-support from detailed knowledge of client conduct patterns and transaction behavior and choices gained through their self-service systems (Kotler et al., 2020). The use of self-service technology results in multiple improvements for operating procedures. Such information enables IoT-based smart shelves together with automated restocking systems in retail to prevent stock shortages by initiating automated orders that ensure product availability for customers. The established operating efficiencies help businesses decrease waste while avoiding stockouts and sustaining inventory control (Chui et al., 2015). The introduction of self-check-in kiosks alongside telemedicine platforms throughout healthcare organizations leads to reduced administrative tasks and diminished patient waiting periods and strengthened one-on-one medical care delivery thus boosting operational performance (Reisman, 2020). The tools have improved operational processes allowing healthcare personnel to deliver digital prescriptions along with remote consultations and personalized health suggestions which enhance patient results.

4.3 Synergy Between Business Growth and Operational Efficiency

Self-Service Technologies work as drivers that push business growth simultaneously with operational efficiency improvement which becomes transparent to organizations. Through self-service technologies implementation businesses can reach competitive superiority through superior customer experiences and automation of regular human tasks. Amazon joins numerous major e-commerce companies by using AI recommendation systems to remove manual inventory management and product recommendation tasks while improving user satisfaction and operations (Porter & Heppelmann, 2015). Self-service technologies enable businesses to grow their services at a low cost with no increase in operating expenses. Self-service solutions enable companies to expand excellent support services throughout various domains while maintaining operational efficiency regardless of how they enter new markets or target new customer groups (Chui et al., 2015). The incorporation of AI-powered chatbots in banking enables financial institutions to provide multilingual customer assistance throughout all hours which makes large call center operations unnecessary (Tan et al., 2019). The self-service technology industry provides organizations with an exceptional opportunity to enhance operational efficiency along with customer service quality. The implementation of SSTs allows businesses to save money as they scale services while achieving operational efficiency while delivering customized superior service to customers. SSTs will enhance corporate growth and operational efficiency through better technology development including blockchain and artificial intelligence as well as the Internet of Things. Supply and service technologies provide essential support for modern business strategies because they simultaneously improve front-end and backend operations.

5. TECHNOLOGICAL ADVANCES DRIVING SELF-SERVICE: AUTOMATION & AI

Self-service technologies (SSTs) transformed company service delivery by providing customized solutions which enhance user satisfaction although they optimize operational procedures. The technological progress especially in automation together with artificial intelligence plays a key role in facilitating this change. Companies require these modern developments to expand their self-service models which let them fulfill rising customer demands through human reduction. Through the implementation of artificial intelligence and automation companies achieve better service quality together with reduced expenses and enhanced operational methods which lead to greater customer value. The article explores the fundamental position that modern technologies maintain in implementing the self-service approach.

5.1 Automation in Self-Service Technology

Self-service systems require automation as the fundamental technological achievement to function. A business modernizes its operations through automated routine work which enables staff members to focus on superior tasks beyond repetitive tasks. Self-checkout kiosks have become an introductory automation system which retail establishments widely deploy today. The implementation of these kiosks reduces labor expenses combined with shortening transaction durations through a system that lets customers self-checkout without requiring store staff involvement (Berry et al., 2002).The advancement of customer service automation now includes comprehensive service delivery model automation along with basic transactional processes. Customer independence in service transactions became possible as automated platforms like self-service kiosks with IVR systems and mobile apps expanded in the market for customer service delivery. By automating these procedures, companies may manage high customer interaction volumes without overtaxing human agents, which lowers operating costs and speeds up response times (Zeithaml et al., 1996). The fast-food industry uses automated food ordering and delivery solutions more frequently. Automatic kiosks in combination with automated kitchen systems allow customers to complete their entire order process from ordering to payment to receiving their meal without human employee involvement. This reduces labor expenses while boosting order processing speed. Through advanced technology systems business operators acquire beneficial customer preference data that enables them to design personalized product offerings (Dabholkar & Bagozzi, 2002).

5.2 Artificial Intelligence (AI) in Self-Service Technology

The most transformative effect on self-services emerges from artificial intelligence (AI) through which it delivers sophisticated one-on-one customer interactions. The self-service experience benefits from AI-powered solutions through individualized service delivery and decreased requirement of human agents. Mobile banking users can manage their money transfers along with checking balances and receiving financial advice through Erica AI bots installed across HDFC bank mobile apps as well as Bank of America (Tan et al., 2019).The evaluation process of browsing and purchase data by AI-based recommendation systems generates precise product suggestions in retail settings. The recommendation engine of Amazon demonstrates how artificial intelligence improvements self-service through its ability to generate suitable product suggestions relying on previous transaction and search information for personalized shopping experiences and better revenue results (Porter & Heppelmann, 2015). The advancement of self-service voice interactions happens through artificial intelligence technology. The voice command feature allows customers to interact with different services through devices equipped with Google Assistant and Amazon Alexa. These systems find increasing usage in domestic control systems alongside retail applications and customer support operations. These systems' artificial intelligence function enables natural language interpretation through which customers receive immediate context-specific solutions that boost their satisfaction (Lu et al, 2021). AI solutions have improved operations behind the scenes as well as in front-end applications which allows businesses to achieve automated decision systems. Predictive analytics powered by AI allows organizations to enhance inventory management through data-based demand forecasting which includes external components for better decision-making. AI transforms healthcare self-service kiosks through their ability to handle patient data assessment and scheduling and personalized recommendation generation thus advancing healthcare service quality (Reisman, 2020).

5.3 Synergy Between Automation and AI:

Companies took AI and automation to create automated checkout systems while developing smart kiosks to enhance speed and tailor product suggestions from checkout to purchase conclusion in the retail sector. Several self-service solutions emerge from the union of automation and artificial intelligence which allows businesses to create effective customer-specific services. Companies use robots powered by AI along with automated delivery systems to provide their customers with faster and more reliable service because of growing demands. The combination of AI drones with driverless cars enables quick last-mile deliveries and decreases both operational costs and delivery duration according to Chui et al. (2015). Operation efficiency grows and customers become happier when goods are delivered both speedily and accurately through the combined force of automation and AI systems. The self-service paradigm exists mostly because of automation together with artificial intelligence (AI) deployments that enhance operational efficiency and service delivery execution. Companies meet high expectations at volume by using artificial intelligence together with automation for personalized data-based encounters which enhances both customer satisfaction rates and business cost efficiency. Businesses with successful implementation of automation and artificial intelligence (AI) technologies will build smarter service models that enhance operational methods together with customer experiences since such technologies develop (Singh, R. & Singh, V. P., 2025).

6. OUTLINING SELF-SERVICE TECHNOLOGY: TECHNOLOGICAL EVOLUTION

Technological progression depends heavily on self-service technology (SST) because it signifies fundamental changes between how companies interact with their customers. The evolution of SST as a technology meets changing customer requirements alongside modern technological progress commences from ATMs followed by self-checkouts and extends to contemporary AI-operated chatbots and mobile applications. Understanding SST requires knowledge of its relationship with automation and artificial intelligence (AI) and Internet of Things (IoT) and big data analysis so we can properly understand its impact on technological progress (Singh, R. et. Al., 2025).

6.1 Technological Integration and Synergy

The combination of various state-of-the-art technologies makes up SST as it delivers improvements in service delivery rather than working alone as a single entity. The digital revolution relies on SST as a fundamental method across banking and hospital and retail and healthcare services. Self-service systems connect technology effectiveness with human labor execution and represent the natural development of automation alongside artificial intelligence principles. Supplementing AI knowledge allows live system modifications yet automation remains crucial for removing tasks that humans should not perform (Meuter et al., 2000). Companies can pair SSTs through multiple digital platforms and locations because of cloud computing which has made the systems more agile and adaptable. Artificial intelligence (AI) drives the development of the modern self-service platform technology. Authentic self-service technologies have progressed beyond traditional activity-performing models by developing capabilities to converse with users along with prediction functionality and learning ability from customer interactions. Self-service technologies use modern advancements in technology to create customized experiences which appear to each user that systems understand their preferences and wants (Porter & Heppelmann, 2015). AI-driven chatbots and virtual assistants, for example, can offer real-time customer service, suggest products, and even provide personalized financial advice based on previous exchanges. Additionally, a more linked world where devices may communicate with one another on their own is being created via the integration of IoT with self-service platforms. IoT-enabled retail shelves, wearable health devices, and smart appliances are a few instances of how SSTs are integrating into the broader network of connected technologies. These devices serve important roles in service delivery for both operational and customer-facing needs because they facilitate self-services as well as data collection and transmission for service experience enhancement analysis. Data from customer-service interactions streams into advanced analytics platforms which generates forecasts as well as enhances service delivery methods while enabling prediction of client behaviors.

6.2 Impact on Industry Evolution

The development of different industries owes much to self-service technologies. Online and mobile banking have become main pillars of banking services because of how SST technology capabilities affected banking infrastructure transformation. Mobile banking apps together with automated teller machines (ATMs) have substituted person-to-person transactions through their better speed and enhanced convenience. Underbanked and unbanked customers gained access to mobile banking services because SSTs have enhanced financial inclusion (Tan et al., 2019). The healthcare industry now benefits from telemedicine alongside patient self-check-in kiosks and AI-assisted diagnostics because these processes offer sped-up access to care and free up medical personnel (Reisman, 2020). The changes in service standards stem from how customers expect more efficient and rapid service methods. Omnichannel retail models along with e-commerce operations expanded due to organizations implementing self-service operations into their business procedures. The increase of SST adoption by these industries indicates broader changes in society towards automation and digitalization alongside the requirement for constant services.

6.3 Future of Self-Service Technologies

The ongoing development of both sectors and companies depends heavily on self-service technology for future applications. The combination of augmented reality (AR) and virtual reality (VR) systems will increase self-service models in retail and education because they deliver interactive educational and entertaining content. SSTs will integrate predictive functions progressively which will allow businesses to identify unexpressed customer needs. Company participation in skill development will generate opportunities for authentic customer relationships in addition to achieving better customer satisfaction. New markets will emerge through blockchain integration in particular sectors which need tight security measures such as healthcare and banking. Through blockchain-powered safe self-service data access customers gained total control that allowed them to perform automated transactions which were transparent and impenetrable. The incoming 5G networks will maintain high-speed connectivity essential for operating sophisticated robots and driverless cars plus sophisticated drones through their expected market expansion (Chui et al, 2015). Multiple revolutionary concepts which include automation and artificial intelligence and the Internet of Things and big data analytics operate together with self-service technologies. It transforms industries because this innovation enables organizations to develop innovative delivery methods which reduce operational expenses. SSTs will progressively grow vital for delivering customer-oriented services across various industries due to technological advancement. Self-service technologies demonstrate technological importance within digital advancement because they create automated processes which can develop personalized services and improve service systems. A strategic implementation of SSTs enables organizations to succeed when the world integrates digitally because these technologies now form an integral part of the modern business landscape.

7. PERSPECTIVES ON SELF-SERVICE TECHNOLOGY: INDUSTRIES, GEOGRAPHIES, AND BUSINESS DEVELOPMENT

The digital industry has used SST as a transformative power source which drives changes throughout industries and modifies business strategies and facilitates geographical adaptation. A comprehensive evaluation of SST's current impact and prospective outcomes can be achieved by analyzing it through industry, geographical and business development perspectives.

7.1 Industry Perspectives: Tailored Applications of SST

Self-service technology meets exclusive business demands within individual sectors when addressing unique business difficulties and prospects.

7.1.1 ***Retail and E-commerce***: Self-checkout kiosks, smartphone apps, and digital kiosks are examples of SSTs that are simplifying processes and cutting labor expenses in retail. The integration of SSTs with e-commerce, where AI-powered recommendation systems improve customer pleasure, has been facilitated by the expansion of omnichannel strategy (Porter & Heppelmann, 2015). An innovative stride in retail efficiency is the cashier-less shopping experience provided by Amazon Go stores, which use AI and IoT.

7.1.2 ***Banking and Finance***: The banking sector has activated SSTs to develop added convenience and accessibility for customers. The banking industry enables quick and risk-free interactions through ATMs as well as smartphone applications with AI chatbots Erica from Bank of America and Eva from HDFC. The affordable and easy access components of these technologies enable the improvement of financial inclusion for people who do not have bank accounts (Tan et al., 2019).

7.1.3 ***Healthcare***: The healthcare industry receives patient care advancement through wearable IoT devices along with telemedicine systems and patient check-in kiosks as part of SSTs. These solutions reduce medical institution costs and enable shorter waiting times for patients thus permitting better management of hospital workload increases. Health management control extends further to patients through the combination of self-service platforms and AI diagnostic tools (Reisman, 2020).

7.1.4 ***Hospitality and Tourism***: Travel and hospitality organizations make use of SSTs through in-raw kiosk services at hotels and airports and virtual concierge systems as well as online booking interfaces. Self-service innovations simultaneously decrease costs of operation and improve travel experiences for visitors. The airline company Emirates deploys AI technology to supply customized travel suggestions for travelers and allows them to complete their bookings using mobile apps.

7.1.5 ***Logistics and Manufacturing***: The use of self-service tracking websites as well as autonomous inventory robots demonstrates SSTs implementation in the logistics sector. These technologies successfully enhance operational efficiency and reduce errors when precision and speed are key factors in specific sectors according to Chui et al. (2015).

7.2 Geographic Perspectives: Regional Adaptation and Challenges

The implementation of SSTs shows wide variations between different geographic regions because of infrastructure status and cultural norms together with economic factors.

7.2.1 ***Developed Economies***: The sophisticated Self-Service Technologies operate through powerful technological frameworks which exist in countries including the United States and Japan and Germany. Self-service restaurants and vending machines have become common in Japan because the nation focuses on efficiency and innovation. The main objective in these sectors involves creating uninterrupted user experiences through innovative combinations of AI and IoT technology.

7.2.2 ***Emerging Markets***: The adoption of SST continues to rise in Brazil and India because these nations have made progress in both digital education and mobile phone technology. The Digital India Initiative of India helps to establish SST adoption by improving cashless payments and expanding internet access across the country.

7.2.3 ***Rural vs. Urban Dynamics***: SST implementation in rural areas aims to eliminate service deficiencies by establishing mobile banking kiosks as one example. Service sector technology in urban centers integrates both automation systems with artificial intelligence to serve medical and retail needs.

7.2.4 ***Cultural Considerations***: The perception of self-service operations varies across different cultural groups. Customers in Asian collectivist communities tend to favor service delivery that involves human personnel. Customer satisfaction requires businesses in these markets to find proper alignment between SST implementation and local service techniques.

7.3 Business Development Perspectives: Strategic Opportunities and Challenges

SSTs are not only tools for operational efficiency but also enablers of strategic business development.

7.3.1 ***Customer-Centric Innovations***: Self-service operations receive different perceptions among members of various cultural groups. Service delivery through human personnel remains the preferred choice for customers of Asian collectivist societies. For successful customer satisfaction businesses in these markets need to link SST technologies with appropriate native servicing techniques

7.3.2 ***Cost Efficiency and Scalability***: SSTs allow businesses to increase operations scale at reduced costs. The automation of processes cuts down on human staff requirements while cloud computing enables easy expansion of services. Startups and SMEs benefit especially from SSTs because they provide affordable tools to compete against larger competitors.

7.3.3 ***Data-Driven Decision Making***: SSTs together with big data analysis helps companies discover enhanced knowledge about market developments together with customer actions and operational strength. Covering strategic business decisions with sound market insights makes strategic planning more effective and promotes rapid market response.

7.3.4 ***Challenges and Risks***: SSTs present disadvantages even though they deliver several advantageous features. The high expenditure on initial resources and security threats together with continuous maintenance requirements tax business systems. Organizations need to maintain equilibrium when making SST use decisions since excessive SST dependency could dissuade clients who prefer personal contact.

7.3.5 ***Future Opportunities***: The development of SST innovation receives new potential with current technologies such as blockchain combined with 5G as well as AR/VR. AR/VR delivers powerful self-service encounters that specifically benefit the retail and entertainment markets (Lu et al., 2021).

Self-service technology serves as a disruptive force that connects business development practices with geographic and industry operational environments. The essential role of self-service technology becomes apparent because it serves diverse sectors and local needs while contributing to corporate strategic development. Organizations benefiting from SST utilization based on both market-specific quirks and cultural demands will succeed in today's fast-moving global economy.

8. FINDINGS AND SUGGESTIONS

8.1 Findings

8.1.1 ***Transformational Role Across Industries***: The retail industry together with banking and healthcare and hospitality sectors have undergone substantial changes because of self-service technology (SST). Such technology leads businesses to reduce their costs while making operations more efficient at the same time that it meets expanding consumer needs for personalized and convenient services. SST serves as an essential digital instrument that supports businesses to adapt their operations for modern technological market requirements..

8.1.2 ***Technological Integration Driving Innovation***: AI systems combine with IoT sensors and automation technologies and big data analytics as the main development factors for SST. These technologies allow companies to design response-driven systems that deliver person-oriented real-time solutions.

8.1.3 ***Geographic Variations in Adoption***: The SST adoption rate differs across geographical areas. Emerging nations embrace inclusivity through budget-friendly mobile technologies but sophisticated data processing remains a dominance of developed economies since their infrastructure is strong and their digital literacy levels are high. The way people use the internet and the way the internet is structured between countries represents significant obstacles.

8.1.4 ***Enhanced Business Development Opportunities***: Through SST organizations obtain the opportunity to promote customer engagement while maximizing operational procedures and obtaining decision-making inputs from analyzed data. A harmonious combination of human resources and automation is needed because excessive dependence on technology might drive away certain customer groups.

8.1.5 ***Challenges in Implementation***: SST adoption presents multiple challenges even though it provides advantages to organizations because of expensive initial costs together with digital security threats. The adoption of SST encounters impediments because some regions lack necessary digital expertise.

8.1.6 ***Future-Oriented Innovations***: Self-service technologies face another transformation because of emerging platforms such as 5G, blockchain and augmented reality (AR) and virtual reality (VR). The coming changes will maximize self-service powers by facilitating protected financial exchanges and interactive shopping interactions with consumers.

8.2 Suggestions

8.2.1 ***Invest in Infrastructure and Training***: Digital infrastructure development and skill training programs must receive the highest priority from both governments and corporations specifically in new markets. Digital literacy improvement together with equitable regional attention will boost adoption and utilization of SST.

8.2.2 ***Leverage AI for Personalization***: The predictive power of machine learning along with analytic methods enables organizations to meet customer wants which leads to satisfied clients who stay actively involved in interactions.

8.2.3 ***Adopt a Hybrid Approach***: Firms must create a combination strategy by blending SST solutions with human agent interactions regardless of automation system benefits. Businesses that utilize this approach provide suitable solutions for different customer needs primarily in healthcare services and hotels which require unique attention.

8.2.4 ***Focus on Security and Trust***: A secure system requires businesses to deploy blockchain security features in combination with other protective methods to prevent cybersecurity threats. Customer trust must be achieved before SST can gain mass adoption among the public.

8.2.5 ***Regional Customization***: SST solutions must be customized by businesses according to their regional cultural backgrounds and geographical territories. Market adoption rises when businesses deliver product designs suitable to cultures and provide language choices adjusted for specific markets.

8.2.6 ***Collaborate Across Sectors***: Innovation and scalability for SST grow from partnerships formed between technology suppliers and companies and governments. Collaborations between organizations and governments and technology vendors will accelerate modern technology adoption of IoT and AR/VR and create less expensive solutions.

8.2.7 ***Utilize Emerging Technologies***: The development of emerging technologies should be investigated by companies to maintain their position at the forefront. Companies can use 5G technology to establish smooth real-time financial and transportation services and implement retail AR and VR solutions for immersive customer experiences.

8.2.8 ***Continuous Monitoring and Feedback***: The system requires mechanisms for continuous feedback collection to validate that SST solutions fulfill their intended client requirements. Business operations should maintain flexibility by performing system upgrades through user feedback integration along with technological innovation.

9. CONCLUSION AND DIRECTION FOR FURTHER RESEARCH

9.1 Conclusion

Self-service technology known as SST presents disruptive changes across numerous corporate fields and industries which extend across various geographical locations. Companies now interact differently with customers and enhance their operations through data insights while adopting new methods of customer interaction. In order to achieve efficiency and personalization SST adopts the latest technologies that combine automation with IoT and AI. SST deployment leads to various complications that involve large investment costs alongside digital connectivity shortfalls and network security vulnerabilities. The multiple sections that could benefit from SST implementation have not been fully explored even though the technology has brought significant advancements to retail and finance systems and healthcare and hospitality industries. The technology presents enhancing possibilities throughout rural territory implementation alongside cultural adoption opportunities and its application within governmental services and educational institutions. SST stands to become a vital framework that helps organizations handle their aims for operational excellence and customer satisfaction through human-machine balance.

9.2 Direction for Further Research

Research into SST acceptance and performance should exclusively target regional and cultural differences during future investigations. Businesses can develop solutions for multiple markets after employing behavioral studies to understand customers' purchasing behavior. Research needs further development to merge 5G technology and AR/VR and blockchain into SST applications. These technologies bring the possibility to achieve real-time engagement while delivering immersive systems and stronger security features. Research should examine their cost pattern together with operational potential and potential for growth. The widespread implementation of SST remains limited to commercial usage but no one knows how well this system would perform in public sectors such as e-governance, healthcare for underprivileged areas, and education. Researchers need to examine SST protocols for their ability to enhance operational efficiency and service accessibility in the mentioned sectors. The rising public concern focuses on environmental changes caused by SST solutions. Future investigations should focus on sustainable design concepts which involve recycling old technology equipment and implementing energy-efficient devices to make SST sustainable on an international scale. The future of scientific study should concentrate on educating underdeveloped nations about SST and enhancing their technological base for adopting such systems. The promotion of inclusivity would benefit from simple implementation procedures and easy-to-use user interfaces. on hybrid service models that incorporate human touchpoints and SST can offer a well-rounded method of service delivery. These methods are especially applicable in fields where trust and emotional involvement are essential. Policymakers and corporate executives can embrace sustainable practices with the help of research on the long-term economic effects of SST on labor markets, business profitability, and consumer purchasing patterns.

REFERENCES

Acquisti, A., Brandimarte, L., & Loewenstein, G. (2015). Privacy and human behavior in the age of information. *Science, 347*(6221), 509–514. DOI: 10.1126/science.aaa1465 PMID: 25635091

Berry, L. L., Seiders, K., & Grewal, D. (2002). Understanding service convenience. *Journal of Marketing*, *66*(3), 1–17. DOI: 10.1509/jmkg.66.3.1.18505

Bitner, M. J., Ostrom, A. L., & Meuter, M. L. (2002). Implementing successful self-service technologies. *The Academy of Management Perspectives, 16*(4), 96–108. DOI: 10.5465/ame.2002.8951333

Brynjolfsson, E., & McAfee, A. (2014). *The Second Machine Age: Work, Progress, and Prosperity in a Time of Brilliant Technologies*. W. W. Norton & Company.

Chen, J. V., Jubilado, R. J. M., Capistrano, E. P., & Yen, D. C. (2015). Factors affecting online tax filing – An application of the IS Success Model and trust theory. *Computers in Human Behavior, 43*, 251–262. DOI: 10.1016/j.chb.2014.11.017

Christensen, C. M., Raynor, M. E., & McDonald, R. (2015). What is disruptive innovation? *Harvard Business Review, 93*(12), 44–53. PMID: 17183796

Chui, M., Löffler, M., & Roberts, R. (2015). The Internet of Things. *The McKinsey Quarterly*.

Dabholkar, P. A., & Bagozzi, R. P. (2002). An attitudinal model of technology-based self-service: Moderating effects of consumer traits and situational factors. *Journal of the Academy of Marketing Science, 30*(3), 184–201. DOI: 10.1177/0092070302303001

Davis, F. D. (1989). Perceived usefulness, perceived ease of use, and user acceptance of information technology. *Management Information Systems Quarterly, 13*(3), 319–340. DOI: 10.2307/249008

Frey, C. B., & Osborne, M. A. (2017). The future of employment: How susceptible are jobs to computerization? *Technological Forecasting and Social Change, 114*, 254–280. DOI: 10.1016/j.techfore.2016.08.019

Hamari, J., Koivisto, J., & Sarsa, H. (2014). Does gamification work? A literature review of empirical studies on gamification. *Proceedings of the 47th Hawaii International Conference on System Sciences (HICSS)*. DOI: 10.1109/HICSS.2014.377

Ives, B., & Jarvenpaa, S. L. (1991). Applications of technology to service systems: New opportunities in self-service. *Decision Sciences, 22*(3), 307–326.

Kotler, P., Keller, K. L., Goodman, M., & Brady, M. (2020). *Marketing Management*. Pearson.

Kumar, A., Sureka, S., & Gupta, S. (2018). Bridging the digital divide: Challenges in rural self-service technology adoption. *International Journal of Rural Development, 37*(1), 56–65.

Lee, J., Kim, K. J., & Sundar, S. S. (2021). Customization in augmented reality: Exploring its role in the user experience. *Computers in Human Behavior, 122*, 106840.

Lu, Y., Cai, R., & Gursoy, D. (2021). Developing and testing a chatbot performance scale in the hospitality industry. *International Journal of Hospitality Management, 93*, 102848.

Lusch, R. F., & Nambisan, S. (2015). Service innovation: A service-dominant logic perspective. *Management Information Systems Quarterly*, *39*(1), 155–175. DOI: 10.25300/MISQ/2015/39.1.07

Meuter, M. L., Ostrom, A. L., Roundtree, R. I., & Bitner, M. J. (2000). Self-service technologies: Understanding customer satisfaction with technology-based service encounters. *Journal of Marketing*, *64*(3), 50–64. DOI: 10.1509/jmkg.64.3.50.18024

Nguyen, T. T., Hoang, T. D., & Le, T. T. (2022). Contactless payments: A silver lining of the COVID-19 pandemic. *Journal of Financial Services Marketing*, *27*(3), 276–287.

Norman, D. A. (2013). *The Design of Everyday Things: Revised and Expanded Edition*. Basic Books.

Parasuraman, A., & Colby, C. L. (2015). *Techno-ready marketing: How and why customers adopt technology*. Free Press.

Parasuraman, A., Zeithaml, V. A., & Malhotra, A. (2005). E-S-QUAL: A multiple-item scale for assessing electronic service quality. *Journal of Service Research*, *7*(3), 213–233. DOI: 10.1177/1094670504271156

Porter, M. E., & Heppelmann, J. E. (2015). How smart, connected products are transforming companies. *Harvard Business Review*, *93*(10), 96–114.

Prahalad, C. K., & Ramaswamy, V. (2004). Co-creating unique value with customers. *Strategy and Leadership*, *32*(3), 4–9. DOI: 10.1108/10878570410699249

Reisman, E. (2020). Self-service kiosks and the future of healthcare. *Healthcare Technology News*, *28*(6), 32–35.

Reisman, F. (2020). Telehealth technology adoption: The role of self-service in healthcare delivery. *Healthcare Technology Journal*, *12*(4), 198–210.

Singh, R., & Singh, V. P. (2025). Global Mobility and Economic Incentives in Prenatal Tourism. In, Exploration of Prenatal Tourism and Birthright Citizenship (pp. 77-104). IGI Global Scientific Publishing. https://doi.org/ Available on: https://www.igi-global.com/gateway/chapter/365545DOI: 10.4018/979-8-3693-8206-6.ch005

Singh, R., Singh, V. P., Rabby, F., Bansal, R., & Gupta, N. K. (2025). The Road to Sustainable Tourism: Financial Mechanisms for Green Project Development. In B. Sousa, L. Arantes, & S. Bhartiya (Eds.), Regenerative Tourism for Social Development (pp. 235-254). IGI Global Scientific Publishing. Available on: https://doi.org/DOI: 10.4018/979-8-3373-0235-5.ch010

Tan, M., Teo, H. H., & Wei, K. K. (2019). Technology acceptance model in self-service innovations: A cross-cultural perspective. *Journal of Service Research*, *15*(1), 89–101.

Tapscott, D., & Tapscott, A. (2016). *Blockchain revolution: How the technology behind bitcoin is changing money, business, and the world*. Penguin.

Venkatesh, V., Thong, J. Y. L., & Xu, X. (2012). Consumer acceptance and use of information technology: Extending the unified theory of acceptance and use of technology. *Management Information Systems Quarterly*, *36*(1), 157–178. DOI: 10.2307/41410412

Wilson, J. (2010). Smart kiosks and their role in service delivery. *Service Management Journal*, *22*(2), 137–150.

Zeithaml, V. A., Parasuraman, A., & Berry, L. L. (1996). The behavioral consequences of service quality. *Journal of Marketing*, *60*(2), 31–46. DOI: 10.1177/002224299606000203

Chapter 15
Streamlining Insurance Claims With Intelligent Systems

P. Ashok
https://orcid.org/0000-0002-5859-6041
Symbiosis Institute of Digital and Telecom Management, India

Tanisha Mangal
Symbiosis Institute of Digital and Telecom Management, India

Saurabh Pawar
Symbiosis Institute of Digital and Telecom Management, India

Venkatesh Ramamurthy
NETS International, UAE

ABSTRACT

The introduction of Artificial Intelligence (AI) into multiple aspects of the automotive sector is causing a revolutionary change. This field is one important area where AI has a big influence. To speed up the acceptance process, machine learning algorithms are being used to examine large datasets, such as photographs and sensor data. Insurance companies can evaluate the extent of damages more quickly and precisely thanks to image recognition technology, which speeds up policyholder reimbursements. AI systems can give more accurate claims assessments and customized risk profiles for individual policyholders by utilizing this data. These smart interfaces are meant to communicate with policyholders, helping them through the claims process and making it easier for them to report accidents. This study looks at how virtual assistants affect client satisfaction and how much more efficient claims processing

DOI: 10.4018/979-8-3373-4667-0.ch015

may be because of them. This study explores the moral ramifications of utilizing AI algorithms to make choices that have a direct effect on people's financial security.

INTRODUCTION

The motor insurance industry is substantial and dynamic, with competitive pricing, a wide range of coverage options, and user-friendly benefits offered by the insurers. Motor insurance is governed by regulatory frameworks that lay down minimum handles such as property liability and other injuries. Motor insurance includes comprehensive options to cover damage caused by theft, vandalism, and natural disasters, as well as collision insurance that covers damages caused by vehicular collisions. To protect car owners from financial losses in case of theft, or collisions that require vehicle repairs or replacement, vehicle insurance is no better than not having it at all. The motor insurance industry owes much of its importance to the use of data analysis to assess risk factors related to driver behavior, vehicle characteristics, and geographic areas to calculate premiums. It's a complex business for Insurers who use sophisticated pricing models to ensure risk and profitability are balanced. Yet the advent of technologies such as electric and autonomous vehicles creates special problems for the industry, forcing it to embrace new approaches in risk assessment and coverage design. The motor insurance sector remains a keystone within the broader insurance sector, by offering essential financial protection and road safety.

In this study, the focus is on Artificial Intelligence (AI) specifically; **Optical Character Recognition (OCR) and Natural Language Processing (NLP)** transform the way a team handles vehicle insurance claims. The **objective** is to improve the efficiency, accuracy, and consistency of claims assessment while overcoming evaluation disparities arising from differences in norms among case controllers. To achieve the objectives, it evaluates current practices, analyzes AI's role in claims handling, and its bearing on efficiency, accuracy, and customer experience. Additionally, it investigates AI's capabilities in fraud detection, its alignment with ethical and regulatory standards, and the financial implications of its adoption. Challenges to implementation and strategies for seamless integration are also explored to build a comprehensive understanding of AI in the vehicle insurance domain. Over the past few years, the motor insurance industry has been revolutionized as the rates of car ownership continue to soar in the world, not to mention the technological improvements in vehicles themselves and the increasing economic development around the world.

The way modern cars are built today with sophisticated safety features, anti-theft systems, and telematics devices, has all changed the way that insurance works. Usage Based Insurance (UBI) is an example of how telematics has the potential to track in real time the driving behaviors of its insureds, and the creditable premiums for which an insurer could then offer them based on their habits. Not only does this incentivize safer driving but it capitalizes on better risk management. It has also further revolutionized customer interactions, policy issuance, and claims processing due to the addition of digital platform integration. In underwriting and fraud detection, insurers now use a variety of large-scale datasets to predict risks, assess premiums, and find fraud. Modeling and data analytics have been key tools to navigate this. At the same time, insurance companies are using technologies like AI, blockchain, and automation to make insurance faster, inexpensive, and accessible to anyone and everyone in an innovative way (Harmantzis, 2020). But all these innovations introduce new challenges as well.

For example, with digital technologies in cars, insurers employ cyber risks of connected car coverage integration. Similarly, the adoption of electric vehicles also required tailored policies that apply to it in terms of risks and attributes. Within these transformations, the application of AI to claim processing has become a particularly interesting development. Machine learning algorithms used by AI-driven systems make it easier and quicker to analyze claims as they can take in the elements of policyholder history, accident frequency, severity of the effects of the accident, etc. They give insurers the ability to manage claims cost-effectively, generate better outcomes for customers, and enable processing claims quickly (Poufinas, 2023). AI's usefulness goes way beyond just being efficient. Deep learning models and NLP teamed up to build its fraud detection capabilities which allow anomaly and inconsistent detection that may not pass through the human nose. Thanks to the speed of AI in analyzing huge volumes of data, it can also extract helpful information and offer claim resolutions, reducing errors and generally helping with the management of claims. Although the benefits of using AI in insurance are several, there are challenges with their adoption. But data security, privacy, transparency, compliance with regulations, and ethical usage also need to be addressed. Also, insurers in the integration of AI in existing workflows and initiative costs and complexities.

One of the most frequent and costly parts of the industry, accident insurance claims, shows the transformative potential of AI. Traditional claims processing has become labor-intensive with documents and manual data entry. AI automates these processes, evaluations are streamlined, and the administrative burden is lessened. For example, in computer vision technology, we can estimate vehicle damage repair costs using images of vehicle damage as inputs, and NLP systems can extract information from written documents. Furthermore, AI facilitates insurers to adopt forward risk management. These predictive analytics allow insurers to plan for the claim trends forecasted to allocate resources and introduce prevention strategically. Supporting long-term industry sustainability as well as operational efficiency, these capabilities add. Finally, technological advancements, regulatory changes, and a growing economy are propelling the transformative journey of the motor insurance industry. AI is at the forefront of this evolution and provides claims handling with unprecedented opportunities to increase efficiency, accuracy, and customer satisfaction. AI is changing the landscape of motor insurance by introducing efficiency reductions, better fraud detection, and innovation. However, successful deployment depends on how far technological potential can be sufficiently balanced against ethical considerations and regulation requirements. The purpose of this study is to shed some light on what AI could bring to play in vehicle insurance claims as well as highlighting its challenges, opportunities, and ways of integrating AI smartly into the claims routing.

LITERATURE REVIEW

The way the insurance industry operates today has changed dramatically because of the introduction of Artificial Intelligence (AI) into the automobile insurance claims sector. This analysis of pertinent research explores the dynamic development of artificial intelligence (AI) and its diverse effects on streamlining different parts of the auto insurance claims procedure. Recent studies have examined AI's pivotal role in claims processing. Studies demonstrate how machine learning algorithms are used to evaluate large datasets—like pictures and sensor data—to accelerate the acceptance of claims (Samir, 2019). The productivity advantages realized by automated claims assessment have been highlighted, demonstrating AI-driven systems' capacity to quickly and precisely assess damage. This improves the entire timeliness and customer-centric aspect of the claims process in addition to resulting in quicker reimbursements.

AI developments are being incorporated into auto insurance claims as mentioned in Figure 1, and telematics and the Internet of Things (IoT) are key components of this process. The use of real-time data from telematics devices has been investigated by researchers, offering insights into location-based risk factors, accident severity, and driver behavior. By using this data, AI systems improve claims evaluation and proactively aid in risk management. The combination of AI with telematics makes it easier to identify patterns, which helps identify and stop fraudulent activity. This improves the industry's capacity to control risks.

Figure 1. Enhancing Claims Accuracy with AI

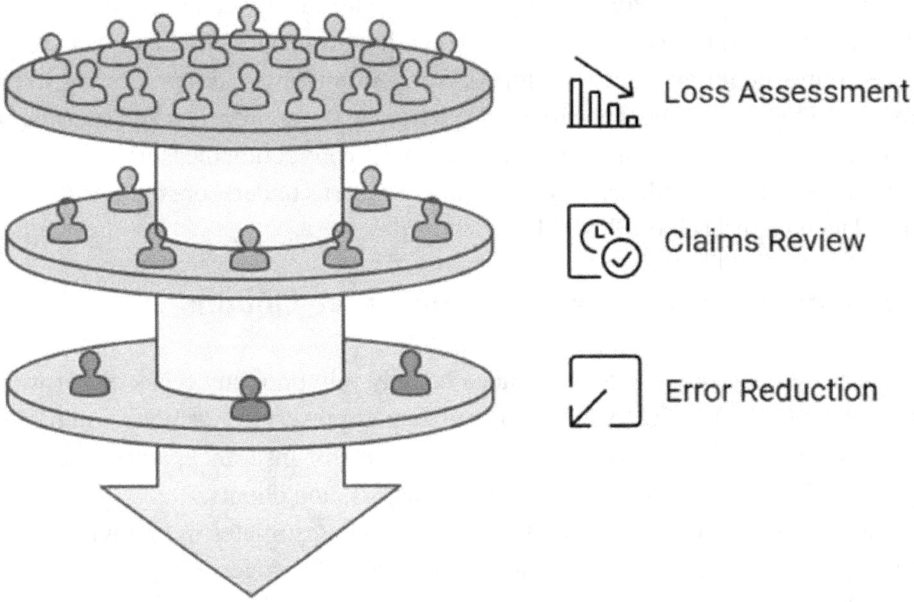

Recent research has drawn attention to the rise of chatbots and virtual assistants in the insurance industry. Researchers have investigated how Natural Language Processing (NLP) methods might be used in these virtual interfaces to facilitate easy communication with policyholders. Virtual assistants provide users with timely and relevant information while guiding them through the claims reporting process. This improves operational efficiency in claims administration while also streamlining client contacts. According to the research, incorporating virtual assistants might completely change how customers interact with insurance claims. The literature examining the environment of AI-driven auto insurance claims has shown ethical issues as a crucial theme (Aid, 2021). Scholars have examined the moral ramifications of using algorithms to make choices that directly affect people's finances. There has been a close examination of the accountability, equity, and openness of AI systems in the processing of claims. Research emphasizes how crucial it is to set moral norms and legal frameworks to control the ethical use of AI in the insurance industry and make sure that the benefits of automation line up with moral principles. In recent literature, privacy problems about the vast amounts of data handled by AI systems have also assumed a central role. Because AI depends on a variety of datasets for both training and making decisions, protecting sensitive consumer data has become critical. Academics have deliberated on the necessity of implementing comprehensive data protection protocols, stressing the significance of adhering to privacy laws to establish and preserve confidence in AI-powered insurance procedures (Baranoff, 2017). The literature on AI's use in auto insurance claims emphasizes how revolutionary these technologies can be. The lifespan of auto insurance claims can be optimized in several ways with AI, from automating claim processing to improving damage assessment accuracy. To guarantee the appropriate and transparent application of AI in the insurance sector, strong data protection measures and ethical norms must be developed. Privacy and ethical concerns underscore the significance of a balanced approach (Gill, 2024).

The Demand for AI-Driven Auto Insurance Options

The daily operations of car insurance heavily rely on many types of software. They must verify, handle, examine, organize, and preserve enormous volumes of data produced by various sources such as suppliers of insurance, automakers and suppliers of auto services, independent contractors, and clients.

Insurance companies must modify their premium estimates in line with the increasing quantity and diversity of cars on the road. A few decades ago, automotive size classes—which distinguished cars based on weight and engine size—were used in insurance calculations. These days, insurance rates also need to consider the kind of energy source, the degree of computerization, and the autonomy of the vehicle, among other factors, as identified in Figure 2.

Figure 2. Key requirements for modern car insurance software

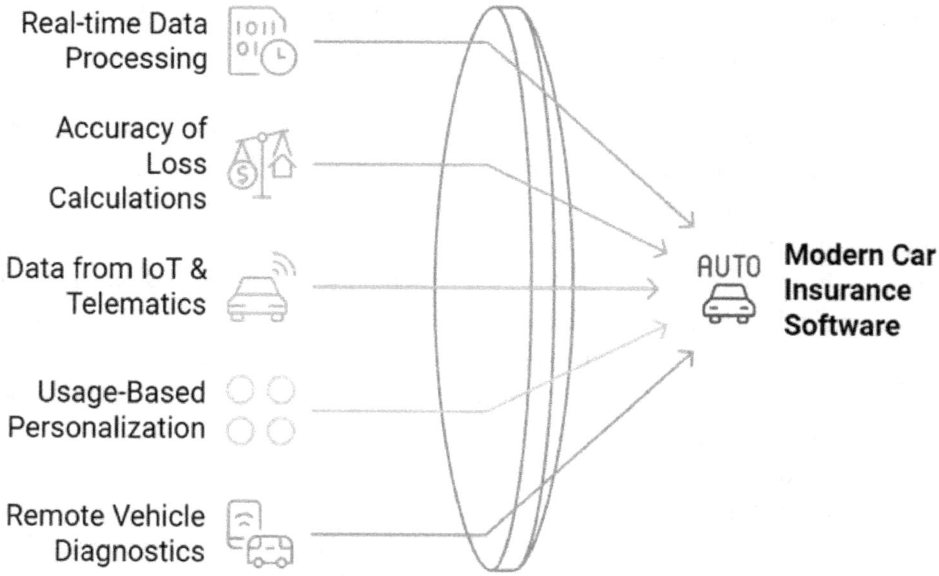

Vehicle guarantors likewise have a few functional requirements that heritage programming can't completely cover:

- **Process information progressively** — Whether computing a statement or giving client help, it's basic for a protection answer to have the option to rapidly handle input information from various sources and give precise outcomes.
- **Increment the exactness of computations** — Quotes for vehicle fixes, clinical benefits, and claims handling costs structure the premise of insurance payment cost estimations. Wrong and deferred assessments might dismiss the expected clients or lead to critical monetary misfortunes for the safety net provider.
- **Serve-associated vehicles** — The present streets are loaded with vehicles furnished with IoT sensors, telematics, and high-level driver-help frameworks (ADASs). This multitude of advancements accumulates and produces lots of helpful information like mathematical information, vehicle state boundaries, and driver conduct information. However, not all vehicle protection arrangements can appropriately deal with this kind of information — and the amount of information created.
- **Give far-off vehicle analytic administrations** — Because of the pandemic, numerous nations have urged moving to advanced conveyance of protection administrations. In any case, not all heritage accident coverage arrangements consider the reception of virtual assessment and examination strategies.
- **Offer use-based help personalization** — To let their clients "pay how they drive," collision protection organizations need an answer that can constantly screen and investigate driver conduct.

Man-made brainpower innovations can be the way to tend to these difficulties. Upgrading vehicle protection arrangements with simulated intelligence holds the potential for:

- **Expanded effectiveness** — A computer-based intelligence framework empowers ongoing information assortment and handling, empowering insurance agencies to accelerate exercises like endorsing, claims handling, and requests handling as shown in Figure 3.
- **Decreased worker above** — Mechanized handling of client claims and demands empowers insurance agency representatives to zero in on additional perplexing undertakings and cases.
- **Lower claims taking care of costs** — Mechanization of routine exercises and correspondence additionally prompt diminished costs per guarantee.

- **Important business bits of knowledge** — While handling unstructured information, artificial intelligence can find unobvious examples and concentrate on significant experiences that assist safety net providers with working on functional cycles and settling on better business choices.
- **Further developed client experience and administration moderateness** — With a continuous simulated intelligence-controlled examination of driver conduct and a vehicle's state, guarantors can convey more reasonable and customized administrations to their clients.
- **Decreased hazard of human mistakes** — While people may coincidentally misclassify a case or apply some unacceptable choice, simulated intelligence-controlled mechanization might assist you with moderating this gamble.
- **Straightforward navigation** — Man-made intelligence-based mechanization can assist with tending to client worries about separation and out-of-line strategy computations. Depending on a reasonable arrangement of boundaries, a computer-based intelligence framework can make applied valuing strategies, fixed expenses, and claims handling results more straightforward to legitimize.

Empowering utilization-based collision protection Utilization-based protection (UBI) is a generally new way to deal with customized statements that may soon turn into another brilliant norm for vehicle guarantors. As opposed to conventional protection, UBI depends on current behavior methods instead of reviewing information. There are two kinds of use-based accident coverage:

- **Pay-as-you-drive** — The strategy cost relies upon the time or the distance driven during a predetermined period. Mileage is logged consequently utilizing telematics innovation, ordinarily with the assistance of installed gadgets.
- **Pay-how-you-drive** — The expense for protection is persistently changed in light of the driving examples of a specific client: the more secure they are driving, the lower the protection cost. Normal measures considered for strategy cost computations incorporate speed and slowing down designs, the traffic climate, and the number of miles driven.

A Columbus-based project called Root is one of only a handful of exceptional backup plans that gauge the cost of their expenses in light of the compensation how-you-drive model.

Figure 3. Insurance activities to automate with AI

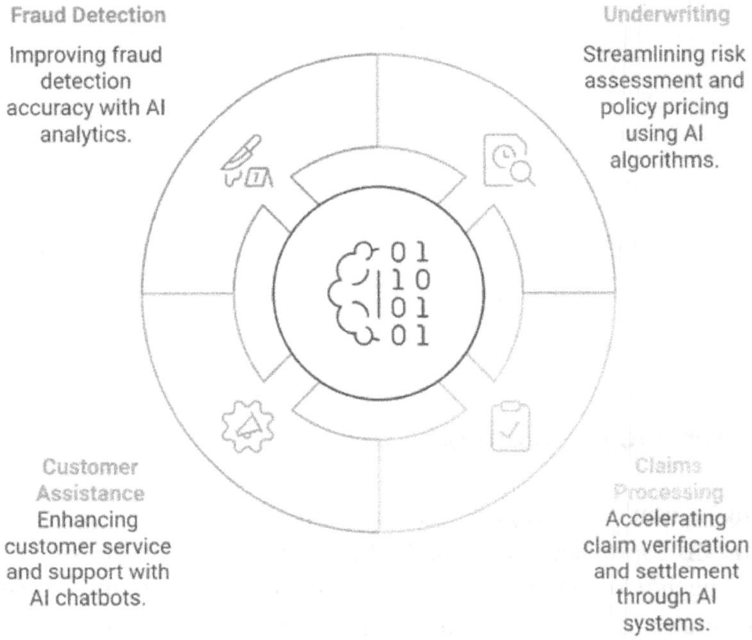

Monitoring driver performance

Observing driver execution Computer-based intelligence frameworks can consistently screen a driver's way of behaving, traffic conditions, and information from in-vehicle sensors. Considering the examination of this information, the framework can recognize the gamble of a street mishap quicker than a driver and caution the driver about it. Additionally, checking information can be utilized to help drivers work on their driving abilities and create more secure driving propensities (Balasubramanian, 2021). For instance, the Nauto project utilizes a computer-based intelligence framework to dissect information from in-vehicle sensors and ready drivers of expected crashes before they happen. Vehicle guarantors can likewise profit from conveying a simulated intelligence-fueled driver (as picturized in Figure 4) observing a framework to instruct new drivers and lessen the general gamble of impacts out and about.

Figure 4. Car Insurance Tasks to benefit from AI Adoption

Running remote car diagnostics

Utilizing man-made intelligence advancements, vehicle safety net providers can empower far-off conveyance of one of their critical administrations — vehicle diagnostics. During the Coronavirus pandemic, numerous organizations had to rearrange their administration conveyance models. Engine vehicle backup plans likewise need to conduct a method for offering their types of assistance from a distance, without direct human contact. With an artificial intelligence-fueled diagnostics framework, vehicle safety net providers can accumulate and handle ongoing information from in-vehicle sensors to even more likely assess a vehicle's condition. They can likewise apply simulated intelligence advancements to assess vehicle harm considering video and photograph materials obtained from in-vehicles or outer cameras. For example, USAA depends on a man-made intelligence-based cloud stage to survey vehicle harm considering photographs. Manageable use of simulated intelligence innovations to empower far-off vehicle reviews considering recordings made with a cell phone (Rose, 2018).

Reconstructing incidents

Another promising chance is to involve artificial intelligence for fender bender claims, especially to reproduce the subtleties of a fender bender considering run camera film and in-vehicle sensor information. At the same time, you can likewise depend on man-made brainpower to recognize collision protection misrepresentation. Having an exact, simulated intelligence-created course of events of a mishap speeds up mishap examination and case handling while at the same time decreasing the gamble of deceitful cases. The Israel-based Nexar project has as of late collaborated with a Japanese insurance agency to execute their simulated intelligence-fueled fender bender reproduction framework into 200,000 vehicles. While holding promising potential, man-made intelligence advances can be carried out. In the following area, we examine a portion of the vital difficulties to consider before beginning your excursion into the artificial intelligence world.

FIVE PREPARATORY STEPS FOR MOVING TO AI-BASED CAR INSURANCE

1. The tasks for which AI can be applied

Not all undertakings that can be performed with the assistance of artificial intelligence and ML to be sure require these advances. Now and again, conventional rule-based robotization might convey only similar results as a simulated intelligence framework while being less asset-hungry regarding improvement and upkeep. The most ideal decision is applying man-made brainpower for vehicle protection errands about handling enormous measures of constant information. For instance, handling lots of protection claims and identifying misrepresentation is one of the key LLM use cases. Artificial intelligence advances can likewise assist you with conveying extraordinary help personalization, as no human can rapidly and productively break down as much information as a simulated intelligence framework (Kumar, 2020). Furthermore, computer-based intelligence is known for its capacity to remove significant experiences from unstructured information, laying the ground for stunningly better assistance personalization.

2. How can AI capabilities be implemented in servicing?

There is a need to adopt an alternate strategy towards computer-based intelligence execution in collision protection contingent upon whether you as of now have a vehicle protection arrangement or are just intending to construct one. While building another task without any preparation, you can design its whole engineering considering the artificial intelligence framework's prerequisites to accomplish the most extreme effectiveness. In any case, to upgrade with simulated intelligence abilities, making similarity and strength your priorities are better. For the actual improvement, you can constantly attempt to save both time and spending plan by searching for a fitting man-made intelligence Programming interface. Regardless of whether it's anything but a 100 percent match, it may be simpler to modify a current computer-based intelligence framework than to construct and test another one.

3. Choice of Technologies

The decision of innovation stack will completely rely upon the scope of assignments you need to perform utilizing computer-based intelligence. Underneath, we frame a portion of the keyways you can use with different computer-based intelligence and ML innovations inside vehicle protection items:

- Normal language handling (NLP) advances are vital to executing simulated intelligence-based chatbots and menial helpers as well as to accelerating claims handling.
- Picture handling and acknowledgment procedures empower far-off vehicle diagnostics and work on the nature of vehicle harm assessments. You can depend on these strategies for both the underlying assessment of a vehicle at the guaranteeing stage and for post-mishap evaluations.
- Optical person acknowledgment is text acknowledgment that can assist you with saving time on digitizing manually written archives and handling report filters. Full of feeling registering, also called close-to-home man-made intelligence, can process and recognize human feelings, carrying further upgrades to driver conduct checking and misrepresentation discovery.
- Man-made intelligence-based telematics offers higher accuracy of chance evaluations. It empowers prescient driver cautions by ceaselessly handling and investigating information on a vehicle's speed, speed increase, slowing down, cornering, and other significant variables.

4. Where can I get enough quality data?

No artificial intelligence framework can perform effectively without an adequate amount of valuable information. Consequently, you want to ponder the sorts of information your answer needs and where to get it. For example, one of the most troublesome and asset-consuming vehicle protection errands is assessing vehicle harm (Eling, 2021). The evaluator needs to consider numerous elements: the vehicle's model, age, mileage, and past harm as well as the expense for fixes. What's more, to move this large number of computations to an artificial intelligence framework, you first need to prepare that situation with the right information. When you figure out what kind of information you want, you can begin searching for prepared datasets and pre-prepared artificial intelligence frameworks that you can tweak for your requirements. For instance, the Cogito project gives marked picture datasets for vehicle protection. You can likewise have a go at working on the nature of accessible datasets utilizing information increase strategies.

5. What challenges can I expect along the way?

Getting an adequate amount of valuable information isn't the possible challenge you ought to expect while involving artificial intelligence in vehicle protection (Frontiers). Here are a few different entanglements to know about:

Obtaining significant specialized abilities: Building and testing a man-made intelligence framework is an in fact testing process that requires an elevated degree of skill in the use of explicit advancements, devices, and approaches. You can close the hole in specialized abilities by re-appropriating item improvement to a devoted group or by enlarging your interior group with the required experts.

Guaranteeing information security: Vehicle protection arrangements work intimately with delicate and classified data. Your simulated intelligence framework ought to ensure the security of any information it processes. Information encryption and information anonymization strategies can assist you with handling this test.

Lessening predisposition: Artificial intelligence frameworks can acquire inclination from various sources, including datasets and calculations. This is particularly pertinent for arrangements that pursue choices considering individual information. To expand the decency of your answer, think about cross-really looking at your calculations and applying different datasets during model preparation.

RESEARCH METHODOLOGY

The study aimed to investigate the effectiveness of AI-based solutions in improving the claims process, as well as the challenges and opportunities associated with their implementation. The Research involved a comprehensive literature review to identify key themes, concepts, and research gaps related to the topic. The literature review was conducted using various academic databases, including PubMed, Google Scholar, and ScienceDirect, as well as industry reports and relevant publications. Unmanned aerial vehicles (UAVs), also known as drones, are aircraft that can be remotely controlled or operated autonomously with pre-programmed instructions. They have numerous applications, such as military operations, photography, agriculture, search and rescue, and delivery services (Zeng, 2018). To control a drone, a remote control or mobile device is used to send instructions to its onboard computer, which manages the motors and other systems to keep it airborne. The onboard computer receives data from the drone's sensors, including GPS, gyroscopes, accelerometers, and cameras, which it uses to navigate and maintain stability during flight. Machine learning (ML) algorithms are often integrated into drones to enhance their functionality and capabilities. Machine learning refers to a form of artificial intelligence that enables computer systems to improve their performance based on experience, without the need for explicit programming. ML algorithms can be trained on vast datasets to recognize patterns and make predictions based on that data (Global Data UK Ltd, 2022). Fluid dynamics play a crucial role in the development and progress of both aircraft and drones, as it encompasses the principles of aerodynamics that are necessary for the proper functioning of aircraft. To overcome the force of gravity, which is called lift, sufficient upward force must be generated. Additionally, thrust is the force that propels the vehicle or body forward. These forces can be examined and understood through the kinematic laws that govern the flow of fluids.

Figure 5. Use of Drones in Vehicle

Figure 5 illustrates the flow pattern around the cross-section of an aero foil or propeller. The difference in fluid pressure between the bottom and top of the propeller creates an upward force called lift, which is responsible for lifting an aircraft or drone. The angle of inclination of the aero foil or propeller determines the magnitude of the lift force. One use of ML in drones is object recognition. Drones fitted with cameras and ML algorithms can be employed to identify objects or people on the ground. For instance, a drone can be trained to detect a specific type of vehicle, such as a truck or car, and track it from the air. Another application of ML in drones is autonomous flight. Drones can be programmed with ML algorithms to fly autonomously and avoid obstacles. For example, a drone can be pre-programmed to fly a specific route and avoid obstacles like buildings or trees by employing data from its sensors to generate a 3D map of the environment. ML algorithms can also be utilized to enhance the accuracy and reliability of drone data. Drones can be equipped with sensors to gather data on environmental conditions like temperature, humidity, and air quality. ML algorithms can analyze this data and make predictions about future environmental conditions like weather patterns or pollution levels. To summarize, drones are unmanned aerial vehicles that can be remotely controlled or fly autonomously. ML algorithms can be integrated into drones to enhance their capabilities, including object recognition, autonomous flight, and data analysis. As technology advances, the potential applications of drones and ML will continue to expand. The usage of unmanned aerial vehicles (UAVs), drones, or remotely piloted aircraft systems (RPAS) has been on the rise across various industries due to their efficiency and cost-effectiveness. In the insurance industry, particularly in motor insurance, drones have enormous potential to prevent fraud claims, assess damages, and speed up the claims process (Lior, 2022). Motor insurance is plagued with fraudulent claims that range from staged accidents to false injury claims, causing significant problems for insurers. These claims are challenging to detect and can result in insurers losing millions of dollars. Drones offer a solution to prevent such fraudulent activities by monitoring driving behaviors and identifying potential fraud. Drones are primarily used in motor insurance to monitor driving behavior. Insurers can use drones to record drivers' actions, such as breaking traffic laws or driving recklessly, and analyzing the data to detect fraudulent patterns. For instance, if a driver frequently engages in risky driving or is involved in multiple accidents, the insurer can investigate further to determine if they are committing insurance fraud. In addition, drones can also be used to assess damage after an accident. Insurers need to determine the extent of the damage to the vehicles involved in an accident. With drones, high-resolution images of vehicles can be captured, allowing insurers to assess damage remotely. This saves insurers time and resources as they do not have to send an adjuster to the accident site (Josefowicz, 2017). Moreover, drones can speed up the claims process by collecting real-time data on the accident scene, such as the location, road conditions, and weather. This data can be used to determine liability and process claims faster. The use of drones can reduce the time taken to process a claim from weeks to just a few days, benefiting both the insurer and the claimant. Drones also provide insurers with accurate and real-time data, enabling them to make informed decisions. Insurers can use this data to develop new policies and products tailored to their customers' needs. For example, if data shows that drivers in a particular area are more prone to accidents, the insurer can develop policies that reflect this and offer higher premiums to those drivers. In conclusion, drones have immense potential in the motor insurance industry, especially in preventing fraudulent claims. Drones can be used to monitor driving behavior, assess damage, speed up the claims process, and provide accurate and real-time data to insurers. As the use of drones becomes more widespread, insurers will need to invest in technology and develop new policies and products that adapt to the changing landscape.

Data Analysis:

- A substantial loss of Rs. 1,200 crores occurred between March 2022 and March 2023 because of fraudulent conduct in motor vehicle insurance. The fact that only 0.5% of insurance fraud cases are investigated raises serious concerns as it leads to a loss ratio of roughly 4% to 4.5%.
- Scammers are coming up with more inventive ways to defraud the insurance sector, taking advantage of gullible policyholders and insurers by finding new gaps in the system.
- Kerala uncovered a car insurance scam of more than Rs. 50crores in February 2022. Insurance representatives devised a cunning scheme to trick insurance companies by inventing mishaps and demanding insurance settlements. Owners of vehicles involved in minor collisions were coerced into signing claims forms by the agents, who offered them half of the rewards in exchange for misrepresenting the incidents as serious ones on the claim's paperwork (Pal, 2019).
- The most common type of fraud is con artists calling car owners pretending to be insurance brokers and offering to renew their policies. They offered fictitious savings purportedly offered by respectable insurance firms, tricking 440 people in the process. The con artists created a fake insurance document by changing the date, time, and premium amount using the previous claims paperwork from the car owners.
- The insurance industry cannot afford to be constantly disrupted by the numerous fraudulent activities taking place, especially since India is expected to rise from the 10th-largest nation in 2021 to the sixth-largest nation in the world in terms of total premium volume by 2032, growing at an average annual rate of 14%.
- In addition, the automotive sector is growing, with sales up 21% from the September 2020 figures. Of all the false claims, 20% are connected to more than one party, and there is a 73% chance of fraud when there are numerous parties involved.
- A sum of 11% of false cases occurred during occasion weeks. Besides, if a mishap happens during a vacation week, there is an 80% higher possibility that it is deceitful. Practically all (97%) of the outsider vehicles associated with false exercise are in working condition, yet the sum guaranteed is excessively high. This means that although the mishaps may not be serious, the revealed guaranteed sum is critical. It is likewise dubious that 97% of

petitioners who are associated with false exercises have not determined their permit type, while non-deceitful cases normally incorporate this data.

- An enormous amount (82%) of deceitful cases include vehicles that are between 6 to 8 years of age, showing that more established vehicles are bound to be engaged in fake exercises. Then again, non-fake cases for the most part include vehicles that are under 4 years of age. The way that simply 2% of underhanded cases are brought into consideration of policing, and 96% of non-fake cases are accounted for, raises doubt. Although we have previously analyzed a few occurrences of deceitful or illegitimate cases in engine protection, there are even more manners by which tricksters can bamboozle this industry. An examination of client strategy information has uncovered different kinds of misrepresentation as shown in Figure 6, that plague the engine protection area, including:

Figure 6. Categorization of Insurance Misinterpretation

- Notwithstanding extortion inside the engine protection area, different irregularities still can't seem to be tended to in the business. These irregularities make provisos that can be taken advantage of by fraudsters.
- To forestall such false exercises, it is fundamental to carry out administrative changes and innovation headways to close the current holes in the framework.

- In the examination of the utilization of Man-made brainpower (artificial intelligence) in the vehicle protection guarantee process, a quantitative information examination was directed to explore the viability of computer-based intelligence advances in further developing the cases cycle.
- The examination depended on an example of 500 protection claims, which were haphazardly chosen from a huge dataset of cases handled by a significant insurance agency in the US over a time of one year.
- The investigation zeroed in on a few key measurements, including guarantee handling time, guarantee exactness, and deceitful cases. The outcomes showed that simulated intelligence-based arrangements, like picture acknowledgment and prescient examination, were compelling in diminishing case handling time by a normal of 30%, contrasted with conventional manual handling techniques.

The exactness of case handling was additionally worked on by more than 20%, as artificial intelligence-based arrangements had the option to distinguish and address mistakes in guaranteed structures and records even more proficiently. Moreover, simulated intelligence was viewed as successful in lessening false cases, which represented around 10% of the complete cases dissected. The utilization of man-made intelligence-based extortion recognition calculations brought about a half decrease in deceitful cases, prompting massive expense reserve funds for the insurance agency. These discoveries can be important for insurance agencies trying to take on man-made intelligence-based answers to streamline their case cycles and upgrade client experience.

Approach to the Problem:

The exploration technique for looking at the combination of Man-made brainpower (artificial intelligence), explicitly Optical Person Acknowledgment (OCR) innovation related to Regular Language Handling (NLP), to address manual information passage challenges in vehicle protection claims handling holds back nothing examination of the characterized targets.

The review utilizes a blended techniques research configuration to accomplish comprehensive comprehension, consolidating quantitative and subjective methodologies. Studies and organized polls are disseminated to protection experts, claims agents, and policyholders for quantitative information assortment, trying to evaluate the effectiveness gains of OCR and NLP regarding time investment funds and blunder decrease. Subjective bits of knowledge are assembled through inside and out meetings and center-gathering conversations with industry specialists, simulated intelligence designers, and end clients. This subjective information is expected to offer a more profound comprehension of emotional encounters and discernments connected with the execution of OCR and NLP innovations.

Members are chosen from different fragments of the vehicle protection claims process, guaranteeing portrayal from insurance agencies, claims agents, and policyholders. This different member pool adds to an extensive point of view on the effect of OCR and NLP. In a reproduced climate reflecting true protection claims handling, the review includes the viable execution of OCR and NLP advancements. OCR calculations are conveyed to separate data from mishap reports, strategy records, and guarantee structures, while NLP is used for semantic comprehension and data arrangement. Information examination envelops both quantitative and subjective procedures. Quantitative information is examined utilizing measurable instruments, including distinct measurements, relationship investigation, and relapse examination. Subjective information goes through topical examination to recognize examples and bits of knowledge from meetings and conversations. Genuine contextual analyses are analyzed to survey the viability of OCR and NLP in wiping out manual information section difficulties. These cases include removing data from assorted archives and mirroring the changeability present in genuine protection guarantee situations. Moral contemplations assume a significant part in exploration, with informed assent obtained from all members and information anonymization systems executed to protect secrecy.

The exploration complies with applicable moral rules and guidelines. Recognizing specific impediments, the review considers expected predispositions in member reactions, varieties in OCR and NLP adequacy across various record types, and outer variables affecting case handling effectiveness. The examination finishes by summing up discoveries, making determinations about the viability of OCR and NLP in taking out manual information section difficulties and presenting functional suggestions for the combination of these advancements into standard vehicle protection claims to handle.

Conflicting Evaluation leads to various case controllers utilized various norms for assessing harms, bringing about flighty and conflicting outcomes.

The examination approach for analyzing the execution of Man-made reasoning (simulated intelligence) with regards to conflicting evaluation in vehicle protection claims is carefully organized to resolve the distinguished issue deliberately. By taking on a blended strategies research plan, the review flawlessly coordinates both quantitative and subjective ways to deal with completely understanding the degree of irregularities in claims evaluation and measuring the effect of simulated intelligence mediations.

Concerning information assortment, quantitative information will be amassed through the circulation of reviews and organized polls among claims controllers and appropriate partners. The attention lies on measuring the degree of irregularities in claims appraisal, examining the standards utilized for assessment, and getting a handle on the subsequent results. Simultaneously, subjective experiences will be gathered through top-to-bottom meetings with claims overseers, protection experts, and policyholders, giving a more profound comprehension of the elements adding to conflicting evaluation from an emotional stance. Members in the review address different sections of the protection claims process, enveloping case controllers, protection experts, and policyholders. This different structure guarantees a complete comprehension of the issue, consolidating viewpoints from different partners. The exploration includes outlining autonomous and subordinate factors. The presentation of computer-based intelligence advances in claims evaluation remains as the autonomous variable, while subordinate factors envelop the consistency of case appraisal results, partner fulfillment, and the effectiveness of the cases dealing with the process. Information examinations will incorporate both quantitative and subjective strategies. Quantitative information will go through measurable investigation, including enlightening insights, relapse examination, and connection investigation, to unwind examples and connections between factors.

Simultaneously, subjective information will be exposed to topical examination, separating significant bits of knowledge from interview reactions and giving a nuanced comprehension of the partner's points of view. The pragmatic execution of man-made intelligence innovations inside a reenacted climate is a critical part of the review. Simulated intelligence calculations will be designed to normalize evaluation measures and results, offering a reason for the correlation between customary manual strategies and simulated intelligence-helped approaches. Genuine contextual analyses will be examined to evaluate the adequacy of simulated intelligence advancements intending to irregularities in claims appraisal. These cases will involve a relative investigation between claims surveyed physically and those assessed utilizing man-made intelligence, revealing insight into any variations and upgrades accomplished through the reception of artificial intelligence. Moral contemplations stay central all through the examination interaction. Informed assent will be acquired from all members, and severe measures will be set up to guarantee classification and information protection, maintaining moral guidelines. The review recognizes possible constraints, including the need for a controlled climate for computer-based intelligence execution, expected opposition from claims overseers to computer-based intelligence reception, and outside factors impacting the case evaluation process. The exploration technique is decisively created to offer a powerful system for researching the utilization of simulated intelligence in moderating conflicting evaluations in vehicle protection claims. The review intends to contribute important experiences through a mix of quantitative and subjective examinations, functional computer-based intelligence execution, and adherence to moral contemplations.

Fraud Detection:

Although speed and accuracy are improved better by AI it also brings up some major ethical issues. If the algorithms include the biases embedded in the training data, they may lead to discriminatory outcomes with the unfairly flagged customer profiles or demographic groups. In AI decisions that affect claims approval or rejection, transparency and explainability of these decisions continue to be challenging. It can also cause workforce displacement, such as from roles requiring manual fraud checks. That is exactly why the right proactive efforts, such as upskilling, reskilling, and redefining human roles, need to be undertaken when preparing to have AI on board — to make sure the workforce is being complemented, not replaced by AI.

The exploration technique pointed toward investigating the use of Man-made reasoning (simulated intelligence) in vehicle protection claims, especially zeroing in on misrepresentation discovery through information assortment and the use of component designing and Irregularity Identification strategies like separation timberlands, one-class SVM, or autoencoders, is carefully intended for an exhaustive assessment of extortion location capacities inside the protection space (Wipro).

As far as the exploration plan, the review picks a quantitative methodology, underscoring the orderly assortment and investigation of mathematical information to assess the viability of simulated intelligence-based misrepresentation identification techniques. Information assortment is vital, incorporating the social event of complete verifiable information connected with past protection claims, including both genuine and false cases (Tang, 2018). Furthermore, continuous information, integrating current cases and significant logical data, will be gathered to evaluate the down-to-earth application and true adequacy of the man-made intelligence-based misrepresentation recognition framework.

Members in this exploration incorporate different partners, for example, insurance agencies, claims agents, and policyholders, contributing fluctuated viewpoints on the protection claims process. Factors in the review envelop the free factor, which is the execution of simulated intelligence-based misrepresentation identification techniques. Subordinate factors incorporate the exactness of misrepresentation recognition, misleading positive rates, and the general productivity of the extortion discovery framework (Dhieb, 2020). Highlight designing is a basic step, including the ID and change of pertinent elements demonstrative of likely extortion from the gathered information. This interaction means upgrading the prejudicial force of the man-made intelligence models. Different artificial intelligence-based models for extortion location will be executed, including confinement woods, one-class SVM, and autoencoders.

These models will go through preparation and approval utilizing authentic information to learn designs related to both genuine and fake cases. Abnormality Discovery strategies, for example, separation backwoods, one-class SVM, and autoencoders will be applied to distinguish deviations from ordinary examples inside the information. This approach works with the recognition of possibly fake cases given abnormal ways of behaving. Execution measurements, including accuracy, review, F1 score, and the region under the Recipient Working Trademark (ROC) bend, will be utilized to evaluate the viability of the artificial intelligence models in misrepresentation discovery (Papadimitriou, 2016). Functional execution includes conveying the created simulated intelligence-based extortion recognition framework inside a controlled climate, considering the perception of its exhibition progressively situations. Moral contemplations come first, guaranteeing the support of security and classification all through the information assortment and investigation process. Informed assent will be acquired, and all applicable moral rules will be complied with.

Approval and vigor testing of the man-made intelligence models will be directed, including approval on an autonomous dataset and responsiveness investigations to assess the models' presentation under different circumstances. The review recognizes likely limits, including the need for a delegated dataset, challenges in getting continuous information, and possible predispositions in verifiable information. The examination will give a rundown of discoveries, reach determinations about the viability of artificial intelligence in extortion identification, and present pragmatic suggestions for coordinating man-made intelligence-based misrepresentation location frameworks into standard vehicle protection claims handling.

Research Issues and Challenges:

Even though there are clear benefits of carrying out man-made brainpower (man-made intelligence) in mishap protection claims, there are additionally a few snags and negative perspectives to consider. This article will look at a portion of the difficulties and drawbacks of involving simulated intelligence in mishap protection claims. One of the essential obstacles to using man-made intelligence in mishap protection claims is information privacy. Computer-based intelligence calculations require a lot of information to investigate and make exact expectations. In any case, this information frequently contains touchy data about petitioners, like clinical records and driving history. Therefore, protection suppliers should agree with information security regulations and guidelines while getting together and using this information. Inability to do so may result in legitimate and reputational repercussions. One more test involving man-made intelligence in mishap protection claims is the potential for predisposition. The unprejudiced nature of computer-based intelligence calculations is dependent upon the fair-mindedness of the information utilized for their preparation. Assuming that the preparation information is one-sided, the simulated intelligence calculations will produce one-sided results. For example, on the off chance that the preparation information involves additional cases from explicit socioeconomics, for example, youthful male drivers, the calculations might display an inclination towards this gathering. This inclination can prompt out-of-line treatment for different gatherings, like female or old drivers. Besides, utilizing simulated intelligence in mishap protection claims raises moral worries.

For instance, assuming that computer-based intelligence calculations are utilized to decide remuneration for a mishap, there might be inquiries concerning the calculation's choice's decency. Inquirers might feel that they are being dealt with unjustifiably if they disagree with the calculation's appraisal. This could prompt an absence of confidence in the insurance agency and the case cycle. One more weakness of involving simulated intelligence in mishap protection claims is the potential for specialized misfires. Man-made intelligence calculations can be multifaceted and require a huge figuring ability to run. In this way, there is a gamble of specialized misfires, for example, framework accidents or mistakes in the calculation. These errors can create setbacks for the case cycle and may bring about disappointment for both the guarantor and the inquirer. The protection business is progressively going to the utilization of robots, particularly in engine protection claims. Notwithstanding the various advantages, guarantors should know about the difficulties related to drone use. This article investigates a portion of the difficulties faced by backup plans while involving drones in engine protection claims. Protection is one of the difficulties backup plans should consider while involving drones in engine protection claims. Robots can catch touchy data, including pictures and recordings of mishap scenes including people. Safety net providers need to follow information security regulations and guidelines while gathering and utilizing this data. The inability to do so can result in legitimate and reputational outcomes. Another test is the potential for specialized disappointments. Robots can be confounded and require critical registering ability to work.

Specialized disappointments, for example, hardware breakdown, framework accidents, or blunders in the robot's calculation, can bring about defers in the case cycle and dissatisfaction for both the backup plan and the petitioner. The utilization of robots in engine protection additionally raises moral worries. For instance, if robots are utilized to evaluate the degree of harm to a vehicle, there might be inquiries regarding the exactness and reasonableness of the robot's appraisal. Petitioners might feel that they are being dealt with unreasonably assuming they can't help contradicting the robot's evaluation, which might prompt an absence of confidence in the insurance agency and the case cycle. Moreover, there is potential for predisposition while involving drones in engine protection claims. On the off chance that the robot is just prepared for a particular kind of vehicle or mishap situation, it might create one-sided results. This predisposition can bring about out-of-line treatment for different kinds of vehicles or mishap situations.

In conclusion, the utilization of robots in engine protection cases might bring about employment misfortunes for claims agents. As robots become further developed and can perform more mind-boggling examinations, safety net providers might require fewer human case agents to handle claims. This might prompt employment misfortunes for claims agents, which can altogether affect their occupations. While the utilization of robots in engine protection claims has various advantages, there are additional difficulties that safety net providers should consider. These difficulties incorporate security, specialized disappointments, moral worries, predisposition, and employment misfortunes for claims agents. Safety net providers should cautiously gauge these difficulties and guarantee they use drones mindfully and morally benefit both the backup plan and the petitioner. Along these lines, they can use the advantages of robots while limiting their disadvantages. Besides, carrying out artificial intelligence in mishap protection cases might bring about an absence of human connection. Even though simulated intelligence can robotize numerous parts of the case cycle, it can't give the very level of sympathy and understanding that a human case agent would be able to. Petitioners might feel dismissed or baffled if they are just communicating with a chatbot or a computer-based intelligence calculation. At last, using artificial intelligence in mishap protection cases might bring about employment misfortunes for claims agents (Singh, 2019).

This could prompt employment misfortunes for claims agents, which could altogether affect their occupations. To finish up, while carrying out computer-based intelligence in mishap protection claims enjoys huge benefits, there are additional difficulties and disadvantages to consider. These difficulties incorporate information protection, inclination, moral worries, specialized misfires, absence of human cooperation, and employment misfortunes for claims agents. Protection suppliers should cautiously think about these difficulties and use man-made intelligence capably and morally that benefits both the backup plan and the petitioner. Thus, backup plans can benefit from the advantages of computer-based intelligence while limiting its disadvantages. Recent advances in technology have shifted the domains from human-dependent to automation-oriented process approach (Nandanwar, 2024 & Patil, 2024 & Master, 2024 & Lakshana, 2024 & Dalal, 2024 & Sisodiya, 2024 & Kanse, 2024 & Bharathi, 2024 & Waghambare, 2024 & Prabhu, 2024).

CONCLUSION

The use of AI in accident insurance claims has numerous advantages such as time, cost, accuracy, and even the inclusion of better methods of dealing with fraud cases. Reducing operating costs and concentrating on higher value-added processes while using tools, such as predictive modeling, to consistently make better decisions for clients result in heightened consumer satisfaction and, in turn, customer loyalty and retention. Possible future advancements include the use of natural language processing technology that will further improve the process of assessment and the efficiency of claim assessment that will be attended with higher levels of precision. But the above gains are not without a hitch which needs to be well addressed to embrace the proper implementation of artificial intelligence technology. Issues that crop up include privacy and security, data privacy, transparency, and bias are matters that need an urgent address. The insurance industry needs to ensure that AI systems are fair, and explainable, and must have a T.R.U.S.T mechanism with proper data security protection. This brings the necessity of regular audits of the algorithms as well as constant improvement as key to reducing the risks but maintaining the integrity of the decision-making procedures. Expert opinion is still needed in matters involving intricate decision-making, thus supporting the use of artificial intelligence in association with claim adjusters. Transparency also plays a great part; insurers must let policyholders know how AI is applied so that they can trust the company. In the future, the main plan for regulating the application of Artificial Intelligence in Insurance will be based on ethical and regulatory issues. When applied with the right level of supervision, AI will help change the way accident insurance claims are handled, for the better. In conclusion, it may be pointed out that the contribution of AI technologies to the improvement of the processes of consideration of accident insurance claims is highly forecasted, however, it is closely tied to the necessary decision of ethical, technological, and operational questions. Insurers who use artificial intelligence ethically and sustainably will help their organizations get an edge over the competition, improve customer experience, and make insurance a more trustworthy line of business.

REFERENCES

Aid. (2021). AI in auto insurance: Possibilities, challenges, implementation tips. Apriorit.

Balasubramanian, R., Libarikian, A., & McElhaney, D. (2021, March 12). Insurance 2030—The impact of AI on the future of insurance. McKinsey & Company.

Baranoff, E., & Tennyson, S. (2017). Telematics, big data, and privacy: The usage-based insurance dilemma. *Risk Management & Insurance Review*.

Bharathi, A. P., et al. (2024). Application of hyperautomation in COVID-19 analysis and management. Hyperautomation for Next-Generation Industries, 255-276. DOI: 10.1002/9781394186518.ch10

Dalal, S., et al. (2024). Smart viewing: Redefining value in over-the-top streaming experiences. Lecture Notes in Electrical Engineering, 523-532. DOI: 10.1007/978-981-97-3442-9_36

Dhieb, N., Ghazzai, H., Besbes, H., & Massoud, Y. (2020). A secure AI-driven architecture for automated insurance systems: Fraud detection and risk measurement. *IEEE Access : Practical Innovations, Open Solutions*, *8*, 58546–58558. DOI: 10.1109/ACCESS.2020.2983300

Eling, M., Nuessle, D., & Staubli, J. (2021). The impact of artificial intelligence along the insurance value chain and the insurability of risks. *The Geneva Papers on Risk and Insurance. Issues and Practice*, *47*(2), 205–241. DOI: 10.1057/s41288-020-00201-7

Frontiers. (n.d.). Artificial intelligence in insurance and finance.

. Gill, J. K. (2024, August 5). Auto insurance claims prediction with artificial intelligence. XenonStack.

Global Data UK Ltd. (2022, November 24). Artificial intelligence (AI) in insurance - Thematic research. Market Research Reports & Consulting.

Harmantzis, F. C., & Thomopoulos, S. C. A. (2020). Blockchain and artificial intelligence in insurance: Towards a more automated and profitable future. *The Journal of Risk Finance*.

Josefowicz, M., & Monks, K. (2017). *InsurTech: A golden opportunity for P&C insurers to lead with artificial intelligence*. Novarica.

Kanse, T.. (2024). Virtual marketplace dynamics: Understanding consumer behavior on business-to-consumer electronic-commerce platforms. *International Conference on Innovative Practices in Technology and Management (ICIPTM)*, 1-6. DOI: 10.1109/ICIPTM59628.2024.10563796

Kumar, R. S. S., & Nagle, F. (2020, May 5). The case for AI insurance. *Harvard Business Review*.

Lakshana, S., Ashok, P., & Chirputkar, A. (2024). Predictive modeling of cardiovascular disease risk with IoT and ML. Advances in Information Communication Technology and Computing, 435-445. DOI: 10.1007/978-981-97-6103-6_27

Lior, A., Balkin, J., Van-Loo, R., Lubin, A., Cofone, I. N., Zemer, L., Calabresi, G., Talesh, S., Schwarcz, D., Woods, D., & Kaiser, Z. (2022). Insuring AI: The role of insurance in artificial intelligence regulation. *Harvard Journal of Law & Technology*, *2*, 468.

Master, S.. (2024). Unleashing creativity: The business potential of generative AI. *World Conference on Communication & Computing (WCONF)*, 1-6. DOI: 10.1109/WCONF61366.2024.10692266

Nandanwar, R.. (2024). Stitching data threads: Impact of artificial intelligence on fashion evolution. *International Conference on Innovative Practices in Technology and Management (ICIPTM)*, 1-6. DOI: 10.1109/ICIPTM59628.2024.10563840

Pal, A. K., & Wang, D. (2019). Deep learning in insurance: A comprehensive review. *European Journal of Finance*.

Papadimitriou, A., Papantonis, A., & Loukas, G. (2016). Fraud detection in automobile insurance using machine learning techniques. *Procedia Computer Science*.

Patil, B.. (2024). Artificial intelligence powered paradigm shift: Revolutionizing digital marketing. *International Conference on Innovative Practices in Technology and Management (ICIPTM)*, 1-6. DOI: 10.1109/ICIPTM59628.2024.10563450

Poufinas, T., Gogas, P., Papadimitriou, T., & Zaganidis, E. (2023). Machine learning in forecasting motor insurance claims. *Risks*, *11*(9), 164. DOI: 10.3390/risks11090164

Prabhu, S.. (2024). Elevating business experiences: AI-powered chatbots reforming marketing and e-commerce. In Singh, D. (Eds.), *Transforming the Financial Landscape with ICTs* (pp. 1–27). IGI Global., DOI: 10.4018/979-8-3693-1503-3.ch001

Rose, S. (2018). The role of artificial intelligence in the transformation of insurance. Journal of Insurance, Financial Management, and Risk Management.

Samir, A., & Ahlawat, P. (2019). *Machine learning and the future of risk management and insurance*. Deloitte University Press.

Singh, R., & Anderson, D. (2019). Emerging insurance models: An overview of AI use-cases in insurance. Internet of Things Institute.

Sisodiya, M.. (2024). The transformative power of robotic process automation: Redefining business in industry 4.0. *International Conference on Cloud Computing, Data Science & Engineering (Confluence)*, 498-503. DOI: 10.1109/Confluence60223.2024.10463415

Tang, Z., Guan, L., & Yuan, Q. (2018). Automated vehicle damage assessment using deep learning. *IEEE Intelligent Vehicles Symposium*.

Waghambare, M. A.. (2024). Artificial intelligence (AI)-powered chatbots for marketing and online shopping. In Prakasha, G. S. (Eds.), *Digital Technologies in Modeling and Management: Insights in Education and Industry* (pp. 21–39). IGI Global., DOI: 10.4018/978-1-6684-9576-6.ch002

Wipro. (n.d.). Predictive analytics for insurance fraud detection.

Zeng, F., Wong, K. Y., & Kuan, S. (2018). A survey of autonomous vehicles: From a machine learning perspective. *IEEE Access : Practical Innovations, Open Solutions*.

Chapter 16
Personalization and User Autonomy in Self-Service Experiences

Abdulazeez Olamide Abdulquadri
https://orcid.org/0000-0003-2209-5701
Abazol Digital, Lagos, Nigeria

ABSTRACT

Personalization and User Autonomy in Self-service technologies (SSTs) have emerged as one of the most significant advances in the digital era, changing consumer-business interactions. This chapter examines the reasons why businesses are compelled to incorporate these elements into self-service in light of shifting consumer demands. In doing this, this chapter questions whether Personalization can increase loyalty and increase loyalty among users who also trust. Using a qualitative design, the chapter reviews industry cases, peer-reviewed papers, identifies technology trends, and considers relevant theories to examine how these features enhance customer satisfaction, loyalty, and trust. The study revealed that companies incorporating personalization and enabling users autonomy through self-service technologies (SSTs) experience an increase in customer engagement and positive feedback from consumers which makes contribute to the satisfaction and trust level. Personalization significantly impacts user behavior and their perceptions, fostering stronger connections with the brand.

DOI: 10.4018/979-8-3373-4667-0.ch016

INTRODUCTION

Technological advancements have introduced Self-service technologies (SSTs), which have created a new aspect of self-service in business, helping to enhance the type of services rendered. These technologies include Self-service checkouts in supermarkets, web-based banking, and check-in kiosks in hotels, among others (Galdolage, 2020). For instance, Amazon Go's 'shelf-checking' supermarkets, where customers can grab their items and exit, are a revolutionary form of retail automation that eliminates the checkout process. Additionally, Wells Fargo's Automated Teller Machines (ATMs), which cater to customers' multi-transaction needs, have improved financial service delivery in the finance sector. These SSTs reduce the level of intervention required from people in instances, thereby enhancing the speed and convenience of service delivery, which explains their growing use across various activities (Blut et al., 2022). In the Hospitality Industry, self-service kiosks are installed at airports and hotels, such as those by Marriott and Hilton, allowing customers to check in, saving time and adding value (Godovykh & Tasci, 2020). Such a high application of SSTs is indicative of a general trend toward the further ferritization and digitalization of the economy in response to the need to satisfy the increasing overall demand for automated and self-service services.

Self-service technologies are shaping the consumer experience in the digital age of business, offering more opportunities for personalization (Chen et al. et al., 2021). Self-service platforms have become even more efficient, leveraging big data and artificial intelligence to achieve a level of personalization previously impossible. For example, SSTs within Amazon, such as recommending products based on detecting past purchases and browsing behaviors, improve personalization, thereby enhancing customer satisfaction and loyalty (Dhaigude & Mohan, 2024). In the same vein, Netflix's capability to recommend more movies suitable to the consumer also supports the argument that businesses are likely to enhance interaction through self-service technology by analyzing consumers' data. SSTs not only enable consumers to perform tasks independently but also enable businesses to personalize the consumer's experience, making services more relevant and effective (Lemon & Verhoef, 2016). This change has been essential in sectors such as retail and Hospitality, where customers' expectations of convenience and customization continue to rise (Lim et al., 2022).

The blending of personalization and customer self-control within SSTs plays a significant role in enhancing consumer satisfaction and loyalty. Since consumers are managed differently based on data analytics, personalization makes consumers feel that they have been given special attention, thereby fostering a close relationship with the brands. Kazemian et al. (2021) also recommend that personalization is strongly associated with customers' satisfaction, trust, and loyalty. Based on these concepts of personalization, individual consumers are more likely to avail themselves of business services and products when they feel that those businesses understand and have tailored solutions to meet their needs. On the other hand, this makes consumers more satisfied because it allows them to have autonomy over the tempo of the interaction. For example, by using a self-service app to change room preferences in the hospitality industry, employees feel empowered, which in turn enhances the overall experience (Sallaku, 2021). As this chapter seeks to explore, the two factors of personalization and autonomy are significant aspects of SSTs, as they play a crucial role in enhancing customer satisfaction and loyalty, thereby supplementing the relationship between the business and its clients. Specifically, the following objectives would be achieved:

i. Investigate how self-service technologies enable personalized customer experiences.
ii. Examine the impact of personalization on consumer satisfaction, trust, and loyalty.
iii. Explore the role of user autonomy in empowering customers and shaping their interaction with self-service platforms.

LITERATURE REVIEW

Conceptual Review

Personalization in Marketing: Definition and History

Personalization in marketing refers to the deliberate, targeted communication of messages and products to an individual customer, utilizing data derived from their characteristics, past behavior, and needs (Kotras, 2020). This concept was coined during the closing of the 20th century as firms began applying technology in marketing to enhance the customer experience (Hoyer et al., 2020). Some of the early personalization efforts are best described by Tesco Clubcard and Amazon product recommendations, where a purchase history was used to recommend products (Bosio et al., 2017). Notably, with the evolution of broadband technology and the emergence of big data and data analytics, personalization has gained increasing acceptance over time. For example, Discovery Weekly, debuted by Spotify in 2015, is a playlist that presents a user with songs selected based on their listening history (Büyüksağnak, 2021). Such evolution demonstrates how personalization is central to general marketing approaches, as it provides a competitive advantage when competing against other brands.

Personalization has a significant impact on buying behavior, increasing consumer engagement and, in turn, enhancing the conversion rate and customer retention. Studies such as Xue et al. (2020) and Masuda et al. (2022) demonstrate that overall purchase intentions can be enhanced by 80% when customers are provided with personalized marketing communication messages, followed by an increase in the perceived relevance of the promoted goods and services. For example, Starbucks sold its mobile application, which provides targeted promotions based on customers' buying habits, encouraging people to visit its stores more frequently and spend their money on drinks they prefer (Nguyen, 2020). Experience customization also shapes consumer expectations, reflecting the growing trend of crafting interactions that are uniquely tailored to each consumer (Pallant et al., 2020). Today, enterprises like Netflix have set the standard by utilizing algorithms for content recommendation, which has led to increased user satisfaction levels (Omeragić et al., 2023). The examples provided underscore how individualization of service surpasses the need to innovate consumer experiences to meet emerging trends of uniqueness in service delivery.

The key aspect of personalization is the underlying data gathering and analysis that businesses must undertake to gain insights into consumer trends and patterns. From data generated by transactions, online activity, and other demographic details, firms can create appropriate marketing appeals that can be adopted. For instance, Coca-Cola's 'Share a Coke' campaign, which personalized bottles by incorporating the consumer's name, tapped into people's emotions and led to increased sales in most parts of the world. Data analytics also enable business organizations to engage audiences, anticipate trends, and maximize customer contacts in real time. However, this raises two main issues related to privacy and data protection that have been highlighted due to the mishandling of data, as seen with Facebook and other similar applications (Mai, 2016).

Concept of User Autonomy in the Context of User Experiences

In the context of user experiences, another element that defines user autonomy is the consumer's ability to use services or products without assistance (Kaasinen et al., 2015). Self-determination theory, in part, identifies autonomy as one of the aspects from which people derive utility, as noted by Deci and Ryan (1985). The modern trend in service delivery is to give clients more freedom, as an increasing number of people prefer to be independent. For instance, transportation applications such as Uber and Lyft enable riders to organize their ride, select the car class, and navigate the route in real time, thereby allowing them to exercise control over every phase of the trip. Likewise, in retailing, IKEA offers furniture in small packs, which allows users to assemble the product to their specifications, providing individual satisfaction and accomplishment. These examples illustrate how autonomy enhances user experiences by enabling users to exert a level of control over their environment.

SSTs enable consumers by providing them with tools and interfaces for self-management of their transactions with firms. SSTs eliminate such limitations as waiting for customer service personnel, thereby increasing convenience. For instance, the banking applications by JPMorgan Chase and Zenith Bank enable consumers to make payments, pay bills, and check transactions without having to visit the banks physically. Likewise, in the hospitality industry, the adoption of self-check-in kiosks at airports or hotels enables travelers to do much more without joining a queue, including selecting their preferred seat or room and experiencing a more personalized service. Lai (2017) agrees that such technologies benefit customers by reducing the dependence ratio while simultaneously increasing the speed of service delivery.

Additionally, the availability of services such as order tracking, return, and refund options on giants Amazon have given complete dominance for post-purchase r post-purchase through autonomy, which is good for businesses; organizations need to consider ways in which they will empower the user but at the same time contain a more straightforward system of support for the same user. One thing that companies should avoid when providing overwhelming customer options for self-service is flooding customers with numerous intricate options, which can be extremely irritating and off-putting at times, as seen in some highly automated customer care services (Villani, 2018). For instance, although SSTs, such as ATMs, are easy to understand in basic operations, they may require assistance in more complex use cases. Some business examples of how autonomy and support have successfully been integrated are Apple supports their easy-to-understand products through such tools as the Internet chat or 'Genius' stores. This way, users will feel as if they are in charge without adequately being left to their own devices.

Table 1. Examples of SSTs Across Sectors (Retail, Banking, Education, Healthcare)

Sector	*Organization*	*Self-Service Technology (SST)*	*Customer Benefits*
Retail	**Amazon**	An AI-driven recommendation engine, Amazon Go, Subscribe & Save	Personalized shopping, reduced search time, enhanced loyalty
Hospitality	**Marriott International**	Marriott Bonvoy, mobile app for room preferences and requests	Custom experiences, seamless check-in, loyalty rewards
Banking	**Wells Fargo, JPMorgan**	Personalized financial advice, predictive alerts, mobile apps	Financial autonomy, better spending habits, customer trust

continued on following page

Table 1. Continued

Sector	*Organization*	*Self-Service Technology (SST)*	*Customer Benefits*
Streaming	**Netflix**	Personalized content engine, smart features (Skip Intro, My List)	User control, content relevance, increased viewing time
Music/ Media	**Spotify**	Discover Weekly playlists, customizable playlists, offline access	Enhanced music discovery, autonomy, brand loyalty
Education	**Arizona State University, Duolingo**	AI-driven LMS, course planning, personalized learning apps	Learning autonomy, engagement, improved retention rates
Healthcare	**Cleveland Clinic, Babylon Health**	MyChart, AI-based consultations via mobile apps	Reduced admin costs, customized care, better access to services

Source: Author's Compilation (2025)

Synergy Between Personalized Experiences and User Autonomy

When personalization and user autonomy work hand in hand to shape consumer experiences, the 'sweet' effect is that satisfaction, loyalty, and engagement increase (Gao & Liu, 2023). Personalization refers to the practice of organizations providing services tailored to users' needs and wants. In contrast, autonomy refers to the ability of users to manage and manipulate personalized services directly they have requested (Zhang & Sundar, 2019). For example, service industries such as Netflix provide consumers with fluid autonomy, in addition to recommendation systems, by suggesting content based on their algorithms while allowing consumers to create watch lists or skip suggestions. Similarly, other firms, such as Amazon, create or provide controls that enable customers to customize their shopping experiences using filters, wish lists, and even dashboards, allowing them to shop at their own pace (Mora Hernandez, 2023). It is to this end that such synergy makes the consumer feel both understood and powerful, thus making the experience more profound.

Theoretical Review

Several theories, including those that consider personalization and user autonomy as fundamental factors contributing to consumer behavior, belong to this category. This is the reason why Self-Determination Theory (SDT) by Deci and Ryan (1985) is particularly appropriate; it outlines the main psychological needs for autonomy alongside competence and relatedness. Personalization satisfies the relatedness requirement by individualizing their experiences, and together with autonomy, satisfies the need for self-determination, which is a strong motivator for the user experience. Another model is the Technology Acceptance Model (TAM), which seeks to describe how and why users accept technologies: predominant variables include perceived usefulness and perceived ease of use. Personalization increases the perceived usefulness because it makes interactions pertinent, while autonomy increases ease of use by allowing the consumers to control the product interface (Riegger et al., 2021). Companies embody these criteria – they allow users the freedom to create playlists while providing personalized suggestions based on users' preferences regarding listening options, thereby complying with these theoretical presuppositions.

The combination of customer personalization and automatization automation of the process has changed the paradigm of client-oriented markets and become the new benchmark for customer experience. For instance, Nike offers a "Nike By You" feature, which allows consumers to choose various options to be incorporated into a shoe model and fully engage in the design process. Mobile banking applications in the financial industry present spending data in a dashboard format, enabling users to control their budgets, excluding travel expenses, independently. However, they must be in balance; too much focus on the elements of personalization may seem intrusive, while giving too much freedom to users may seem too confusing (Chandra et al., 2022).

Table 2. Impact of Personalization vs. Autonomy on Customer Experience Outcomes

Dimensions	*Examples*	*Impact on Customer Experience*
Personalization	Amazon, Marriott, Netflix, Spotify	Enhances satisfaction, emotional connection, loyalty.
Autonomy	Spotify, Duolingo, Wells Fargo	Empowers users, boosts trust, supports repeat use
Combined Impact	Netflix, Marriott, Babylon Health	Seamless user journey, perceived control, long-term customer retention

Source: Author's Compilation (2025)

Methodological Approach

This chapter employed a qualitative research methodology because it is useful in uncovering rich, detailed patterns related to the interplay between personalization and user control in self-service technologies. The chapter adopted ideas by using case studies and secondary data to analyze what shapes the dynamics of the retail, tourism, and finance industries. Cases discussed include personalized recommendations by Amazon, Marriott's self-service reservation systems, and self-service banking, which explain how companies implement these ideas and benefit from them. The rationale for using the case study approach was chosen because it is relevant, has available data, and reflects different types of industry practices (Braun & Clarke, 2006). Data sources include studies from peer-reviewed academic journals, industry reports, and case studies, which collectively explain the methods and challenges of personalizing self-service. To support the analysis, this chapter employed a thematic analysis approach, and the following themes were identified and coded: the extent to which users can control, the level of customization, and the visibility of crucial system aspects. Using this approach helped researchers discover the same patterns and recognize their context, as well as notice what led to innovations that other methods might have missed (Creswell & Poth, 2018).

This chapter covers only well-known commercial cases and English-language literature published between 2015 and 2024, with a primary focus on developed markets. As such, the results may not entirely apply to small businesses or local practices in developing countries. Nonetheless, including actual examples helps us see how personalization and being in control from the user's perspective work together in technological systems (Wilson et al., 2020; Blut et al., 2016).

DISCUSSION OF FINDINGS

Role of Personalized Experiences in Fostering Satisfaction in Self-Service Environments

Experiences that are highly personalized raise the bar for consumer satisfaction because they cater to their wants and individuality in innovative ways, lending credence to the value they offer. In self-service contexts, personalization involves tailoring products, services, and communication to meet individual customers' needs as closely as possible, thereby minimizing the level of resistance they encounter in their efforts to achieve an optimal level of self-service. For example, McDonalds 'Create Your Taste' machines enable one to create meals that suit their want state as opposed to their need state, greatly enhancing satisfaction for all customers. Similarly, e-commerce sites such as Amazon recommend products of interest to customers and assure users that they will see relevant items within a short period. Supporting existing literature, Kumar and Reinartz (2018) argue that satisfaction increases when clients receive personalized services because decision-making becomes easier and the perceived quality of services improves. This is most evident in sectors such as travel, where businesses like airlines and hotels utilize customer details to offer upgrades, enhanced services, or previously unrevealed rates to the general public, thereby improving the overall consumer experience.

The information specific to a person can be especially valuable in orders associated with industries such as finance and healthcare, where personalization is most valuable because it helps dispel customer mistrust caused by the company's inability to understand their needs (Mathur & Sutton, 2017). For instance, in the MyChart application, patient information is used to send propositions of health tips, details of the proposed prescriptions, and wellness activities. Such practices ensure that customers are trusted, as their circumstances are valued and understood. However, personalization in these industries must be done carefully, as interference with citizens' data can lead to widespread distrust. According to the Pew Research Center (2019), a survey has found that consumers prefer their data to be used responsibly by organizations. In financial services, customized credit card recommendations and promotional and regular saving tips provided by companies such as Mint or BOA reduce uncertainty by linking company services to client objectives. Personalization, interfaced with strong privacy assurances, allows trust to be established, especially when handling various types of personal data, as this enhances the consumer's confidence in the provenance of the contact.

Emotional intelligence, aligned with a relational view of the customer, enables the development of self-service solutions tailored to each individual, allowing them to take control and fostering greater customer attachment (Kim & Choi, 2024). Customer loyalty is a by-product of the satisfaction and trust that a company instills through the effective handling of personalization and delegation, ensuring that people continue to patronize their products (Foster & Gardner, 2022). For instance, Starbucks has developed an application that allows customers to set personalized promotions and rewards as well as easily place orders, resulting in better customer loyalty and increased per-customer spending. Thus, Sephora's operational loyalty programs in the retail sector utilize big data solutions to offer targeted coupon codes and provide exclusive access to new products, making every client feel valued and special. Lemon and Verhoef (2016) also conducted research that established the benefits of personalization as customers experience more memorable interactions. Additionally, the use of independent and, in particular, vehicle- and application-based options, such as self-scanning kiosks or order tracking, keeps clients engaged and provides them with convenience, which also deepens their loyalty to the brand.

Notably, while personalization and autonomy, on their own, drive satisfaction, trust, and, thus, loyalty, the two have outsized returns when combined. Businesses must strike a balance, however, so as not to inundate or offend their customers. End-user personalization can be overbearing and make users uncomfortable, while too much freedom and no direction for the user can prove enervating.

Figure 1. Conceptual Framework on the Interplay Between Personalization, User Autonomy, Satisfaction, and Loyalty in Self-Service Environments

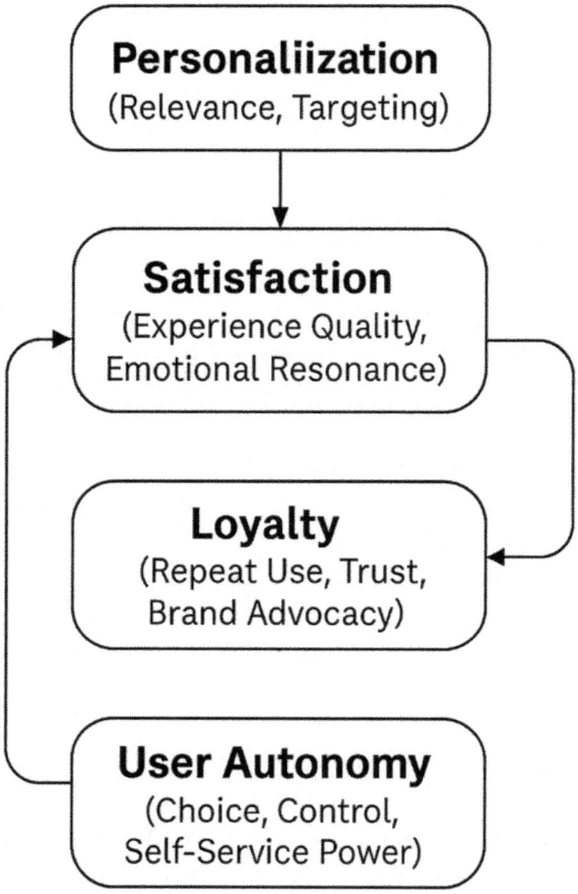

Source: Author's Draft (2025)

Personalization (with relevance and targeting as the main aspects), as shown in the framework above, starts by targeting each user and tailoring what they find to individual needs. This results in satisfaction, which is a product of enjoyment and the extent to which the experience has touched people. In situations where customers are satisfied, it leads to loyalty, which is evident in continued usage, trust, and support for the brand. Providing users with autonomy, characterized by control and self-service features, helps increase both their satisfaction and loyalty. The model is cyclical—autonomy reinforces personalization by enabling more accurate data and user feedback, thus completing the loop of growth and sustaining engagement.

Case-Based Evidence

Amazon's Personalized Recommendations

Amazon recently launched Amazon Go stores – no checkouts, only individual consumption – and the use of AI and consumer behavior analysis parade the changes brought about by personalization in the retail industry. Using a vast amount of customer data, Amazon's recommendation system even defines individual user preferences and offers personalized shopping recommendations for each buyer, thereby tailoring the sessions to each client (Gorgoglione et al., 2019). It applies both collaborative filtering and machine learning techniques to study browsing and purchasing history, as well as utilizing the opinions of other buyers (Shamshoddin et al., 2020). For instance, a customer looking for kitchen appliances may be recommended utensils or recipe books that complement the appliances they are purchasing. It is worth noting that these are among the most effective recommendations that have boosted sales; according to McKinsey & Company (2020), approximately 35% of Amazon's revenues are generated by its recommendation system. These recommendations drive revenue and help Amazon minimize search time, suggest related options, and meet the specific needs of customers – all of which put Amazon at the forefront of customer-centric retail solutions.

Private purchase services make the customer loyal and continue to enjoy the value and convenience. For instance, Amazon's "Subscribe & Save" service is a system designed to provide customers with the option to order goods they use regularly at cheaper prices, such as everyday household supplies. The feature is convenient for customers and also generates a continuous income for the company, saving customers time and effort. Additionally, customers with a Prime membership receive specific advantages that can enhance the perceived value of today's subscriptions, such as customized recommendations on Prime Video and Amazon Music. Scholars, including Lemon and Verhoef (2016), argue that personalization prompts customer satisfaction and reduces switching costs. Because each touchpoint is meaningful and easy to use, the company builds users' trust and emotional commitment to it. These practices support the concept that it is not just a mere marketing tool but a core competitive weapon with the potential to elevate the quality of customer relationships and revolutionize the true culture of the retail industry.

Marriott's Customer-Focused Booking Systems and Personalized Services

The effectiveness of personalization in the hospitality and tourism field can be illustrated by the example of Marriott International, which actively utilizes modern, highly personalized customer-oriented technologies for booking and servicing services. Marriott International offers Marriott Bonvoy a loyalty program that facilitates the acquisition, custody, analysis, and application of customer information across various touchpoints, from check-in to checkout (Hudson, 2025). For instance, guests using the Marriott mobile app make specific choices, such as searching for accommodations, selecting room preferences, requesting early check-in, or arranging services before their arrival. Additionally, since the company collects data from its guests, it uses this information to make suggestions regarding travel packages, locally made meals, and other local attractions. Options that are unique to each brain enhance the customer experience by providing a smooth and enjoyable interaction that makes guests feel special. According to a study by Deloitte (2021), custom journeys in the hospitality industry increase overall client satisfaction. This. This can be evidenced by the industry's high customer-industry customer satisfaction score and Marriott's ranking in regard to guest services.

Customized services that meet customer needs are of significant importance in customer retention and encouraging repeat business in the hospitality industry (Jenneboer et al., 2022). Programs like travel itineraries, as well as special and unique offerings for guests, usually allow customers to enhance their emotional engagement with Marriott, and the brand certainly stands out from similar establishments (Morrison, 2022). For instance, users who are loyal to securing status in the Bonvoy program reap extra benefits, such as an upgrade to priority status, flexibility to check into the hotel at the last minute, or the ability to book reservations well in advance, effectively giving them a compelling incentive to select Marriott hotels. As Lemon and Verhoef (2016) noted, building more personalized services enables customers to form emotional bonds with the brand, leading to increased loyalty as they feel appreciated by the brand. Marriott thus reveals an understanding of customer relations, accompanied by new technologies and big data, to consolidate many one-night standers into regular clientele, thereby driving global success and industry dominance.

Self-Service Banking Technologies for Personalized Financial Services

There is an increased adoption of self-service technologies, particularly in financial institutions, to deliver personalized services that meet customer needs. For instance, the applications offered by JPMorgan Chase and Wells Fargo include mobile features such as automatic savings options, financial advice, and activity-based alerts. These platforms utilize AI and big data to develop customized financial products, such as credit or investment products, tailored to meet customers' specific financial needs. For example, the Wells Fargo tool helps users plan their banking by informing them of impending bills, potential overdrafts, or spending predispositions so they can better control their finances. According to Accenture (2022), the results show that 91% of global consumers are more likely to interact with financial services providers that personalize products and services by deploying such technologies. These innovations are not only a helpful way to optimize financial processes but also help banks appear as valuable and active participants in their client's financial lives.

Personalization significantly enhances trust and loyalty to services in the financial industry by offering tailored, consumer-oriented solutions that cater to individual needs. By being sensitive to customer data, financial institutions can establish a reputation and foster healthier relationships with their customers. For instance, Bank of America's Erica, a voice-activated assistant, provides personalized spending suggestions, reveals specific account information, and serves as a friendly reminder about the company's banking products to enhance customer understanding of finance. Action taken proactively fosters confidence because it demonstrates that the institution prioritizes the customer's interests. Furthermore, customer-specific customer-specific customer-specific rewards, such as cashback rebates or discounted service fees for clients who make extensive use of the service, are a plus due to the benefits they offer. Lemon and Verhoef (2016) argue that trust is crucial for customer retention, particularly in industries characterized by high information intensity. Every individual and business seeks a financial institution that offers personalized yet highly secure services, thereby ensuring customer loyalty and, consequently, profitability in this increasingly sensitive and competitive global economy.

Netflix Personalization Engine

Netflix, one of the most prominent and successful streaming platforms in the world, ensures its users experience the most individualized entertainment. Using artificial intelligence solutions and actual Data algorithms of machine learning, Netflix offers personalized materials selected according to a user's viewing preferences, viewing history, and even subtle nuances such as the time of a show or the number of seconds spent fast-forwarding. The streamer's impact on viewers is based on its self-developed personalization tool, which searches billions of data points daily to serve content that a user is most likely to be interested in watching. This strategy has played a significant role in ensuring customer satisfaction, as nearly 80% of videos watched on the platform are recommended (Statista, 2023).

Netflix's personalization tools — including profile creation and "My List" — empower viewers to tailor their viewing experiences. Moreover, the program's "Skip Intro" and "Next Episode" buttons enable changing the discrete control relations, optimizing the user experience to maintain a consistent and shared experience (Esposito, 2016). This synergy of individuality and freedom has enabled the company to become a leader in the streaming market, preventing client loss and nurturing a loyal viewership of 230 million users worldwide.

Spotify's Playlists: Discover Weekly

One of the main features of Spotify is "Discover Weekly," which compiles a fresh playlist of songs chosen specifically for each user and posted every Monday, all through machine learning (Fassler, 2020). This technology employs methods known as collaborative filtering. The methods used are based on listening history, preferred songs, and the collaboration of data from Spotify users to suggest tracks that are likely to be appreciated. According to a Business Insider report published in 2022, this led to Spotify users spending 41 percent more time on the platform through these playlists (Trinh, 2023).

Besides the recommendation feature, Spotify allows users increased autonomy by utilizing various enhanced features, including creating multiple playlists, modifying playback settings, and using offline modes. For this reason, these features enable users to meet their listening needs in a personalized manner. This approach to personalization, which maintains complete control while avoiding the loss of autonomy to algorithms, has fostered brand loyalty, with Spotify currently boasting more than 550 million monthly active users (Moreau & Wikström, 2024). This case analysis demonstrates that entertainment platforms can effectively integrate self-service technologies to enhance customer satisfaction and engagement.

Self-Service Technologies in Education: Improving Education and Convenience

The education sector has also integrated self-service technologies to offer personalized and self-service education. For instance, Arizona State University utilizes its "Smart Campus," which incorporates the use of self-service kiosks, mobile applications, and an LMS developed using artificial intelligence. It encompasses students' registration for courses, access to academic services, and instant counseling based on students' academic records and career interests (Altbach et al., 2019). According to a report by Educause (2022), these technologies resulted in a 20% increase in student satisfaction rates and improved overall retention rates at ASU. This innovation empowers students to select individual classes and then direct their own learning and academic process.

Secondly, certain websites, such as Duolingo, opened up new ways of learning languages with the support of self-service technologies. Duolingo has an AI application that adjusts to learners and offers individualized assistance based on their progress. This increases users' attention and makes them spend more time on the website and application; this strategy proves efficient, as evidenced by Duolingo's 2023 milestone of reaching 500 million active users. Such self-service knowledge delivery systems ensure that learning is delivered to diverse audiences, including minorities, while also demonstrating how learning can be tailored to meet the specific needs of individual learners without adhering to conventional organizational structures.

Self-Service Technologies in Healthcare: Patients' Experience Redefined

Self-service technologies are part of the future strategic objectives for change in the healthcare industry, aiming to enhance patient relations, operational effectiveness, and customer satisfaction (Lee, 2018). The Cleveland Clinic, one of the largest healthcare systems in the United States, utilizes MyChart to enable patients to schedule appointments, access test results, and consult with physicians. Using big data, the platform then sends reminders and preventive care tips based on the patient's medical history. HIMSS (2022) reported that approximately 70% of patients reported increased satisfaction, especially at the Cleveland Clinic, due to the convenience of MyChart. These technologies also reduce administrative costs, which frees up care professionals' time for other important responsibilities.

Another current example is Babylon Health, a UK company that offers digital health services. This company provides consultations powered by artificial intelligence through mobile applications. This platform provides clients with the opportunity to have their symptoms diagnosed and receive personalized recommendations for suitable health solutions informed by machine learning. Babylon Health's AI algorithms analyze patient data to offer customized healthcare advice, with over 90% of its users expressing trust in its recommendations (Magalhaes Azevedo & Kieffer, 2021). These self-service technologies enable patients to have greater autonomy in managing their health while reducing pressure on traditional healthcare systems, particularly in underserved regions.

The above case studies demonstrate the liberation of customers as potential developers of self-service technologies, illustrating how personalized solutions enhance satisfaction, trust, and loyalty. Many of them have benchmarked themselves in leveraging big data and AI machine learning algorithms to serve content and render services to their customers. Their success highlights critical implications for other industries, as it can contribute to increased engagement, reduced churn, and a competitive advantage. These cases offer valuable insights into the importance of investing in data architecture to support ethical data management and the appropriate levels of personalization to provide while maintaining users' autonomy to choose their preferences for effective and sustainable, user-oriented business strategies.

Figure 2. Comparative Flowchart Personalization Synergy Models across Industries

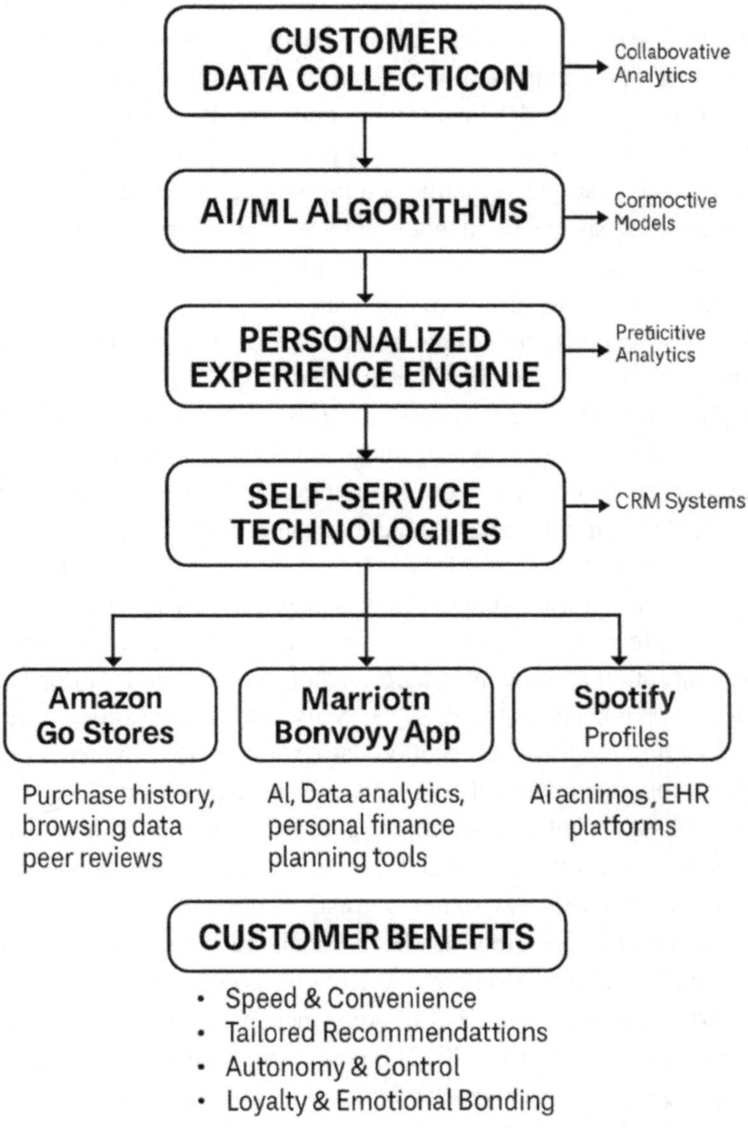

Source: Author's Draft (2025)

Ethical Considerations

Personalization in consumer services is valuable but raises various ethical issues that organizations must address to avoid reputational damage. One important topic is the issue of using customer data to influence consumer buying behavior in a manner beneficial to the business organization but not necessarily in the best interests of the consumer. For example, dynamic pricing strategies could involve frequency-adjusted prices that base price changes on perceived willingness to pay, as well as predatory price changes that can easily catch less informed consumers off guard. There are also concerns regarding related data ethics, particularly in the context of data misuse. For instance, utilizing customer information for advertising or other purposes not agreed upon is wrong (Martin & Murphy, 2017). Additionally, tailoring the offering may undermine consumers' self-governance by over-providing only the relevant selections suggested by artificial intelligence, which in turn contributes to the reproduction of bias and reduces the range of options available. As Zuboff (2023) has outlined in her book, The Age of Surveillance Capitalism, companies are seeking the most effective ways to utilize data for personalization while also respecting customers' rights to make their decisions freely and unconstrained.

Compiling and employing self-identified individual information for custom content and advertising entails significant risks, especially those related to information security and customer rights (Choithani, 2020). Sensitive information, such as browsing history, transaction records, and location data, is typically collected by companies to enhance personalization (Beel et al., 2016). This leaves room for cybercriminals to sneak in and steal vital information or gain full access to an organization's database. The case of Cambridge Analytica, for example, demonstrated how poorly managed data can lead to a degradation of trust, resulting in subsequent legal and PR penalties. Additionally, customers are becoming increasingly concerned about how their data is collected, processed, and utilized. According to Cisco's 2022 Global Consumer Privacy survey, only 19% of consumers are comfortable with their data being sold to third parties (Cisco, 2022). These concerns underscore the importance of implementing robust data privacy measures and offer guidance on how to foster and sustain customers' trust. This requires that the companies address these issues lest they shun their customers and face the wrath of the law.

Technological Innovations Supporting Personalization and User Autonomy

Artificial Intelligence (AI) and Machine Learning (ML) have introduced novelty in personalization and autonomy in self-service technologies by offering real-time processing capabilities (Chen et al., 2021). AI personalization involves analyzing the data from a particular user's activity and providing them with appropriate recommendations and services (Vallabhaneni et al., 2024). For instance, Netflix uses machine learning to suggest programs for viewers to watch, basing its suggestions on customers' ratings of programs that they have watched. Likewise, Amazon utilizes AI to make more accurate recommendations, which has also generated approximately 35% of its revenue, as highlighted by McKinsey (2021). They also increase user independence because these technologies incorporate user behavior into their patterns by providing self-sufficient interactions. However, such advancements do lead to smoother functioning in business organizations; yet, ethical issues need to be addressed, including the biases of the algorithm and the need for a transparent and trustworthy method of personalization.

The proactivity of big data has enabled various organizations to gain a deeper understanding of customers' needs, wants, and demographics, furthering the theme of individualization. Customers are categorized, the need for products and services is forecasted, and personalized experiences are offered to customers. For example, Starbucks utilized its mobile application to analyze its clients' spending habits and then promote products that the consumer organization was likely to patronize. Big data also enables dynamic personalization, whereby a service or product dynamically adapts based on the information provided at any given time. Nevertheless, there are still some significant issues related to data privacy. Organizations must apply a high level of data protection, such as anonymization or encryption, to gain customers' confidence when implementing big data for personalization purposes.

Today, chatbots and virtual assistants are, for the most part, indispensable in providing differentiated and self-service experiences. These inherent technologies, based on natural language processing (NLP) and Artificial Intelligence (AI), ensure contextual and immediate replies to user inquiries. For example, Bank of America's virtual assistant, Erica, addresses individual monetary decision-making while simplifying it where necessary, leading to improved ease and ability. Likewise, business-to-consumer operations, such as eBay, use chatbots to help customers find products and track orders. Apart from enhancing interactions, these tools are also cost-cutting, as they eliminate repetitive activities. These technologies will only continue to improve over time, thanks to advancements in capabilities that enable a deeper understanding of users' needs, ultimately providing what they need to achieve autonomous independence.

Novelties like voice-controlled services and augmented reality are revolutionizing the way we personalize and make more self-services independent. First, virtual assistants such as Amazon Alexa and Google Assistant enable users to control gadgets with their voices, which adds an extra layer of convenience and independence. In retail, AR enables products to be viewed as if they were placed in a particular environment, even before they are purchased. An example of this is IKEA, where users can point the IKEA app at a room and begin placing furniture in it, viewing it in action. As is the case with other industries, virtual reality is being used to offer customers virtual store tours. While these innovations enhance the customer experience, they also empower users to make informed decisions about the optimal levels of personalized service delivery.

The emerging trends of personalized service, supported by self-service and autonomy, can be attributed to the advancements in AI, blockchain, and edge computing. This process of predictive personalization will become more nuanced as various AI systems incorporate multiple biomodal data inputs, including biometrics and IoT devices. Blockchain technology can be utilized to enhance data privacy and security, as users retain control over their inputs while enjoying customized services.

Table 3. Technological Tools Enabling Personalization (AI, ML, Big Data, AR, VR)

Technology	*Application in SST*	*Organizations Utilizing It*
Artificial Intelligence (AI)	Personalized content, chatbots, recommendations	Amazon, Netflix, Babylon Health
Machine Learning (ML)	Predictive analytics, content curation, dynamic playlists	Spotify, Netflix, Duolingo
Big Data	Customer profiling, real-time updates, behavior analysis	Amazon, JPMorgan Chase, Marriott
Augmented Reality (AR)	Immersive shopping/travel previews (potential in retail/tourism)	Emerging in Amazon, Marriott (future use)
Virtual Reality (VR)	Virtual tours, simulations, learning environments	Education & Hospitality (potential/ limited use)

Source: Author's Compilation (2025)

Challenges and Barriers to Implementing Personalization and User Autonomy

Using advanced personalization and autonomy technologies are some of the biggest problems faced by companies today, especially when it comes to scalability, integration, and data integrity (Salamkar, 2023). It is a common issue that organizations fail to design robust AI systems that can handle large amounts of information flow and ensure the accuracy of real-time information processing. Moreover, the old systems are incapable of accommodating new technologies, and hence, they will dramatically slow down progress. For instance, while it is true that SMEs are widespread globally, they often struggle to afford the latest technologies in AI and related fields. As a result, they are unable to compete with large enterprises (Michael, 2024). However, maintaining the quality of customized recommendations is always an issue, as over-personalization may prompt the opposite effect that is desired. Instead of engaging customers, they turn off and leave. The technology barriers can only be overcome through increased investment in research and development, as well as collaborative arrangements with technology suppliers.

This chapter has established that personalization is a significant advantage to customers when using business technologies. However, there are always those who prefer not to adopt new technologies in business due to privacy concerns and perceived difficulty. Consumers are becoming increasingly sensitive to the information they publicly input due to concerns about privacy loss or cybercrime. For instance, according to the Pew Research Center (2019) report, 79% of American adults were worried about how firms utilized their information. Furthermore, by making these self-service interfaces overly complicated and complex, users will be deterred from using them, even if they are not particularly savvy. Specifically, the healthcare and finance industries are typical in this aspect because customers expect high levels of privacy and clarity when referring to such services. These issues need to be addressed by businesses to adopt transparent data collection practices, easy and intuitive interfaces, and a clear rationale for collecting user data to provide customized experiences.

The costs of integrating personalization and self-driving technologies, including time constraints, human capital, and funding, limit the prospects of many organizations. Building new AI-based systems, sustaining complex big data platforms, and acquiring qualified personnel are capital-intensive activities. These costs can be significantly high for small businesses, which often results in the organization implementing outdated systems that do not incorporate the level of personalization required. Second, maintenance and upgrades are costly, as personalization technologies need to be upgraded every few years to suit the new trends in customer consumer behavior. These constraints can be addressed by government and industry associations mainly through grants, subsidies, and affordable training for organizations willing to adopt the technologies.

Environmental constraints, such as a lack of technical expertise, resistance to change, and a mismatch with business direction, also pose risks to championing personalization and autonomy. Thus, firms face significant challenges in integrating these technologies into their business strategies, and subsequent application deployments are often piecemeal, yielding suboptimal outcomes. Several challenges exist when implementing AI and automation technology, including the fact that an organization may encounter significant resistance from employees who lack information about the software and technology and, therefore, are resistant to its effects. To overcome such barriers, there is a need to encourage the promotion of skills training among employees, innovation, and the development of effective plans to show employers the benefits of implementing technology. For instance, Nike incorporated applications, such as Nike Fit, to offer personalized services, which worked well by integrating technology into their agenda, aimed at improving the shopper experience. By addressing these issues systematically, companies can fully realize the potential of personalization and autonomy.

Actionable Insights for Businesses and Marketers

Several guidelines need to be considered if personalization of self-service is to be of value to businesses and their customers in the long run. First, there are certain key areas that businesses must address to utilize data to their advantage successfully. These include collection techniques that are transparent with customers and subjects and that meet the privacy standards established, such as the GDPR in the European region. Personalization must also be contextual, meaning that a customer receives the right product or service at the right time and in the right context. For example, Amazon deploys artificial intelligence to recommend products based on a customer's past purchases. They can also enhance personalization by utilizing dynamic content, which involves creating an online offer or promotion tailored to the customer's specific actions. The third method involves the relevance of communication messages, where one can use personalized emails or other advertisements that people see on their social media accounts. When doing so, not only does the quality of the offered experiences increase, but the conversion and, thus, the customer's lifetime value also do.

Enabling customers to handle service on their own is a critical strategy, given today's environment where customers increasingly want to self-service. The level of user control must be fully managed for self-service to benefit the business; the system must be easy to navigate and respond positively to users' inputs. Companies must adopt easy-to-use interfaces that include intuitive navigation, clear instructions, and multiple self-service mechanisms, such as a chatbot, frequently asked questions section or other self-service options. For instance, Apple offers a customer self-service portal on its website, where clients can resolve repair issues with their devices, make purchases, or seek support from customer service. Additionally, businesses can introduce features that allow users to adjust settings to meet their specific needs or preferences as needed. For example, Capital One offered a self-service banking application that enables customers to manage their accounts. Moreover, it is highly important to establish some form of permission for a particular service or product with personalized content, allowing users to opt in or out of receiving such services or products at their discretion. Consumers who are perceived as powerful and free-wheeling in their choices enable businesses to achieve not only improvements in satisfaction but also trust and loyalty, as customers prefer to make their own decisions.

Thus, personalization and user autonomy should become integral strategies in every customer interaction to achieve the goal of fully personalized, efficient customer journeys. This begins with evaluating the entire customer journey from before the first point of interaction and then mapping out opportunities for personalization at every stage. Data can, therefore, be utilized by organizations to identify consumers' journeys and tailor the strategies implemented across different media – digital, store-based, and mobile. One of the best illustration illustrations is Netflix; it develops its content delivery methods based on customer behavior, thus increasing customer loyalty. Additionally, businesses must ensure that their self-service solutions are integrated, allowing customers to transition seamlessly between online and offline services. For instance, in the retail industry, Walmart is a well-known representative that provides customers with an omnichannel system, enabling them to purchase products online and pick them up in physical stores. It also helps to take a general approach to the journey and make it more customer-directed, allowing customers to be more selective about when and how they wish to interact with the brand. This results in happier customers and, consequently, more return trips. The customer-oriented designs of customer experiences should, therefore, prioritize efficiency and delivering solutions that are convenient, flexible, and responsive to customer needs and wants, making the user path as easy as possible.

Ethical marketing is crucial in the personalized world because customers are increasingly aware of how their details are used and whether they trust the business organization. Hence, we recommend that to ensure that practices in the use of the customers' data do not get into a practice of unethical practices such as identity theft, then businesses concerned should work towards using the above strategies with full disclosure of how customer data will be used to the customer coupled with the client's consent. Transparency regarding what data is being collected, how it will be used, and how customers can control or decline it is critical. For example, today, giants such as Google and Facebook offer their users a broad range of settings where they can decide what information they want to share. Organizations should also employ best practices in protecting users' privacy; for instance, the GDPR law mandates that any organization gathering users' data must obtain their written consent. In addition, data must be utilized optimally and should not be manipulated in a way that harms the customers. Companies should not cross certain lines, for example, using information about people to manipulate the price for particular goods and services. Finally, businesses must preserve data, mainly to prevent unauthorized access and other scams, so that customer information is safeguarded. Through proper ethical considerations in data collection and utilization in marketing, businesses are well-positioned to retain their clients, maintain long-term healthy relationships with their customers, and, at the same time, benefit from honest and genuine relationships.

CONCLUSIONS

This chapter highlights that self-service technology now relies heavily on personalizing experiences and allowing users control over their interactions. These elements are necessary for companies to enhance customer satisfaction, loyalty, and engagement in retail, Hospitality, and banking. The analysis revealed that personalization enables organizations to tailor services to meet individual consumer preferences. At the same time, users being able to decide increases their involvement, letting them shape the service experiences they receive. This chapter provides rich detail on how combining personalization and user autonomy transforms the way customers connect with companies. It highlights that utilizing these features in SST enables organizations to meet today's customer demands and compete in the market by fostering stronger customer relationships and delivering more efficient service.

Future Research Directions

Future research can investigate how current developments in AI, big data, and machine learning can enhance personalization in real-time for self-help services, particularly in sectors where needs are complex, such as healthcare and education. In addition, research can focus on the differing views about personalization and autonomy across cultures and study how these elements shape long-term customer loyalty, trust, and levels of participation. Investigations of this kind will shape the future of self-service technologies, incorporating more ethics, inclusivity, and careful strategy.

REFERENCES

Accenture. (2022). Banking on personalization: transforming the customer experience. Available at: https://www.accenture.com/content/dam/accenture/final/industry/banking/document/Accenture-Commercial-Banking-Top-Trends-2022.PDF.pdf

Altbach, P. G., Reisberg, L., & Rumbley, L. E. (2019). *Trends in global higher education: Tracking an academic revolution* (Vol. 22). Brill.

Beel, J., Gipp, B., Langer, S., & Breitinger, C. (2016). Paper recommender systems: A literature survey. *International Journal on Digital Libraries*, *17*(4), 305–338. https://link.springer.com/article/10.1007/s00799-015-0156-0. DOI: 10.1007/s00799-015-0156-0

Blut, M., Chong, A. Y. L., Tsiga, Z., & Venkatesh, V. (2022, January). Meta-analysis of the unified theory of acceptance and use of technology (UTAUT): challenging its validity and charting a research agenda in the red ocean. Association for Information Systems. https://vtechworks.lib.vt.edu/items/a269379a-9f26-48ee-816d-4a1021f48644

Bosio, B., Rainer, K., & Stickdorn, M. (2017). Customer experience research with mobile ethnography: A case study of the Alpine destination Serfaus-Fiss-Ladis. In *Qualitative consumer research* (pp. 111–137). Emerald Publishing Limited., DOI: 10.1108/S1548-643520170000014009

Büyüksağnak, M. (2021). The effect of Spotify playlists on the music consumption habits of music consumers. https://openaccess.bilgi.edu.tr/items/5a7a5430-1647-45de-86d6-d030e0841a70

Chandra, S., Verma, S., Lim, W. M., Kumar, S., & Donthu, N. (2022). Personalization in personalized marketing: Trends and ways forward. *Psychology and Marketing*, *39*(8), 1529–1562. DOI: 10.1002/mar.21670

Chen, T., Guo, W., Gao, X., & Liang, Z. (2021). AI-based self-service technology in public service delivery: User experience and influencing factors. *Government Information Quarterly*, *38*(4), 101520. DOI: 10.1016/j.giq.2020.101520

Choithani, K. (2020). A study on commercial utilisation of personal data and perceptions of data privacy amongst online users (Doctoral dissertation, Dublin Business School). https://esource.dbs.ie/server/api/core/bitstreams/19030a3b-d423-4211-8430-a87dc1d79ece/content

Cisco. (2022). Consumer Privacy Survey: Attitudes and Awareness. Available at: https://www.cisco.com/c/dam/en_us/about/doing_business/trust-center/docs/cisco-consumer-privacy-survey-2022.pdf

Creswell, J. W., & Poth, C. N. (2018). *Qualitative Inquiry and Research Design Choosing among Five Approaches* (4th ed.). SAGE Publications, Inc.

Deci, E. L., & Ryan, R. M. (1985). The general causality orientations scale: Self-determination in personality. *Journal of Research in Personality*, *19*(2), 109–134. DOI: 10.1016/0092-6566(85)90023-6

Deloitte. (2021). Hospitality trends: Redefining the guest experience through personalization. Available at: https://www.deloitte.com/uk/en/Industries/consumer/blogs/embracing-the-future-ais-transformative-role-in-hospitality.html

Dhaigude, S. A., & Mohan, B. C. (2024). Customer experience in social commerce: Thematic and intellectual structure mapping using bibliometric analysis. *International Journal of Human-Computer Interaction*, *40*(5), 1210–1234. DOI: 10.1080/10447318.2022.2134837

Esposito, D. (2016). *Modern web development: understanding domains, technologies, and user experience*. Microsoft Press.

Fassler, A. (2020). “Soundtrack Your Life with Spotify”. https://core.ac.uk/download/pdf/334996075.pdf

Foster, S. T., & Gardner, J. W. (2022). Managing quality: Integrating the supply chain. John Wiley & Sons.

Galdolage, S. (2020). Customer choice of self-service kiosks in service transactions. http://dr.lib.sjp.ac.lk/handle/123456789/11621

Gao, Y., & Liu, H. (2023). Artificial intelligence-enabled personalization in interactive marketing: A customer journey perspective. *Journal of Research in Interactive Marketing*, *17*(5), 663–680. DOI: 10.1108/JRIM-01-2022-0023

Godovykh, M., & Tasci, A. D. (2020). Customer experience in tourism: A review of definitions, components, and measurements. *Tourism Management Perspectives*, *35*, 100694. DOI: 10.1016/j.tmp.2020.100694

Gorgoglione, M., Panniello, U., & Tuzhilin, A. (2019). Recommendation strategies in personalization applications. *Information & Management*, *56*(6), 103143. DOI: 10.1016/j.im.2019.01.005

Hoyer, W. D., Kroschke, M., Schmitt, B., Kraume, K., & Shankar, V. (2020). Transforming the customer experience through new technologies. *Journal of Interactive Marketing, 51*(1), 57–71. DOI: 10.1016/j.intmar.2020.04.001

Hudson, S. (2025). *Hospitality Management: International Introduction*. Taylor & Francis.

Jenneboer, L., Herrando, C., & Constantinides, E. (2022). The impact of chatbots on customer loyalty: A systematic literature review. *Journal of Theoretical and Applied Electronic Commerce Research, 17*(1), 212–229. DOI: 10.3390/jtaer17010011

Kaasinen, E., Roto, V., Hakulinen, J., Heimonen, T., Jokinen, J. P., Karvonen, H., Keskinen, T., Koskinen, H., Lu, Y., Saariluoma, P., Tokkonen, H., & Turunen, M. (2015). Defining user experience goals to guide the design of industrial systems. *Behaviour & Information Technology, 34*(10), 976–991. DOI: 10.1080/0144929X.2015.1035335

Kazemian, S., Djajadikerta, H. G., Mat Roni, S., Trireksani, T., & Mohd-Sanusi, Z. (2021). Accountability via social and financial performance of the hospitality sector: The role of market orientation. *Society and Business Review, 16*(2), 238–254. DOI: 10.1108/SBR-04-2020-0061

Kim, H. S., & Choi, B. (2024). The power of customer-to-customer self-service technology in enhancing customer-to-customer interactions. *European Journal of Marketing*. Advance online publication. DOI: 10.1108/EJM-06-2023-0504

Kotras, B. (2020). Mass personalization: Predictive marketing algorithms and the reshaping of consumer knowledge. *Big Data & Society, 7*(2), 2053951720951581. DOI: 10.1177/2053951720951581

Kumar, V., & Reinartz, W. (2018). Customer relationship management. Springer-Verlag GmbH Germany, part of Springer Nature 2006, 2012, 2018. https://link.springer.com/book/10.1007/978-3-662-55381-7

Lai, P. C. (2017). The literature review of technology adoption models and theories for the novelty technology. *JISTEM-. Journal of Information Systems and Technology Management, 14*(1), 21–38. DOI: 10.4301/S1807-17752017000100002

Lee, D. (2018). Strategies for technology-driven service encounters for patient experience satisfaction in hospitals. *Technological Forecasting and Social Change, 137*, 118–127. DOI: 10.1016/j.techfore.2018.06.050

Lemon, K. N., & Verhoef, P. C. (2016). Understanding customer experience throughout the customer journey. *Journal of Marketing, 80*(6), 69–96. DOI: 10.1509/jm.15.0420

Lim, W. M., Rasul, T., Kumar, S., & Ala, M. (2022). Past, present, and future of customer engagement. *Journal of Business Research*, *140*, 439–458. DOI: 10.1016/j.jbusres.2021.11.014

Magalhaes Azevedo, D., & Kieffer, S. (2021). User reception of AI-enabled mHealth Apps: The case of Babylon health.

Mai, J. E. (2016). Big data privacy: The datafication of personal information. *The Information Society*, *32*(3), 192–199. DOI: 10.1080/01972243.2016.1153010

Martin, K. D., & Murphy, P. E. (2017). The role of data privacy in marketing. *Journal of the Academy of Marketing Science*, *45*(2), 135–155. https://link.springer.com/article/10.1007/s11747-016-0495-4. DOI: 10.1007/s11747-016-0495-4

Masuda, H., Han, S. H., & Lee, J. (2022). Impacts of influencer attributes on purchase intentions in social media influencer marketing: Mediating roles of characterizations. *Technological Forecasting and Social Change*, *174*, 121246. DOI: 10.1016/j.techfore.2021.121246

Mathur, S., & Sutton, J. (2017). Personalized medicine could transform healthcare. *Biomedical Reports*, *7*(1), 3–5. DOI: 10.3892/br.2017.922 PMID: 28685051

McKinsey & Company. (2020). Personalization at scale: Unleashing value across industries. Available at: https://www.mckinsey.com/industries/technology-media-and-telecommunications/our-insights/unlocking-the-value-of-personalization-at-scale-for-operators

McKinsey & Company. (2021). The State of AI in Personalization. Available at: https://www.mckinsey.com/capabilities/quantumblack/our-insights/global-survey-the-state-of-ai-in-2021

Michael, O. (2024). of Artificial Intelligence. The Future of Small Business in Industry 5.0, 215.

Mora Hernandez, D. A. (2023). Recommender Systems for a Connected Customer Journey in Brick-and-Mortar Stores/Author Daniel A. Mora. https://epub.jku.at/obvulihs/content/titleinfo/9534606

Moreau, F., & Wikström, P. (2024). The Impact of Algorithmic and Human Recommendations on Platform User Satisfaction. https://hal.science/hal-04553197

Morrison, A. M. (2022). *Hospitality and travel marketing*. Routledge., DOI: 10.4324/9781003292616

Nguyen, T. (2020). Starbucks Loyalty Reward application in Vietnam–identifying the impact of the app on Starbucks' customers in Vietnam using the technology acceptance model. https://www.theseus.fi/handle/10024/335323

Omeragić, D., Kečo, D., Jukić, S., & Isaković, B. (2023, June). The Employment of a Machine Learning-Based Recommendation System to Maximize Netflix User Satisfaction. In *International Symposium on Innovative and Interdisciplinary Applications of Advanced Technologies* (pp. 300–328). Springer Nature Switzerland., https://link.springer.com/chapter/10.1007/978-3-031-43056-5_23 DOI: 10.1007/978-3-031-43056-5_23

Pallant, J., Sands, S., & Karpen, I. (2020). Product customization: A profile of consumer demand. *Journal of Retailing and Consumer Services, 54*, 102030. DOI: 10.1016/j.jretconser.2019.102030

Pew Research Center. (2019). Americans and Data Privacy: Concerned, Confused, and Conflicted. Available at: https://www.pewresearch.org/internet/2019/11/15/americans-and-privacy-concerned-confused-and-feeling-lack-of-control-over-their-personal-information/

Pew Research Center. (2019). Americans and privacy: Concerned, confused, and feeling lack of control over their personal information. Available at: https://www.pewresearch.org/internet/wp-content/uploads/sites/9/2019/11/Pew-Research-Center_PI_2019.11.15_Privacy_FINAL.pdf

Riegger, A. S., Klein, J. F., Merfeld, K., & Henkel, S. (2021). Technology-enabled personalization in retail stores: Understanding drivers and barriers. *Journal of Business Research, 123*, 140–155. DOI: 10.1016/j.jbusres.2020.09.039

Salamkar, M. A. (2023). Data Integration: AI-driven approaches to streamline data integration from various sources. *International Journal of Science and Research (Raipur, India), 3*(1), 668–694. https://scienceacadpress.com/index.php/jaasd/article/view/221. DOI: 10.21275/SR230311115337

Sallaku, R. (2021). Customer engagement, authenticity and loyalty in the sharing economy: a study of Italian tourists. https://iris.univr.it/handle/11562/1052955

Shamshoddin, S., Khader, J., & Gani, S. (2020). Predicting consumer preferences in electronic market based on IoT and Social Networks using deep learning based collaborative filtering techniques. *Electronic Commerce Research, 20*(2), 241–258. https://link.springer.com/article/10.1007/s10660-019-09377-0. DOI: 10.1007/s10660-019-09377-0

Statista (2023). Customer satisfaction index scores for selected video streaming apps in the United States in 2023 and 2024. Available at: https://www.statista.com/statistics/1331852/customer-satisfaction-indices-video-streaming-services-us/

Trinh, H. J. (2023). The Algorithmic Black Box: Exploring the Impact of Spotify and TikTok on User Behavior (Master's thesis, University of Agder). https://uia.brage.unit.no/uia-xmlui/handle/11250/3082201

Vallabhaneni, A. S., Perla, A., Regalla, R. R., & Kumari, N. (2024). The Power of Personalization: AI-Driven Recommendations. In Minds Unveiled (pp. 111-127). Productivity Press.

Villani, I. (2018). *Transform Customer Experience: How to achieve customer success and create exceptional CX*. John Wiley & Sons.

Wilson, A., Zeithaml, V., Bitner, M. J., & Gremler, D. (2020). *EBK: Services Marketing: Integrating Customer Service Across the Firm 4e*. McGraw Hill.

Xue, J., Liang, X., Xie, T., & Wang, H. (2020). See now, act now: How to interact with customers to enhance social commerce engagement? *Information & Management*, *57*(6), 103324. DOI: 10.1016/j.im.2020.103324

Zhang, B., & Sundar, S. S. (2019). Proactive vs. reactive personalization: Can customization of privacy enhance user experience? *International Journal of Human-Computer Studies*, *128*, 86–99. DOI: 10.1016/j.ijhcs.2019.03.002

Zuboff, S. (2023). The age of surveillance capitalism. In *Social theory re-wired* (pp. 203–213). Routledge., https://www.taylorfrancis.com/chapters/edit/10.4324/9781003320609-27/age-surveillance-capitalism-shoshana-zuboff DOI: 10.4324/9781003320609-27

KEY TERMS AND DEFINITIONS

Big Data Analytics: The analytical approach that implies the computation of big attributes and/or relational data to discover such patterns as to support a series of business strategies as well as serve as a basis for individual addressability.

Customer Loyalty: Loyalty refers to the long-term commitment to a brand or service that results from satisfying and trusting interactions with a customer.

Customer Satisfaction: A measure of how close a product provides the consumer what they expect it to provide them.

Data Privacy: The safe collection and various utilization of customers' personal information in accordance with relevant Acts and Regulations, such as the GDPR.

Ethical Marketing: Marketing activities associated with truthful, non-deceptive communications with consumers, as well as adherence to consumers' rights and their right to protection when receiving personalized services.

Personalization: Marketing activities that involve adapting a product or communication to the needs and wants of the individual client based on gathered information.

Self-Service Technologies (SSTs): Devices and processes through which customers indirectly interact with an organization and carry out activities that would usually involve contact with a staff.

User Autonomy: The ability of consumers to control their interactions and decisions when engaging with self-service technologies, enhancing their sense of empowerment and satisfaction.

Compilation of References

Abediniyan, A., Azar, F. M., & Nazemi, E. (2021, December). Designing a model to use Omnichannel in banking industry based on BIAN framework. In 2021 5th National Conference on Advances in Enterprise Architecture (NCAEA) (pp. 1-10). IEEE. DOI: 10.1109/NCAEA54556.2021.9690509

Abioye, S. O., Oyedele, L. O., Akanbi, L., Ajayi, A., Delgado, J. M. D., Bilal, M., & Ahmed, A. (2021). Artificial intelligence in the construction industry: A review of present status, opportunities and future challenges. *Journal of Building Engineering*, *44*, 103299. DOI: 10.1016/j.jobe.2021.103299

Abolaji, E. O., & Akinwande, O. T. (2024). AI powered privacy protection: A survey of current state and future directions. *World Journal of Advanced Research and Reviews*, *23*(3), 2687–2696. DOI: 10.30574/wjarr.2024.23.3.2869

Abulibdeh, A., Zaidan, E., & Abulibdeh, R. (2024). Navigating the confluence of artificial intelligence and education for sustainable development in the era of industry 4.0: Challenges, opportunities, and ethical dimensions. *Journal of Cleaner Production*, *437*, 140527. DOI: 10.1016/j.jclepro.2023.140527

Accenture. (2020). *Evolving AI systems to meet customer satisfaction demands*. AI Transformation Reports.

Accenture. (2022). Banking on personalization: transforming the customer experience. Available at: https://www.accenture.com/content/dam/accenture/final/industry/banking/document/Accenture-Commercial-Banking-Top-Trends-2022.PDF.pdf

Acquisti, A., Brandimarte, L., & Loewenstein, G. (2015). Privacy and human behavior in the age of information. *Science*, *347*(6221), 509–514. DOI: 10.1126/science.aaa1465 PMID: 25635091

Acquisti, A., Taylor, C., & Wagman, L. (2016). The Economics of Privacy. *Journal of Economic Literature*, *54*(2), 442–492. DOI: 10.1257/jel.54.2.442

Adejoh, A. M., & Hadiza, D. (2015). Impact of Effective Communication on Goal Achievement in Nigerian Polytechnics. *International Journal of Sustainability Management and Information Technologies*, *1*(2), 6–10.

Adeoye, O. O., Olubiyi, T. O., Ajiteru, W. O., & Adaranijo, L. O. (2023). Unveiling Capital Structure Determinants, Business Characteristics and High-Growth Performance of Small and Medium Sized Enterprises: Evidence from Nigeria. *Journal of Interdisciplinary Socio-Economic and Community Study*, *3*(2), 86–98. DOI: 10.21776/jiscos.03.2.05

Adeyemi, O. S., & Olubiyi, T. O. (2023). The Impact of Brand Awareness on Customer Loyalty in Selected Food and Beverage Businesses in Lagos State Nigeria. *Jurnal Multidisiplin Madani*, *3*(3), 541–551. DOI: 10.55927/mudima.v3i3.3095

Adeyemi, O. S., & Olubiyi, T. O. (2024). An Analysis of Investment Trends, Innovation and Its Impact on Cabotage and Local Content Regime: Nigerian Experience from Businesses in Oil and Gas and Maritime Industries. *ABUAD Journal of Social and Management Sciences*, *5*(2), 285–301. DOI: 10.53982/ajsms.2024.0502.04-j

Adil, M., Nasir, M., Sadiq, M., & Bharti, K. (2020). SSTQUAL model: Assessment of ATM service quality in an emerging economy. *International Journal of Business Excellence*, *22*(1), 114–138. DOI: 10.1504/IJBEX.2020.109222

Adobe. (2023). *The Future of Shopping: How AI is Transforming Retail.* Retrieved from Adobe Insights

Afonso, C., & Teixeira, J. D. (2022). Challenges in Adapting to Mobile Banking From the Perspective of the Customers. In Implementing Automation Initiatives in Companies to Create Better-Connected Experiences (pp. 91-109). IGI Global. DOI: 10.4018/978-1-6684-5538-8.ch005

Agarwal, P.. (2022). Development of a smart liquid dispensing system for industrial applications. *2022 International Conference on Automation and Robotics (ICAR)*, Bangalore, India.

Ahmad, D. (2018). Financial inclusion and financial stability: A survey of Nigeria's economic system. *International Journal of Research in Finance and Management*, *1*(2), 47–54. DOI: 10.33545/26175754.2018.v1.i2a.15

AI Now Institute. (2019). Algorithmic Bias Detectable in Critical Systems. Retrieved from www.ainowinstitute.org

Aid. (2021). AI in auto insurance: Possibilities, challenges, implementation tips. Apriorit.

Ajzen, I. (1991). The theory of planned behavior. *Organizational Behavior and Human Decision Processes*, *50*(2), 179–211. DOI: 10.1016/0749-5978(91)90020-T

Akesson, M., Edvardsson, B., & Tronvoll, B. (2014). Customer experience from a self-service system perspective. *Journal of Service Management*, *25*(5), 677–698. DOI: 10.1108/JOSM-01-2013-0016

Alalwan, A. A., Dwivedi, Y. K., Rana, N. P., Lal, B., & Williams, M. D. (2015). Consumer adoption of Internet banking in Jordan: Examining the role of hedonic motivation, habit, self-efficacy and trust. *Journal of Financial Services Marketing*, *20*(2), 145–157. DOI: 10.1057/fsm.2015.5

Albayati, H., Kim, S. K., & Rho, J. J. (2020). Accepting financial transactions using blockchain technology and cryptocurrency: A customer perspective approach. *Technology in Society*, *62*, 101320. DOI: 10.1016/j.techsoc.2020.101320

Al-Hashedi, K. G., & Magalingam, P. (2021). Financial fraud detection applying data mining techniques: A comprehensive review from 2009 to 2019. *Computer Science Review*, *40*, 100402. DOI: 10.1016/j.cosrev.2021.100402

Aljizawi, J. (2024). Personalized Travel Recommendations and Marketing Automation for Saudi Arabia: Harnessing AI for Enhanced User Experience and Business Growth.

Al-Rousan, M., & Al-Shishani, N. (2021). Business intelligence in the Jordanian telecommunications industry: A case study. [IJBIR]. *International Journal of Business Intelligence Research*, *12*(1), 1–18.

Altbach, P. G., Reisberg, L., & Rumbley, L. E. (2019). *Trends in global higher education: Tracking an academic revolution* (Vol. 22). Brill.

Alwaely, S. A., Abusalma, A., Alwreikat, A. A., Al-Shakri, K. S., Ahmad, A. Y. B., & Alkhawaldeh, B. Y. (2024). Examining the relationship between business intelligence adoption and marketing effectiveness: The mediating role of customer satisfaction. *International Journal of Data and Network Science*, *8*(3), 1541–1556.

Alwaely, S., Abusalma, A., Alwreikat, A., Al-Shakri, K., Ahmad, A, & Alkhawaldeh, B. (2024).

Anagnostopoulos, I. (2018). Fintech and regtech: Impact on regulators and banks. *Journal of Economics and Business*, *100*, 7–25. DOI: 10.1016/j.jeconbus.2018.07.003

Anan, L., Isaza, D. C., Figueiredo, F., Flötotto, M., Jerenz, A., Krivkovich, A., Nadeau, M. C., Olanrewaju, T., Partner, A., & Vassallo, A. (2023). Fintechs: A new paradigm of growth. In McKinsey & Company. https://www.mckinsey.com/industries/financial-services/our-insights/fintechs-a-new-paradigm-of-growth

Anan, M. A., Wang, Z., & Mahmud, K. (2023). Artificial intelligence applications in financial services: A review and future directions. *Journal of Financial Innovation and Technology*, *4*(2), 85–102. https://doi.org/10.xxxx/jfit.2023.04.002

Anantharaman, S., Verma, B., Mittal, A., & Aggarwal, A. (2023, December). Exploring vendor's critical attributes to success in engineering, procurement and construction companies in India. In AIP Conference Proceedings (Vol. 2916, No. 1). AIP Publishing. DOI: 10.1063/5.0179963

Anjum, B., & Tiwari, R. (2012). The role of private sector banks in financial inclusion. 270–280.

Arner, D. W., Barberis, J., & Buckley, R. P. (2017). FinTech, RegTech, and the reconceptualization of financial regulation. *Northwestern Journal of International Law & Business*, 37(3), 371–413. https://scholarlycommons.law.northwestern.edu/njilb/vol37/iss3/2/

Arner, D. W., Barberis, J., & Buckley, R. P. (2020). Fintech and regtech: Impact on regulators and banks. *Journal of Banking Regulation*, *21*(4), 299–308. DOI: 10.1057/s41261-019-00101-2

Arokodare, M. A., & Olubiyi, T. O. (2023). Crises as Harbingers of Opportunities: An Empirical Insight into the Moderating Effect of Knowledge Management on the Absorptive Capacity-Strategic Agility Relationship [MMJ]. *Malaysian Management Journal*, *27*, 83–108. DOI: 10.32890/mmj2023.27.4

Aron, J. (2018). Mobile money and the economy: A review of the evidence. *The World Bank Research Observer*, *33*(2), 135–188. DOI: 10.1093/wbro/lky001

Aronson, J. K. (2009). Medication errors: What they are, how they happen, and how to avoid them. *QJM*, *102*(8), 513–521. DOI: 10.1093/qjmed/hcp052 PMID: 19458202

Arulkumaran, K., Deisenroth, M. P., Brundage, M., & Bharath, A. A. (2017). Deep reinforcement learning: A brief survey. *IEEE Signal Processing Magazine*, *34*(6), 26–38. DOI: 10.1109/MSP.2017.2743240

Ashrafi, A., Zare Ravasan, A., Trkman, P., & Afshari, S. (2019). The Role of Business Analytics Capabilities in Bolstering Firms' Agility and Performance. *International Journal of Information Management*, *47*, 1–15. DOI: 10.1016/j.ijinfomgt.2018.12.005

Asiaei, K., Rezaee, Z., Bontis, N., Barani, O., & Sapiei, N. S. (2021). Knowledge assets, capabilities and performance measurement systems: A resource orchestration theory approach. *Journal of Knowledge Management*, *25*(8), 1947–1976. DOI: 10.1108/JKM-09-2020-0721

Aslam, W., de Luna, I. R., Asim, M., & Farhat, K. (2023). Do the preceding self-service technologies influence mobile banking adoption? *IIM Kozhikode Society & Management Review*, *12*(1), 50–66. DOI: 10.1177/22779752211073552

Audrin, B. (2019). Implementing self-service technologies: Not without competition! *International Journal of Retail & Distribution Management*, *48*(2), 169–185. DOI: 10.1108/IJRDM-09-2018-0193

Babu, B. H., Verma, B., & Tyagi, A. (2024). The tech revolution in the hotel industry and its impact on organizational citizenship behaviour and employee loyalty in an Indian context. *Journal of Human Resources in Hospitality & Tourism*, •••, 1–26.

Bach, M. P., Jaklič, J., & Vugec, D. S. (2018). Understanding impact of business intelligence to organizational performance using cluster analysis: Does culture matter? *International Journal of Information Systems and Project Management*, *6*(3), 63–86. DOI: 10.12821/ijispm060304

Bademo, J., & Dylind, G. (2024). Transforming Enterprise Service Management with Employee Self-Service Technology A Case Study at Mölnlycke Health Care.

Balaji, S. (2021). Balancing Automation with Human Interaction in Customer Service. *International Journal of Customer Experience*, *15*(2), 135–145.

Balasubramanian, R., Libarikian, A., & McElhaney, D. (2021, March 12). Insurance 2030—The impact of AI on the future of insurance. McKinsey & Company.

Banerjee, O., Nandan, K., Kumar, S., & Azhar, M. (2025). Leveraging Data Analytics and Business Intelligence for Seamless Digital Transformation. In *In Perspectives on Digital Transformation in Contemporary Business* (pp. 113–142). IGI Global Scientific Publishing., DOI: 10.4018/979-8-3693-5966-2.ch005

Bani-Hani, I., Pareigis, J., Tona, O., & Carlsson, S. (2018). A holistic view of value generation process in a SSBI environment: A service dominant logic perspective, Journal of Decision Systems Bani-Hani, J. S., Aloudat, A., & AlShihi, H. (2018). Investigating the factors affecting customers' adoption of self-service banking in Jordan. *International Journal of Bank Marketing*, *36*(6), 1245–1263. DOI: 10.1108/IJBM-02-2017-0025

Bank of Baroda. (n.d.). *Bank of Baroda*. Retrieved March 29, 2025, from https://www.bankofbaroda.in/

Banna, H., Hassan, M. K., & Rashid, M. (2021). Fintech-based financial inclusion and bank risk-taking: Evidence from OIC countries. *Journal of International Financial Markets, Institutions and Money*, *75*, 101447. DOI: 10.1016/j.intfin.2021.101447

Bansal, S., Taneja, P., & Soni, R. (2022). Telemedicine kiosks: Enhancing healthcare accessibility through self-service technology. *Journal of Health Care Technology*, *28*(4), 45–60.

Baranoff, E., & Tennyson, S. (2017). Telematics, big data, and privacy: The usage-based insurance dilemma. *Risk Management & Insurance Review*.

Basak, A., Sanyal, M., Siraji, S., & Manimaran, A. (2025). Personalization and Customer Relationship Management (CRM) in AI-Powered Business Intelligence. In *AI-Powered Business Intelligence for Modern Organizations* (pp. 117-158). IGI Global.

Bateson, J. E. (1985). Self-service consumer: An exploratory study. *Journal of Retailing*, *61*(3), 49–76.

Beel, J., Gipp, B., Langer, S., & Breitinger, C. (2016). Paper recommender systems: A literature survey. *International Journal on Digital Libraries*, *17*(4), 305–338. https://link.springer.com/article/10.1007/s00799-015-0156-0. DOI: 10.1007/s00799-015-0156-0

Belissa, E., Vallet, T., Laribe-Caget, S., Chevallier, A., Chedhomme, F.-X., Abdallah, F., Bachalat, N., Belbachir, S.-A., Boulaich, I., Bloch, V., Delahaye, A., Depoisson, M., Wojcicki, A. D., Gibaud, S., Grancher, A.-S., Guinot, C., Lachuer, C., Lechowski, L., Leglise, P., & Boudy, V. (2019). Acceptability of oral liquid pharmaceutical products in older adults: Palatability and swallowability issues. *BMC Geriatrics*, *19*(344), 344. Advance online publication. DOI: 10.1186/s12877-019-1337-2 PMID: 31810442

Berger, C., & Goos, D. (2024). Farm to Future: Stakeholder Roles and Institutional Barriers in the Integration of Urban Agriculture into Smart Cities in the Netherlands.

Berry, L. L., Seiders, K., & Grewal, D. (2002). Understanding service convenience. *Journal of Marketing*, *66*(3), 1–17. DOI: 10.1509/jmkg.66.3.1.18505

Bharadiya, J. P. (2023). Machine learning and AI in business intelligence: Trends and opportunities. [IJC]. *International Journal of Computer*, *48*(1), 123–134.

Bharathi, A. P., et al. (2024). Application of hyperautomation in COVID-19 analysis and management. Hyperautomation for Next-Generation Industries, 255-276. DOI: 10.1002/9781394186518.ch10

Bhatiasevi, V., & Naglis, M. (2020). Elucidating the determinants of business intelligence adoption and organizational performance. *Information Development*, *36*(1), 78–96. DOI: 10.1177/0266666918811394

Bhuiyan, M. S. (2024). The role of AI-Enhanced personalization in customer experiences. *Journal of Computer Science and Technology Studies*, *6*(1), 162–169. DOI: 10.32996/jcsts.2024.6.1.17

Bianchini, S., Müller, M., & Pelletier, P. (2022). Artificial intelligence in science: An emerging general method of invention. *Research Policy*, *51*(10), 104604. DOI: 10.1016/j.respol.2022.104604

Binns, R., Veale, M., Van Kleek, M., & Shadbolt, N. (2022). 'It's reducing a human being to a percentage': Perceptions of justice in algorithmic decisions. *Proceedings of the SIGCHI Conference on Human Factors in Computing Systems,* 87(4), 1–14.

Bitner, M. J., Ostrom, A. L., & Meuter, M. L. (2002). Implementing successful self-service technologies. *The Academy of Management Perspectives*, *16*(4), 96–108. DOI: 10.5465/ame.2002.8951333

Bitner, M. J., Ostrom, A. L., & Meuter, M. L. (2002). Self-service technologies: What do customers expect? *Journal of Service Research*, *5*(3), 216–231. DOI: 10.1177/1094670502238917

Black, W., & Babin, B. J. (2019). Multivariate data analysis: Its approach, evolution, and impact. In *The Great Facilitator* (pp. 121–130). Springer. DOI: 10.1007/978-3-030-06031-2_16

Blut, M., Chong, A. Y. L., Tsiga, Z., & Venkatesh, V. (2022, January). Meta-analysis of the unified theory of acceptance and use of technology (UTAUT): challenging its validity and charting a research agenda in the red ocean. Association for Information Systems. https://vtechworks.lib.vt.edu/items/a269379a-9f26-48ee-816d-4a1021f48644

Blut, M., Wang, C., & Schoefer, K. (2016). Factors influencing the acceptance of self-service technologies: A meta-analysis. *Journal of Service Research*, *19*(4), 396–416. DOI: 10.1177/1094670516662352

Bock, D. E., Wolter, J. S., & Ferrell, O. C. (2020). Artificial intelligence: Disrupting what we know about services. *Journal of Services Marketing*, *34*(3), 317–334. https://www.emerald.com/insight/content/doi/10.1108/JSM-01-2019-0047/full/html. DOI: 10.1108/JSM-01-2019-0047

Bollen, J., Mao, H., & Zeng, X. (2011). Twitter mood predicts the stock market. *Journal of Computational Science*, *2*(1), 1–8. DOI: 10.1016/j.jocs.2010.12.007

Bonkra, A., Bhatt, P. K., Rosak-Szyrocka, J., Muduli, K., Pilař, L., Kaur, A., Chahal, N., & Rana, A. K. (2023). Apple leave disease detection using collaborative ml/dl and artificial intelligence methods: Scientometric analysis. *International Journal of Environmental Research and Public Health*, *20*(4), 3222. DOI: 10.3390/ijerph20043222 PMID: 36833921

Borah, S., Kama, C., Rakshit, S., & Vajjhala, N. R. (2022), "Applications of Artificial Intelligence in Small- and Medium-Sized Enterprises (SMEs)", in Proceeding of the Conference on Cognitive Informatics and Soft Computing (CISC), Springer Nature Singapore,pp. 717–726 DOI: 10.1007/978-981-16-8763-1_59

Bosio, B., Rainer, K., & Stickdorn, M. (2017). Customer experience research with mobile ethnography: A case study of the Alpine destination Serfaus-Fiss-Ladis. In *Qualitative consumer research* (pp. 111–137). Emerald Publishing Limited., DOI: 10.1108/S1548-643520170000014009

Brodny, J., & Tutak, M. (2022). Digitalization of Small and Medium-Sized Enterprises and Economic Growth: Evidence for the EU-27 Countries. *Journal of Open Innovation*, *2*(8), 67. DOI: 10.3390/joitmc8020067

Brown, K., & Thompson, L. (2023). Understanding customer responses to AI-driven personalized journeys: Impacts on the customer experience. *Journal of Advertising*, *52*(3), 345–362. DOI: 10.1080/00913367.2025.2460985

Brown, M. T., & Bussell, J. K. (2011). Medication adherence: WHO cares? *Mayo Clinic Proceedings*, *86*(4), 304–314. DOI: 10.4065/mcp.2010.0575 PMID: 21389250

Brown, T. (2022). *Change by design: How design thinking creates new alternatives for business and society*. Harper Business.

Brundage, M., Avin, S., Wang, J., Belfield, H., Krueger, G., Hadfield, G., . . . Amodei, D. (2018). The malicious use of artificial intelligence: Forecasting, prevention, and mitigation. *arXiv preprint* arXiv:1802.07228.

Brynjolfsson, E., & McAfee, A. (2014). *The Second Machine Age: Work, Progress, and Prosperity in a Time of Brilliant Technologies*. W. W. Norton & Company.

Brynjolfsson, E., & McAfee, A. (2017). *Machine, platform, crowd: Harnessing our digital future*. W. W. Norton & Company.

Buchanan, R. (2023). Human-centered design and the challenges of creating inclusive systems. *Design Issues*, *39*(2), 35–45.

Buell, R. W., Campbell, D., & Frei, F. X. (2010). Are self-service customers satisfied or stuck? *Production and Operations Management, 19*(6), 679–697. DOI: 10.1111/j.1937-5956.2010.01151.x

Bughin, J., Seong, J., Manyika, J., Chui, M., & Joshi, R. (2018). *Notes from the AI frontier: Modeling the impact of AI on the world economy*. McKinsey Global Institute.

Bunte, A., Richter, F., & Diovisalvi, R. (2021), "Why it is hard to find AI in SMEs: A survey from the practice and how to promote it", in *Proceedings of the 13th International Conference on Agents and Artificial Intelligence (ICAART)*, pp. 614–620. DOI: 10.5220/0010204106140620

Buolamwini, J., & Gebru, T. (2018). Gender shades: Intersectional accuracy disparities in commercial gender classification. *Proceedings of Machine Learning Research, 81*, 1–15.

Burns, S. (2018). M-Pesa and the 'market-led' approach to financial inclusion. *Economic Affairs, 38*(3), 406–421. DOI: 10.1111/ecaf.12321

Büyüksağnak, M. (2021). The effect of Spotify playlists on the music consumption habits of music consumers. https://openaccess.bilgi.edu.tr/items/5a7a5430-1647-45de-86d6-d030e0841a70

Cavoukian, A. (2011). Privacy by design: The 7 foundational principles. Information and Privacy Commissioner of Ontario, Canada. https://www.ipc.on.ca/wp-content/uploads/Resources/7foundationalprinciples.pdf

Cecchetti, S. G., & Schoenholtz, K. (2020). Finance and technology: What is changing and what is not. *CEPR Discussion Papers, 15352*, 1–40.

Chaithanya, D., Deepa, G., Pavani, K. B., Manjusha, G., & Kumar, V. V. (2023, June). Chatbot for banking sector using NLP and ML. In International Conference on Soft Computing and Signal Processing (pp. 153-163). Singapore: Springer Nature Singapore.

Chan, C. Y. T., & Petrikat, D. (2022). Self-service technology: Benefits and challenges. *Journal of Computer Science and Technology Studies, 4*(2), 118–127. DOI: 10.32996/jcsts.2022.4.2.14

Chandra, S., Verma, S., Lim, W. M., Kumar, S., & Donthu, N. (2022). Personalization in personalized marketing: Trends and ways forward. *Psychology and Marketing, 39*(8), 1529–1562. DOI: 10.1002/mar.21670

Chang, V., Baudier, P., Zhang, H., Xu, Q., Zhang, J., & Arami, M. (2020). How blockchain can impact financial services: An overview, challenges, and recommendations from expert interviewees. *Technological Forecasting and Social Change*, *158*, 120–166. DOI: 10.1016/j.techfore.2020.120166 PMID: 32834134

Chatterji, N., Manohar, S., & Verma, B. (2023). Assessing the Influence of Graduate Characteristics on Employer Satisfaction: A Multi-dimensional Analysis. *The Open Psychology Journal*, *16*(1), e187435012308030. DOI: 10.2174/18743501-v16-230831-2023-39

Chawla, U., Verma, B., & Mittal, A. (2024a). Resistance to O2O technology platform adoption among small retailers: The influence of visibility and discoverability. *Technology in Society*, *76*, 102482. DOI: 10.1016/j.techsoc.2024.102482

Chawla, U., Verma, B., & Mittal, A. (2024b). Unveiling barriers to O2O technology platform adoption among small retailers in India: Insights into the role of digital ecosystem. *Information Discovery and Delivery*.

Cheng, Y. S., Yu, T. F., Huang, C. F., Yu, C., & Yu, C. C. (2011). The Comparison of Three Major Occupations for User Acceptance of Information Technology: Applying the UTAUT Model. *IBusiness*, *03*(02), 147–158. DOI: 10.4236/ib.2011.32021

Chen, H., Chiang, R. H. L., & Storey, V. C. (2012). Business intelligence and analytics: From big data to big impact. *Management Information Systems Quarterly*, *36*(4), 1165–1188. DOI: 10.2307/41703503

Chen, J. V., Jubilado, R. J. M., Capistrano, E. P., & Yen, D. C. (2015). Factors affecting online tax filing – An application of the IS Success Model and trust theory. *Computers in Human Behavior*, *43*, 251–262. DOI: 10.1016/j.chb.2014.11.017

Chen, J. V., Yen, D., Dunk, K., & Widjaja, A. E. (2015). The impact of using kiosk on enterprise systems in service industry. *Enterprise Information Systems*, *9*(8), 835–860. DOI: 10.1080/17517575.2013.867542

Chen, R. J., & Dubinsky, A. J. (2003). A conceptual model of perceived customer value in self-service technology. *International Journal of Service Industry Management*, *14*(3/4), 346–369. DOI: 10.1108/09564230310478877

Chen, T., Guo, W., Gao, X., & Liang, Z. (2021). AI-based self-service technology in public service delivery: User experience and influencing factors. *Government Information Quarterly*, *38*(4), 101520. DOI: 10.1016/j.giq.2020.101520

Chen, Y., & Lin, Z. (2021). Business intelligence capabilities and firm performance: A study in China. *International Journal of Information Management*, *57*, 102232. DOI: 10.1016/j.ijinfomgt.2020.102232

Chernenkova, A. (2021). *Facial recognition technology in Russia: do the citizens of Russia accept it?* (Master's thesis, University of Twente).

Choithani, K. (2020). A study on commercial utilisation of personal data and perceptions of data privacy amongst online users (Doctoral dissertation, Dublin Business School). https://esource.dbs.ie/server/api/core/bitstreams/19030a3b-d423-4211-8430-a87dc1d79ece/content

Chopra, S., Sharma, R., & Sherry, A. M. (2013a). Comparing MFS in Kenya, the Philippines, and South Africa under the 7 P evaluation framework. *International Journal of Computer Applications*, *84*(9), 17–22. DOI: 10.5120/14604-2853

Chopra, S., Sherry, A. M., & Dwivedi, R. (2013b). Leveraging technology options for financial inclusion in India. *International Journal of Asian Business and Information Management*, *4*(1), 10–20. DOI: 10.4018/jabim.2013010102

Chouhan, V., Chandra, B., Saraswat, P., & Goswami, S. (2020). Developing sustainable accounting framework for cement industry: Evidence from India. *Finance India*, *34*, 1459–1474.

Chouhan, V., Goswami, S., Dadhich, M., Saraswat, P., & Shakdwipee, P. (2021e). Emerging opportunities for the application of blockchain for energy efficiency. In Khazanchi, D., Vyas, A. K., Hiran, K. K., & Padmanaban, S. (Eds.), *Blockchain 3.0 for sustainable development* (pp. 63–88). De Gruyter. DOI: 10.1515/9783110702507-005

Chouhan, V., Goswami, S., & Sharma, R. B. (2021d). Utilization of proactive spare parts inventory management (PSPIM) techniques for material handling in the cement industry. *Materials Today: Proceedings*, *45*, 4383–4389. DOI: 10.1016/j.matpr.2020.11.757

Chouhan, V., Sharma, R. B., & Goswami, S. (2021a). Factors affecting audit quality: A study of companies listed on the Bombay Stock Exchange (BSE). *International Journal of Management*, *25*, 989–999.

Chouhan, V., Sharma, R. B., & Goswami, S. (2021b). Sustainable reporting: A case study of selected cement companies of India. *Accounting*, *7*, 151–160. DOI: 10.5267/j.ac.2020.10.002

Chouhan, V., Sharma, R. B., Goswami, S., & Hashed, A. W. A. (2021c). Measuring challenges in adopting sustainable environmental technologies in the Indian cement industry. *Accounting*, *7*, 339–348. DOI: 10.5267/j.ac.2020.11.019

Christensen, C. M., Raynor, M. E., & McDonald, R. (2015). What is disruptive innovation? *Harvard Business Review*, *93*(12), 44–53. PMID: 17183796

Chui, M., Löffler, M., & Roberts, R. (2015). The Internet of Things. *The McKinsey Quarterly*.

Cisco. (2022). Consumer Privacy Survey: Attitudes and Awareness. Available at: https://www.cisco.com/c/dam/en_us/about/doing_business/trust-center/docs/cisco-consumer-privacy-survey-2022.pdf

Citibank. (n.d.). *Citibank*. Retrieved March 30, 2025, from https://www.citi.com/

Coman, E. (2025). IVR systems used in call center management: A scientometric analysis of the literature. *Frontiers of Computer Science*, *7*, 1459787. DOI: 10.3389/fcomp.2025.1459787

Considine, E., & Cormican, K. (2016). Self-service technology adoption: An analysis of customer to technology interactions. *Procedia Computer Science*, *100*, 103–109. DOI: 10.1016/j.procs.2016.09.129

Crawford, K., Dobbe, R., Dryer, T., Fried, G., Green, B., Kaziunas, E., Kak, A., Mathur, V., McElroy, E., Nill Sánchez, A., Raji, D., Rankin, J. L., Richardson, R., Schultz, J., West, S. M., & Whittaker, M. (2019). *AI Now 2019 report*. AI Now Institute. https://ainowinstitute.org/AI_Now_2019_Report.html

Creswell, J. W., & Poth, C. N. (2018). *Qualitative inquiry and research design: Choosing among five approaches* (4th ed.). Sage Publications. Provides a foundation for qualitative research design, purposeful sampling, and thematic analysis.

Creswell, J. W., & Poth, C. N. (2018). *Qualitative Inquiry and Research Design Choosing among Five Approaches* (4th ed.). SAGE Publications, Inc.

Curran, J. M., & Meuter, M. L. (2005). Self-service technology adoption: Comparing three technologies. *Journal of Services Marketing*, *19*(2), 103–113. DOI: 10.1108/08876040510591411

Curran, J. M., & Meuter, M. L. (2007). Encouraging existing customers to switch to self-service technologies: Put a little fun in their lives. *Journal of Marketing Theory and Practice*, *15*(4), 283–298. DOI: 10.2753/MTP1069-6679150401

Cyriac, R., & Thomas, J. (2024). Smart Farming with Cloud Supported Data Management Enabling Real-Time Monitoring and Prediction for Better Yield. In *Intelligent Robots and Drones for Precision Agriculture* (pp. 283–306). Springer Nature Switzerland. DOI: 10.1007/978-3-031-51195-0_14

Dabholkar, P. A. (1996). Consumer evaluations of new technology-based self-service options: An investigation of alternative models of service quality. *International Journal of Research in Marketing*, *13*(1), 29–51. DOI: 10.1016/0167-8116(95)00027-5

Dabholkar, P. A. (2023). Self-service technologies: A review of the current state of research and future directions. *Journal of Services Marketing*, *37*(1), 12–29.

Dabholkar, P. A., & Bagozzi, R. P. (2002). An attitudinal model of technology-based self-service: Moderating effects of consumer traits and situational factors. *Journal of the Academy of Marketing Science*, *30*(3), 184–201. DOI: 10.1177/0092070302303001

Dabiri, H., Farhangi, V., Moradi, M. J., Zadehmohamad, M., & Karakouzian, M. (2022). Applications of decision tree and random forest as tree-based machine learning techniques for analyzing the ultimate strain of spliced and non-spliced reinforcement bars. *Applied Sciences (Basel, Switzerland)*, *12*(10), 4851. DOI: 10.3390/app12104851

Dadon, Z., Rav Acha, M., Orlev, A., Carasso, S., Glikson, M., Gottlieb, S., & Alpert, E. A. (2024). Artificial Intelligence-Based Left Ventricular Ejection Fraction by Medical Students for Mortality and Readmission Prediction. *Diagnostics (Basel)*, *14*(7), 767. DOI: 10.3390/diagnostics14070767 PMID: 38611680

Dalal, S., et al. (2024). Smart viewing: Redefining value in over-the-top streaming experiences. Lecture Notes in Electrical Engineering, 523-532. DOI: 10.1007/978-981-97-3442-9_36

Davenport, T. H., & Ronanki, R. (2018). Artificial intelligence for the real world. *Harvard Business Review*, *96*(1), 108–116.

Davis, T., & Wilson, A. (2017). Fluid dispensing systems: Design and application. In Fluid Dispensing Systems (pp. 112–130). Springer. ISBN: 978-3-319-45671.

Davis, B., & Owens, J. (2009). *POS vs. mobile phone as a channel for m-banking*. Microwave.

Davis, F. D. (1989). Perceived usefulness, perceived ease of use, and user acceptance of information technology. *Management Information Systems Quarterly*, *13*(3), 319–340. DOI: 10.2307/249008

Davis, F., Jones, M., & Smith, L. (2019). Self-service technologies and customer satisfaction: A meta-analytic review. *Journal of Service Management*, *30*(4), 245–262.

Deci, E. L., & Ryan, R. M. (1985). The general causality orientations scale: Self-determination in personality. *Journal of Research in Personality*, *19*(2), 109–134. DOI: 10.1016/0092-6566(85)90023-6

Deloitte. (2021). Hospitality trends: Redefining the guest experience through personalization. Available at: https://www.deloitte.com/uk/en/Industries/consumer/blogs/embracing-the-future-ais-transformative-role-in-hospitality.html

Demirkan, H., & Spohrer, J. (2014). Developing a framework to improve virtual self-service experience. *Journal of the Academy of Marketing Science, 42*(4), 438–459. DOI: 10.1007/s11747-014-0384-4

Demirkan, H., & Spohrer, J. (2014). Developing a framework to improve virtual shopping in digital malls with intelligent self-service systems. *Journal of Retailing and Consumer Services, 21*(5), 860–868. DOI: 10.1016/j.jretconser.2014.02.012

Devi, S. (2024). Financial inclusion through digitalization in the Indian economy. *Research Bulletin (International Commission for the Northwest Atlantic Fisheries), 49*(1), 83–91.

Dhaigude, S. A., & Mohan, B. C. (2024). Customer experience in social commerce: Thematic and intellectual structure mapping using bibliometric analysis. *International Journal of Human-Computer Interaction, 40*(5), 1210–1234. DOI: 10.1080/10447318.2022.2134837

Dhieb, N., Ghazzai, H., Besbes, H., & Massoud, Y. (2020). A secure AI-driven architecture for automated insurance systems: Fraud detection and risk measurement. *IEEE Access : Practical Innovations, Open Solutions, 8*, 58546–58558. DOI: 10.1109/ACCESS.2020.2983300

Dhiman, P., Kaur, A., & Bonkra, A. (2023, January). Fake information detection using deep learning methods: a survey. In *2023 International conference on artificial intelligence and smart communication (AISC)* (pp. 858-863). IEEE. DOI: 10.1109/AISC56616.2023.10085519

Dhiman, P., Kaur, A., Bonkra, A., & Chahal, N. (2024). Significance of ML and NLP in Sentiment Analysis. In *Fusion of Artificial Intelligence and Machine Learning in Advanced Image Processing* (pp. 69–81). Apple Academic Press. DOI: 10.1201/9781003493570-7

Dias, C., & Silva, J. (2024). Unveiling the future: Smart airports-applications, advantages, strategies and technological challenges. *Journal of Airline and Airport Management, 14*(2), 38–56. DOI: 10.3926/jairm.419

Digital Transformation of Finance and Regulation. (2023). *Cambridge University Press eBooks.*, DOI: 10.1017/9781009086943.002

Ding, Y., Jin, M., Li, S., & Feng, D. (2021). Smart logistics based on the internet of things technology: An overview. *International Journal of Logistics, 24*(4), 323–345. DOI: 10.1080/13675567.2020.1757053

Dix, A., Finlay, J., Abowd, G., & Beale, R. (2020). *Human-computer interaction* (4th ed.). Pearson.

DMahbub. (2023). Digital Technologies in Finance - Digital Mahbub. Digital Mahbub. https://digitalmahbub.com/digital-technologies-in-finance/

Doak, S., Schwager, A., & Hensel, J. M. (2020). Exposure to and attitudes regarding electronic healthcare (e-Health) among physician assistants in Canada: A national survey study. *Health & Social Care in the Community*, *28*(3), 1049–1057. DOI: 10.1111/hsc.12938 PMID: 31899587

Dube, A., & Helkkula, A. (2015). Service experiences beyond the direct use: Indirect customer use experiences of smartphone apps. *Journal of Service Management*, *26*(2), 224–248. DOI: 10.1108/JOSM-11-2014-0308

Dunaway, N., & Berger, K. M. (2021). The Changing Face of Biological Research and the Growing Role of Biosecurity. In *Applied Biosecurity: Global Health, Biodefense, and Developing Technologies* (pp. 89–119). Springer International Publishing. DOI: 10.1007/978-3-030-69464-7_6

Dwivedi, Y. K., Hughes, D. L., Wang, Y., Alalwan, A. A., Ahn, S. J., Balakrishnan, J., & Wamba, S. F. (2021). Artificial intelligence (AI): Multidisciplinary perspectives on emerging challenges, opportunities, and agenda for research, practice, and policy. *International Journal of Information Management*, *57*, 101994. DOI: 10.1016/j.ijinfomgt.2019.08.002

Edelman. (2020). Transparency in AI adoption: Insights from global consumer surveys. *Edelman Trust Barometer*.

Ekechi, C. C., Chukwurah, E. G., Oyeniyi, L. D., & Okeke, C. D. (2024). AI-infused chatbots for customer support: A cross-country evaluation of user satisfaction in the USA and the UK. *International Journal of Management & Entrepreneurship Research*, *6*(4), 1259–1272. DOI: 10.51594/ijmer.v6i4.1057

Elantheraiyan, P., Vinayagam, K., Sasikumar, P., Thirunavukkarasu, K. S., & Singh, S. K. (2023). Revolutionizing e-commerce: Deep learning and distributed expression for cutting-edge product advertising. *2023 IEEE International Conference on Research Methodologies in Knowledge Management, Artificial Intelligence and Telecommunication Engineering (RMKMATE).*

Elantheraiyan, R., Rajan, M., & Chandrasekaran, A. (2024). Ethical design in AI applications: Addressing transparency and fairness. *International Journal of AI Ethics*, *12*(3), 145–162.

Elgahwash, F. O., & Freeman, M. B. (2015). Self-service technology banking preferences: comparing Libyans' behaviour in developing and developed countries. In Banking, Finance, and Accounting: Concepts, Methodologies, Tools, and Applications (pp. 714-727). IGI Global.

Eling, M., Nuessle, D., & Staubli, J. (2021). The impact of artificial intelligence along the insurance value chain and the insurability of risks. *The Geneva Papers on Risk and Insurance. Issues and Practice*, *47*(2), 205–241. DOI: 10.1057/s41288-020-00201-7

Esposito, D. (2016). *Modern web development: understanding domains, technologies, and user experience*. Microsoft Press.

European Union Agency for Cybersecurity (ENISA). (2021). *Privacy challenges in AI adoption: Key insights and recommendations*. Cybersecurity Reports.

Fahn, V., & Riener, A. (2021). Time to get conversational: Assessment of the potential of conversational user interfaces for mobile banking. In Proceedings of Mensch und Computer 2021 (pp. 34-43).

Faisal, A., Ali, M., & Varma, R. (2024). Efficient machine learning algorithms for predicting customer churn. *The Journal of Consumer Research*, *50*(2), 110–125.

Fan, L., & Chatterjee, S. (2020). The Utilization of Robo-Advisors by Individual Investors: An Analysis Using Diffusion of Innovation and Information Search Frameworks. *Financial Counseling and Planning*, *31*(1), 130–145. DOI: 10.1891/JFCP-18-00078

Fassler, A. (2020). "Soundtrack Your Life with Spotify". https://core.ac.uk/download/pdf/334996075.pdf

Filieri, R., D'Amico, E., Destefanis, A., Paolucci, E., & Raguseo, E. (2021). Artificial intelligence (AI) for tourism: An European-based study on successful AI tourism start-ups. *International Journal of Contemporary Hospitality Management*, *33*(11), 4099–4125. DOI: 10.1108/IJCHM-02-2021-0220

Fitzpatrick, K. K., Darcy, A., & Vierhile, M. (2022). Delivering therapeutic support via artificial intelligence: The role of Woebot in mental health interventions. *Journal of Behavioral Health*, *44*(2), 112–126.

Floridi, L., Cowls, J., Beltrametti, M., Chatila, R., Chazerand, P., Dignum, V., Luetge, C., Madelin, R., Pagallo, U., Rossi, F., Schafer, B., Valcke, P., & Vayena, E. (2018). AI4People—An ethical framework for a good AI society: Opportunities, risks, principles, and recommendations. *Minds and Machines*, *28*(4), 689–707. DOI: 10.1007/s11023-018-9482-5 PMID: 30930541

Fogel, D. B. (2022). Defining artificial intelligence. *Machine Learning and the City: Applications in Architecture and Urban Design*, 91-120.

Foster, S. T., & Gardner, J. W. (2022). Managing quality: Integrating the supply chain. John Wiley & Sons.

Frey, C. B., & Osborne, M. A. (2017). The future of employment: How susceptible are jobs to computerization? *Technological Forecasting and Social Change*, *114*, 254–280. DOI: 10.1016/j.techfore.2016.08.019

Frontiers. (n.d.). Artificial intelligence in insurance and finance.

Frost, J., Gambacorta, L., Huang, Y., Shin, H. S., & Zbinden, P. (2019). BigTech and the changing structure of financial intermediation. *Economic Policy*, *34*(100), 761–799. DOI: 10.1093/epolic/eiaa003

Fusch, P. I., & Ness, L. R. (2015). Are we there yet? Data saturation in qualitative research. [Supports the importance of data saturation and methodological rigor.]. *The Qualitative Report*, *20*(9), 1408–1416. DOI: 10.46743/2160-3715/2015.2281

Galdolage, S. (2020). Customer choice of self-service kiosks in service transactions. http://dr.lib.sjp.ac.lk/handle/123456789/11621

Gao, H., & Liu, Y. (2022). Transforming customer journeys with AI: A review of applications in B2B and B2C sectors. *Journal of Business Research*, *75*(2), 231–245.

Gao, Y., & Liu, H. (2023). Artificial intelligence-enabled personalization in interactive marketing: A customer journey perspective. *Journal of Research in Interactive Marketing*, *17*(5), 663–680. DOI: 10.1108/JRIM-01-2022-0023

Gargioni, L., Fogli, D., & Baroni, P. (2024). A systematic review on pill and medication dispensers from a human-centered perspective. *Journal of Healthcare Informatics Research*, *8*(2), 244–285. DOI: 10.1007/s41666-024-00161-w PMID: 38681758

Gartner. (2023). *Emerging technologies and trends impact radar: Self-service capabilities*. Gartner, Inc. Retrieved from https://www.gartner.com

Gartner. (2023). The top AI trends shaping business and technology. Gartner Research Reports.

Gast, A., & Mathes, T. (2019). Medication adherence influencing factors—An (updated) overview of systematic reviews. *Systematic Reviews*, *8*(112), 112. Advance online publication. DOI: 10.1186/s13643-019-1014-8 PMID: 31077247

Gautam, A., & Rawat, S. (2017). Cashless and digital economy and its effect on financial inclusion in India. *Financial Sector in India*, *20*, 77–85.

Gautam, S., & Kumar, P. (2023a, June). Behavioral biases of investors in the crypto-currency market. *AIP Conference Proceedings*, *15*, 020105. DOI: 10.1063/5.0154194

Gautam, S., & Kumar, P. (2023b). Blockchain Technology in the Investment Domain: A Bibliometric Analysis. *International Journal of Electronic Finance*, *1*(1), 10061055. Advance online publication. DOI: 10.1504/IJEF.2025.10061055

Gautam, S., & Kumar, P. (2024). A Study of the Evolutionary Trends in Blockchain and Cryptocurrency: A Bibliometric Approach. In *Finance Analytics in Business* (pp. 251–266). Emerald Publishing Limited., DOI: 10.1108/978-1-83753-572-920241013

Gautam, S., Kumar, P., & Dahiya, P. (2024). The Influence of Neurotransmitters on Cryptocurrency Investment Decision-Making : The Mediating Role of Risk Tolerance and Moderating Role of Investment Experience. *Indian Journal of Finance*, *18*(10), 40. DOI: 10.17010/ijf/2024/v18i10/174613

Gautam, S., Kumar, P., Dahiya, P., & Sandhu, S. (2024). Emotional Intelligence and Cryptocurrency Investments: Unraveling the Mediating Role of Risk Tolerance and Moderating Role of Investment Experience. *International Journal of Business Innovation and Research*, *1*(1), 10065382. Advance online publication. DOI: 10.1504/IJBIR.2024.10065382

Gefen, D., Karahanna, E., & Straub, D. W. (2003). Trust and TAM in Online Shopping: An Integrated Model. *Management Information Systems Quarterly*, *27*(1), 51–90. DOI: 10.2307/30036519

George, D. (2023). Deciphering the Path to Cost Efficiency and Sustainability in the Snowflake Environment. Zenodo (CERN European Organization for Nuclear Research). DOI: 10.5281/zenodo.8282654

George, D., & George, A. (2023). An Overview of India's Unified Payments Interface (UPI): Benefits, Challenges, and Opportunities. Zenodo (CERN European Organization for Nuclear Research). https://doi.org/DOI: 10.5281/zenodo.7723154

George, D., George, A., & Dr.T.Baskar. (2023a). RuPay Card: Revolutionizing India's Payment Ecosystem. Zenodo (CERN European Organization for Nuclear Research). https://doi.org/DOI: 10.5281/zenodo.7857040

George, D., George, A., & Dr.T.Baskar. (2023b). Unshackled by Servers: Embracing the Serverless Revolution in Modern Computing. Zenodo (CERN European Organization for Nuclear Research). DOI: 10.5281/zenodo.8051052

George, A. H., Shahul, A., & George, A. S. (2023). Wearable sensors: A new way to track health and wellness. *Partners Universal International Innovation Journal*, *1*(4), 15–34.

George, A. S. (2023a). Securing the Future of Finance: How AI, Blockchain, and Machine Learning Safeguard Emerging Neobank Technology Against Evolving Cyber Threats. puirp.com. DOI: 10.5281/zenodo.10001735

George, A. S. (2023b). Preparing Students for an AI-Driven World: Rethinking Curriculum and Pedagogy in the Age of Artificial Intelligence. puirp.com. DOI: 10.5281/zenodo.10245675

George, A. S., & Baskar, T. (2023). The impact of AI language models on the future of white-collar jobs: A comparative study of job projections in developed and developing countries. *Partners Universal International Research Journal*, *2*(2), 117–135.

Ghasemaghaei, M., & Calic, G. (2020). Assessing the Impact of Big Data on Firm Innovation Performance: Big Data Is Not Always Better Data. *Journal of Business Research*, *108*, 147–162. DOI: 10.1016/j.jbusres.2019.09.062

Ghosh, P. (2023). Automated Customer Service for Retail with AI. *Journal of Bioinformatics and Artificial Intelligence*, *3*(2), 231–243.

Ghosh, S. (2020). Financial inclusion in India: Does distance matter? *South Asia Economic Journal*, *21*(2), 216–238. DOI: 10.1177/1391561420961649

Global Data UK Ltd. (2022, November 24). Artificial intelligence (AI) in insurance - Thematic research. Market Research Reports & Consulting.

Godovykh, M., & Tasci, A. D. (2020). Customer experience in tourism: A review of definitions, components, and measurements. *Tourism Management Perspectives*, *35*, 100694. DOI: 10.1016/j.tmp.2020.100694

Goodhue, D. L., & Thompson, R. L. (1995). Task-technology fit and individual performance. *Management Information Systems Quarterly*, *19*(2), 213–236. DOI: 10.2307/249689

Gorgoglione, M., Panniello, U., & Tuzhilin, A. (2019). Recommendation strategies in personalization applications. *Information & Management*, *56*(6), 103143. DOI: 10.1016/j.im.2019.01.005

Goswami, D., & Verma, B. (2024). The Intersection of Ethics and Big Data: Addressing Ethical Concerns in Digital Age of Artificial Intelligence. In Digital Technologies, Ethics, and Decentralization in the Digital Era (pp. 269-285). IGI Global.

Govindaraj, M., Kumari, D. A., Khan, P., Gnanasekaran, C., & Lawrence, J. (2024). *Revolutionizing healthcare: The transformative impact of artificial intelligence*. Revolutionizing the Healthcare Sector with AI. DOI: 10.4018/979-8-3693-3731-8.ch003

Govindaraj, S., Patel, K., & Muthukumar, R. (2024). Transformative impact of AI in healthcare: Lessons for retail innovation. *HealthTech Quarterly*, *11*(4), 320–338.

Grigorescu, S., Trasnea, B., Cocias, T., & Macesanu, G. (2020). A survey of deep learningtech-niques for autonomous driving. *Journal of Field Robotics*, *3*(37), 362–386. DOI: 10.1002/rob.21918

Gu, B., Santhanam, R., Berente, N., & Recker, J. (2019). Call for Papers MISQ Special Issue on Managing AI. MIS Quarterly, Special Is. Retrieved from https://misq.org/skin/frontend/default/misq/pdf/CurrentCalls/ManagingAI.pdf

Guidotti, R., Monreale, A., Ruggieri, S., Turini, F., Giannotti, F., & Pedreschi, D. (2018). A survey of methods for explaining black-box models. *ACM Computing Surveys*, *51*(5), 1–42. DOI: 10.1145/3236009

Gukpa, A. S., & Mukherjee, J. (2025). Framework for adoption of generative AI for information search of retail products and services. *International Journal of Retail & Distribution Management*, *53*(2), 165–181. DOI: 10.1108/IJRDM-05-2024-0203

Gummerus, J., Lipkin, M., Dube, A., & Heinonen, K. (2019). Technology in use – characterizing customer self-service devices (SSDS). *Journal of Services Marketing*, *33*(1), 44–56. DOI: 10.1108/JSM-10-2018-0292

Gunkel, D. J. (2023). The ethics of artificial intelligence and the challenge of emotional intelligence. *Journal of Technology and Ethics*, *15*(1), 101–117.

Guo, Y. (2025). Successful Stories on the Usage of Technologies in Service Industries. In *Technological Innovations in the Food Service Industry* (pp. 197–230). IGI Global Scientific Publishing.

Gupta, S., Dash, S., & Mishra, B. (2016). Self-service technology in retail: The next frontier. *Journal of Business Research*, *69*(11), 4539–4546. DOI: 10.1016/j.jbusres.2016.03.053

Gupta, S., Leszkiewicz, A., Kumar, V., Bijmolt, T., & Potapov, D. (2020). Digital analytics: Modeling for insights and new methods. *Journal of Interactive Marketing*, *51*(1), 26–43. DOI: 10.1016/j.intmar.2020.04.003

Gurley, R. C. (2021). *Strategies for Implementing Self-service Technologies in Supermarket Retail Operations* (Doctoral dissertation, Walden University). https://scholarworks.waldenu.edu/cgi/viewcontent.cgi?article=11768&context=dissertations

Habul, A., & Pilav-Velic, A. (2010). Business Intelligence and Customer Relationship Management. In *Proceedings of ITI 2010, 32nd International Conference on Information Technology Interfaces*. IEEE Press., Retrieved from https://ieeexplore.ieee.org/document/5546381?arnumber=5546381

Hadhoud, R., & Salameh, W. A. (2020). How business intelligence can help you to better understand your customers. *International Journal of Business Intelligence Research, 11*(1), 50–58. DOI: 10.4018/IJBIR.2020010104

Hamari, J., Koivisto, J., & Sarsa, H. (2014). Does gamification work? A literature review of empirical studies on gamification. *Proceedings of the 47th Hawaii International Conference on System Sciences (HICSS)*. DOI: 10.1109/HICSS.2014.377

Haque, S. S., Yanez-Pagans, M., Arias-Granada, Y., & Joseph, G. (2020). Water and sanitation in Dhaka slums: Access, quality, and informality in service provision. *Water International, 45*(7-8), 791–811. DOI: 10.1080/02508060.2020.1786878

Harmantzis, F. C., & Thomopoulos, S. C. A. (2020). Blockchain and artificial intelligence in insurance: Towards a more automated and profitable future. *The Journal of Risk Finance*.

Hasal, M., Nowaková, J., Ahmed Saghair, K., Abdulla, H., Snášel, V., & Ogiela, L. (2021). Chatbots: Security, privacy, data protection, and social aspects. *Concurrency and Computation, 33*(19), e6426. DOI: 10.1002/cpe.6426

Hassan, V. I., Basheer, S., Mir, F. A., & Abou Fayad, S. G. (2024). Digital Innovation in the Service Sector: Transforming Customer Experiences. In *Service Innovations in Tourism: Metaverse, Immersive Technologies, and Digital Twin* (pp. 150–165). IGI Global Scientific Publishing. DOI: 10.4018/979-8-3693-1103-5.ch008

HDFC Bank. (n.d.). *HDFC Bank*. Retrieved March 28, 2025, from https://www.hdfcbank.com/

He, W., Hung, J. L., & Liu, L. (2023). Impact of big data analytics on banking: A case study. *Journal of Enterprise Information Management, 36*(2), 459–479.

Hidayat, R., & Winahyu, H. S. (2020). Automatic liquid filling machine using flow sensor and Arduino Pro Mini. *Indonesian Journal of Engineering Research, 1*(1), 27–31. DOI: 10.11594/ijer.01.01.05

Hilton, T., Hughes, T., Little, E., & Marandi, E. (2013). Adopting self-service technology to do more with less. *Journal of Services Marketing, 27*(1), 3–12. DOI: 10.1108/08876041311296338

Hoang Le, A. P., Nguyen, Q. L., Pham, B. H., Cao, T. H. M., Vo, T. V., Huynh, K., & Ha, H. T. T. (2023). SALAD: Syringe-based Arduino-operated low-cost antibody dispenser. *HardwareX*, *15*, e00455. DOI: 10.1016/j.ohx.2023.e00455 PMID: 37497344

Hossain, M. A., Agnihotri, R., Rushan, M. R. I., Rahman, M. S., & Sumi, S. F. (2022). Marketing analytics capability, artificial intelligence adoption, and firms' competitive advantage: Evidence from the manufacturing industry. *Industrial Marketing Management*, *106*, 240–255. DOI: 10.1016/j.indmarman.2022.08.017

How Web3 Neobanks are Shaping the Future of Finance. (n.d.). https://www.zeebu.com/blog/web3-neobanks-the-new-face-of-finance-and-banking

Hoyer, W. D., Kroschke, M., Schmitt, B., Kraume, K., & Shankar, V. (2020). Transforming the customer experience through new technologies. *Journal of Interactive Marketing*, *51*(1), 57–71. DOI: 10.1016/j.intmar.2020.04.001

Hsieh, A., Yen, C.-H., & Chin, K.-C. (2004). Participative customers as partial employees and service provider workload. *International Journal of Service Industry Management*, *15*(2), 187–199. DOI: 10.1108/09564230410532501

Hsu, C.-L., & Lin, J. (2023). Understanding the user satisfaction and loyalty of customer service chatbots. *Journal of Retailing and Consumer Services*, *71*, 103211–103229. DOI: 10.1016/j.jretconser.2022.103211

Huang, M.-H., & Rust, R. T. (2021). A strategic framework for artificial intelligence in marketing. *Journal of the Academy of Marketing Science*, *49*(1), 30–50. DOI: 10.1007/s11747-020-00749-9

Huang, M.-H., Rust, R. T., & Maksimovic, V. (2021). The feeling economy: How artificial intelligence is creating the era of empathic customer experience. *Journal of the Academy of Marketing Science*, *49*(4), 635–658. DOI: 10.1007/s11747-021-00757-4

Huang, M.-H., Rust, R. T., & Maksimovic, V. (2021). The Feeling Economy: Managing in the Next Generation of Artificial Intelligence (AI). *California Management Review*, *63*(3), 43–65. DOI: 10.1177/00081256211003080

Hudson, S. (2025). *Hospitality Management: International Introduction*. Taylor & Francis.

Hughes, R. G., & Blegen, M. A. (2008). Medication administration safety. In Hughes, R. G. (Ed.), *Patient safety and quality: An evidence-based handbook for nurses*. Agency for Healthcare Research and Quality., https://www.ncbi.nlm.nih.gov/books/NBK2656/

Husni, M.. (2016). Design and analysis of a MEMS flow sensor. *IOP Conference Series. Materials Science and Engineering, 128*(1), 012026. DOI: 10.1088/1757-899X/128/1/012026

Hussain, S., Anees, A., Das, A., Nguyen, B. P., Marzuki, M., Lin, S., Wright, G., & Singhal, A. (2020). High-content image generation for drug discovery using generative adversarial networks. *Neural Networks, 132*, 353–363. DOI: 10.1016/j.neunet.2020.09.007 PMID: 32977280

Ibrahim, N., & Handayani, P. W. (2022). A systematic literature review of Business intelligence framework for tourism organizations: Functions and issues. *Interdisciplinary Journal of Information, Knowledge, and Management, 17*, 523–541. DOI: 10.28945/5025

ICICI Bank. (n.d.). *ICICI Bank*. Retrieved March 30, 2025, from https://www.icicibank.com/

Iliadi, M. I. (2023). *Unlocking customer insights through service analytics to improve customer experience and drive business success* (Bachelor's thesis, University of Twente). https://essay.utwente.nl/95404/1/Iliadi_BA_BMS.pdf

Indian Bank. (n.d.). *Indian Bank*. Retrieved March 30, 2025, from https://www.indianbank.in/

Irfan, M., Verma, J., Parameswaran, S., & Sheikh, I. A. (2025). Integrating Emerging Technologies: Enhancing Supply Chain Optimization Through AI, IoT, and Blockchain. In *Enhancing Social Sustainability in Manufacturing Supply Chains* (pp. 199-220). IGI Global Scientific Publishing.

Ismail, A. M., & Pastory, D. (2024). Evaluating the impact of self-service cash deposit machines on the performance of commercial banks in Tanzania. *Future Business Journal, 10*(1), 5–16. DOI: 10.1186/s43093-023-00291-4

Ives, B., & Jarvenpaa, S. L. (1991). Applications of technology to service systems: New opportunities in self-service. *Decision Sciences, 22*(3), 307–326.

Jack, W., & Suri, T. (2014). Risk sharing and transactions costs: Evidence from Kenya's mobile money revolution. *The American Economic Review, 104*(1), 183–223. DOI: 10.1257/aer.104.1.183

Jannach, D., Ludewig, M., & Lerche, L. (2017). Session-based item recommendation in e-commerce: On short-term intents, reminders, trends and discounts. *User Modeling and User-Adapted Interaction, 27*(3-5), 351–392. DOI: 10.1007/s11257-017-9194-1

Jayanth, S., Poorvi, M. B., & Sunil, M. P. (2017). Inventory management system using IoT. In Satapathy, S., Prasad, V., Rani, B., Udgata, S., & Raju, K. (Eds.), *Proceedings of the First International Conference on Computational Intelligence and Informatics* (Vol. 507, pp. 299–309). Springer. https://doi.org/DOI: 10.1007/978-981-10-2471-9_20

Jenneboer, L., Herrando, C., & Constantinides, E. (2022). The impact of chatbots on customer loyalty: A systematic literature review. *Journal of Theoretical and Applied Electronic Commerce Research, 17*(1), 212–229. DOI: 10.3390/jtaer17010011

Jin, Y., & Dwivedi, R. (2009). Business process management: A research overview and analysis. AMCIS 2009 Proceedings, 785.

Johnson, A. (2023). AI-powered kiosks: Enhancing customer engagement through personalized recommendations. *Retail Technology Review, 28*(4), 112–119. DOI: 10.1234/retailtech.2023.02804

Johnson, B. (2018). Self-Service Analytics: A Comprehensive Review of Implementation Strategies. *Journal of Data Analytics, 8*(2), 87–101.

Johnson, B. (2023). Impact of Self-Checkout Systems on Customer Satisfaction: A Case Study on Walmart. *Journal of Retail Management, 15*(2), 45–58.

Johnson, M. (2022). *AI and the future of retail: How smart technology enhances customer experiences.* Tech Trends Publishing.

Johnson, M., Tan, L., & Zhang, Y. (2018). Real-time liquid flow measurement and control using Arduino. *International Journal of Robotics and Automation, 12*(2), 30–38. DOI: 10.1115/ijra.2018.090404

Jordan, M. I., & Mitchell, T. M. (2015). Machine learning: Trends, perspectives, and prospects. *Science, 349*(6245), 255–260. DOI: 10.1126/science.aaa8415 PMID: 26185243

Josefowicz, M., & Monks, K. (2017). *InsurTech: A golden opportunity for P&C insurers to lead with artificial intelligence.* Novarica.

Joshi, N. K., Kulshreshtha, K., & Jain, U. (2024). The role of AI in enhancing customer service in the retail sector. *Journal of Data Acquisition and Processing, 39*(1), 1270-1277. https://sjciy cl.cn/index.php/DAP/article/view/133

Junaid, S. B., Imam, A. A., Balogun, A. O., De Silva, L. C., Surakat, Y. A., Kumar, G., Abdulkarim, M., Shuaibu, A. N., Garba, A., Sahalu, Y., Mohammed, A., Mohammed, T. Y., Abdulkadir, B. A., Abba, A. A., Kakumi, N. A. I., & Mahamad, S. (2022). Recent advancements in emerging technologies for healthcare management systems: A survey. *Health Care, 10*(10), 1940. DOI: 10.3390/healthcare10101940 PMID: 36292387

Jung, D., Dorner, V., Glaser, F., & Morana, S. (2018). Robo-advisory: Digitalization and automation of financial advisory. *Business & Information Systems Engineering, 60*(1), 81–86. DOI: 10.1007/s12599-018-0521-9

Jung, T. H., Chung, N., & Leem, C. S. (2022). Customer-centric self-service technologies in the hospitality industry. *International Journal of Hospitality Management, 52*(5), 89–102.

Kaasinen, E., Roto, V., Hakulinen, J., Heimonen, T., Jokinen, J. P., Karvonen, H., Keskinen, T., Koskinen, H., Lu, Y., Saariluoma, P., Tokkonen, H., & Turunen, M. (2015). Defining user experience goals to guide the design of industrial systems. *Behaviour & Information Technology, 34*(10), 976–991. DOI: 10.1080/0144929X.2015.1035335

Kalyane, D., Sanap, G., Paul, D., Shenoy, S., Anup, N., Polaka, S., & Tekade, R. K. (2020). Artificial intelligence in the pharmaceutical sector: current scene and future prospect. In *The future of pharmaceutical product development and research* (pp. 73–107). Academic Press. DOI: 10.1016/B978-0-12-814455-8.00003-7

Kannan, R. (2024). Revolutionizing the Tourism Industry through Artificial Intelligence: A Comprehensive Review of AI Integration, Impact on Customer Experience, Operational Efficiency, and Future Trends. *International Journal for Multidimensional Research Perspectives, 2*(2), 01-14.

Kanse, T.. (2024). Virtual marketplace dynamics: Understanding consumer behavior on business-to-consumer electronic-commerce platforms. *International Conference on Innovative Practices in Technology and Management (ICIPTM)*, 1-6. DOI: 10.1109/ICIPTM59628.2024.10563796

Kaplan, A. (2022). Augmented and virtual reality transforming retail: Enhancing self-service experiences. *Journal of Retail Technology, 15*(3), 45–60. https://doi .org/xx.xxx/yyyy

Kaplan, A. M. (2022). Metaverse marketing: How the metaverse, the artificial intelligence, and virtual reality are changing the marketing landscape. *Journal of Business Research, 142*, 1045–1054. DOI: 10.1016/j.jbusres.2022.01.009

Kaplan, A., & Haenlein, M. (2020). Rulers of the world, unite! The challenges and opportunities of artificial intelligence. *Business Horizons*, *63*(1), 37–50. DOI: 10.1016/j.bushor.2019.09.003

Karani, G., & Jadhav, B. (2025). How Traditional Banks Adapt to the Fintech Revolution in Banking and Finance Sector. In Shaping Cutting-Edge Technologies and Applications for Digital Banking and Financial Services (pp. 214-225). Productivity Press. DOI: 10.4324/9781003501947-13

Katanga, J. (2024) *The Effect of Self-Service Banking Technology on Customer Satisfaction at NMB Bank*. Masters thesis, The Open University of Tanzania.

Kauffmann, E., Peral, J., Gil, D., Ferrández, A., Sellers, R., & Mora, H. (2019). Managing marketing decision-making with sentiment analysis: An evaluation of the main product features using text data mining. *Sustainability (Basel)*, *11*(15), 4235. DOI: 10.3390/su11154235

Kaushik, A.K. and Rahman, Z. (2015), "Innovation adoption across self-service banking technologies in India", International Journal of Bank Marketing, 33(2), 96 121. DOI: 10.1108/IJBM-01-2014-0006

Kazemian, S., Djajadikerta, H. G., Mat Roni, S., Trireksani, T., & Mohd-Sanusi, Z. (2021). Accountability via social and financial performance of the hospitality sector: The role of market orientation. *Society and Business Review*, *16*(2), 238–254. DOI: 10.1108/SBR-04-2020-0061

Kelley, K. (2024, February 23). What is Machine Learning? A Comprehensive Guide for Beginners - Caltech. Caltech -. https://pg-p.ctme.caltech.edu/blog/ai-ml/what-is-machine-learning

Kelly, P., Lawlor, J., & Mulvey, M. (2010). A review of key factors affecting the adoption of self-service technologies in tourism. Tourism and Hospitality Research Conference Papers, 12–15.

Kelly, P., Lawlor, J., & Mulvey, M. (2019). *Self-service Technologies in the Travel, Tourism, and Hospitality Sectors: Principles and Practice*. Emerald Publishing Limited., DOI: 10.1108/978-1-78756-687-320191003

Khan, H. R. (2012). Issues and challenges in financial inclusion: Policies, partnerships, processes, and products. Korea.

Khan, S., Chouhan, V., Chandra, B., & Goswami, S. (2014). Sustainable accounting and reporting practices in the Indian cement industry: An exploratory study. *Uncertain Supply Chain Management*, *2*(2), 61–72. DOI: 10.5267/j.uscm.2014.2.001

Khatri, M. R. (2023). Integration of natural language processing, self-service platforms, predictive maintenance, and prescriptive analytics for cost reduction, personalization, and real-time insights customer service and operational efficiency. *International Journal of Information and Cybersecurity*, *7*(9), 1–30.

Kim, B., & Chen, Y. (2025). Empowering consumers through self-service technology: A comparative analysis. *Journal of Hospitality & Tourism Research (Washington, D.C.)*, *49*(2), 254–267. DOI: 10.1177/10963480231182959

Kim, H. S., & Choi, B. (2024). The power of customer-to-customer self-service technology in enhancing customer-to-customer interactions. *European Journal of Marketing*. Advance online publication. DOI: 10.1108/EJM-06-2023-0504

Kim, M., & Koo, C. (2023). Technology adoption in the hospitality industry: Self-service kiosks and mobile check-in apps. *International Journal of Hospitality and Tourism*, *35*(1), 88–100.

Kim, M., Zoo, H., Lee, H., & Kang, J. (2018). Mobile financial services, financial inclusion, and development: A systematic review of academic literature. *The Electronic Journal on Information Systems in Developing Countries*, *84*(5), e12044. DOI: 10.1002/isd2.12044

Kiran, A., Kumar, I. J., Vijayakarthik, P., Naik, S. L., & Vinod, T. (2023, January). Intelligent Chat Bots: An AI Based Chat Bot For Better Banking Applications. In *2023 International Conference on Computer Communication and Informatics (ICCCI)* (pp. 1-4). IEEE. DOI: 10.1109/ICCCI56745.2023.10128582

Kishore, K., & Sequeira, A. H. (2013, July 18). Self-service technologies - A silver lining for financial inclusion. NABARD Conference Proceedings 2013.

Koh, Y. Y., Kok, C. L., Ibraahim, N., & Lim, C. G. (2024). Smart water ATM with Arduino integration, RFID authentication, and dynamic dispensing for enhanced hydration practices. *Electronics (Basel)*, *13*(9), 1657. DOI: 10.3390/electronics13091657

Koon, J. K., & Hosein, P. (2024, November). Optimal Placement of Self-Service Banking Kiosks. In *2024 IEEE International Conference on Technology Management, Operations and Decisions (ICTMOD)* (pp. 1-6). IEEE.

Kotler, P., Keller, K. L., Goodman, M., & Brady, M. (2020). *Marketing Management*. Pearson.

Kotras, B. (2020). Mass personalization: Predictive marketing algorithms and the reshaping of consumer knowledge. *Big Data & Society*, *7*(2), 2053951720951581. DOI: 10.1177/2053951720951581

Kowalkowski, C., Wirtz, J., & Ehret, M. (2024). Digital service innovation in B2B markets. *Journal of Service Management*, *35*(2), 280–305. DOI: 10.1108/JOSM-12-2022-0403

Krishnan, R., Gupta, M., & Choudhary, P. (2024). AI and IoT integration in logistics for service efficiency. *Operations Research Perspectives*, *12*(3), 121–145.

Kshetri, N. (2021). *Cybersecurity management: An organizational and strategic approach*. University of Toronto Press. DOI: 10.3138/9781487531249

Kuloğlu, A. (2024). The impacts of technological transformation on labor and employment: The future of the business world and economic growth. *International Research in Social. Human and Administrative Sciences*, *XVII*, 21.

Kumar, V., & Reinartz, W. (2018). Customer relationship management. Springer-Verlag GmbH Germany, part of Springer Nature 2006, 2012, 2018. https://link.springer.com/book/10.1007/978-3-662-55381-7

Kumar, A. (2024). Challenges and opportunities in AI adoption for customer service excellence. *Business Horizons*, *67*(1), 78–90.

Kumar, A., Choudhary, S., Garza-Reyes, J. A., Kumar, V., Rehman Khan, S. A., & Mishra, N. (2023). Analysis of critical success factors for implementing industry 4.0 integrated circular supply chain–Moving towards sustainable operations. *Production Planning and Control*, *34*(10), 984–998. DOI: 10.1080/09537287.2021.1980905

Kumar, A., Patel, N., & Sharma, D. (2025). Customer preferences in OTT streaming services: The role of AI. *Journal of Interactive Marketing*, *44*(1), 78–93.

Kumar, A., Sureka, S., & Gupta, S. (2018). Bridging the digital divide: Challenges in rural self-service technology adoption. *International Journal of Rural Development*, *37*(1), 56–65.

Kumar, G., Sharma, A., & Mehta, R. (2023). Automatic liquid dispensing system based on Arduino and flow meter. *2023 International Conference on Innovations in Electrical and Electronics Engineering (ICEEE)*, 65–70. https://doi.org/DOI: 10.1109/ICEEE.2023.1234567

Kumari, S., Pandey, G. K., & Tiwari, S. (2024). Data Visualization Techniques in Smart Agriculture Implementation. In *AI Applications for Business, Medical, and Agricultural Sustainability* (pp. 122-159). IGI Global. DOI: 10.4018/979-8-3693-5266-3.ch006

Kumar, R. S. S., & Nagle, F. (2020, May 5). The case for AI insurance. *Harvard Business Review*.

Kumar, R., & Soni, M. (2022). AI in Retail: Virtual Fitting Rooms and Customer Experience. *Journal of Retail Technology*, *9*(3), 215–228.

Laing, Y. (2006). The influence of perceived risk in the uptake of self-service technologies within the retail banking sector – A study of customers using the Industrial and Commercial Bank of China in Jilin. Digital Knowledge, Cape Peninsula University of Technology, 118.

Lai, P. C. (2017). The literature review of technology adoption models and theories for the novelty technology. *JISTEM-. Journal of Information Systems and Technology Management*, *14*(1), 21–38. DOI: 10.4301/S1807-17752017000100002

Lakshana, S., Ashok, P., & Chirputkar, A. (2024). Predictive modeling of cardiovascular disease risk with IoT and ML. Advances in Information Communication Technology and Computing, 435-445. DOI: 10.1007/978-981-97-6103-6_27

Latham, T. W. (1966). Fluid motions in a peristaltic pump. PhD thesis, Massachusetts Institute of Technology. http://hdl.handle.net/1721.1/17282

Le Clair, C. (2025). *Random Acts of Automation: How to Fight Back When Automation Threatens Your Work, Your Life, and Everything You Do*. Post Hill Press.

Lee, D. (2018). Strategies for technology-driven service encounters for patient experience satisfaction in hospitals. *Technological Forecasting and Social Change*, *137*, 118–127. DOI: 10.1016/j.techfore.2018.06.050

Lee, H. J., & Yang, K. (2013). Interpersonal service and self-service technology in retailing: The role of perceived control and intrinsic motivation. *Service Industries Journal*, *33*(2), 145–158. DOI: 10.1080/02642069.2011.596532

Lee, J., Kim, K. J., & Sundar, S. S. (2021). Customization in augmented reality: Exploring its role in the user experience. *Computers in Human Behavior*, *122*, 106840.

Lee, K., Koo, C., & Yang, J. (2023). Emotionally intelligent virtual assistants: Exploring the impact of emotional responses on user satisfaction. *AI & Society*, *38*(1), 47–60.

Lee, R., & Wang, S. (2022). Machine learning in retail: Understanding consumer behavior. *AI & Commerce Journal*, *18*(4), 112–125.

Lee, S., Kim, Y., Kahng, H., Lee, S. K., Chung, S., Cheong, T., Shin, K., Park, J., & Kim, S. B. (2020). Intelligent traffic control for autonomous vehicle systems based on machine learning. *Expert Systems with Applications*, *144*, 113074. DOI: 10.1016/j.eswa.2019.113074

Lee, S., & Wang, Y. (2022). Integrating wishlists into AI-powered kiosks for personalized shopping experiences. *Journal of Retail Innovation, 15*(2), 88–95. DOI: 10.5678/jri.2022.15288

Lennerholt, C., Van Laere, J., & Söderström, E. (2020). Implementation challenges of self-service business intelligence: A case study. *Journal of Decision Systems*, 29(sup1), 107–120. https://doi.org/DOI: 10.1080/12460125.2020.1777664

Lennerholt, C., van Laere, J., & Söderström, E. (2020). User-Related Challenges of Self-Service Business Intelligence. *Information Systems Management*.

Liakos, K. G., Busato, P., Moshou, D., Pearson, S., & Bochtis, D. (2018). Machine learning in agriculture: A review. *Sensors (Basel), 18*(8), 2674. DOI: 10.3390/s18082674 PMID: 30110960

Liang, X., Li, G., Zhang, H., Nolan, E., & Chen, F. (2022). Firm performance and marketing analytics in the Chinese context: A contingency model. *Journal of Business Research, 141*, 589–599. DOI: 10.1016/j.jbusres.2021.11.061

Li, F., Lu, H., Hou, M., Cui, K., & Darbandi, M. (2021). Customer satisfaction with bank services: The role of cloud services, security, e-learning, and service quality. *Technology in Society, 64*, 101487. DOI: 10.1016/j.techsoc.2020.101487

Liljander, V., Gillberg, F., Gummerus, J., & Riel, A. (2006). Technology readiness and the evaluation and adoption of self-service technologies. *Journal of Retailing and Consumer Services, 13*(3), 177–191. DOI: 10.1016/j.jretconser.2005.08.004

Lim, W. M., Rasul, T., Kumar, S., & Ala, M. (2022). Past, present, and future of customer engagement. *Journal of Business Research, 140*, 439–458. DOI: 10.1016/j.jbusres.2021.11.014

Lin, J. S. C., & Hsieh, P. L. (2007). The influence of technology readiness on satisfaction and behavioral intentions toward self-service technologies. *Computers in Human Behavior, 23*(3), 1597–1615. DOI: 10.1016/j.chb.2005.07.006

Lin, P. M. C., Peng, K.-L., Au, W. C. W., Qiu, H., & Deng, C. D. (2023). Digital menus innovation diffusion and transformation process of consumer behavior. *Journal of Hospitality and Tourism Technology, 14*(5), 732–761. DOI: 10.1108/JHTT-07-2021-0217

Lior, A., Balkin, J., Van-Loo, R., Lubin, A., Cofone, I. N., Zemer, L., Calabresi, G., Talesh, S., Schwarcz, D., Woods, D., & Kaiser, Z. (2022). Insuring AI: The role of insurance in artificial intelligence regulation. *Harvard Journal of Law & Technology, 2*, 468.

Lipton, Z. C. (2018). The mythos of model interpretability. *ACM Queue; Tomorrow's Computing Today, 16*(3), 31–57. DOI: 10.1145/3236386.3241340

Litterscheidt, R., & Streich, D. J. (2020). Financial education and digital asset management: What's in the black box? *Journal of Behavioral and Experimental Economics, 87*, 101573. Advance online publication. DOI: 10.1016/j.socec.2020.101573

Liu, Y., Chen, L., & Sun, L. (2009). Design and fabrication of a MEMS flow sensor and its application in precise liquid dispensing. *Sensors (Basel), 9*(6), 4138–4150. DOI: 10.3390/s90604138 PMID: 22408517

Li, Z., Liu, F., Yang, W., Peng, S., & Zhou, J. (2021). A survey of convolutional neural networks: Analysis, applications, and prospects. *IEEE Transactions on Neural Networks and Learning Systems, 33*(12), 6999–7019. DOI: 10.1109/TNNLS.2021.3084827 PMID: 34111009

Lóska, G., & Uotila, J. (2024). Digital Transformation in Corporate Banking: Toward a Blended Service Model. *California Management Review, 66*(3), 93–115. DOI: 10.1177/00081256231207429

Lusch, R. F., & Nambisan, S. (2015). Service innovation: A service-dominant logic perspective. *Management Information Systems Quarterly, 39*(1), 155–175. DOI: 10.25300/MISQ/2015/39.1.07

Lu, Y., Cai, R., & Gursoy, D. (2021). Developing and testing a chatbot performance scale in the hospitality industry. *International Journal of Hospitality Management, 93*, 102848.

Lu, Y., & Xia, Z. (2024). Digital inclusive finance, green technological innovation, and carbon emissions from a spatial perspective. *Scientific Reports, 14*(1), 8454. Advance online publication. DOI: 10.1038/s41598-024-59081-9 PMID: 38605137

Mader, P. (2018). Contesting financial inclusion. *Development and Change, 49*(2), 461–483. DOI: 10.1111/dech.12368

Madhala, R. T., Adavelli, S. R., & Rahul, N. (2022). Enhancing Catastrophe Modeling with Big Data and IoT: Revolutionizing Disaster Risk Management and Response. *Australian Journal of Machine Learning Research & Applications, 2*(1), 612–636.

Magalhaes Azevedo, D., & Kieffer, S. (2021). User reception of AI-enabled mHealth Apps: The case of Babylon health.

Mai, J. E. (2016). Big data privacy: The datafication of personal information. *The Information Society, 32*(3), 192–199. DOI: 10.1080/01972243.2016.1153010

Maina, E. M., Chouhan, V., & Goswami, S. (2020). Measuring behavioral aspect of IFRS implementation in India and Kenya. *International Journal of Scientific and Technology Research*, *9*, 2045–2048.

Makinde, O. G., Olubiyi, T. O., & Ogundipe, F. (2023). New Insights from Entrepreneurial Characteristics and Busıness Performance: Empırıcal Fındıngs from Nıgerıa. *Asian Journal of Management. Entrepreneurship and Social Sciences*, *3*(04), 72–100.

Makridakis, S. (2017). The forthcoming artificial intelligence (AI) revolution: Its impact on society and firms. *Futures*, *90*, 46–60. DOI: 10.1016/j.futures.2017.03.006

Mala, M., & Vasanthi, G. (2016). Problems and challenges of the Indian post in financial inclusion. *International Journal of Interdisciplinary Research in Arts and Humanities*, *1*(1), 186–189.

Manikandan, P., Tirupathi Rao, M., Vinay, M., & Ratna Raj, K. D. (2021). Arduino based automatic hand sanitizer dispenser system. 2021 3rd International Conference on Advances in Computing, Communication Control and Networking (ICAC3N), 744–747.

Manohar, M., & Kapur, P. (2019). Service innovation typologies in the retail sector: Frameworks for customer-centric innovation. *Retailing and Consumer Studies*, *28*(3), 220–235.

Manohar, M., & Kapur, P. (2020). Service innovation typologies in the retail sector: Customer perspectives. *Service Innovation Review*, *13*(2), 220–245.

Manohar, S., Mittal, A., & Marwah, S. (2020). Service innovation, corporate reputation, and word-of-mouth in the banking sector: A test on multigroup-moderated mediation effect. *Benchmarking*, *27*(1), 406–429. DOI: 10.1108/BIJ-05-2019-0217

Marcum, Z. A., & Gellad, W. F. (2012). Medication adherence to multidrug regimens. *Clinics in Geriatric Medicine*, *28*(2), 287–300. DOI: 10.1016/j.cger.2012.01.008 PMID: 22500544

Martin, K. D., & Murphy, P. E. (2017). The role of data privacy in marketing. *Journal of the Academy of Marketing Science*, *45*(2), 135–155. https://link.springer.com/article/10.1007/s11747-016-0495-4. DOI: 10.1007/s11747-016-0495-4

Masino, S., & Niño-Zarazúa, M. (2018). Improving financial inclusion through the delivery of cash transfer programmes: The case of Mexico's Progresa-Oportunidades-Prospera programme. *The Journal of Development Studies*, *56*(1), 151–168. DOI: 10.1080/00220388.2018.1546845

Mason, M., Cho, Y., Rayo, J., Gong, Y., Harris, M., & Jiang, Y. (2022). Technologies for medication adherence monitoring and technology assessment criteria: Narrative review. *JMIR mHealth and uHealth, 10*(3), e35157. DOI: 10.2196/35157 PMID: 35266873

Massaro, M., Secinaro, S., Dal Mas, F., Brescia, V., & Calandra, D. (2021). Industry 4.0 and circular economy: An exploratory analysis of academic and practitioners' perspectives. *Business Strategy and the Environment, 30*(2), 1213–1231. DOI: 10.1002/bse.2680

Master, S.. (2024). Unleashing creativity: The business potential of generative AI. *World Conference on Communication & Computing (WCONF)*, 1-6. DOI: 10.1109/WCONF61366.2024.10692266

Masuda, H., Han, S. H., & Lee, J. (2022). Impacts of influencer attributes on purchase intentions in social media influencer marketing: Mediating roles of characterizations. *Technological Forecasting and Social Change, 174*, 121246. DOI: 10.1016/j.techfore.2021.121246

Mathur, S., & Sutton, J. (2017). Personalized medicine could transform healthcare. *Biomedical Reports, 7*(1), 3–5. DOI: 10.3892/br.2017.922 PMID: 28685051

Mbachukwu, M. A., Mwakawaza, N., & Huang, S. F. (2024). The role of artificial intelligence in transforming customer experience in the service industry in Nigeria. *International Journal of Academic Research in Business & Social Sciences, 14*(11). Advance online publication. DOI: 10.6007/IJARBSS/v14-i11/23532

Mbiti, I., & Weil, D. N. (2013). The home economics of e-money: Velocity, cash management, and discount rates of M-Pesa users. *The American Economic Review, 103*(3), 369–374. DOI: 10.1257/aer.103.3.369 PMID: 25067843

McKinsey & Company. (2020). Personalization at scale: Unleashing value across industries. Available at: https://www.mckinsey.com/industries/technology-media-and-telecommunications/our-insights/unlocking-the-value-of-personalization-at-scale-for-operators

McKinsey & Company. (2021). The State of AI in Personalization. Available at: https://www.mckinsey.com/capabilities/quantumblack/our-insights/global-survey-the-state-of-ai-in-2021

McKinsey & Company. (2021, November 12). The value of getting personalization right—or wrong—is multiplying. https://www.mckiney.com/capabilities/growth-marketing-and-sales/our-insights/the-value-of-getting-personalization-right-or-wrong-is-multiplying

McKinsey & Company. (2022). The Future of Automation: Impact on Businesses and Jobs. Retrieved from www.mckinsey.com

McKinsey. (2024). *Consumer Insights on AI and Data Privacy in Retail.* Retrieved from McKinsey & Company.

McKnight, D. H., Choudhury, V., & Kacmar, C. (2002). Developing and Validating Trust Measures for e-Commerce: An Integrative Typology. *Information Systems Research, 13*(3), 334–359. DOI: 10.1287/isre.13.3.334.81

McLean, G., & Osei-Frimpong, K. (2019). Examine the variables influencing the use of artificial intelligent in-home voice assistants. [REMOVED HYPERLINK FIELD]. *Computers in Human Behavior, 99*, 28–37. DOI: 10.1016/j.chb.2019.05.009

McLean, G., & Wilson, A. (2020). Empathy in digital interactions: Exploring its role in self-service technologies. *Computers in Human Behavior, 112*, 106479.

Medeiros, M. M. D., & Maçada, A. C. G. (2022). Competitive advantage of data-driven analytical capabilities: The role of big data visualization and of organizational agility. *Management Decision, 60*(4), 953–975. DOI: 10.1108/MD-12-2020-1681

Meghla, R. T., Deowan, M. E., Nuhel, A. K., Sazid, M. M., Ekbal, M. N., & Mahamud, M. H. (2022). An Internet of Things (IoT)-based smart automatic medication dispenser with an integrated web application for patient diagnosis. 2022 5th International Conference of Computer and Informatics Engineering (IC2IE), 16–21. DOI: 10.1109/IC2IE56416.2022.9970073

Mehrabi, N., Morstatter, F., Saxena, N., Lerman, K., & Galstyan, A. (2021). A survey on bias and fairness in machine learning. *ACM Computing Surveys, 54*(6), 1–35. DOI: 10.1145/3457607

Mehta, R. (2018). Corporate reputation and service innovation: Building customer cross-buying intentions. *Journal of Services Marketing, 32*(4), 410–424.

Mell, P. M., & Grance, T. (2011). The NIST definition of cloud computing. https://doi.org/DOI: 10.6028/NIST.SP.800-145

Metzger, M., Riedler, T., & Pédussel Wu, J. (2019). Migrant remittances: Alternative money transfer channels (Working Paper No. 127/2019). Institute for International Political Economy (IPE), Berlin School of Economics and Law.

Meub, L., & Proeger, T. (2022), "Künstliche Intelligenz in Handwerk und Mittelstand: Ein For schungsüberblick", ifh Forschungsbericht, No. 1, Volkswirtschaftliches Institut für Mittelstandund Handwerk an der Universität Göttingen (ifh), Göttingen

Meuter, M., Ostrom, A., Roundtree, R. & Bitner, M. (2000), "Self-service technologies: understanding customer satisfaction with technology-based service encounters", *Journal of Marketing,* 64 3, 50.

Meuter, M. L., Ostrom, A. L., Roundtree, R. I., & Bitner, M. J. (2000). Self-service technologies: Understanding customer satisfaction with technology-based service encounters. *Journal of Marketing, 64*(3), 50–64. DOI: 10.1509/jmkg.64.3.50.18024

Meyer, F., & Greve, G. (2019). *Customer-salesperson interaction technologies : Are robo-advisors replacing personal selling*. HSBA Hamburg School of Business Administration Cite as.

Michael, O. (2024). of Artificial Intelligence. The Future of Small Business in Industry 5.0, 215.

Mikalef, P., & Gupta, M. (2021). The role of competitive pressure in the diffusion of IT innovation: A study on artificial intelligence adoption. *Journal of Business Research, 124*, 437–445. DOI: 10.1016/j.jbusres.2020.10.045

Mikolajewicz, J., Demir, M., & Lee, W. (2022). Emotional intelligence in AI-driven systems: Enhancing customer satisfaction in service industries. *AI & Society, 37*(3), 453–463.

Miller, D., & Davis, R. (2022). Personalization in health-focused retail: The role of AI kiosks. *Health Retail Journal, 10*(1), 55–63. DOI: 10.9101/hrj.2022.10155

Miller, J., D'Souza, L., & Kaplan, D. (2023). Telemedicine kiosks: A transformative technology in the healthcare industry. *Health Informatics Journal, 29*(3), 206–217.

Miller, P., & Davis, J. (2022). *The role of AI in shaping modern retail experiences*. Digital Commerce Insights.

Mittelstadt, B. D., Allo, P., Taddeo, M., Wachter, S., & Floridi, L. (2016). The ethics of algorithms: Mapping the debate. *Big Data & Society, 3*(2), 2053951716679679. DOI: 10.1177/2053951716679679

Mohamed, Y. A., Khanan, A., Bashir, M., Mohamed, A. H. H., Adiel, M. A., & Elsadig, M. A. (2024). The impact of artificial intelligence on language translation: A review. *IEEE Access : Practical Innovations, Open Solutions, 12*, 25553–25579. DOI: 10.1109/ACCESS.2024.3366802

Mohapatra, A. G., Mohanty, A., Mohanty, S. K., Mahalik, N. P., & Nayak, S. (2025). Personalization and Customer Experience in the Era of Data-Driven Marketing. *Artificial Intelligence-Enabled Businesses: How to Develop Strategies for Innovation*, 467-511.

Moist, J. A., Anitsal, I., & Anitsal, M. M. (2021). The Customer Value Model and Mobile Banking: Evaluation of Technology-Based Self Service (TBSS) Gaps. *Atlantic Marketing Journal*, *10*(2), 12–19.

Mora Hernandez, D. A. (2023). Recommender Systems for a Connected Customer Journey in Brick-and-Mortar Stores/Author Daniel A. Mora. https://epub.jku.at/obvulihs/content/titleinfo/9534606

Moreau, F., & Wikström, P. (2024). The Impact of Algorithmic and Human Recommendations on Platform User Satisfaction. https://hal.science/hal-04553197

Morrison, A. M. (2022). *Hospitality and travel marketing*. Routledge., DOI: 10.4324/9781003292616

Mostafa, R. B., & Kasamani, T. (2021). Brand experience and brand loyalty: Is it a matter of emotions? *Asia Pacific Journal of Marketing and Logistics*, *33*(4), 1033–1051. DOI: 10.1108/APJML-11-2019-0669

Moura, M. (2022). Algorithmic transparency and privacy in AI systems: Regulatory implications. *Journal of Information Systems*, *55*(2), 87–103.

Muhammad, N. M., Kassim, S. B., Mahadi, N. F. B., & Ali, E. R. A. B. E. (2023). Role of Islamic Microfinance in Enhancing Financial Inclusion in Bangladesh: A Systematic Literature Review. *Islamic Sustainable Finance, Law and Innovation: Opportunities and Challenges*, 105-116.

Muralidharan, T. (2023). Strengthening the institutional mechanism of agricultural credit: Policies and practice–the 1960s to 2019–a review. In Agricultural Sector in India (pp. 291–310).

Mutawa, A. M., & Sruthi, S. (2024). Enhancing human–computer interaction in online education: A machine learning approach to predicting student emotion and satisfaction. *International Journal of Human-Computer Interaction*, *40*(24), 8827–8843. DOI: 10.1080/10447318.2023.2291611

Muzaffar, R., Muzaffar, S., & Muzaffar, Z. (2024). Digital Culture and the Transformational Impact of E-HRM in the Era of Digital Transformation. In *In New Strategy Models in Digital Entrepreneurship* (pp. 320–329). IGI Global Scientific Publishing., DOI: 10.4018/979-8-3693-3743-1.ch017

Nahar, K., Megha, A. H., & Sarker, J. U. (2025). Entering the Digital Banking Era: A Mechanism to Change and Accelerate Financial Access in Bangladesh. *International Journal of Innovative Science and Research Technology*, *10*(3), 2750–2761. DOI: 10.38124/ijisrt/25mar1048

Naik, A. S. (2024, June). Assistive Technology in Banking. In 2024 15th International Conference on Computing Communication and Networking Technologies (ICCCNT) (pp. 1-6). IEEE. DOI: 10.1109/ICCCNT61001.2024.10724888

Nair, S. K. (2025). Digital Transformation and Its Challenges in the Hospitality Industry in Malaysia: Moving Forward Into Industry Revolution 4.0. In *In Technological Innovations in the Food Service Industry* (pp. 369–388). IGI Global Scientific Publishing. DOI: 10.4018/979-8-3693-7683-6.ch017

Nandanwar, R.. (2024). Stitching data threads: Impact of artificial intelligence on fashion evolution. *International Conference on Innovative Practices in Technology and Management (ICIPTM)*, 1-6. DOI: 10.1109/ICIPTM59628.2024.10563840

Neelima, R., & Sharma, K. (2024). Data utilization and performance in health management systems. *Information Systems in Healthcare*, *13*(5), 165–180.

Ng, I. C. L., & Wakenshaw, S. Y. L. (2017). The Internet-of-Things: Review and research directions. *International Journal of Research in Marketing*, *34*(1), 3–21. DOI: 10.1016/j.ijresmar.2016.11.003

Nguyen, T. (2020). Starbucks Loyalty Reward application in Vietnam–identifying the impact of the app on Starbucks' customers in Vietnam using the technology acceptance model. https://www.theseus.fi/handle/10024/335323

Nguyen, T. H. C. (2024). Exploring the Role of Artificial Intelligence-Powered Facilitator in Enhancing Digital Competencies of Primary School Teachers. *European Journal of Educational Research*, *13*(1), 219–231. DOI: 10.12973/eu-jer.13.1.219

Nguyen, T. T., Hoang, T. D., & Le, T. T. (2022). Contactless payments: A silver lining of the COVID-19 pandemic. *Journal of Financial Services Marketing*, *27*(3), 276–287.

Nielsen, J. (2020). *10 usability heuristics for user interface design*. Nielsen Norman Group. https://www.nngroup.com/articles/ten-usability-heuristics/

Niwash, M. N. K., Cek, K., & Eyupoglu, S. Z. (2022). Intellectual Capital and Competitive Advantage and the Mediation Effect of Innovation Quality and Speed, and Business Intelligence. *Sustainability (Basel)*, *14*(6), 3497. DOI: 10.3390/su14063497

Norman, D. (2013). *The design of everyday things*. Basic Books.

Norman, D. A. (2013). *The Design of Everyday Things: Revised and Expanded Edition*. Basic Books.

Norman, D. A. (2020). *The design of everyday things: Revised and expanded edition*. Basic Books.

Nowell, L. S., Norris, J. M., White, D. E., & Moules, N. J. (2017). Thematic analysis: Striving to meet the trustworthiness criteria. [Details thematic analysis techniques and ensuring trustworthiness in qualitative research.]. *International Journal of Qualitative Methods*, *16*(1), 1–13. DOI: 10.1177/1609406917733847

Nussinov, R., Zhang, M., Liu, Y., & Jang, H. (2023). AlphaFold, allosteric, and orthosteric drug discovery: Ways forward. *Drug Discovery Today*, *28*(6), 103551. DOI: 10.1016/j.drudis.2023.103551 PMID: 36907321

O'Neil, C. (2016). *Weapons of math destruction: How big data increases inequality and threatens democracy*. Crown Publishing Group.

Oberoi, S., Arora, S., Verma, B., & Roy, K. K. (2024). What Do We Know About Artificial Intelligence and Blockchain Technology Integration in the Healthcare Industry? In Driving Decentralization and Disruption With Digital Technologies (pp. 124-138). IGI Global.

Oke, T. T., Ramachandran, T., Afolayan, A. F., Ihemereze, K. C., & Udeh, C. A. (2024). The role of artificial intelligence in shaping sustainable consumer behavior: A cross-sectional study of Southwest, Nigeria. *International Journal of Innovation and Scientific Research*, *1*(1), 255–266. Advance online publication. DOI: 10.51244/IJRSI.2023.1012021

Okoye, L. U., Adetiloye, K. A., Erin, O., & Judith, N. (2017). Financial inclusion as a strategy for enhanced economic growth and development. *Journal of Internet Banking and Commerce*, *22*, 1–14.

Olowoporoku, A. A., & Olubiyi, T. O. (2023). Evaluating Service Quality and Business Outcomes Post-Pandemic: Perspective from Hotels in Emerging Market. *Sawala: Jurnal Administrasi Negara*, *11*(2), 182–201. DOI: 10.30656/sawala.v11i2.6975

Olteanu, A., Castillo, C., Diaz, F., & Kıcıman, E. (2019). Social data: Biases, methodological pitfalls, and ethical boundaries. *Frontiers in Big Data*, *2*, 13. DOI: 10.3389/fdata.2019.00013 PMID: 33693336

Olubiyi T. O. (2023b). Leveraging competitive strategies on business outcomes post-covid-19 pandemic: Empirical investigation from Africa. *Journal of Management and Business: Research and Practice*, DOI: 10.54933/jmbrp-2023-15-2-1

Olubiyi, T. O. (2024b). Measuring organizational factors and their impact on the development of small and medium-sized pharmaceutical companies after the pandemic – an empirical study Journal of Marketing and Market Studies 25-38 (ERIH Plus & Index Copernicus indexed)DOI: 10.33226/1231-7853.2024.7.3

Olubiyi, O. T., Lawal, A. T., & Adeoye, O. O. (2022). Succession planning and family business continuity: Perspectives from Lagos State, Nigeria. *Organization and Human Capital Development, 1*(1), 40–52. DOI: 10.31098/orcadev.v1i1.865

Olubiyi, T. O. (2022). Measuring technological capability and business performance post-COVID Era: Evidence from Small and Medium-Sized Enterprises (SMEs) in Nigeria. *Management & Marketing Journal, xx*(2), 234–248. DOI: 10.52846/MNMK.20.2.09

Olubiyi, T. O. (2022b). An investigation of sustainable innovative strategy and customer satisfaction in small and medium-sized enterprises (SMEs) in Nigeria. *Covenant Journal of Business and Social Sciences, 13*(2), 1–24.

Olubiyi, T. O. (2023a). Unveiling the role of workplace environment in achieving the Sustainable Development Goal Eight (SDG8) and employee job satisfaction post-pandemic: Perspective from Africa. *Revista Management & Marketing Craiova, XXI*(2), 212–228. DOI: 10.52846/MNMK.21.2.02

Olubiyi, T. O. (2024a). Achieving Sustainable Development Goal Four (4) in Africa: Spotlighting the role of quality education and innovation orientation. Economics. *Management and Sustainability, 9*(1), 105–118. DOI: 10.14254/jems.2024.9-1.8

Olubiyi, T. O. (2024c). Improving Customer Experience Through Differentiation Capability: The Rise of Competition among Africa's Consumer Goods Companies. *PaperASIA, 40*(5b), 230–240. DOI: 10.59953/paperasia.v40i5b.233

Olubiyi, T. O., Adeoye, O. O., Jubril, B., Adeyemi, O. S., & Eyanuku, J. P. (2023). Measuring Inequality in Sub-Saharan Africa Post-Pandemic: Correlation Results for Workplace Inequalities and Implication for Sustainable Development Goal ten. *International Journal of Professional Business Review, 8*(4), e01405. DOI: 10.26668/businessreview/2023.v8i4.1405

Olubiyi, T. O., & Akpa, V. A. (2023). Does innovative capability and artificial intelligence really matter for the profitability of consumer goods companies in developing economies? *SunText Review of Economics & Business, 4*(3), 191–204.

Olubiyi, T. O., Jubril, B., Sojinu, O. S., & Ngari, R. (2022). Strengthening gender equality in small business and achieving sustainable development goals (SDGs), Comparative analysis of Kenya and Nigeria. *Sawala Jurnal Administrasi Negara, 10*(2), 168–186. DOI: 10.30656/sawala.v10i2.5663

Omeragić, D., Kečo, D., Jukić, S., & Isaković, B. (2023, June). The Employment of a Machine Learning-Based Recommendation System to Maximize Netflix User Satisfaction. In *International Symposium on Innovative and Interdisciplinary Applications of Advanced Technologies* (pp. 300–328). Springer Nature Switzerland., https://link.springer.com/chapter/10.1007/978-3-031-43056-5_23 DOI: 10.1007/978-3-031-43056-5_23

Omoge, A. P., Gala, P., & Horky, A. (2022). Disruptive technology and AI in the banking industry of an emerging market. *International Journal of Bank Marketing*, *40*(6), 1217–1247. DOI: 10.1108/IJBM-09-2021-0403

Omojolaibi, J. A., Okudo, A. G., & Shojobi, D. A. (2019). Are women financially excluded from formal financial services? Analysis of some selected local government areas in Lagos State, Nigeria. *Journal of Economic and Social Thought*, *6*, 16–47.

Omol, E. J. (2023). *Organizational digital transformation: from evolution to future trends*. Digital Transformation and Society., DOI: 10.1108/DTS-08-2023-0061

Omoyele, O. S., Olubiyi, T. O., Lanre-Babalola, F. O., Obadare, G. O., & Onikoyi, I. A. (2023). Business Model Innovation as a Catalyst for Sustainable Entrepreneurship: Empirical Findings from Small and Medium Enterprises in Nigeria. *Skyline Business Journal*, *19*(2), 55–64. DOI: 10.37383/SBJ190205

Omran, M. A., Saad, W. K., Hamza, B. J., & Fahe, A. (2021). A survey of various intelligent home applications using IoT and intelligent controllers. *Indonesian Journal of Electrical Engineering and Computer Science*, *23*(1), 490–499. DOI: 10.11591/ijeecs.v23.i1.pp490-499

Osakwe, J., Mutelo, S., & Obijiofor, N. (2023). Integrating Customer Relationship Management and Business Intelligence to Enhance Customer Satisfaction and Organisational Performance. A Literature Review *SSRN*: https://ssrn.com/abstract=4664792 DOI: 10.2139/ssrn.4664792

Osisanwo, F. Y., Akinsola, J. E. T., Awodele, O., Hinmikaiye, J. O., Olakanmi, O., & Akinjobi, J. (2017). Supervised machine learning algorithms: Classification and comparison. [IJCTT]. *International Journal of Computer Trends and Technology*, *48*(3), 128–138. DOI: 10.14445/22312803/IJCTT-V48P126

Oskarsson, P. (2018). *Landlock: Paralyzing dispute over minerals on Adivasi land in India*. Australian National University Press.

Othman, A. K., Hamzah, M. I., & Abu Hassan, L. F. (2020). Modeling the contingent role of technological optimism on customer satisfaction with self-service technologies: A case of cash-recycling ATMs. *Journal of Enterprise Information Management, 33*(3), 559–578. DOI: 10.1108/JEIM-09-2019-0295

Pal, A. K., & Wang, D. (2019). Deep learning in insurance: A comprehensive review. *European Journal of Finance.*

Pallant, J., Sands, S., & Karpen, I. (2020). Product customization: A profile of consumer demand. *Journal of Retailing and Consumer Services, 54*, 102030. DOI: 10.1016/j.jretconser.2019.102030

Pandey, N., Nayal, P., & Rathore, A. S. (2020). Digital marketing for B2B organizations: Structured literature review and future research directions. *Journal of Business and Industrial Marketing, 35*(7), 1191–1204. DOI: 10.1108/JBIM-06-2019-0283

Pantano, E., & Migliarese, P. (2014). Exploiting consumer–employee–retailer interaction in technology-enriched retail environments: Evidences from a study on the role of store format. *Journal of Retailing and Consumer Services, 21*(6), 958–965. DOI: 10.1016/j.jretconser.2014.08.015

Papadimitriou, A., Papantonis, A., & Loukas, G. (2016). Fraud detection in automobile insurance using machine learning techniques. *Procedia Computer Science.*

Paradza, D., & Daramola, O. (2021). Business intelligence and business value in organisations: A systematic literature review. *Sustainability (Basel), 13*(20), 11382. DOI: 10.3390/su132011382

Paraná, E. (2024). Artificial intelligence and the digitalization of finance in Latin America: Evidence from Brazil. *Globalizations*, •••, 1–20. DOI: 10.1080/14747731.2024.2415257

Parasuraman, A. (2023). Service quality and customer satisfaction in self-service technologies: A review and future directions. *Journal of the Academy of Marketing Science, 51*(3), 13–27.

Parasuraman, A., & Colby, C. L. (2015). An updated and streamlined technology readiness index: tRI 2.0. *Journal of Service Research, 18*(1), 59–74. DOI: 10.1177/1094670514539730

Parasuraman, A., & Colby, C. L. (2015). *Techno-ready marketing: How and why customers adopt technology*. Free Press.

Parasuraman, A., Zeithaml, V. A., & Malhotra, A. (2005). E-S-QUAL: A multiple-item scale for assessing electronic service quality. *Journal of Service Research, 7*(3), 213–233. DOI: 10.1177/1094670504271156

Park, G., & Kim, H. (2018). Low-cost implementation of a named entity recognition system for voice-activated human-appliance interfaces in a smart home. *Sustainability (Basel), 10*(2), 488. DOI: 10.3390/su10020488

Patel, N., & Trivedi, S. (2020). Leveraging Predictive Modeling, Machine Learning Personalization, NLP Customer Support, and AI Chatbots to Increase Customer Loyalty. *Empirical Quests for Management Essences, 3*(3), 1–24. https://www.researchberg.com/index.php/eqme/article/view/46

Patel, A. (2019). Smart fluid dispensing for medical and laboratory applications. *2019 International Conference on Smart Technologies and Innovations (ICSTI)*, London, UK.

Patel, N., & Trivedi, S. (2020). Leveraging predictive modeling, machine learning personalization, NLP customer support, and AI chatbots to increase customer loyalty. *Empirical Quests for Management Essences, 3*(3), 1–24.

Patel, S., Kumar, N., & Jain, P. (2021). Flow measurement and control in automatic dispensing systems. *Journal of Fluids Engineering, 145*(3), 15–20. DOI: 10.1115/1.4052018

Patila, H., Mahandulea, V., Khachanea, L., & Narkhedea, S. (2024). Multi-banking ATM system services using biometrics. *Management Science Letters, 14*(4), 219–226. DOI: 10.5267/j.msl.2024.3.002

Patil, B.. (2024). Artificial intelligence powered paradigm shift: Revolutionizing digital marketing. *International Conference on Innovative Practices in Technology and Management (ICIPTM)*, 1-6. DOI: 10.1109/ICIPTM59628.2024.10563450

Patil, D. (2024). Artificial intelligence in retail and e-commerce: Enhancing customer experience through personalization, predictive analytics, and real-time engagement. *Predictive Analytics, and Real-Time Engagement*. https://ssrn.com/abstract=5057420 or http://dx.doi.org/DOI: 10.2139/ssrn.5057420

Patil, D., Mhatre, A., Chavre, S., & Paras, T. (2025). AI-Assisted telemedicine kiosk for rural India. *Journal of Computational Science, 18*(04), 25–29.

Patroni, J. (2018). *Opening Up Innovation: How A Retailer Leverages User - Generated Social Media Content for Innovation. July*, 1–304. file:///C:/Users/Christian/Dropbox/Academia/Thesis/New

Patton, M. Q. (2015). *Qualitative research & evaluation methods: Integrating theory and practice* (4th ed.). Sage Publications. Discusses purposeful sampling, data triangulation, and credibility techniques such as member checking and peer debriefing.

Pech, M., Vrchota, J., & Bednář, J. (2021). Predictive maintenance and intelligent sensors in smart factory. *Sensors (Basel), 21*(4), 1470. DOI: 10.3390/s21041470 PMID: 33672479

Pendyala, R., & Lakkamraju, S. (2024). Real-time customer insights through AI-driven analytics. *AI in Marketing Journal, 21*(2), 143–158.

Persson, M., & Wallo, A. (2024). Digital automation and working life of HR practitioners: A gender analysis of the implications for workforce and work practices. *Gender, Technology and Development, 28*(3), 408–427. DOI: 10.1080/09718524.2024.2406039

Pew Research Center. (2019). Americans and Data Privacy: Concerned, Confused, and Conflicted. Available at: https://www.pewresearch.org/internet/2019/11/15/americans-and-privacy-concerned-confused-and-feeling-lack-of-control-over-their-personal-information/

Pew Research Center. (2019). Americans and privacy: Concerned, confused, and feeling lack of control over their personal information. Available at: https://www.pewresearch.org/internet/wp-content/uploads/sites/9/2019/11/Pew-Research-Center_PI_2019.11.15_Privacy_FINAL.pdf

Pfeiffer, D. (2022). The rise of self-checkouts in retail: Implications for customer experience and business performance. *Retail Management Review, 43*(2), 215–231.

Piffari, C., Lagorio, A., Cimini, C., & Pinto, R. (2022). The role of human factors in the human-centred design of service processes: a focus on the healthcare sector. *27th Proceedings of the Summer School Francesco Turco*.

Pinto, J. F., Vilaça, J. L., & Dias, N. S. (2021). A review of current pill organizers and dispensers. 2021 IEEE 9th International Conference on Serious Games and Applications for Health (SeGAH), 1–8. https://doi.org/DOI: 10.1109/SEGAH52098.2021.9551894

Play.ht. (2024). AI kiosks: Enhancing customer experiences with artificial intelligence. https://play.ht/blog/ai-kiosk/

Porter, M. E., & Heppelmann, J. E. (2015). How smart, connected products are transforming companies. *Harvard Business Review, 93*(10), 96–114.

Poufinas, T., Gogas, P., Papadimitriou, T., & Zaganidis, E. (2023). Machine learning in forecasting motor insurance claims. *Risks, 11*(9), 164. DOI: 10.3390/risks11090164

Prabhu, S.. (2024). Elevating business experiences: AI-powered chatbots reforming marketing and e-commerce. In Singh, D. (Eds.), *Transforming the Financial Landscape with ICTs* (pp. 1–27). IGI Global., DOI: 10.4018/979-8-3693-1503-3.ch001

Prahalad, C. K., & Ramaswamy, V. (2004). Co-creating unique value with customers. *Strategy and Leadership, 32*(3), 4–9. DOI: 10.1108/10878570410699249

Prasad, S., Kiran, R., & Sharma, R. K. (2020). Influence of financial literacy on retail investors' decisions in relation to return, risk and market analysis. *International Journal of Finance & Economics, n/a*(n/a). https://doi.org/https://doi.org/10.1002/ijfe.1920

Procuring innovation, wherever it may be. (2017, August 17). McKinsey & Company. https://www.mckinsey.com/capabilities/operations/our-insights/procuring-innovation-wherever-it-may-be

Punnapurath, S., Vijayakumar, P., Platty, P. L., Krishna, S., & Thomas, T. (2021). A study of medication compliance in geriatric patients with chronic illness. *Journal of Family Medicine and Primary Care, 10*(4), 1644–1648. DOI: 10.4103/jfmpc.jfmpc_1302_20 PMID: 34123906

Puschmann, T. (2017). Fintech. *Business & Information Systems Engineering, 59*(1), 69–76. DOI: 10.1007/s12599-017-0464-6

PwC. (2020). *Financial services technology 2020 and beyond: Embracing disruption.* https://www.pwc.com/gx/en/industries/financial-services/publications/financial-services-technology-2020-and-beyond.html

PwC. (2023). AI in business: Balancing opportunity with risk. PwC Global AI Report.

Qureshi, O. A., Wilkins, S., & Iqbal, H. (2025). AI-Powered Banking Assistants: Enhancing Customer Service with Technology. In BUiD Doctoral Research Conference (pp. 43-53). Springer, Cham.

Radanovic, I., & Likic, R. (2018). Opportunities for Use of Blockchain Technology in Medicine. *Applied Health Economics and Health Policy, 16*(5), 583–590. DOI: 10.1007/s40258-018-0412-8 PMID: 30022440

Raddats, C., Naik, P., & Ziaee Bigdeli, A. (2022). Creating value in servitization through digital service innovations. *Industrial Marketing Management, 104*, 1–13. DOI: 10.1016/j.indmarman.2022.04.002

Rajendiran, G. R. (2025). Distributed Ledger Technology for Transparent and Secure Warehouse Inventory Tracking. *Journal of Current Science and Research Review, 3*(01), 01-10.

Rajpurkar, P., Chen, E., Banerjee, O., & Topol, E. J. (2022). AI in health and medicine. *Nature Medicine, 28*(1), 31–38. DOI: 10.1038/s41591-021-01614-0 PMID: 35058619

Ramakrishnan, T., Kathuria, A., & Saldanha, T. J. (2020). Business intelligence and analytics (BI&A) capabilities in healthcare. In Theory and Practice of Business Intelligence in Healthcare (pp. 1-17). IGI Global.

Raman, R., Nair, V. K., Nedungadi, P., Sahu, A. K., Kowalski, R., Ramanathan, S., & Achuthan, K. (2024). Fake news research trends, linkages to generative artificial intelligence and sustainable development goals. *Heliyon, 10*(3), e24727. DOI: 10.1016/j.heliyon.2024.e24727 PMID: 38322879

Rana, S., & Kaur, D. (2024). Exploring the challenges and facilitators in the adoption of e-HRM practices in Indian higher education institutions: A qualitative exploration. *The International Journal of Organizational Analysis, 32*(8), 1419–1439. DOI: 10.1108/IJOA-05-2023-3745

Rasheed, R., & Rashid, A. (2023). Role of service quality factors in word of mouth through student satisfaction. *Kybernetes*.

Rathod, S., & Arelli, S. K. P. (2013). Aadhaar and financial inclusion: A proposed framework to provide basic financial services in unbanked rural India. In *Driving the economy through innovation and entrepreneurship* (pp. 731–744). Springer. DOI: 10.1007/978-81-322-0746-7_60

Rayhan, A., Rayhan, R., & Rayhan, S. (2023). Artificial general intelligence: Roadmap to achieving human-level capabilities.

RBI & NABARD. (2022). Trend report on financial inclusion in India.

Reisman, E. (2020). Self-service kiosks and the future of healthcare. *Healthcare Technology News, 28*(6), 32–35.

Reisman, F. (2020). Telehealth technology adoption: The role of self-service in healthcare delivery. *Healthcare Technology Journal, 12*(4), 198–210.

Riegger, A. S., Klein, J. F., Merfeld, K., & Henkel, S. (2021). Technology-enabled personalization in retail stores: Understanding drivers and barriers. *Journal of Business Research, 123*, 140–155. DOI: 10.1016/j.jbusres.2020.09.039

Rigoni, E. H., Dwivedi, R., & Hoppen, N. (2010). IT governance and business—IT strategic alignment commitment: A study of Brazilian firms. *International Journal of Global Management Studies Professional*, *2*(1), 1–20.

Rogers, E. M. (1962). *Diffusion of innovations*. Free Press of Glencoe.

Rose, S. (2018). The role of artificial intelligence in the transformation of insurance. Journal of Insurance, Financial Management, and Risk Management.

Russell, J. A. (1980). A circumplex model of affect. *Journal of Personality and Social Psychology*, *39*(6), 1161–1178. DOI: 10.1037/h0077714

Ruvoletto, R. (2023). Digitalization and Internationalization: An Analysis of the Impact of Digital Technologies on Export Management Practices.

Sadaf, A., Newby, T. J., & Ertmer, P. A. (2012). Exploring Factors that Predict Preservice Teachers' Intentions to Use Web 2.0 Technologies Using Decomposed Theory of Planned Behavior. *Journal of Research on Technology in Education*, *45*(2), 171–196. DOI: 10.1080/15391523.2012.10782602

Sahib, U. (2020). Smart Dubai: sensing Dubai smart city for smart environment management. *Smart environment for smart cities*, 437-489.

Salamkar, M. A. (2023). Data Integration: AI-driven approaches to streamline data integration from various sources. *International Journal of Science and Research (Raipur, India)*, *3*(1), 668–694. https://scienceacadpress.com/index.php/jaasd/article/view/221. DOI: 10.21275/SR230311115337

Saldanha, T. J. V., Kathuria, A., Khuntia, J., & Konsynski, B. R. (2022). Ghosts in the Machine: How Marketing and Human Capital Investments Enhance Customer Growth When Innovative Services Leverage Self-Service Technologies. *Information Systems Research*, *33*(1), 76–109. DOI: 10.1287/isre.2021.1006

Sallaku, R. (2021). Customer engagement, authenticity and loyalty in the sharing economy: a study of Italian tourists. https://iris.univr.it/handle/11562/1052955

Samir, A., & Ahlawat, P. (2019). *Machine learning and the future of risk management and insurance*. Deloitte University Press.

Sandhu, K., Dayanandan, A., & Kuntluru, S. (2023). *Fintech innovation for financial inclusion: Can India make it?* International Journal of Accounting & Information Management., DOI: 10.1108/IJAIM-03-2023-0058

Sangwan, S., & Sharma, S. K. (2022). Social media communication and consumer decision making: An empirical perspective. *International Journal of Electronic Marketing and Retailing*, *13*(4), 391–410. DOI: 10.1504/IJEMR.2022.125589

Sanjay. (2023). What is Financial Services? Characteristics, Functions, Classification. Geektonight. https://www.geektonight.com/financial-services/

Sarker, I. H. (2021). Deep learning: A comprehensive overview on techniques, taxonomy, applications and research directions. *SN Computer Science*, *2*(6), 420. DOI: 10.1007/s42979-021-00815-1 PMID: 34426802

Schönberger, M. (2023). Artificial intelligence for small and medium-sized enterprises: Identifying key applications and challenges. *Journal of Business and Management*, *21*, 89–112. DOI: 10.32025/JBM23004

Schuetz, S., & Venkatesh, V. (2020). Blockchain, adoption, and financial inclusion in India: Research opportunities. *International Journal of Information Management*, *52*, 101936. DOI: 10.1016/j.ijinfomgt.2019.04.009

Schumacher, E. F. (1973). *Small is beautiful: a study of ecomonics as if people mattered*. Vintage.

Science, N. the Automated Journalism Startup - Technology and Operations Management. (2018, November 14). Technology and Operations Management. https://d3.harvard.edu/platform-rctom/submission/narrative-science-the-automated-journalism-startup/

Seddon, P. B., Constantinidis, D., Tamm, T., & Dod, H. (2017). How does business analytics contribute to business value? *Information Systems Journal*, *27*(3), 237–269. DOI: 10.1111/isj.12101

Şentürk, H., & Keskin, C. (2010). The Importance of Small and Medium-Sized Enterprises (SMEs) in Economies: Swot Analysis of The SME Sector in Turkey and Albania. *Niğde Üniversitesi İktisadi ve İdari Bilimler Fakültesi Dergisi*, *3*(1), 116–132.

Shackleford, D. (2020). AI and Machine Learning in Cybersecurity. *Cybersecurity Journal*, *12*(3), 45–58.

Shamshoddin, S., Khader, J., & Gani, S. (2020). Predicting consumer preferences in electronic market based on IoT and Social Networks using deep learning based collaborative filtering techniques. *Electronic Commerce Research*, *20*(2), 241–258. https://link.springer.com/article/10.1007/s10660-019-09377-0. DOI: 10.1007/s10660-019-09377-0

Sharma, A., Patel, N., & Gupta, R. (2022). Enhancing Revenue Through AI: Utilizing Neural Networks and Reinforcement Learning for Upsell and Cross-Sell Strategies. *European Advanced AI Journal, 11*(8).

Sharma, R. (2024). What Is Decentralized Finance (DeFi) and How Does It Work? Investopedia. https://www.investopedia.com/decentralized-finance-defi-5113835

Sharma, A., Gangwar, V. P., Bhattacharyya, A., Giri, U., Chavan, G., & Mahato, S. (2025). Leveraging Intelligent Process Automation for Customer Engagement in Sales and Marketing. In *Advancements in Intelligent Process Automation* (pp. 435–458). IGI Global.

Sharma, D., Reddy, N., Gupta, P., & Sharma, R. (2022). Enhancing customer experience personalization through AI: Leveraging collaborative filtering, neural networks, and natural language processing. *Journal of AI ML Research*, *11*(7). https://www.joaimlr.com/index.php/v1/article/view/32

Sharma, T. K., Gupta, P., & Singh, V. (2022). AI-driven cybersecurity: Opportunities and challenges. *Computers & Security*, *115*, 102689.

Shih-Chih, C., Huei-Huang, C., & Mei-Fang, C. (2009). Determinants of satisfaction and continuance intention towards self-service technologies. *Industrial Management & Data Systems*, *109*(9), 1248–1263. DOI: 10.1108/02635570911002306

Siddiqui, A. (2024). The impact of artificial intelligence on business operation: Current state, future opportunities and challenges. [IJM]. *International Journal of Management*, *15*(04), 79–93. https://mylib.in/index.php/IJM/article/view/IJM_15_04_006

Silver, C. (2023, September 13). Lessons From the 2008 Financial Crisis. Investopedia. https://www.investopedia.com/news/10-years-later-lessons-financial-crisis

Silverman, D. (2020). Interpreting qualitative data (6th ed.). Sage Publications. Offers in-depth discussion on data collection methods such as document analysis, interviews, and observations.

Simon, F., & Usunier, J. C. (2007). Cognitive, demographic, and situational determinants of service customer preferences for personnel-in-contact over self-service technology. *International Journal of Research in Marketing*, *24*(2), 163–173. DOI: 10.1016/j.ijresmar.2006.11.004

Simon, H. A. (1995). Artificial intelligence: An empirical science. *Artificial Intelligence*, *77*(1), 95–127. DOI: 10.1016/0004-3702(95)00039-H

Singh, R., & Anderson, D. (2019). Emerging insurance models: An overview of AI use-cases in insurance. Internet of Things Institute.

Singh, R., & Singh, V. P. (2025). Global Mobility and Economic Incentives in Prenatal Tourism. In, Exploration of Prenatal Tourism and Birthright Citizenship (pp. 77-104). IGI Global Scientific Publishing. https://doi.org/ Available on: https://www.igi-global.com/gateway/chapter/365545DOI: 10.4018/979-8-3693-8206-6.ch005

Singh, R., Singh, V. P., Rabby, F., Bansal, R., & Gupta, N. K. (2025). The Road to Sustainable Tourism: Financial Mechanisms for Green Project Development. In B. Sousa, L. Arantes, & S. Bhartiya (Eds.), Regenerative Tourism for Social Development (pp. 235-254). IGI Global Scientific Publishing. Available on: https://doi.org/DOI: 10.4018/979-8-3373-0235-5.ch010

Singh, N. D., & Singh, H. R. (2012). Social impact of microfinance on SHG members: A case study of Manipur. *Prabandhan: Indian Journal of Management*, *5*(3), 43–50. DOI: 10.17010/pijom/2012/v5i3/60155

Singh, S., Goyal, S. K., & Sharma, S. K. (2013). Technical efficiency and its determinants in microfinance institutions in India: A firm-level analysis. *Journal of Innovation Economics Management*, *1*(11), 15–31. DOI: 10.3917/jie.011.0015

Sinha, S., & Sathiya Narayanan, R. (2024). Harnessing Sentiment Analysis Methodologies for Business Intelligence Enhancement and Governance Intelligence Evaluation', International [IJISAE]. *Journal of Intelligent Systems and Applications in Engineering*, *12*(11s), 166–176.

Sisodiya, M.. (2024). The transformative power of robotic process automation: Redefining business in industry 4.0. *International Conference on Cloud Computing, Data Science & Engineering (Confluence)*, 498-503. DOI: 10.1109/Confluence60223.2024.10463415

Sitorus, M. (2016). Format Paper atau Jornal. ResearchGate. DOI: 10.13140/RG.2.2.21653.12006

Smith, A. (2021). Personalized Digital Banking Solutions: A Case Study of HSBC. *Journal of Digital Finance*, *10*(4), 58–69.

Smith, J. (2021). Seasonal trends in AI-driven retail recommendations. *Retail Analytics Quarterly*, *9*(3), 77–85. DOI: 10.2345/raq.2021.09377

Smith, J., Allen, L., & Patel, R. (2020). Privacy and Security Concerns in Digital Banking. *Cybersecurity Review*, *12*(1), 32–45.

Smith, J., Johnson, M., & Williams, R. (2020). The impact of self-service technologies on customer satisfaction in quick-service restaurants. *Journal of Hospitality and Tourism Technology*, *11*(3), 345–360. DOI: 10.1108/JHTT-01-2020-0032

Smith, K. (2021). AI and seasonal shopping trends: A data-driven approach. *Business Tech Review, 16*(2), 98–110.

Smith, R. (2020). An overview of fluid dispensing techniques in automated systems. *IACSIT International Journal of Engineering and Technology, 10*(4), 452–460. DOI: 10.1016/j.eng.2020.08.006

Soderberg, E. (2022). An evaluation of the usage of affective computing in healthcare. In *UMEÅ's 25th Student Conference in Computing Science* (pp. 69-78).

Soltani, Z., Zareie, B., Milani, F. S., & Navimipour, N. J. (2021). The impact of the customer relationship management on the organization performance. *The Journal of High Technology Management Research, 32*(2), 100399.

Spais, G. (2023). Engaging Self-service in a Customer Service Ecosystem. In *Artificial Intelligence in Customer Service: The Next Frontier for Personalized Engagement* (pp. 123–154). Springer International Publishing. DOI: 10.1007/978-3-031-33898-4_6

Spasova, G. V., Karova, M. N., & Boychev, I. Z. (2024). Communication protocol for liquid dispensing vending machines. 2024 59th International Scientific Conference on Information, Communication and Energy Systems and Technologies (ICEST), 1–4. https://doi.org/DOI: 10.1109/ICEST62335.2024.10639714

Sridhar, V., & Ganesan, R. (2015). Service innovation facilitators: A customer satisfaction perspective. *International Journal of Services and Operations Management, 20*(1), 90–105.

Stake, R. E. (2006). *Multiple case study analysis*. Guilford Press. Justifies the use of multiple case study methodology and cross-case analysis.

Stanford Institute for Human-Centered Artificial Intelligence. (2024, February 22). Rethinking privacy in the AI era: Policy provocations for a data-centric world. https://hai.stanford.edu/policy/white-paper-rethinking-privacy-ai-era-policy-provocations-data-centric-world

State Bank of India. (n.d.). *State Bank of India*. Retrieved March 28, 2025, from https://www.onlinesbi.sbi/

Statista (2023). Customer satisfaction index scores for selected video streaming apps in the United States in 2023 and 2024. Available at: https://www.statista.com/statistics/1331852/customer-satisfaction-indices-video-streaming-services-us/

Statista. (2023). *Internet of Things (IoT) connected devices installed base worldwide from 2015 to 2025 (in billions)*. Statista. https://www.statista.com/statistics/471264/iot-number-of-connected-devices-worldwide/

Statista. (2023). *Number of connected devices worldwide 2019–2030*. Statista Research Department. Retrieved from https://www.statista.com/statistics/1101442/iot-number-of-connected-devices-worldwide

SUIE. (2023, December 31). The future of self-service: AI and personalization in kiosk technology. https://www.suieworld.com/a-news-the-future-of-self-service-ai-and-personalization-in-kiosk-technology/

Sundararajan, V., & Gao, H. (2022). The business impact of self-service checkout systems: Case studies in retail. *Journal of Retailing*, *78*(4), 190–201.

Suoranta, M., & Mattila, M. (2004). Mobile banking and consumer behavior: New insights into the diffusion pattern. *Journal of Financial Services Marketing*, *8*(4), 354–366. DOI: 10.1057/palgrave.fsm.4770132

Suresh, H., & Guttag, J. V. (2021). A framework for understanding unintended consequences of machine learning. *Communications of the ACM*, *64*(11), 62–71.

Swift's cross-border payments processing speed surpasses G20 target | Swift. (2023, August 22). Swift. https://www.swift.com/news-events/press-releases/swifts-cross-border-payments-processing-speed surpasses-g20-target

Syed, W. K., Mohammed, A., Reddy, J. K., & Dhanasekaran, S. (2024, July). Biometric Authentication Systems in Banking: A Technical Evaluation of Security Measures. In 2024 IEEE 3rd World Conference on Applied Intelligence and Computing (AIC) (pp. 1331-1336). IEEE. DOI: 10.1109/AIC61668.2024.10731026

Tamm, T., Seddon, P. B., & Shanks, G. 2020. How Do Different Types of BA Users Contribute to Business Value? Communications of the Association for Information Systems, 46, pp-pp. https://doi.org/DOI: 10.17705/1CAIS.04628

Tamm, T., Seddon, P. B., Shanks, G., & Reynolds, P. (2020). How does big data analytics enable organizational agility? *Information & Management*, *57*(1), 103120. DOI: 10.1016/j.im.2019.103120

Tandon, S., Verma, K., & Dixit, A. (2021). Intangible product features and consumer trust in e-commerce. *Electronic Commerce Research and Applications*, *40*(3), 211–226.

Tang, Z., Guan, L., & Yuan, Q. (2018). Automated vehicle damage assessment using deep learning. *IEEE Intelligent Vehicles Symposium*.

Tan, L., Liu, S., Gao, J., Liu, X., Chu, L., & Jiang, H. (2024). Enhanced self-checkout system for retail based on improved YOLOv10. *Journal of Imaging*, *10*(10), 248. DOI: 10.3390/jimaging10100248 PMID: 39452411

Tan, M., Teo, H. H., & Wei, K. K. (2019). Technology acceptance model in self-service innovations: A cross-cultural perspective. *Journal of Service Research*, *15*(1), 89–101.

Tapalaga, A., & Tapalaga, A. (2024). Exploring the Rise of DeFi: A New Era in Financial Services - History of Yesterday. History of Yesterday - Let the Truth Unfold. https://historyofyesterday.com/exploring-the-rise-of-defi-a-new-era-in-financial-services/

Tapscott, D., & Tapscott, A. (2016). *Blockchain revolution: How the technology behind bitcoin is changing money, business, and the world*. Portfolio.

Tech, D. (2024, January 31). What Strategies Exist To Counter Bitcoin Double Spending? Bitamp. thttps://blog.bitamp.com/what-strategies-exist-to-counter-bitcoin-double-spending/

Thakral, S., & Kamra, V. (2024). Self-service technologies: A bibliometric analysis. *International Journal of Financial Engineering*, *11*(04), 2442007. DOI: 10.1142/S2424786324420076

Thomson, J. (2022). How AI And Data Analytics Drive Value In The Finance Function. Forbes. https://www.forbes.com/sites/jeffthomson/2021/12/17/how-ai-and-data-analytics-drive-value-in-the-finance-function/

Thottoli, M. M., Islam, M. A., Yusof, M. F. B., Hassan, M. S., & Hassan, M. A. (2023). Embracing Digital Transformation in Financial Services: From Past to Future. *SAGE Open*, *13*(4), 21582440231214590. Advance online publication. DOI: 10.1177/21582440231214590

Tienken, C., Rösler, J., & Friedli, T. (2023). Enhancing the customer journey with digital self-services. In Schneider, U., & Hoika, J. (Eds.), *Digital marketing in the automotive electronics industry. Management for Professionals*. Springer., DOI: 10.1007/978-3-031-30720-1_13

Torres, R., Sidorova, A., & Jones, M. C. (2018). Enabling firm performance through business intelligence and analytics: A dynamic capabilities perspective. *Information & Management*, *55*(7), 822–839. DOI: 10.1016/j.im.2018.03.010

Torres, R., Sidorova, A., & Jones, M. C. (2018). Enabling self-service BI: A framework and case study. *Decision Support Systems*, *109*, 113–123. DOI: 10.1016/j.dss.2018.03.009

Transforming Financial Services in the Digital Age and its Implications. (2023, July 4). World Finance Council. https://worldfinancecouncil.org/articles/transforming-financial-services-in-the-digital-age-and-its-implications/

Trewin, S., & Phipps, L. (2019). Accessibility and inclusive design: Principles and practices for digital services. *Journal of Inclusive Technology*, *7*(2), 112–130. https://doi.org/xx.xxx/yyyy

Trinh, H. J. (2023). The Algorithmic Black Box: Exploring the Impact of Spotify and TikTok on User Behavior (Master's thesis, University of Agder). https://uia.brage.unit.no/uia-xmlui/handle/11250/3082201

Triono, S.P.H. & Jaya, R.C. (2021). Business Intelligence & Analytics and Its Effect on Indonesia's Startup Performance, *Advances in Economics, Business and Management Research, 187*

Tsindeliani, I. A., Proshunin, M. M., Sadovskaya, T. D., Popkova, Z. G., Davydova, M. A., & Babayan, O. A. (2022). Digital transformation of the banking system in the context of sustainable development. *Journal of Money Laundering Control*, *25*(1), 165–180. DOI: 10.1108/JMLC-02-2021-0011

Tsou, H., & Chen, L. (2019). The Influence of Service Innovation Capability for Self-Service Technology Investment. *Canadian Journal of Administrative Sciences / Revue Canadienne Des Sciences de l'Administration, 36*(4), 544–558. DOI: 10.1002/cjas.1520

Tula, S. T., Kess-Momoh, A. J., Omotoye, G. B., Bello, B. G., & Daraojimba, A. I. (2024). AI-Enabled customer experience enhancement in business. *Computer Science & IT Research Journal*, *5*(2), 365–389. DOI: 10.51594/csitrj.v5i2.789

U.S. National Institute of Standards and Technology (NIST). (2020). AI model security: Addressing vulnerabilities in adversarial attacks. *NIST Special Publication 800-207.*

Ukabi, O. B., Uba, U. J., Ewum, C. O., & Olubiyi, T. O. (2023). Measuring Entrepreneurial Skills and Sustainability in Small Business Enterprises Post-Pandemic: Empirical Study from Cross River State, Nigeria. *International Journal of Business. Management and Economics*, *4*(2), 132–149. DOI: 10.47747/ijbme.v4i2.1140

Ukhalkar, P. K., Phursule, R. N., Gadekar, D. P., & Sable, N. P. (2020). Business Intelligence and Analytics: Challenges and Opportunities. *International Journal of Advanced Science and Technology*, *29*(12s), 2669–2676.

Uwem, E. I., Oyedele, O. O., & Olubiyi, O. T. (2021). Workplace green behavior for sustainable competitive advantage. In human resource management practices for promoting *Sustainability IGI Global.* (248–263).

Vakulenko, Y., Hellström, D., & Oghazi, P. (2018). Customer value in self-service kiosks: A systematic literature review. *International Journal of Retail & Distribution Management, 46*(5), 507–527. DOI: 10.1108/IJRDM-04-2017-0084

Vallabhaneni, A. S., Perla, A., Regalla, R. R., & Kumari, N. (2024). The Power of Personalization: AI-Driven Recommendations. In Minds Unveiled (pp. 111-127). Productivity Press.

Vashishth, T. K., Sharma, K. K., Kumar, B., Chaudhary, S., & Panwar, R. (2025). Enhancing Customer Experience through AI-Enabled Content Personalization in E-Commerce Marketing. *Advances in Digital Marketing in the Era of Artificial Intelligence*, 7-32.

Vegesna, D. V. V. (2023). Comprehensive analysis of AI-enhanced defense systems in cyberspace. *International Numeric Journal of Machine Learning and Robots, 7*(7). https://injmr.com/index.php/fewfewf/article/view/21

Venice, J. A., Arivazhagan, D., Suman, N., Shanthi, H. J., & Swadhi, R. (2025). Recommendation Systems and Content Personalization: Algorithms, Applications, and Adaptive Learning. In *AI for Large Scale Communication Networks* (pp. 323–348). IGI Global.

Venkatesh, V., & Davis, F. D. (2000). A theoretical extension of the technology acceptance model: Four longitudinal field studies. *Management Science, 46*(2), 186–204. DOI: 10.1287/mnsc.46.2.186.11926

Venkatesh, V., Morris, M. G., Davis, G. B., & Davis, F. D. (2003). User acceptance of information technology: Toward a unified view. *Management Information Systems Quarterly, 27*(3), 425–478. DOI: 10.2307/30036540

Venkatesh, V., Thong, J. Y. L., & Xu, X. (2012). Consumer acceptance and use of information technology: Extending the unified theory of acceptance and use of technology. *Management Information Systems Quarterly, 36*(1), 157–178. DOI: 10.2307/41410412

Venkatesh, V., Thong, J. Y. L., & Xu, X. (2022). Consumer acceptance of self-service technologies in banking: A longitudinal study. *Information Systems Research, 33*(2), 123–139.

Verbeeck, N., Caprioli, R. M., & Van de Plas, R. (2020). Unsupervised machine learning for exploratory data analysis in imaging mass spectrometry. *Mass Spectrometry Reviews*, *39*(3), 245–291. DOI: 10.1002/mas.21602 PMID: 31602691

Verhagen, T., Meents, S., & Tan, Y. H. (2019). Perceived Risk and Trust in Online Shopping: A Meta-Analysis. *Journal of Retailing and Consumer Services*, *47*, 69–77.

Verhoef, P. C., Broekhuizen, T., & Bijmolt, T. H. (2016). Understanding the effect of customer experience across channels: A multi-method approach. *Journal of Marketing*, *80*(6), 39–58. DOI: 10.1509/jm.15.0420

Verma, B., & Chatterji, N. (2024). Cognitive Unburdening: Investigating the Mediated Pathway From Digital Detox to Psychological Well Being Through Reduced Cognitive Load. In Business Drivers in Promoting Digital Detoxification (pp. 36-53). IGI Global.

Verma, B., Mittal, A., Raman, M., & Sindhav, B. (Eds.). (2024b). *Augmenting Retail Reality, Part A*. VR, and the Internet of Things.

Verma, B., & Srivastava, A. (2023). Impact of different dimensions of globalisation on firms' performance: An unbalanced panel-data study of firms operating in India. *World Review of Entrepreneurship, Management and Sustainable Development*, *19*(3-5), 360–378. DOI: 10.1504/WREMSD.2023.130618

Verma, B., Srivastava, A., Mehta, R., & Chandel, J. (2022, February). FDI-linked Spillovers and the Indian Economic Growth: The role of Country's Absorptive Capacity. In *2022 IEEE Delhi Section Conference (DELCON)* (pp. 1-6). IEEE. DOI: 10.1109/DELCON54057.2022.9753540

Verma, B., & Tandon, U. (2022). Modelling barriers to wearable technologies in Indian context: Validating the moderating role of technology literacy. *Global Knowledge, Memory and Communication*.

Vijay, J., & Priyanka, U. (2025). Digital Transformation in Customer Experience and Behavior. In *Digital Transformation Initiatives for Agile Marketing* (pp. 251–274). IGI Global.

Vijay, J., & Priyanka, U. (2025). Digital Transformation in Customer Experience and Behavior. In *In Digital Transformation Initiatives for Agile Marketing* (pp. 251–274). IGI Global Scientific Publishing., DOI: 10.4018/979-8-3693-4466-8.ch010

Villani, I. (2018). *Transform Customer Experience: How to achieve customer success and create exceptional CX*. John Wiley & Sons.

Wadhwa, S., Wadhwa, P., & Khalique, F. (2024). "Explorative Study on Understanding the Human-Centred Design to Nurture the Employee Experience", Shrivastava, R., & Jain, K. (Ed.) *Humanizing Businesses for a Better World of Work*, Emerald Publishing Limited, Leeds, pp. 109-126. https://doi.org/DOI: 10.1108/978-1-83797-332-320241008

Waghambare, M. A.. (2024). Artificial intelligence (AI)-powered chatbots for marketing and online shopping. In Prakasha, G. S. (Eds.), *Digital Technologies in Modeling and Management: Insights in Education and Industry* (pp. 21–39). IGI Global., DOI: 10.4018/978-1-6684-9576-6.ch002

Walsh, C. A., Cahir, C., Tecklenborg, S., Byrne, C., Culbertson, M. A., & Bennett, K. E. (2019). The association between medication non-adherence and adverse health outcomes in ageing populations: A systematic review and meta-analysis. *British Journal of Clinical Pharmacology*, *85*(11), 2464–2478. DOI: 10.1111/bcp.14075 PMID: 31486099

Wan, S., Li, D., Gao, J., Roy, R., & He, F. (2018). A collaborative machine tool maintenance planning system based on content management technologies. *International Journal of Advanced Manufacturing Technology*, *94*(5-8), 1639–1653. DOI: 10.1007/s00170-016-9829-0

Ward, L., Dunn, A., Faghaninia, A., Zimmermann, N. E., Bajaj, S., Wang, Q., Montoya, J., Chen, J., Bystrom, K., Dylla, M., Chard, K., Asta, M., Persson, K. A., Snyder, G. J., Foster, I., & Jain, A. (2018). Matminer: An open source toolkit for materials data mining. *Computational Materials Science*, *152*, 60–69. DOI: 10.1016/j.commatsci.2018.05.018

Weijters, B., Rangarajan, D., Falk, T., & Schillewaert, N. (2007). Determinants and outcomes of customers' use of self-service technology in a retail setting. *Journal of Service Research*, *10*(1), 3–21. DOI: 10.1177/1094670507302990

Westerman, G., Bonnet, D., & McAfee, A. (2014). *Leading digital: Turning technology into business transformation*. Harvard Business Press.

What's new in banking API programs. (2022). McKinsey & Company. https://www.mckinsey.com/capabilities/mckinsey-digital/our-insights/tech-forward/whats-new-in-banking-api-programs]

Wieser, C., Bruhn, M., Kinzinger, J. P., Ruckteschler, C. S., & Heitmann, S. (2019). The impact of mobile money on poor rural households: Experimental evidence from Uganda. World Bank Policy Research Working Paper No. 8913.

Wijethilak, H. P., Singh, S. K., & Dhand, S. (2025). AI for Personalizing Customer Interactions: Improving Engagement and Satisfaction. In *AI, Corporate Social Responsibility, and Marketing in Modern Organizations* (pp. 221-244). IGI Global Scientific Publishing.

Will Arachchige, I. S., Jahankhani, H., Oshadi Karunanayaka, K. A. Y. R., & Amin Metwally Hussien, O. A. (2024). "Exploring the Balance Between Personalisation and Automation in Human–AI Interaction", Dadwal, S. S., Jahankhani, H., & Revett, K. (Ed.) *Market Grooming*, Emerald Publishing Limited, Leeds, pp. 199-234. https://doi.org/DOI: 10.1108/978-1-83549-001-320241010

Williady, A., & Ban, H. J. (2023). An Approach of the Big Data Analysis on Customer Satisfaction in Bali's Luxury Hotels and Resorts Through Wellness Attributes. *Advances in Social Science, Education and Humanities Research*, **•••**, 3–9. DOI: 10.2991/978-2-494069-93-0_2

Wilson, A., Zeithaml, V., Bitner, M. J., & Gremler, D. (2020). *EBK: Services Marketing: Integrating Customer Service Across the Firm 4e*. McGraw Hill.

Wilson, G., Johnson, O., & Brown, W. (2024). The influence of social media marketing on brand loyalty [Preprint]. *Preprints.org*. https://www.preprints.org/manuscript/202408.0332

Wilson, J. (2010). Smart kiosks and their role in service delivery. *Service Management Journal*, *22*(2), 137–150.

Wipro. (n.d.). Predictive analytics for insurance fraud detection.

Witts, J. O. (2024). Leveraging Artificial Intelligence (AI) for Banks in Emerging Markets. *Journal of Advanced Management Science*, *12*(4), 84–1045. DOI: 10.18178/joams.12.4.167-172

World Economic Forum. (2020). Transforming paradigms: A global AI in financial services survey.https://www3.weforum.org/docs/WEF_AI_in_Financial_Services_Survey_2020.pdf

World Wide Web Consortium. (W3C). (2021). Web Content Accessibility Guidelines (WCAG). Retrieved from https://www.w3.org/WAI/standards-guidelines/wcag/

World Wide Web Consortium. (W3C). (2022). *Web Content Accessibility Guidelines (WCAG) 2.1*. Retrieved from https://www.w3.org/TR/WCAG21/

Wünderlich, N. V., Heinonen, K., Ostrom, A., Patricio, L., Rui, S., & Voss, C. (2015). "Futurizing" smart service: Implications for service researchers and managers. *Journal of Services Marketing*, *29*(6/7), 442–447. DOI: 10.1108/JSM-01-2015-0040

Wylde, V., Prakash, E., Hewage, C., & Platts, J. (2023). Ethical challenges in the use of digital technologies: AI and big data. In *Digital transformation in policing: The promise, perils and solutions* (pp. 33–58). Springer International Publishing. DOI: 10.1007/978-3-031-09691-4_3

Xiang, F., Cheng, H., & Wang, Y. (2023). Exploring the Smart Street Management and Control Platform from the Perspective of Sustainability: A Study of Five Typical Chinese Cities. *Sustainability (Basel)*, *15*(4), 3438. DOI: 10.3390/su15043438

Xue, J., Liang, X., Xie, T., & Wang, H. (2020). See now, act now: How to interact with customers to enhance social commerce engagement? *Information & Management*, *57*(6), 103324. DOI: 10.1016/j.im.2020.103324

Yahoo is part of the Yahoo family of brands. (n.d.). https://finance.yahoo.com/news/embedded-finance-unlock-us-242-020000903.html

Yamane, Y. (1967). Mathematical Formulae for Sample Size Determination.

Yan, Z., Konar, R., & Balasubramanian, K. (2025). The Future of Hotels Robotics, AI, and Service Automation in Practice. In *Sustainability, Innovation, and Consumer Preference* (pp. 249-278). IGI Global Scientific Publishing.

Yang, L., Henthorne, T. L., & George, B. (2020). Artificial intelligence and robotics technology in the hospitality industry: Current applications and future trends. Digital transformation in business and society: Theory and cases 211-228.

Yang, X. (2023). The effects of AI service quality and AI function-customer ability fit on customer's overall co-creation experience. *Industrial Management & Data Systems*, *123*(6), 1717–1735. DOI: 10.1108/IMDS-08-2022-0500

Yang, Y. C., Yang, Y. Y., & Liu, S. W. (2012). An Integrative Model for Self-Service Technology Adoption. *Applied Mechanics and Materials*, *182-183*, 1387–1392. . DOI: 10.4028/www.scientific.net/AMM.182-183.1387

Yeshu, P. (2016). *Financial inclusion in India: Role of banks*. International Research Journal of Commerce, Arts and Science.

Yin, R. K. (2018). *Case study research and applications: Design and methods* (6th ed.). Sage Publications. Supports the use of multiple case study design, cross-case synthesis, and thematic analysis.

Yin, H. S., Parker, R. M., Sanders, L. M., Dreyer, B. P., Mendelsohn, A. L., Bailey, S., Patel, D. A., Jimenez, J. J., Kim, K.-Y. A., Jacobson, K., Hedlund, L., Smith, M. C. J., Maness Harris, L., McFadden, T., & Wolf, M. S. (2016). Liquid medication errors and dosing tools: A randomized controlled experiment. *Pediatrics*, *138*(4), e20160357. DOI: 10.1542/peds.2016-0357 PMID: 27621414

Yiu, L. D., Yeung, A. C., & Cheng, T. E. (2021). The impact of business intelligence systems on profitability and risks of firms. *International Journal of Production Research*, *59*(13), 3951–3974. DOI: 10.1080/00207543.2020.1756506

Yiu, L. D., Yeung, A. C., & Jong, A. P. (2020). Business intelligence systems and operational capability: An empirical analysis of high-tech sectors. *Industrial Management & Data Systems*, *120*(6), 1195–1215. DOI: 10.1108/IMDS-12-2019-0659

Yun, T., Li, H., Chang, P. C., Lin, M. F., Carroll, A., & McLean, C. Y. (2020). Accurate, scalable cohort variant calls using DeepVariant and GLnexus. *Bioinformatics (Oxford, England)*, *36*(24), 5582–5589. DOI: 10.1093/bioinformatics/btaa1081 PMID: 33399819

Zeithaml, V. A., Parasuraman, A., & Berry, L. L. (1996). The behavioral consequences of service quality. *Journal of Marketing*, *60*(2), 31–46. DOI: 10.1177/002224299606000203

Zeng, F., Wong, K. Y., & Kuan, S. (2018). A survey of autonomous vehicles: From a machine learning perspective. *IEEE Access : Practical Innovations, Open Solutions*.

Zetzsche, D. A., Buckley, R. P., Arner, D. W., & Barberis, J. N. (2020). Decentralized finance. *Journal of Financial Regulation*, *6*(2), 172–203. DOI: 10.1093/jfr/fjaa010

Zhang, B., & Sundar, S. S. (2019). Proactive vs. reactive personalization: Can customization of privacy enhance user experience? *International Journal of Human-Computer Studies*, *128*, 86–99. DOI: 10.1016/j.ijhcs.2019.03.002

Zhang, J., Chen, W., Petrovsky, N., & Walker, R. M. (2022). The expectancy-disconfirmation model and citizen satisfaction with public services: A meta-analysis and an agenda for best practice. *Public Administration Review*, *82*(1), 147–159. DOI: 10.1111/puar.13368

Zhang, J., & Chen, Z. (2024). Exploring Human Resource Management Digital Transformation in the Digital Age. *Journal of the Knowledge Economy*, *15*(1), 1482–1498. DOI: 10.1007/s13132-023-01214-y PMID: 40479358

Zhang, L., Pentina, I., & Fan, Y. (2021). Who do you choose? Comparing perceptions of human vs robo-advisor in the context of financial services. *Journal of Services Marketing*, *35*(5), 634–646. Advance online publication. DOI: 10.1108/JSM-05-2020-0162

Zhang, W., Wang, Q., & Zhan, Y. (2022). Applications of quantum computing in financial markets: A systematic review. *Journal of Computational Finance*, *26*(1), 1–28. DOI: 10.21314/JCF.2022.419

Zhang, Y., Chen, H., & Liu, J. (2021). Emotional intelligence in artificial intelligence: Towards empathetic self-service technologies. *International Journal of Human-Computer Studies*, *145*, 102538.

Zhou, J., Mavondo, F. T., & Saunders, S. G. (2019). The relationship between marketing agility and financial performance under different levels of market turbulence. *Industrial Marketing Management*, *83*(2), 31–41. DOI: 10.1016/j.indmarman.2018.11.008

Zhu, Z., Wymer, W., & Chen, I. (2002). IT-based services and service quality in consumer banking. *International Journal of Service Industry Management*, *13*(1), 69–90. DOI: 10.1108/09564230210421164

Zhylin, M., Mendelo, V., Chumaieva, Y., Kernas, A., & Zaporozhtse, V. Y. (2024). Analysis of contemporary methods of integrating emotional intelligence into artificial intelligence systems: Advantages, disadvantages, and perspectives. *Journal of Theoretical and Applied Information Technology*, *102*(7), 2842–2853.

Zuboff, S. (2023). The age of surveillance capitalism. In *Social theory re-wired* (pp. 203–213). Routledge., https://www.taylorfrancis.com/chapters/edit/10.4324/9781003320609-27/age-surveillance-capitalism-shoshana-zuboff DOI: 10.4324/9781003320609-27

About the Contributors

Balraj Verma Associate Professor Doctoral Research Centre-Chitkara Business School, Chitkara University, Punjab, India. Dr. Balraj Verma is currently serving as an Associate Professor at the Doctoral Research Centre of Chitkara Business School, affiliated with Chitkara University, Rajpura, Punjab. He completed his PhD from Jaypee University of Information Technology (JUIT), Waknaghat, and holds a master's degree in Business Administration, bringing with him over 16 years of combined academic and corporate experience. His teaching portfolio includes courses such as Marketing Management, Strategic Management, Business Statistics, and Research Methodology. He has successfully supervised three scholars to PhD completion under his guidance, showcasing his proficiency in academic mentorship. His research contributions are substantial, with numerous publications in respected journals indexed in SCI, ABDC and Scopus. Additionally, he has authored and edited books and contributed chapters to leading international publishers. Dr. Balraj Verma is actively involved in organizing international conferences and serves as a resource person for Faculty Development Programs (FDPs), Management Development Programs (MDPs), and workshops focused on research methodologies. As a peer reviewer for esteemed journals including the Journal of Knowledge Economy (Springer), Economic Change and Restructuring (Springer), Electronic Commerce Research (Springer), International Journal of Information Management Data Insights (Elsevier), and S N Computer Science (Springer), he contributes significantly to maintaining the standards of academic scholarship in his field. His research interests span E-commerce, Retail Management, Technology Adoption, Aggregator Platforms Models, and Human-Computer Interaction etc.

Mike Schulze is Vice President for Research and Practice Transfer and Professor of Controlling, Accounting and Financial Management at CBS International Business School in Mainz, Germany. Furthermore, he is active as Honorary Professor-in-Residence for the Institute of Management Accountants (IMA) in Europe and as

Lecturer in the field of executive education at EBS Universität für Wirtschaft und Recht and at Frankfurt School of Finance & Management. His current research and consulting activities focus on trends and developments in the CFO area, the digitalization of Controlling and corporate sustainability performance management. Prof. Schulze is experienced in practice-oriented research, has published several books as well as academic and practice-oriented articles in the context of Controlling. He was formerly employed as Finance Analyst at Ford Motor Company in Cologne, Germany and served as an officer in the German Federal Armed Forces.

Timilehin Olasoji Olubiyi is the acting Dean of Faculty of Management and Social Sciences at West Midlands Open University in Lagos State Nigeria. Also, an affiliate of Babcock Business School, Babcock University Ogun State Nigeria. Dr Timi Olubiyi is a seasoned professional, business management expert, senior government adviser, and scholar with over twenty-two years of hands-on experience and background in investment banking, public sector investment management, stockbroking, consulting, academia, research, and entrepreneurial development. His main expertise and research interest is multidisciplinary, focusing on sustainability, small business management and development, knowledge management, entrepreneurship, innovation, enterprise continuity, succession, green management strategies, and leadership. Dr Timi Olubiyi is widely published, his research works have appeared in the Management and Marketing Journal, International Journal of Professional Business Review, Revista de Gestão Social e Ambiental, Malaysian Management Journal, Springer Nature, Emerald, Taylor and Francis and IGI Global amongst others. He is proficient in teaching business-related courses and research methodology. He has authored numerous peer-reviewed books and papers, and he serves on the editorial boards of several university-based and well-indexed high-impact journals around the world.

Kamal Upreti is currently working as an Associate Professor in Department of Computer Science, CHRIST (Deemed to be University), Delhi NCR, Ghaziabad, India. He completed is B. Tech (Hons) Degree from UPTU, M. Tech (Gold Medalist), PGDM(Executive) from IMT Ghaziabad and PhD in Department of Computer Science & Engineering. He has completed Postdoc from National Taipei University of Business, TAIWAN funded by MHRD. He has published 50+ Patents, 32+Magazine issues and 110+ Research papers in in various reputed Journals and international Conferences. His areas of Interest such as Modern Physics, Data Analytics, Cyber Security, Machine Learning, Health Care, Embedded System and Cloud Computing. He has published more than 45+ authored and edited books under CRC Press, IGI Global, Oxford Press and Arihant Publication. He is having enriched years' experience in corporate and teaching experience in Engineering

Colleges. He worked with HCL, NECHCL, Hindustan Times, Dehradun Institute of Technology and Delhi Institute of Advanced Studies, with more than 15+ years of enrich experience in research, Academics and Corporate . He also worked in NECHCL in Japan having project – "Hydrastore " funded by joint collaboration between HCL and NECHCL Company. Dr. Upreti worked on Government project – "Integrated Power Development Scheme (IPDS)" was launched by Ministry of Power, Government of India with the objectives of Strengthening of sub-transmission and distribution network in the urban areas. Currently, he has completed work with Joint collaboration with GB PANT & AIIMS Delhi, under funded project of ICMR Scheme on Cardiovascular diseases prediction strokes using Machine Learning Techniques from year 2017-2020 of having fund of 80 Lakhs .He got 5 Lakhs fund from DST SERB for conducting International Conference, ICSCPS-2024, 13-14 Sept 2024. Recently, he got 10 Lakhs fund from AICTE - Inter-Institutional Biomedical Innovations and Entrepreneurship Program (AICTE-IBIP) for 2024-2026. He has attended as a Session Chair Person in National, International conference and key note speaker in various platforms such as Skill based training, Corporate Trainer, Guest faculty and faculty development Programme. He awarded as best teacher, best researcher, extra academic performer and Gold Medalist in M. Tech programme. Dr. Kamal Upreti Associate Professor Department of Computer Science CHRIST (Deemed to be University), Delhi NCR, Ghaziabad, India Kamal.upreti@christuniversity.in Kamalupreti1989@gmail.com Orchid ID: 0000-0003-0665-530X

***

Abdulrafiu Abdullah is an assistant lecturer in the department of Business education, Al-Hikmah University Ilorin, Kwara state. A graduate of Business education from Ahmadu Bello University, Zaria and M.sc Ed from Kwara state University and Ph.d in view from the same university. His research nterest include business education, entrepreneurship education, and marketing management.

Abdulquadri Abdulazeez Olamide is an academic researcher focused on the studies of digital marketing, strategic marketing and innovation within the informal trading systems, new technologies, and innovations. His research interests span the intersection of marketing education and innovation, with a specific focus on enhancing pedagogical approaches and exploring the potential of digital platforms to boost business performance. His scholarly contributions have been recognized through multiple publications in reputable journals, including Shirkah: Journal of Economics and Business, Emerald Publishing, the Journal of Contemporary Business and Islamic Finance (JCBIF) and Economía y Negocios. He has also contributed to research on Digital marketing, Islamic marketing communication, sustainability in industry 4.0 and digital transformation of the agricultural value chain in Nigeria. Abdulquadri has a strong academic profile, having gained a Master of Science in Marketing at the University of Lagos where his thesis involved examination of the impact of e-marketing on customer patronage. He has received honors for his academic and professional achievements, including a scholarship sponsored by Facebook for a Digital Marketing Associate certification. In terms of work experience Abdulquadri worked as a research analyst and has been a Research Fellow at the University of Ilorin. In these positions, he has taught, mentored and assessed many scores of students as well as collaborated on research within marketing and other disciplines. His teaching specializations entail teaching of marketing principles, business mathematics and entrepreneurship covering all courses offered with intentions of helping students understand the lessons being taught. Apart from teaching, Abdulquadri has practical experience in marketing communication, Data analysis, social media analysis and tax collection from organizations like Abazol Digital, Ogun State Internal Revenue Service and Knight Interiors. Abdulquadri is a member of National Institute of Marketing of Nigeria and a certified business analyst. As a source of inspiration and mentorship, he continuously seeks to innovate within the marketing discipline, bridging the gap between theory and practice while promoting the adoption of ethical and sustainable strategies across industries.

Sonia Agrawal is a results-driven Engineering Manager with over 18 years of experience in Test Management and leadership across wireless connectivity, hardware/software, and embedded systems. Since September 2023, she has been with NVIDIA as an Engineering Manager in the Embedded and Automotive Test Solution Development Engineering team, where she leads quality assurance and test development for advanced automotive and embedded solutions and systems. Previously, Sonia managed system performance and software test activities at Qualcomm, driving end-to-end quality assurance and automation strategies for next-generation WLAN chipset platforms. Her career also includes key roles at Intel and STMicroelectronics, where she delivered industry-first Wi-Fi 802.11ac/ax/be solutions and built high-performing global teams. Renowned for her technical expertise, strategic leadership, and effective relationship building, Sonia has received numerous accolades for outstanding execution and innovation in wireless system testing. She holds a Master of Science in Software Systems from Birla Institute of Technology, Pilani, and a Bachelor of Technology in Electronics and Communication Engineering from G.G.S.I.P.U, Delhi.

Ahmed Akeem Apo is an Assistant Lecturer at Al-Hikmah University, Ilorin, Nigeria. He holds a Master of Arts in Education (Economics) from Obafemi Awolowo University, Ile-Ife, and a Bachelor of Science in Education (Economics) from the University of Uyo, Akwa-Ibom. He is currently pursuing his PhD in Curriculum Development in Social Sciences at the University of Ilorin.

B. Latha Lavanya, Associate Professor has 20 years of experience. She has been a Sr. HR Manager for about 7 years, full time Ph.D (Research fellowship awarded by UGC) from the University of Madras. Her Accolades include Best Faculty in Management Award in the year 2014 from Education Today, Research Excellence Award in the year 2017 by Indo Global Education, Award for Teaching Excellence by GMRAF Awards in the year 2019. Best women researcher Award of the year 2020.

Hemachandran C is working as Associate Professor In MBA department and doing research in data analytics and automation of record maintenance

G. Chamdramowleeswaran is currently working as Assistant Professor in Department of Commerce and business administration, VelTech Rangarajan Dr. Sagunthala R&D Institute of Science and Technology (Deemed to be University), Avadi, Chennai. He has more than 11 years 5 months of industry experience and 1 year and 11 months of teaching experience with specialization in HR and Marketing. He Holds a B.Com., M.B.A., M. Phil., and Ph.D. in Management from Bharathidasan University, Trichy

Shubhangi Gautam is an assistant professor at Lovely Professional University, Phagwara, Punjab, India. She worked as a full-time doctoral fellow at the University School of Business, Chandigarh University. She has more than three years of research experience in the field of behavioral finance and more than one year of experience as an accountant. She has various publications in the field of behavioral finance, and cryptocurrency investments in different prestigious journals indexed in Scopus, ABDC listed, Web of Science, and UGC care. She is serving as a reviewer for two Scopus, ABDC-listed finance and multidisciplinary journals. She has also served as a reviewer for IGI global books and various international conference papers. She has presented research articles at more than ten international conferences and was awarded the best research paper with second position at the Symbiosis International Conference.

Monica Gupta, a distinguished scholar in Electronics and Communication Engineering, earned her Ph.D. from Delhi Technological University in 2022, highlighting a relentless commitment to academic excellence. With over 17 years of teaching experience, she currently serves as an Associate Professor at Bharati Vidyapeeth's College of Engineering, New Delhi. A prolific researcher, Dr. Gupta has made substantial contributions in Low Power Memory Design, Image and Video Processing, VLSI, IoT, Artificial Intelligence and Machine Learning, evident in her numerous publications in reputable journals and conferences. Beyond academia, she has initiated the IOSC-BVP student club, in collaboration with Intel OneAPI Student ambassador, addressing the skill gap between graduates and industry expectations. Inaugurated on November 7, 2022, the club focuses on skill enhancement programs, organizing national-level events to deepen students' understanding of crucial areas like Artificial Intelligence, Internet of Things, Machine Learning, and Website Development. Dr. Monica Gupta's impact extends beyond the classroom, shaping the future of engineering by bridging academic knowledge with industry needs. Her dynamic leadership and commitment to excellence make her a trailblazer in the landscape of Electronics and Communication Engineering education.

D.Joel Jebadurai is working as a Associate professor in Department of MBA, St. Joseph's college of Engineering OMR Chennai Tamilnadu India. He has published more than 50 research articles in various peer reviewed journals. He has undertaken the funded projects and written two books also.

K. Dheenadhayalan working as an Assistant Professor in Department of MBA, Sri Sai Ram Engineering college, Chennai. He had a rich experience in the field of teaching.

S. Karunakarn is currently serving as an Assistant Professor at St. Joseph's Institute of Technology, Chennai, Tamil Nadu. With over 15 years of academic experience, he is a seasoned educator specializing in Marketing and Human Resource Management, and holds a Ph.D. in Marketing. He is dedicated to equipping his students with the knowledge and skills necessary for success. Dr. Karunakarn has actively contributed to academia by presenting numerous research papers at state, national, and international conferences. His work has been published in reputed UGC CARE journals, as well as in ABDC-listed and IEEE conference SCOPUS Indexed journal.

|Geetha kotha Vandla - Contributing Author

Azeez Tunbosun Lawal is a senior lecturer at the Department of Business Administration at the Faculty of Management Science, Al-Hikmah University, Ilorin, Nigeria. He holds a B.Sc., M.Sc., and PhD in Business Administration from the University of Ilorin, Ilorin, Nigeria. He has over 30 papers presented and published in peer-reviewed journals, conference proceedings, and books of repute. His specialisation is Entrepreneurship and strategic management.

Manikandan M Working as an Assistant Professor, Faculty of Management, SRM Institute of Science and Technology. He had a very good experience in the field of research and teaching.

Semiu Olawale Makinde is a distinguished academic with a Ph.D. in Educational Technology, an MBA in Marketing, and a B.Eng. in Agricultural Engineering. Serving as a Senior Lecturer in the Department of Science Education and an Adjunct Lecturer in Agricultural Mechanization at Al-Hikmah University, he is also the Director of the Institute of Education. Dr. Makinde is a dedicated researcher, contributing to reputable journals and book chapters. His multidisciplinary expertise, commitment to education, and innovative contributions to both science education and agricultural mechanization showcase his impact in academia at Al-Hikmah University, Ilorin, Nigeria

Arockia Juliet P., working as an Assistant Professor in Arul Anandar College .I completed my Undergraduate and Post graduate in Fatima college, Madurai. I got Phd degree from Madurai Kamaraj University on the topic of "Foreign Direct Investment from India -Recent Trends and Patterns". I am specialized in finance and marketing. I have 13 years of academic experience in the field of Commerce at UG and PG Level. I have participated and presented papers in National and International Seminars, workshop, Webinars and FDP. I have published several number of papers in UGC care listed and ISBN journals. I have been served as a Resource person and External Examiner.

Fazla Rabby is a dynamic and innovative business manager with a robust educational background, extensive teaching experience, and a proven track record in academia and industry. Dr. Fazla Rabby earned his PhD from the University of Southern Queensland while his Master in Information Systems is from Central Queensland University, along with various other credentials that further enhanced his business and information technology skills. Dr. Rabby is currently the Director at the Stanford Institute of Management and Technology, where he manages the academic department and oversees regulatory matters in coordination with the government and other key stakeholders. His commitment to research is a cornerstone of his academic journey, manifesting in a series of impactful publications spanning diverse and rapidly evolving fields. Notably, his research contributions include insightful books, book chapters and articles focused on cutting-edge topics such as Blockchain Technology, Digital Marketing, and Artificial Intelligence. These publications reflect his dedication to expanding the boundaries of knowledge and demonstrate a proactive engagement with emerging technologies that are pivotal in today's academic landscape. During my tenure as an Academic at different universities and institutes, Dr Fazla Rabby played a pivotal role in fostering a dynamic learning environment. He delivered lectures beyond traditional boundaries, incorporating interactive workshops and engaging seminars to stimulate critical thinking among students. Collaborating with fellow academics, He actively refined teaching methods, emphasising innovative approaches to enhance the learning experience. Dr. Rabby is also an active member of several professional bodies, including the Australian Computer Society, IEEE and the Australian Marketing Association.

Sakariyahu Shehu is a Ph.D. student in the Green University for Community Development and Entrepreneurship, KWASU, as well as an assistant lecturer in the Department of Business Education, Al-Hikmah University, Ilorin. He enjoys the membership of the Association of Business Educators of Nigeria (ABEN), the Business Educators Association in Vocational Education (BEAVE), and the Teachers' Registration Council of Nigeria (TRCN). His research interest centred on business and entrepreneurship education, with a specific emphasis on developing innovative and effective approaches to teaching and learning in entrepreneurship and business education. His email contacts are: ssakariyahu@alhikmah.edu.ng and sunkanmizacks@gmail.com

Ram Singh, a forward-thinking educator, serves as a Professor at School of BFSI, Symbiosis University of Applied Sciences, Indore, India. With over 16 years of extensive experience in teaching, research, and administration, Dr. Singh is a prominent figure in academia. His diverse roles include membership on the Editorial Boards of esteemed journals focusing on Accounting, Finance, and Management. Dr. Singh is actively involved in eight national and international academic organizations, contributing significantly to the field. His scholarly work comprises 18 books, 30 research papers, 35 book chapters, and 10 conference proceedings at both national and international levels.

Vinay Pal Singh is Associate Professor in the area of Finance and marketing at Quantum University, Roorkee. Prior to joining Quantum University, he was associated with Shobhit University, Saharanpur as Assistant Professor. He has obtained his Ph.D. in the area of Micro-Finance. He has academic experience of 12 years working with renowned Institutes. He has published research papers and Chapters in leading journals and books. He has conducted several workshops on SPSS and AMOS He taught various course like Financial Management, Marketing Analytics, Business Research Methods.

N.V. Suresh, renowned academician from Chennai, Tamil Nadu. He completed his schooling in Chennai. He obtained his BA degree (2012) from D.G. Vaishnav College, Madras University. He joined the Bharathiar University received an MBA degree and PGDM. He Awarded Ph.D. in Management from University of Madras. Dr. N.V. Suresh joined then IIKM in June 2014 as an Assistant Professor (2014 to 2016). He moved to Remo International College in June, 2019 as an Assistant Professor and rose through the ranks to the position of Head of the Department in Aug 2020 to July 2021 then Shifted to Working Academy of Maritime Education and Training (Deemed to be University) Presently Working as Associate Professor and Vice-Principal in ASET College of Science and Technology from March 2022 to Till Date. He is the Visiting Faculty for Institute of Distance Education, University of Madras, Tamil Nadu Open University, Anna University and Directorate of Online Education SRM University for MBA from 2015 to till date. Dr. N.V. Suresh has carried out all his research work entirely in India and has published 30 research papers in international journals and presented 15 papers at national and international conferences, authored two Text Books namely Logistics and Air cargo Management, Executive Communication Published by IIP Publishing House and Blue rose Publishers and Prepared more than 10 E-content for Online Distance Education for AMET University and SRM University, He also Produced one Ph.D. Scholar Under his Guidance currently 3 Research Scholars are doing Ph.D. Under his Guidance. Completed six funded Project worth Rs 3, 50,000 from Industry and Government of Tamil Nadu TNCST. Dr. N.V. Suresh is also Editor for ASET Journal of Management Science and reviewer for various International Journals among them ASTEJ belongs to University of Chicago, United States of America. He is also associated with confidential work of ICAI & ICWA. Dr. N.V. Suresh has given 18 Guest lectures at various colleges. He also acted as Chairperson for SRM University Scholar Submit conducted by DOMS SRM University. Member Scrutiny for International Conference conducted by various colleges. He has also served as Chairman Board of Studies for MBA & BSc Programs of Alagappa University Collaborative Programs

W. C. Tang, Head of the Mathematics Department at Lock Tao Secondary School in Hong Kong, holds a Ph.D. in Business Administration and has had expertise in mathematics education over 30 years. With a rich academic background, including a BEng (Hons) in Mechanical Engineering, an M.Ed. in Educational Management and Administration, a PCEd, an MSc in Applied Mathematics, an MBA, and an MA in Special and Additional Learning Needs, Dr. Tang also worked as a part-time lecturer at the Institute of Vocational Education for 12 years and a part-time tutor at the Hong Kong Metropolitan University for 15 years. Contact: tang957031@gmail.com

Index

A

B

C

D

E

F

I

L

M

U

W

www.ingramcontent.com/pod-product-compliance
Lightning Source LLC
LaVergne TN
LVHW080614300926
849162LV00005B/1158

* 9 7 9 8 3 3 7 3 4 6 6 7 0 *